Machining with Abrasives

Mark J. Jackson • J. Paulo Davim
Editors

Machining with Abrasives

Springer

Editors
Mark J. Jackson
Purdue University
MET, Center for Advanced Manufacturing
College of Technology
West Lafayette, IN 47907-2021
USA
jacksomj@purdue.edu

J. Paulo Davim
University of Aveiro
Department of Mechanical Engineering
3810-193 Aveiro, Portugal
pdavim@mec.ua.pt

ISBN 978-1-4419-7301-6 e-ISBN 978-1-4419-7302-3
DOI 10.1007/978-1-4419-7302-3
Springer New York Dordrecht Heidelberg London

Printed on acid-free paper

Springer is part of Springer Science+Business Media (www.springer.com)

Preface

Machining with abrasives is a widely used process for producing intricate shapes on components and artifacts and for providing precise tolerances that are required especially for the manufacture of engineering components. Grinding technology is associated with the current developments in this area and this book contains chapters on abrasive materials, bonding systems, grinding wheel design, micro and nanogrinding, control of grinding processes, abrasive grinding, abrasive water jet grinding, and dressing of grinding wheels. The chapters are written by subject matter experts from around the world and provide an authoritative text on the current developments in machining with abrasive tools. Chapter 1 focuses on new developments in abrasive materials and associated bonding systems ranging from conventional abrasives to superabrasive materials. Chapter 2 provides an interesting prose on heat treatment and the performance of vitrified bonds, while Chap.3 takes a look at the design of grinding wheels and wheel safety. The dressing of grinding wheels is discussed in Chap.4 and leads into a chapter on the on the surface integrity in ground surfaces (Chap.5). The control aspect of grinding is explained lucidly in Chap.6, while Chap.7 focuses on the exciting new area of nanogrinding. The final chapters deal with polishing with loose abrasive particles and an authoritative overview of impact abrasive machining.

The structure of the book is based on matter provided by many colleagues and the author wishes to thank the contributors of this book for helping construct a source of knowledge and information on machining with abrasives and for granting the editors permission to use such material. The editors also acknowledge the help and support of Inderscience Publishers for allowing the chapter contributors and editors to reproduce their work that was originally published in refereed Inderscience journals and other refereed journals where permissions have been acknowledged.

Dr. Mark J. Jackson
Dr. J. Paulo Davim

About the Editors

Dr. Mark J. Jackson
C. Eng., Engineering Council of London, U.K., 1998
M. A. Status, Natural Sciences, University of Cambridge, U.K., 1998
Ph.D., Mechanical Engineering, Liverpool, U.K., 1995
M.Eng., Mechanical & Manufacturing Engineering, Liverpool, U.K., 1991
O.N.D., Mechanical Engineering, Halton College, U.K., 1986
O.N.C. Part I, Mechanical Engineering, Halton College, U.K., 1984

Doctor Jackson began his engineering career in 1983 when he studied for his O.N. C. part I examinations and his first-year apprenticeship-training course in mechanical engineering. After gaining his Ordinary National Diploma in Engineering with distinctions and I.C.I. prize for achievement, he read for a degree in mechanical and manufacturing engineering at Liverpool Polytechnic and spent periods in industry working for I.C.I. Pharmaceuticals, Unilever Industries, and Anglo Blackwells. After graduating with a Master of Engineering (M. Eng.) degree with Distinction under the supervision of Professor Jack Schofield, M.B.E., Doctor Jackson subsequently read for a Doctor of Philosophy (Ph.D.) degree at Liverpool in the field of materials engineering focusing primarily on microstructure-property relationships in vitreous-bonded abrasive materials under the supervision of Professor Benjamin Mills. He was subsequently employed by Unicorn Abrasives' Central Research & Development Laboratory (Saint-Gobain Abrasives' Group) as materials technologist, then technical manager, responsible for product and new business development in Europe, and university liaison projects concerned with abrasive process development. Doctor Jackson then became a research fellow at the Cavendish Laboratory, University of Cambridge, working with Professor John Field, O.B.E., F.R.S., on impact fracture and friction of diamond before becoming a lecturer in engineering at the University of Liverpool in 1998. At Liverpool, Dr. Jackson established research in the field of micromachining using mechanical tools, laser beams, and abrasive particles. At Liverpool, he attracted a number of research grants concerned with developing innovative manufacturing processes for which he was jointly awarded an Innovative Manufacturing Technology Center from the Engineering and Physical Sciences Research Council in November 2001. In 2002, he became associate professor of mechanical engineering and faculty associate in the Center for Manufacturing Research, and Center for Electric Power at Tennessee Technological University (an associated university of Oak Ridge National Laboratory), and a faculty associate at Oak Ridge National Laboratory. Dr. Jackson was the academic adviser to the Formula SAE Team at Tennessee Technological University. In 2004 he moved to Purdue University as Professor of Mechanical Engineering in the College of Technology.

Doctor Jackson is active in research work concerned with understanding the properties of materials in the field of microscale metal cutting, micro- and nanoabrasive machining, and laser micro machining. He is also involved in developing next generation manufacturing processes. Doctor Jackson has directed, co-directed, and managed research grants funded by the Medical Research Council, Engineering and Physical Sciences Research Council, The Royal Society of London, The Royal Academy of Engineering (London), European Union, Ministry of Defense (London), Atomic Weapons Research Establishment, National Science Foundation, N. A.S.A., U. S. Department of Energy (through Oak Ridge National Laboratory), Y12 National Security Complex at Oak Ridge, Tennessee, and Industrial Companies, which has generated research income in excess of $15 million. Dr. Jackson has organized many conferences and served as General Chairman of the International Surface Engineering Congress and is Deputy President of the World Academy of Materials and Manufacturing Engineering. He has authored and co-authored over 250 publications in archived journals and refereed conference proceedings, has written a book on "micro and nanomanufacturing", is guest editor to a number of refereed journals, and has edited a book on "commercializing micro- and nanotechnology products". He is the co-editor of the "*Journal of Manufacturing Technology Research*", associate editor of the "*International Journal of Molecular* Engineering", and is on the editorial boards of the "*International Journal of Machining and Machinability of Materials*", "*International Journal of Computational Materials Science and Surface Engineering*", "*International Journal of Nanomanufacturing*", "*International Journal of Nano and Biomaterials*", and the "*International Journal of Manufacturing Research*".

Dr. J. Paulo Davim
Aggregation, Mechanical Engineering, University of Coimbra, Portugal, 2005
Ph.D., Mechanical Engineering, University of Porto, Portugal, 1997
M.Sc., Mechanical Engineering (Materials & Manufacturing Processes), University of Porto, Portugal, 1991
Graduation (5 years), Mechanical Engineering, University of Porto, Portugal, 1986

Doctor J. Paulo Davim received his Ph.D. degree in Mechanical Engineering from the University of Porto in 1997 and the Aggregation from the University of Coimbra in 2005. Between, 1986/96, he was lecturer in University of Porto. Currently, he is an Aggregate Professor at the Department of Mechanical Engineering of the University of Aveiro and the Head of MACTRIB - Machining and Tribology Research Group. He has more 24 years of teaching and research experience in manufacturing, materials and mechanical engineering. His main research interests are the machining and tribology and more recently the application of computational and statistical methods in manufacturing as well as all aspects of sustainable manufacturing. He was coordinator of several research projects and supervisor of 2 Ph.D. thesis and 10 M.Sc. thesis. He is Guest Editor of journals, book Series Editor and Scientific Advisory for many international journals and conferences. Presently, he is an Editorial Board member of 15 international journals and acts as reviewer for than 45 prestigious ISI web Science journals. He has also published more than 20 book chapters and 300 articles in journals and conferences (more 150 articles in ISI Web Science, h-index 15).

He is the Editor of the "*International Journal of Machining and Machinability of Materials*", "*International Journal of Manufacturing, Materials and Mechanical Engineering*", "*Journal of Machining and Forming Technologies*", "*Journal of Modern Manufacturing Technology*" *and Co Editor* of the "*International Journal of Surface Science and Engineering*", "*International Journal of Mechatronics and Manufacturing Systems*", "*International Journal of Materials Engineering Innovation*", and the "*Journal of Manufacturing Technology Research*".

Contents

Contributors

Yasser M. Ali
School of Mechanical and Manufacturing Engineering, The University of New South Wales, Sydney, NSW 2052, Australia

J. Paulo Davim
Department of Mechanical Engineering, University of Aveiro, 3810-193 Aveiro, Portugal

Michael P. Hitchiner
Saint-Gobain Abrasives, Inc., 27588 Northline Road, Romulus, MI 48174, USA

Han Huang
School of Mechanical and Mining Engineering, The University of Queensland, Brisbane, QLD 4072, Australia

Mark J. Jackson
MET, Center for Advanced Manufacturing, Purdue University, College of Technology, 401 North Grant Street, West Lafayette, IN 47907-2021, USA

Jian Lu
Department of Mechanical, Materials and Aerospace Engineering, University of Central Florida, 4000 Central Florida Blvd, Orlando, FL 32816, USA

Abdolreza Rasifard
Competence Center of Grinding Technology and Superfinishing, Hochschule Furtwangen University, Villingen-Schwenningen, Germany

Taghi Tawakoli
Department of Mechanical and Process Engineering, Hochschule Furtwangen University, Villingen-Schwenningen, Germany

Jun Wang
School of Mechanical and Manufacturing Engineering, The University of New South Wales, Sydney, NSW 2052, Australia

Chengying Xu
Department of Mechanical, Materials and Aerospace Engineering, University of Central Florida, 4000 Central Florida Blvd, Orlando, FL 32816, USA

Ling Yin
School of Engineering, James Cook University, Brisbane, QLD 4811, Australia

L.C. Zhang (Liangchi Zhang)
School of Mechanical and Manufacturing Engineering, The University of New South Wales, Sydney, NSW 2052, Australia

Libo Zhou
Department of Intelligent Systems Engineering, Ibaraki University, Hitachi, Japan

Chapter 1
Abrasive Tools and Bonding Systems

Mark J. Jackson and Michael P. Hitchiner

Abstract The manufacture of high performance grinding wheels is heavily dependent upon the selecting the correct abrasive grain in terms of cutting ability, fracturing ability and the shape of the grain that controls metal removal rate, dressing interval, coolant delivery to the wheel-workpiece interface and chip evacuation. This chapter explains the recent developments in abrasive grains and bonding systems and should allow the reader to better select the correct abrasive grain and bonding system to economically machine materials.

Keywords Abrasive tools · Bonding · Grinding wheels

1.1 Abrasive Grain Characteristics

1.1.1 Grain Shape

Grain shape makes an enormous impact on grain strength, grinding performance, and packing characteristics that impact wheel formulation and manufacture. Shape will affect the *r* term in the undeformed chip thickness calculation $t' = \{[Vw/(Vs \cdot Cr)] \cdot (d/De)^{1/2}\}^{1/2}$ where r is the ratio of undeformed chip width to chip depth. This in turn impacts grinding power, finish and force/grit. Shape and size are interlinked especially for particles of indeterminate shape i.e. not perfectly spherical, cubic, etc. For synthetic diamond particles, for example, there exists an

M.J. Jackson (✉)
MET, College of Technology, Purdue University, 401 North Grant Street, West Lafayette, IN 47907, USA
e-mail: jacksomj@purdue.edu

M.J. Jackson and J.P. Davim (eds.), *Machining with Abrasives*,
DOI 10.1007/978-1-4419-7302-3_1, © Springer Science+Business Media, LLC 2011

infinite combination of particle shapes derived from the transition between octahedral and cubic forms, as shown in Fig. 1.1. Moreover, crystal imperfections, and polycrystalline particles, further add to the variety of diamond forms.

A blocky round grain will in general be far stronger than an angular, sharp-cornered grain. Quantifying exactly what "blocky" and "angular" mean, and defining the characteristics key to shape have been the sources of considerable study both for grinding performance and batch to batch quality control of wheel manufacture. A variety of parameters for describing the shape of particle projections, classified according to the salient feature of the measurement, is shown in Table 1.1. The boxed parameters are those considered to be of most interest in a study by De Pellegrin et al. [1, 2], whereby the shape characteristics of diamond grain were studied in relation to grinding performance using optical microscopy and image analysis.

Two key diametrical dimensions are the major and minor diameters, d_a and d_b respectively as illustrated in Fig. 1.2, provide a fundamental measure of particle size. It may be noted that, depending on the orientation of the caliper lines relative to the given projection, many different diameters may be obtained. Although size is an important feature, it is shape that governs particle abrasiveness. Algebraic combinations of linear dimensions provide measures of shape.

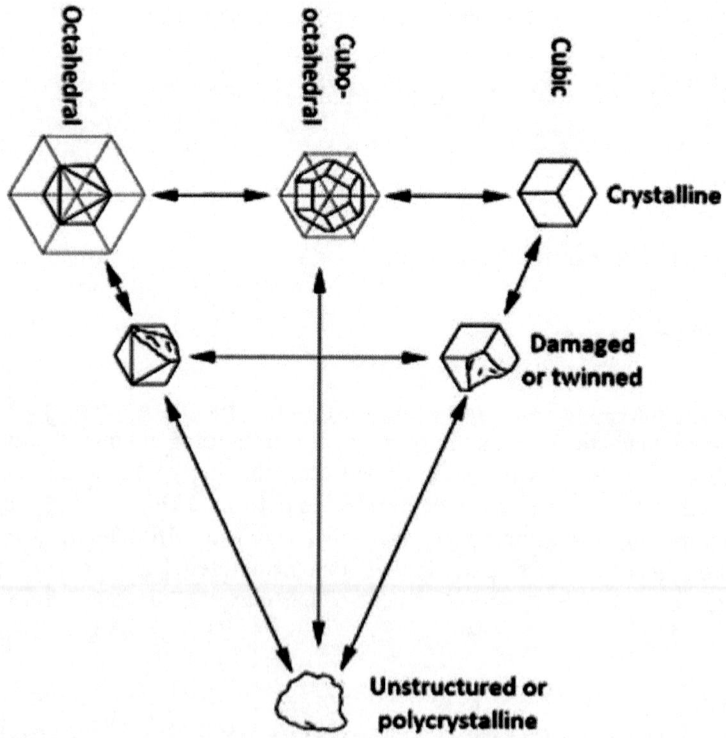

Fig. 1.1 Diamond particle shape map

Table 1.1 Classification of parameters for measuring particle shape

Span	Area	Perimeter	Ratio	Average	Composite
Major diameter	**Projection area**	Silhouette perimeter	**Aspect ratio**	Average radius	Spike parameter
Minor diameter	Area moment	Perimeter moment	**Convexity**	Average diameter	Fractal dimension
Minimum diameter	Hull area	Hull perimeter	Circularity	Mean angle	**Sharpness**
Feret diameter	Equivalent radius (circle area based)	Equivalent radius (circle perimeter based)	Shape factor	Average curvature	Fourier/Radance
	Equivalent ellipse		Ellipticity		Roughness
					Angularity

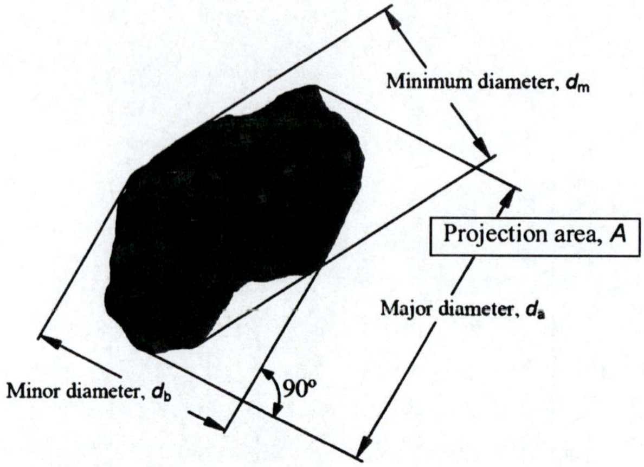

Fig. 1.2 Dimensional definitions for a grain 2D projection

Examples of such calculated values from this projection include:

(a) **Aspect ratio**, the ratio of the major to minor diameter d_a/d_b. A useful parameter to describe grain elongation.
(b) **Projection area**, the area enclosed by the boundary of its projection. It is an indirect measure of size and bulk of the particle. It is important as part of the calculation of grain convexity.
(c) **Convexity** is a characteristic that strongly relates to the strength of the grain and its abrasive potential. A grain is convex if an idealized elastic membrane stretched across its projection leaves no space between itself and the grain's surface. The degree of convexity correlates with lower mechanical integrity but higher abrasive aggressiveness with the grain being, on average, less blocky. Convexity also correlates with the notion of grain irregularity.
Convexity C as a parameter is defined as

$$C = \frac{A_f \pm A_p}{A_p}$$

where A_p is the projected area of the grain, A_f is the fill area between the grain projection and the idealized elastic membrane stretched across the projection (Fig. 1.3).
(d) Grain "sharpness" is a parameter that has been developed specifically for the characterization of abrasive grains based on chip formation modeling whereby the abrasive rate is governed by the degree of penetration into the workpiece, as illustrated in Fig. 1.4.

The functional relationship between the two orthogonal areas, Ω and Λ, is known as the "groove function", see Fig. 1.5. Averaging applied to numerous particle

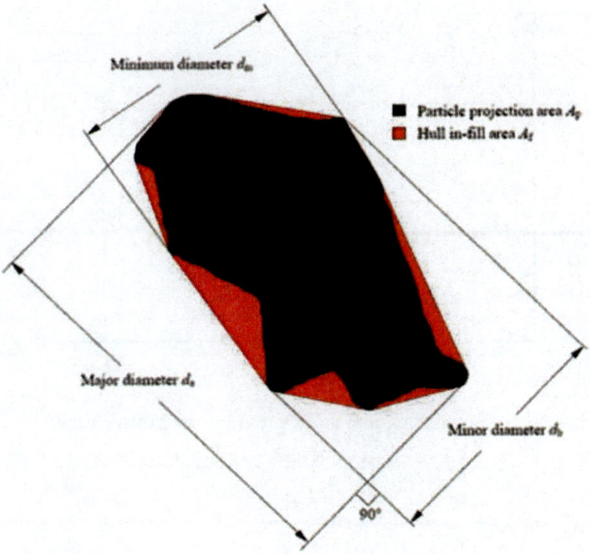

Fig. 1.3 Illustration of fill and projection areas for a grain 2D projection

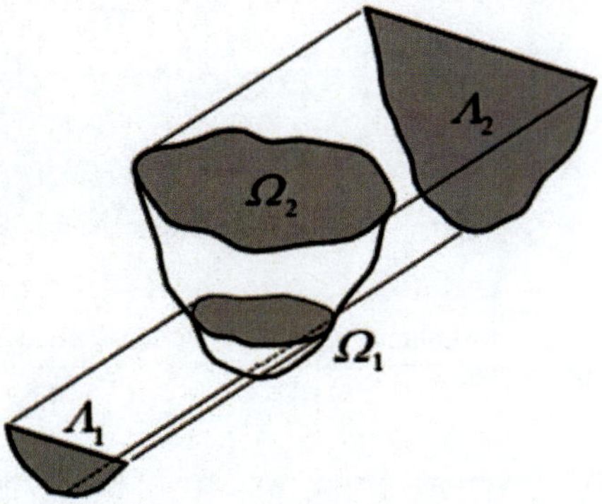

Fig. 1.4 Orthogonal grain projections relative to penetration depth

projections, and to the different particle orientations, yields the "average groove function" [1]. This function embodies the abrasive characteristics of an agglomeration of particles as might be found in a grinding wheel or coated abrasive paper.

From the groove function it may be seen that particle abrasiveness depends not only on the shape of individual particles, but also the complex distribution of the particles that constitute an abrasive surface [3]. The groove function embodies far greater information than any single parameter. The convenience of a single parameter led to the synthesis of the groove function into the "sharpness" parameter. The sharpness parameter involves rescaling the axes of the groove function by

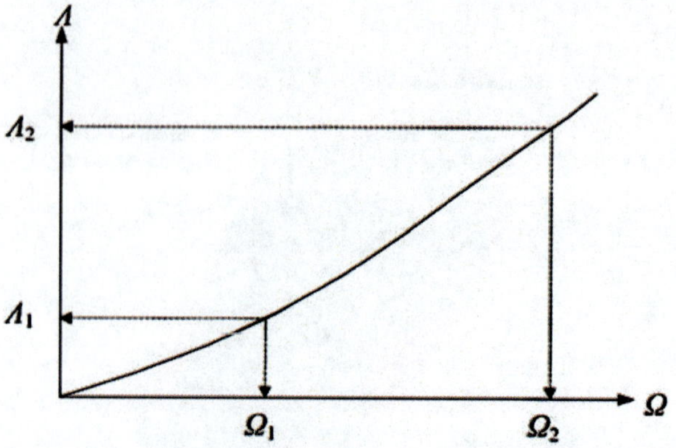

Fig. 1.5 Relationship between orthogonal areas provides the groove function

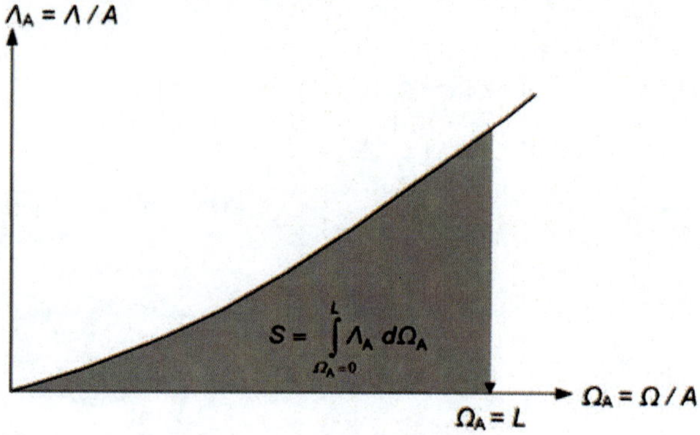

Fig. 1.6 Integration of the dimensionless average groove function, to obtain the sharpness parameter, S

the average projection area, A, of the particle sample. In this way, the groove function becomes non-dimensional – in terms of Ω_A and Λ_A – and particle samples of nominally different size can be compared solely in terms of their abrasiveness, which is governed by the slope of the curve. Sharper and more abrasive particles exhibit overall steeper slopes, and on this basis, integration of the dimensionless average groove function with respect to Ω_A, between zero and an arbitrary limit, L, yields the single parameter, sharpness (S), as illustrated in Fig. 1.6. The choice of L is governed by the extent to which it is expected that the particles will penetrate the workpiece. A value of 0.3 was chosen because it represents a realistic average penetration during abrasion; i.e. the average indentation area, Ω, is 0.3 of the average projection area, A.

Having developed the sharpness parameter, De Pellegrin et al. [1] went on to compare it to convexity and aspect ratio for six different types of commercially

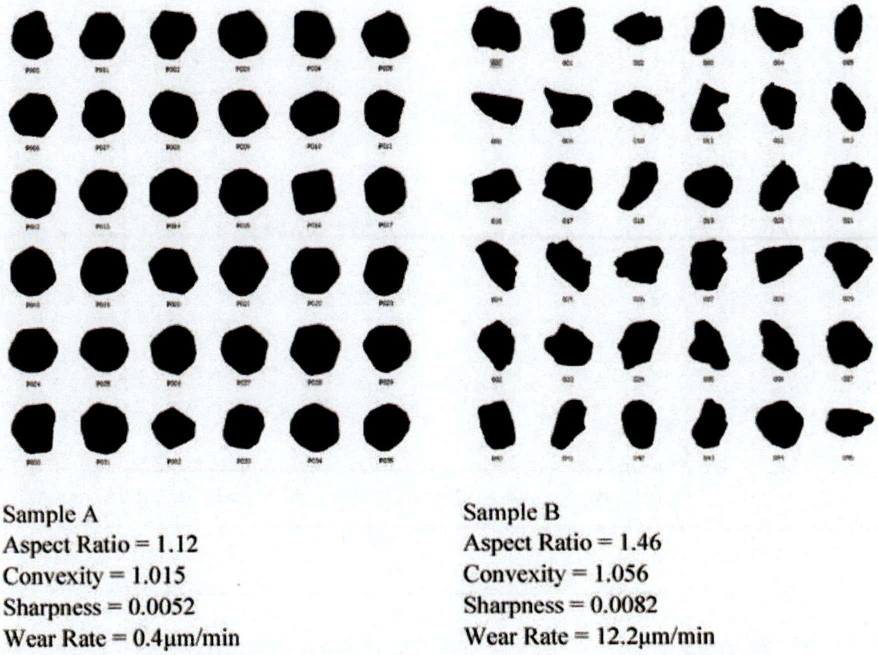

Fig. 1.7 Diamond grain projection sample extremes with associated shape characteristics and wear rates [1]

available diamond grains (Fig. 1.7). The 2D projections of the two extreme-most samples (blocky → angular) are shown in Fig. 1.4. The difference in shape is visually apparent, but the strength of the sharpness technique is that it is able to quantify much subtler differences in particle abrasiveness. Wear rate measurements were made for each of the six samples. Wear tests consisted of abrading polyurethane blocks with constant load and grinding wheels specially fabricated from the candidate particles. A 0.99 correlation coefficient was observed between wear rate and sharpness, and a 0.98 correlation coefficient with convexity. Aspect ratio only managed a modest 0.82. While sharpness correctly ordered all particle types in terms of wear rate, it was surprising to find that convexity offered almost the same performance as a predictor of wear rate, as illustrated in Fig. 1.8.

This interesting result suggests that convexity, a much more convenient parameter to calculate than sharpness, might be imbued with the qualities that make it a sound predictor of particle abrasiveness. This hypothesis needs further testing and verification. (For further detail on the various parameters discussed above see [1, 4–10].)

Particle shape is fundamental to the performance of abrasives, however, the mechanical and thermal integrity of the grains can be just as important when grinding tenacious materials, as treated in the next section.

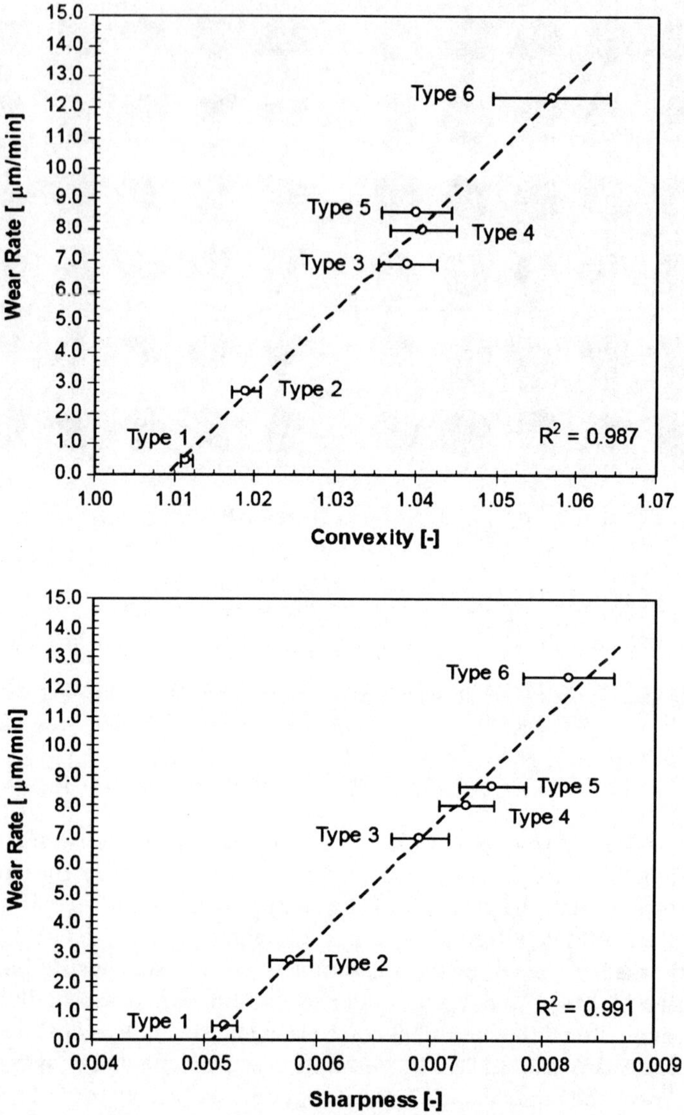

Fig. 1.8 The correlation of convexity and sharpness with wear rate experiments

1.1.2 Attritious Wear Factors

Attritious wear is wear that occurs "atom by atom" by interactions of the grain with the workpiece. These interactions are both physical and chemical, and can be quite complex. They can involve mechanical fracture (abrasion) at the microscopic scale and plastic deformation. Heat from friction and chip formation can lead to localized diffusion, chemical degradation and decomposition of the grain, and even melting. The clean surfaces exposed by the creation of a chip are highly reactive and can drive chemical reactions that would normally occur at much higher temperatures. Even the presence of oxygen in the atmosphere has a profound effect by neutralizing the clean steel chip surface. Grinding in a vacuum will generally lead to high levels of loading from metal to metal and grain to metal welding.

Hardness is the key factor in controlling attritious wear characterized by mechanical micro-fracture and plastic deformation. In general a grain has to be at least 20% harder than the workpiece to be suitable as an abrasive. Temperature plays an important factor as localized temperatures can easily exceed several 100°C and hardness of abrasives such as alumina drops markedly with temperature. The Pechiney's Abral™ Al-O-N abrasive was developed recently in part because of its high temperature hardness compared to alumina (Fig. 1.9) [11]. (It should also be noted that the Abral™ grain is not wetted by molten steel, suggesting an additional benefit of a lower tendency for welding leading to loading during the grind.)

The impact of hardness and other attritious wear controlling factors can be seen by the comparison of typical G Ratio values for the major abrasive type diamond and CBN, alumina and silicon carbide grinding various industrial workpieces (Table 1.2).

Grinding alumina the wear is essentially all mechanical for each abrasive type especially at low wheel speeds where heat generation is minimal. The effect in this case of hardness is very apparent. Diamond with its superior hardness provides a G Ratio of typically 100 times greater than the second hardest abrasive, CBN, at slightly over half the hardness. Similarly, the G Ratio for CBN is about 100 times higher than for Silicon Carbide with a similar proportional reduction in hardness. Alumina abrasive, with a hardness approaching that of the workpiece, gives a very poor G Ratio and is in effect non-functional as an abrasive.

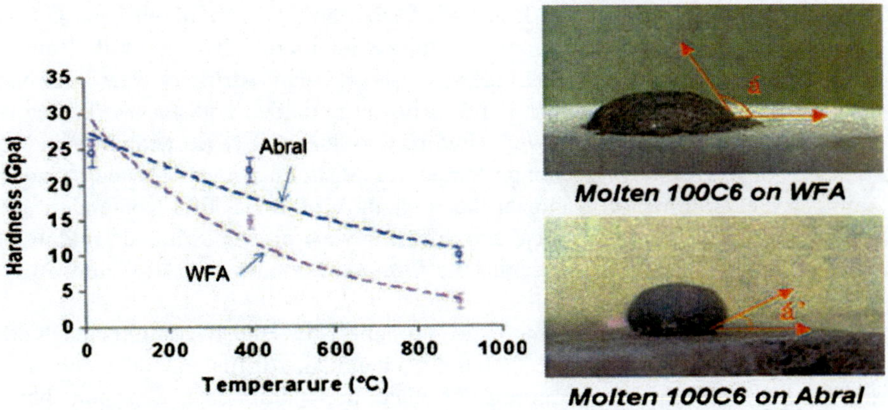

Fig. 1.9 Hardness and wetting characteristics of white fused alumina (WFA) and Al-O-N (Abral) abrasive

Table 1.2 Typical relative wear resistance values for major abrasive grain types during grinding

	Grinding alumina	Grinding steel	Grinding nickel	Grinding titanium
Diamond 9,000 Hv	100,000	1,000	100	500
CBN 4,500 Hv	1,000	10,000	5,000	100
Alox 1,800 Hv	<1	5–10	10	1
SiC 2,800 Hv	10	1–5	1	10

The situation with steel and nickel is very different. Diamond has a strong affinity for iron and even stronger affinity for nickel, both metals being true solvents of diamond at high temperature. CBN shows no such affinity. Therefore even though the mechanical wear should be in the favor of diamond by 100:1 in fact the wear rate of diamond is at least ten times higher than for CBN. This is also seen in turning even at low speeds with relatively little heat generation. For example, a diamond tool turning mild steel wears 10,000 times faster than the same diamond tool turning aluminum – silica alloy of the same hardness. The reactions can be modified; for example a diamond tool turning pure nickel wears 10,000 times faster than turning electroless nickel containing 10% phosphorus. The phosphorus appears to form a non-reactive nickel phosphide phase. Diamond is used to CNC profile the faces of large electroless nickel coated mirrors to nanometer accuracies for satellite telescopes. The reactivity of diamond with iron can also be reduced in the absence of high temperatures by the presence of free carbon in, for example, pearlitic cast iron. Grinding cast iron at high speed and removal rates, CBN will out-perform diamond by two orders of magnitude. However when honing cast iron cylinder blocks at 2–3 m/s with free graphite present, diamond will out-perform CBN by 10:1. If the carbon is present as carbide rather than graphite however, the performance of diamond and CBN is reversed. Diamond is also used to burnish hardened steel and, with minimal heat and no clean metal created to make contact with the diamond, the life of the tool is extraordinarily long. Silicon carbide also shows high attritious wear grinding ferrous materials due to chemical and diffusion reactions with the workpiece to form carbides and silicides. Even aluminum oxide will react with iron in the presence of oxygen at high enough temperatures although this is generally not a significant enough factor to impact the economics of wheel life. Komanduri and Shaw [12] has identified chemical and diffusion wear processes in the grinding of ferrous alloys with SiC, in particular the formation of carbides (Ni_3C and Co_3C) grinding superalloys.

Finally, the grinding titanium is probably the most complicated with numerous wear mechanisms including chemical, diffusion, adhesion and mechanical impacting the economics of wheel life v grain cost. Titanium shows a chemical affinity for SiC and alumina grain with reactions of the form [13].

$$\begin{aligned} \text{Silicon Carbide}: \ & SiC + O_2 \rightarrow SiO_2 + C \\ & SiC + 2O_2 \rightarrow SiO_2 + CO_2 \\ & Ti + SiO_2 + O_2 \rightarrow TiO_2 + SiO_2 \\ & Ti + SiC + O_2 \rightarrow TiC + SiO_2 \\ \text{Aluminum Oxide}: \ & 2Al_2O_3 + Ti \rightarrow TiC + SiO_2 \end{aligned}$$

Diamond also shows some affinity with titanium and will form the TiC phase at high enough temperatures but the reaction rate is not on the same order of magnitude as for the true solvent metals (e.g., Fe, Ni) with carbon. Instead we might expect some impact from high adhesion forces. CBN, or more precisely the oxide layer formed on the surface of the CBN grains from reaction with oxygen in the air or water in the coolant, is also somewhat reactive with titanium but these reactions only appear to happen at high temperatures. Again adhesion forces may be expected to be more significant. What studies have been made of wear mechanisms of CBN on titanium have generally been in turning. Here, the primary chemical wear mechanism was with the binder between the CBN grains in compact tools. With recently developed binderless CBN tools the attritious wear mechanisms were primarily a mixture of mechanical, adhesion and diffusion with no dominating chemical wear factor [14].

Kugemai et al. [13] showed that, when grinding titanium, diamond abrasive gave the highest G Ratio, lowest grinding forces and lowest grinding temperatures of all the standard abrasive types including alumina-zirconias while alumina based grains gave the worst. Kumar [15] found that a medium toughness diamond abrasive could grind titanium alloys more efficiently than CBN or SiC. Nevertheless, from the standpoint of economics, SiC is still the abrasive of choice in industry based in part on abrasive cost but more on ease of use, especially dressing of forms critical for aerospace and medical applications, while being able to generate acceptable workpiece quality. This will become an area of increased research with the commercialization of titanium intermetallics such as γ-TiAl for aerospace, power generation and diesel turbochargers.

1.1.3 Grain Fracture Toughness

Whereas hardness provides a measure for the tendency of grain to wear by attrition on the atomic scale, fracture toughness (or it's inverse term "friability") provides a measure for the loss of abrasive due to breakdown by fracturing or splintering of the grain typically at the micron level (micro-fracture) or greater (macro-fracture). The degree of fracture is in large degree dependent on grain properties such as crystal size and morphology, impurities, inclusions and pre-existing cracks, and shape. It is also very dependent on the level and nature of the forces applied to the grain during grinding and from factors in the grinding environment such as thermal shock from coolant. Attritious wear leads to the creation of wear flats that dramatically increases the force exerted on the grain and in turn leads to increased levels of fracture.

Fracture toughness, particularly of superabrasive grain, is most commonly evaluated by a vibration – impact test. A grain sample of a known particle size distribution is placed in a tube with steel ball bearings and shaken with a fixed amplitude and frequency for a given length of time. The grain particle size distribution is then re-measured to assess the level of breakdown. The grain is either measured as received to give a Toughness Index (TI) value; or after processing at high temperatures, typical of those seen in wheel manufacturing process or use, to give a Thermal Toughness Index (TTI) value. The high temperature processing can occur either in a vacuum, or in the same atmosphere as that used in wheel manufacturing, or even after mixing with wheel bond e.g. glass or vitrified frit, which is dissolved with HF acid subsequent to heat treatment. In general the TTI will be less than the TI as temperature causes the expansion of inclusions, reactions with the atmosphere, and infiltration of surface flaws with glass bond. Where the grain has previously been through a significant degree of crush processing, especially for fused alumina grain, high temperature calcining can actually increase the TTI by annealing existing cracks. Crush strength measurements are also made on single grains.

Hagiwara et al. [16] have developed a method of evaluating the grain strength from fly cut measurements using single grains. They evaluated the grain strengths in terms of a fracture onset probability but also categorized the shape of the fractured grains. They proposed ten primary edge models for the morphology of fractured grains (Fig. 1.10). Their study reports measurements made on populations of grains using examples of alumina and SiC abrasives.

Probability of survival of a grain is given by $P_t = 1 - e^{-\gamma t}$ where γ is the fracture onset coefficient. Values for γ and primary fracture modes are given in Table 1.3.

The technique readily distinguishes between alumina and SiC, showing the highly friable nature of the latter, as well as more subtle differences between various grades within a grain family. Grain toughness must be matched to both the wheel bond characteristics and the grinding conditions (Fig. 1.11).

Ideally the grain should fracture creating the loss of relatively fine particles typically at the micron or sub-micron level; a process termed "micro-fracturing". The remaining portion of the grain should remain sharp and able to cut. If the grain is too tough relative to the bond holding it, or the grinding force/grain is extremely high, then the grain can undergo total break-out or loss without doing any useful work. If the bond is strong enough to hold the grain but there are high grinding forces/grain, and/or the grain crystallite size is large, then the fracture is often more one of coarse loss of grain by "macro-fracturing" still without the full amount of possible useful work being obtained. On the other hand if the grain is much weaker than the bond and/or prone to high attritious wear due to mechanical, heat or chemical wear, then "glazing" occurs resulting in the creation of wear flats, high

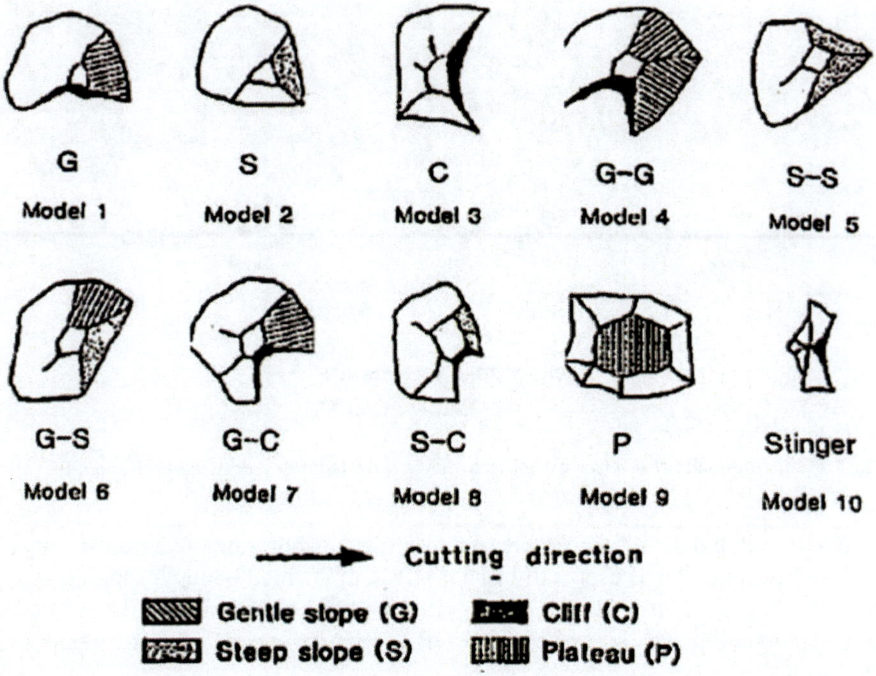

Fig. 1.10 Primary edge models for the morphology of fractured grains [16]

Table 1.3 Fracture onset and mode of fracture data from fly cut measurements [16]

Abrasive type	Fracture onset coefficient, mean value γ_{mean}	Fracture modes, most common
Alumina/2.3% titania (brown alumina)	0.046	1,10.5
White alumina	0.174	10,3,8
Single grain, temperature annealed, alumina	0.142	1,4,5
Black silicon carbide	0.407	10,2
Green silicon carbide	0.317	10,3,5

grinding forces and increased interface temperature. Higher forces will lead in turn to more fracture.

The ideal stable state for wheel wear is a limited amount of attritious wear controlled by micro-fracture. The maximum amount of wear flat area is set by the onset of thermal damage. For ferrous materials this limit is about (1–2)% of the wheel surface when using alumina or SiC abrasives, and about (4–5)% when using CBN or diamond abrasives due to their higher thermal diffusivity or ability to remove heat from the grind zone.

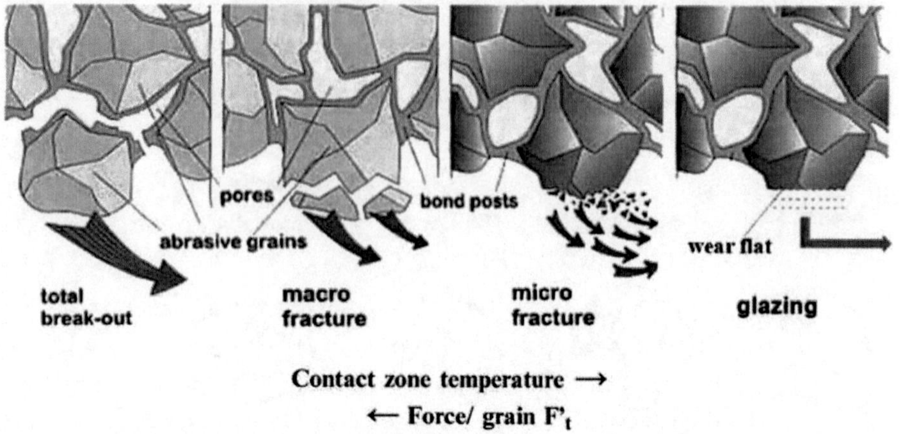

Fig 1.11 Grain/bond breakdown modes in grinding wheels (after a drawing by Rappold)

Fracture behavior is also important in terms of the grains reaction to impact during dressing. This is especially true of CBN in vitrified bonds for high production grinding. Ishikawa et al. [17] reported a study of dressing vitrified bonded wheels containing coarse grade (80#) GE 1 abrasive using rotary diamond disc dress rolls.

It was determined that at a modest crush ratio of +0.2 there was a definite shift from predominantly a micro-fracture regime at a depth of dress of 1 μm to a macro-fracture regime at a depth of dress of 3 μm (Fig. 1.12). Changes to dress depth of as small as 0.5 μm had a significant effect on power and finish As the crush ratio was increased from +0.2 to +0.8 the level of macro-fracture increased dramatically to dominate the process. These results are specific to a particularly friable grade and size of CBN typical of that used in resin and weaker vitrified bonds. It is therefore to be expected that a tougher grade of CBN or a finer grit size would require either a higher crush ratio and/or deeper depth of cut to achieve the same degree of fracture. Evidence for this is suggested in the work by Takaga et al. [18] who found that when dressing 80# GE 500 abrasive at a depth of dress of 5 μm micro-fracture dominated even at a crush ratio of +0.5 but at a crush ratio of +0.9 was replaced by macro-fracture. GE 500 is a particularly tough grade used primarily in single layer plated applications.

The control of the level of micro-fracture and the resulting wheel surface morphology is especially critical to dressing of vitrified CBN wheels with rotary diamond dressers for high production grinding. A major challenge with vitrified CBN is the initial rapid changes in grinding characteristics immediately after dress especially for relatively weak systems or burn sensitive grind operations. The effect is illustrated in Fig. 1.13 with examples by Jakobuss et al. [19] of normal force changes for different dress crush ratios and wheel speeds. The problem is in the first perhaps 5% of possible grinding between dress cycles. The effect is also seen with conventional abrasives to some extent but is fleeting being essentially complete

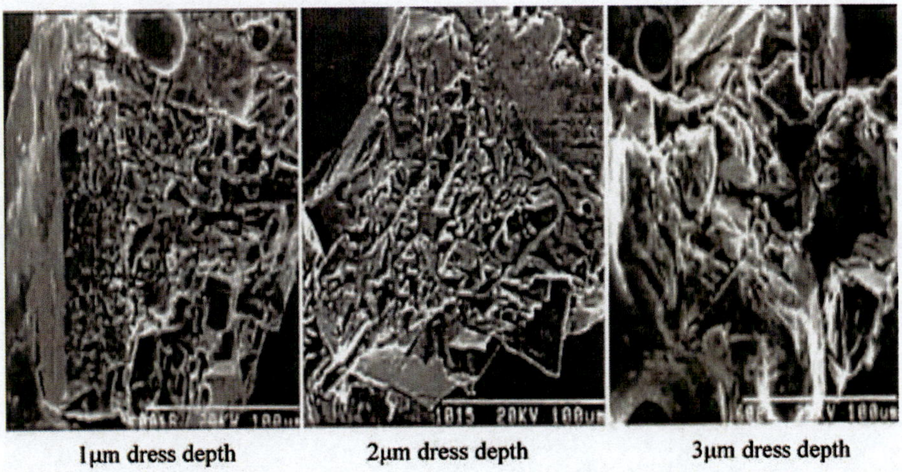

Fig. 1.12 Micro-to-macro-fracture trend rotary diamond dressing CBN at increasing dress depth

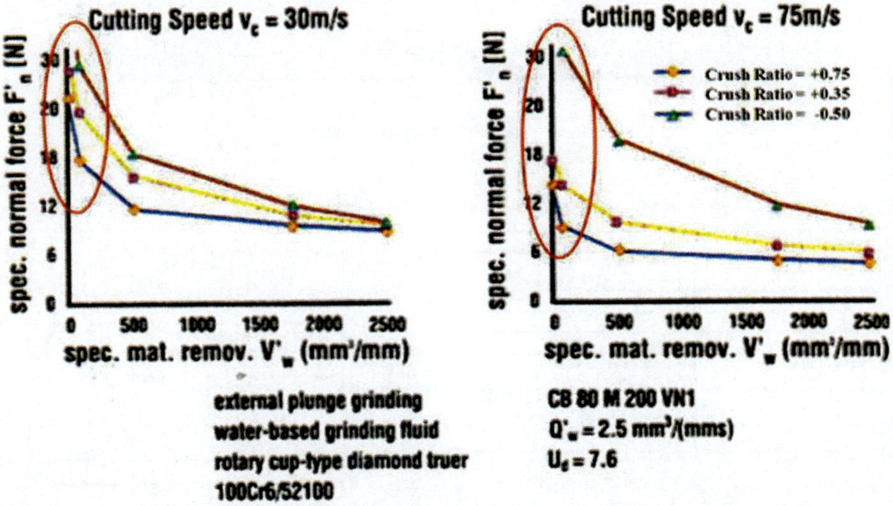

This figure will be printed in b/w

Fig. 1.13 Influence of crush ratio on normal grinding force when rotary diamond dressing of vitrified CBN wheels

before the first workpiece component has been finished ground. For a CBN ground process with several hundred workpieces being ground between dresses this rapid change in grinding force can lead to taper and burn issues or the need for special programming routines to reduce feed rates to compensate.

Fujimoto et al. [20] made a detailed study of the wear of grain in the surface of a vitrified CBN wheel by means of three dimensional multiprobe SEM and fractal dimension analysis. Grinding conditions along with force, finish and wear data are given in Fig. 1.14. They identified three stages in a complicated sequence of events as the wheel wore. Immediately after dress they observed a break-in period with higher wear and the characteristics drop in grinding forces. Observations of individual grains Fig. 1.15a, b found this drop was associated with a loss of unstable grain edges and the formation of new sharp cutting edges. Analysis of three dimensional profiles reveals that the cutting edge density is reduced (Fig. 1.16) affecting the grain density in the wheel surface to a depth of at least 20 μm. After

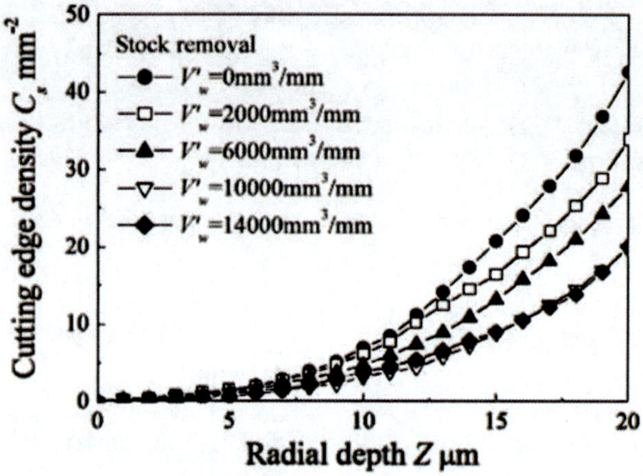

Experimental conditions

Grinding method	Surface plunge grinding, Up-cut
Grinding wheel	cBN80L100V Dimensions: ϕ200 × t10 mm
Peripheral wheel speed v_s	33 m/s
Work speed v_w	0.15 m/s
Wheel depth of cut a	10 μm
Grinding fluid	Soluble type (JIS W-2-2) 2% dilution
Workpiece	High speed tool steel (JIS SKH51) Hardness: HRC65 Dimensions: 100^l×5^t×30^h mm

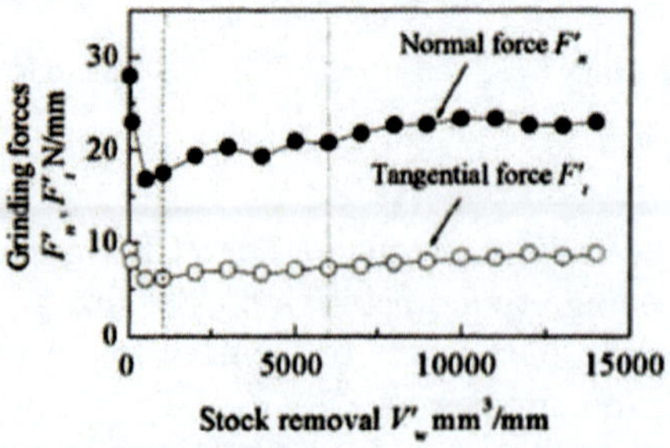

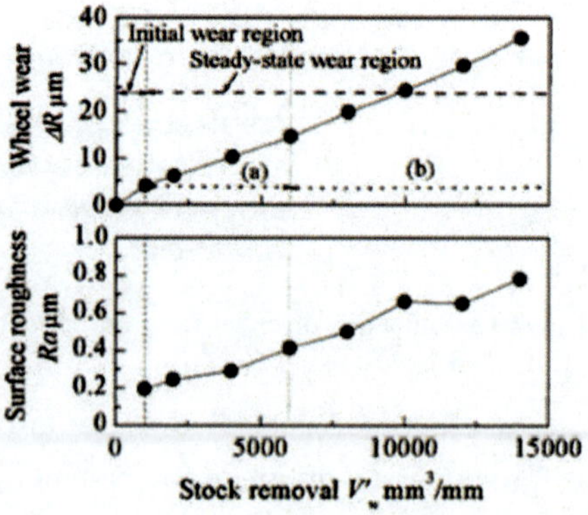

Fig. 1.14 Grinding conditions, force, wear and finish from study on CBN wheel wear [20]

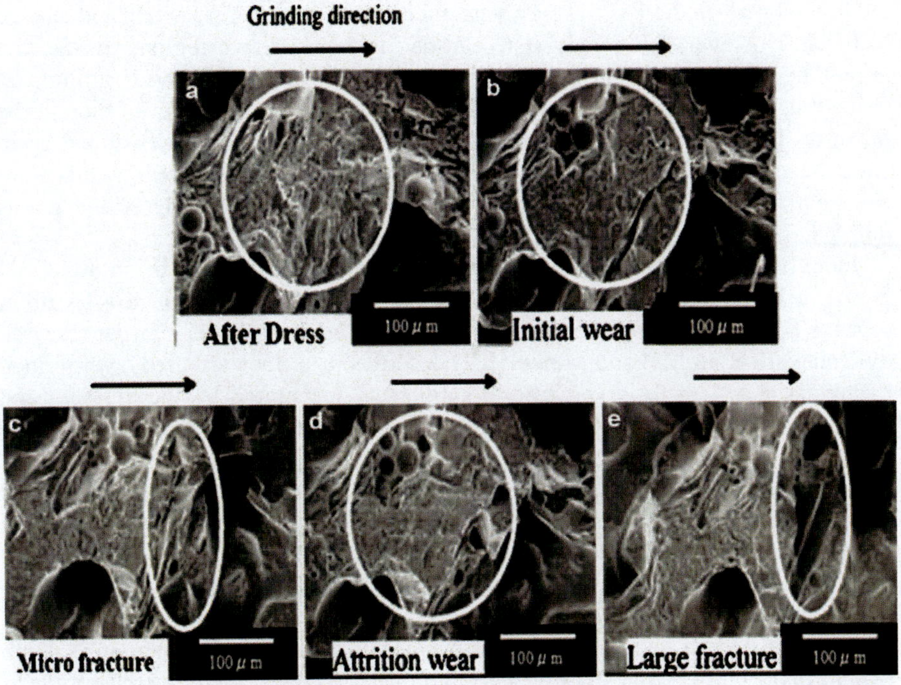

Fig. 1.15 Photographic study of CBN grain breakdown for conditions shown in Fig. 1.14 [20]

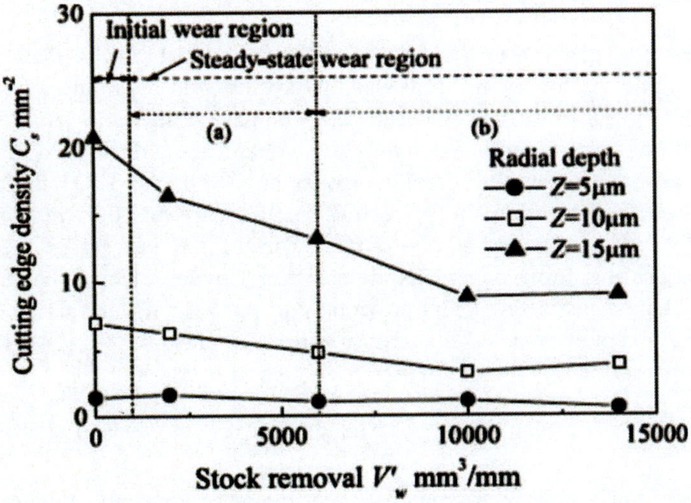

Fig. 1.16 Grain cutting edge density variation with workpiece material removed after dress and with depth below the wheel surface [20]

the initial break-in there is a steady state condition in terms of wear rate and change in finish. However, this steady state can be divided into two regions. In the first region there is micro-fracturing of the grains and signs of attritious wear building on the cutting edges. Interestingly the grinding forces climb slightly even though the finish is climbing too indicating major changes in cutting edge density and grain edge shape. The second region begins after about 15 μm of wheel wear; at this point the wear flats on the grains are pronounced and macro-fracturing of the grains dominates. The grinding forces now remain constant.

Interestingly, in a high production manufacturing environment using 80# CBN grain, the amount of wear between dresses is usually limited to about 10–15 μm in order to keep the process under control over repeated dress cycles. Earlier research by Yonekura et al. [21] and Mindek [22] described a surface affected layer termed "Tsukidashiryo", or "Active Surface Roughness" generated by the dressing and grinding processes, and varying in depth from a few microns to over 30. After the initial dress and grind cycles the surface is conditioned such that the dress amount of the second dress cycle is now critical. Over-dressing, that is removal of total depths of >20 μm, will result in a closed wheel similar to the first dress; under-dressing, that is removal of total depths of <5 μm (for 80# grain size), will result in fewer parts/dress as the cutting edge density in now too low. Much of the optimization involved with vitrified CBN processes is in the selection of the appropriate dress depth per pass to control the level of micro-fracture, and the total dress depth to control the cutting edge density. This in turns limits the amount the magnitude of the break-in period while maintaining an optimum parts/dress.

1.2 Silicon Carbide

Silicon carbide (SiC) was the first of the synthetic abrasives that ushered in twentieth century manufacturing. It was first synthesized in commercial quantities around 1891 by Dr Edward G Acheson [23] who gave it the trade name "Carborundum", and was initially produced in only small quantities and sold as a substitute for diamond powder for lapping precious stones at $880/lb (at 1,891 dollar value!). With process optimization the price fell precipitously $0.10/lb in 1938. Today (2010), the price is about $0.80/lb. The heart of the process is the Acheson resistance heating furnace, an adaption of the Cowles electric batch smelting furnace patented just a few years previously in 1885, in which quartz silica sand and petroleum coke is reacted at a temperature of around 2,400°C [24]. The overall reaction is described by the carbothermic reduction equation:

$$SiO_2 + 3C \rightarrow SiC + 2CO$$

The furnace is prepared by placing a large carbon resistor rod on a horizontal bed or trough of raw materials to which a heavy current is applied. The raw material also includes sawdust to add porosity to help release the CO, and salt to remove iron impurities. The whole process takes from 36 h to 10 days and yields typically 10–50 tons of product. From the time it is formed the SiC remains a solid as no melting occurs (SiC sublimates at 2,700°C).

Pure SiC is colorless. Two grades of SiC are produced for abrasive applications – "green" and "black". Green SiC is the purer produced from a virgin mix of sand and coke; black SiC is produced from recycled feed including amorphous SiC from previous furnacing cycles; the black coloration comes from iron impurities. Green and black products are also sorted in terms of distance from the carbon rod, with some green material being obtained closer to the electrode even with recycled feed.

Figure 1.17 shows photographs of a Saint-Gobain SiC plant in Norway. The first photograph is a view of the filled bed of an Acheson furnace, the second photograph shows the product after it has been removed from the furnace with the carbon rod still embedded in the center of the ingot. After withdrawal of the rod and

Fig. 1.17 SiC Acheson furnaces and raw product as removed from the furnace (courtesy Saint-Gobain)

removal of the surrounding amorphous SiC the remaining mass is 98% SiC to be further processed.

The Acheson process has remained essentially unchanged for many decades. As such the primary driver for the location of manufacturing has been cheap, easily available electric power – most commonly hydro-electric power. The original Acheson furnaces were driven by power from Niagara Falls although North American production is now severely challenged by operating costs; global manufacturing is now dominated by China with almost half the market. Other countries with significant grain production include Brazil, Russia and Vietnam. More recently Czech, Spain and even Bhutan have come on line although much of this product is for other applications than abrasives.

There has been a large upsurge in interest in SiC for applications such as tank and body armor, heat resistant bodies for kiln ware, high temperature electronic devices, and wire saw applications for electronics spurring research into both manufacturing processes and SiC material properties. Silicon carbide is the hardest of the conventional abrasives with a Knoop hardness of 2,500 and a surface Knoop microhardness of (2,900–3,100) kg mm^2 at room temperature. Microhardness falls off with temperature as shown in Fig. 1.18 [25, 26].

Green Silicon Carbide is the higher purity silicon carbide manufactured with typically >98.5% of SiC. The crystal type is alpha phase silicon carbide in the form of hexagonally shaped platelets. Black Silicon Carbide is of a lower purity (95–98%) and consists of the alpha phase with both hexagonal and rhombohedral forms. The green is the slightly harder but more friable and angular. For this reason green silicon carbide is used for grinding hard metals such as chilled cast iron rolls, titanium, and metal and ceramic cutting materials. Black silicon carbide is used more for grinding of soft non-ferrous metals and non-metallics like rubber, wood, ceramics and glass. Both SiC grades are more friable than fused alumina grain (Fig. 1.19). SiC does show reactivity or solubility with iron, limiting its use grinding ferrous materials. It is also susceptible to oxidation at higher temperatures. Thermal properties are shown in Fig. 1.20 [27].

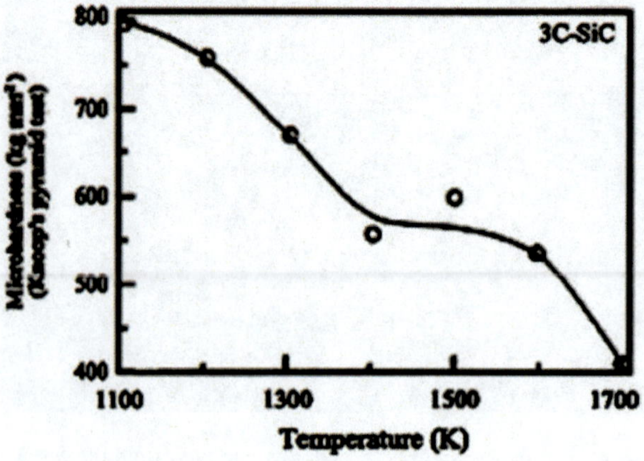

Fig. 1.18 Hardness variation with temperature for SiC

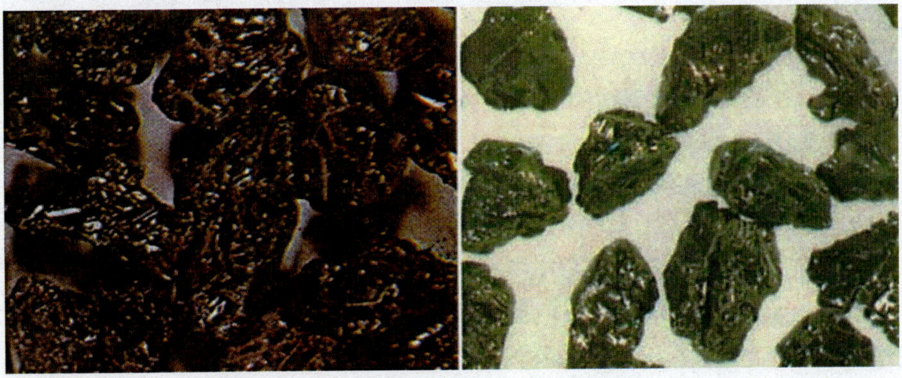

Fig. 1.19 Examples of *black* and *green* SiC

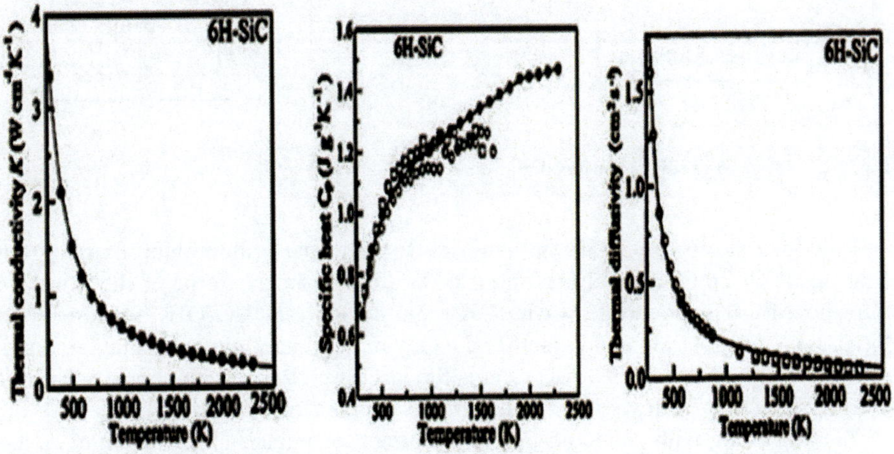

Fig. 1.20 Thermal properties – κ, C_p and $(\kappa C_p)^{1/2}$ of SiC

1.3 Fused Alumina

The second great innovation in abrasive technology at the turn of the twentieth century, after the Acheson process for SiC synthesis, was the invention of the Higgins electric arc furnace for the production of electro-fused alumina (or "Alundum") by Aldus C. Higgins of the Norton Company in 1904 [28]. Prior to this, wheel makers had used naturally occurring aluminum oxide in the forms of the minerals emery and corundum but the variability in chemical and mechanical properties made controlling wheel formulations difficult. Emery abrasive use now is limited to coated paper.

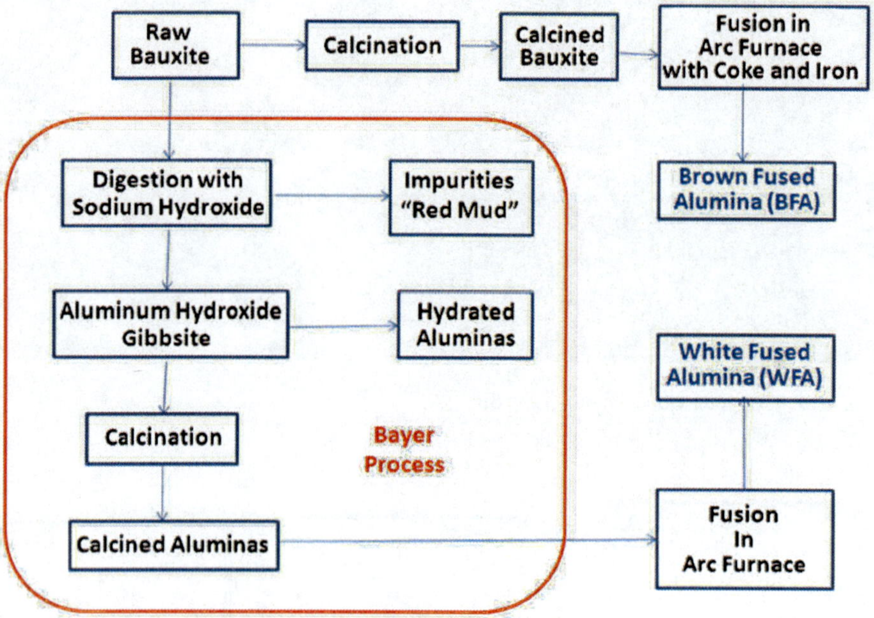

Fig. 1.21 Processing routes for fused alumina

Raw bauxite is the starting material for all fused aluminum oxide grain (Fig. 1.21). It consists of up to about 60% alumina in the form of the minerals gibbsite $Al(OH)_3$, boehmite γ-AlO(OH), and diaspore α-AlO(OH), together with the iron oxides goethite and hematite, the clay mineral kaolinite and small amounts of anatase and titania, TiO_2. Australia is the largest producer with almost a third of world production, followed by China, Brazil, Guinea, and Jamaica [29].

Bauxite along with coke and iron is the direct feed material for fusion to create the Brown Fused Alumina (BFA) a family of abrasives containing controlled amount of up to 4% titania. Bauxite can also be purified prior to fusion by the Bayer Process invented in 1887 by Karl Bayer in Russia. In this case, bauxite is heated in a pressure vessel with sodium hydroxide solution at 150–200°C. After separation of the iron based residue (red mud) by filtering, pure gibbsite is precipitated by cooling the liquid and seeding with fine grained aluminium hydroxide. Gibbsite is then converted into aluminium oxide by calcining. The Bayer process removes almost all of the natural impurities present in the raw bauxite, but leaves behind 0.1–0.4% soda (Na_2O) in the purified calcined alumina. This is the feed for the production of White Fused Alumina (WFA) and its family of abrasives. The Bayer Process increases the cost of WFA feed material by about a factor of 5 compared with bauxite feed for BFA. A Higgins electric arc furnace consists of a thin steel or aluminum shell on a heavy metal hearth (Fig. 1.22). A wall of water running over the outside of the shell cools it sufficiently to maintain the shell integrity in combination with a thin layer of aluminum oxide that forms on

Fused Higgins shell in water cooling

Electrode arm assembly

5 MVA Higgins EAF in operation

An electrode of Higgins EAF under power

Fig. 1.22 Examples of Higgins furnaces and equipment details (courtesy Whiting Equipment, Canada Inc.)

the inside due its extremely poor thermal conductivity. Steel was historically the normal shell material, as it has a relatively high melting point, but aluminum is now preferred especially for WFA fusions to prevent discoloration from iron rust contamination Feed material is poured into the bottom of the furnace and a carbon starter rod laid on it. Two or three large vertical carbon rods are then brought down to touch and a heavy current applied. The starter rod is rapidly consumed but the heat generated melts the bauxite, which then becomes an electrolyte. Feed material is added continually over the next several hours to build up the volume of melt to as much as 20 tons. Current flow is controlled by adjusting the height of the electrodes which are eventually consumed in the process. The reaction conditions of a BFA fusion as a result of the added coke which reacts with the oxygen in the impurities to produce carbon monoxide, reducing the silica to silicon and iron oxide to iron which combine along with the added iron to form a heavy, highly fluid, ferrosilicon phase. Silica is also lost as fume due to the high temperatures. In addition titania levels may be adjusted by reduction to titanium that precipitates out with the ferrosilicon. A typical fusion of 30 tons may take around 20 h to completely fill and melt the contents of the furnace pot, cooling time may be up to 4 days. The cooling is very directional; the insulating outer layer of the melt is quenched next to the pot and highly microcrystalline. There is then a large crystalline, dendritic, growth region in a radial direction with solidification in towards the center of the pot as heat flows from the center out. The pot has a high profile with an aspect ratio of about 1:1. Impurities will concentrate in the liquid phase in the center and towards the bottom of the forming ingot. After cooling the ingot must be broken up and hand sorted to remove the primary concentration of impurities. Additional iron and ferrosilicon are subsequently removed with magnetic separators during crushing (Fig. 1.23).

The reaction conditions for a WFA fusion are in general not considered reducing in that the only carbon present is from the electric arc and starter rods. The biggest concern is the conversion of the residual soda from the Bayer process feed into Sodium β-Alumina which crystallizes as soft hexagonal plates in the alumina. As Sodium β-Alumina has a lower melting point than alumina it will again concentrate in the in portions of the ingot that solidify last. The Higgins furnace has evolved from an original design with a small capacity of 1–5 tons to furnaces today of up to 40 tons with pot diameters of up to 3.5 m (12 ft) and a power supply of up to 4 MVA. It requires 2.2 MVA h to produce 1 tonne of BFA and 1.5 MVA h to produce 1 tonne of WFA. However, with increasing capacity and efficiency demands there has been a move towards even larger tilting furnaces up to 6 m in diameter that can pour the molten alumina into pots with water cooled hearths. These furnaces use a power supply of as much as 10 MVA or even greater and can pour up to 24 tons every 4 h whilst maintaining a more consistent batch to batch chemistry.

Pour pot design for use with tilt furnaces can have a major impact on grain structure and chemistry (Fig. 1.24). For example for a WFA fusion pour into a high profile pot the cooling process and output is similar to a Higgins furnace i.e. a large alumina crystallite size with dendritic growth and a very low sodium β alumina content after sorting. However pouring into a low profile pot (aspect ratio $\ll 1$) onto a cold hearth results in a much faster cooling rate, a fine crystallite alumina structure and a much more evenly dispersed – but higher – sodium β-alumina content. The high thermal gradient when cooling a WFA ingot in a deep profile pot causes crystallization of α-alumina in a dendritic habit made up of inter-grown rhombohedra extending along the thermal gradient. This type of crystal is the result of the edges of the rhombohedron growing much faster than the faces (Fig. 1.25). Abrasive grain made from this will tend to fracture in relatively large fragments along well defined planes but be very self-sharpening.

Crystallization in low profile pots will show structures with less directional growth and that are finer in crystal size resulting in smaller fragments during grain fracture; the material is also about 10% softer from higher sodium β-alumina contamination. For BFA fusions in tilt pour furnaces about 25% of the furnace content is poured each time while most of the ferrosilicon collects at the bottom where it can accumulate over many regular pours until it is removed in a "deep

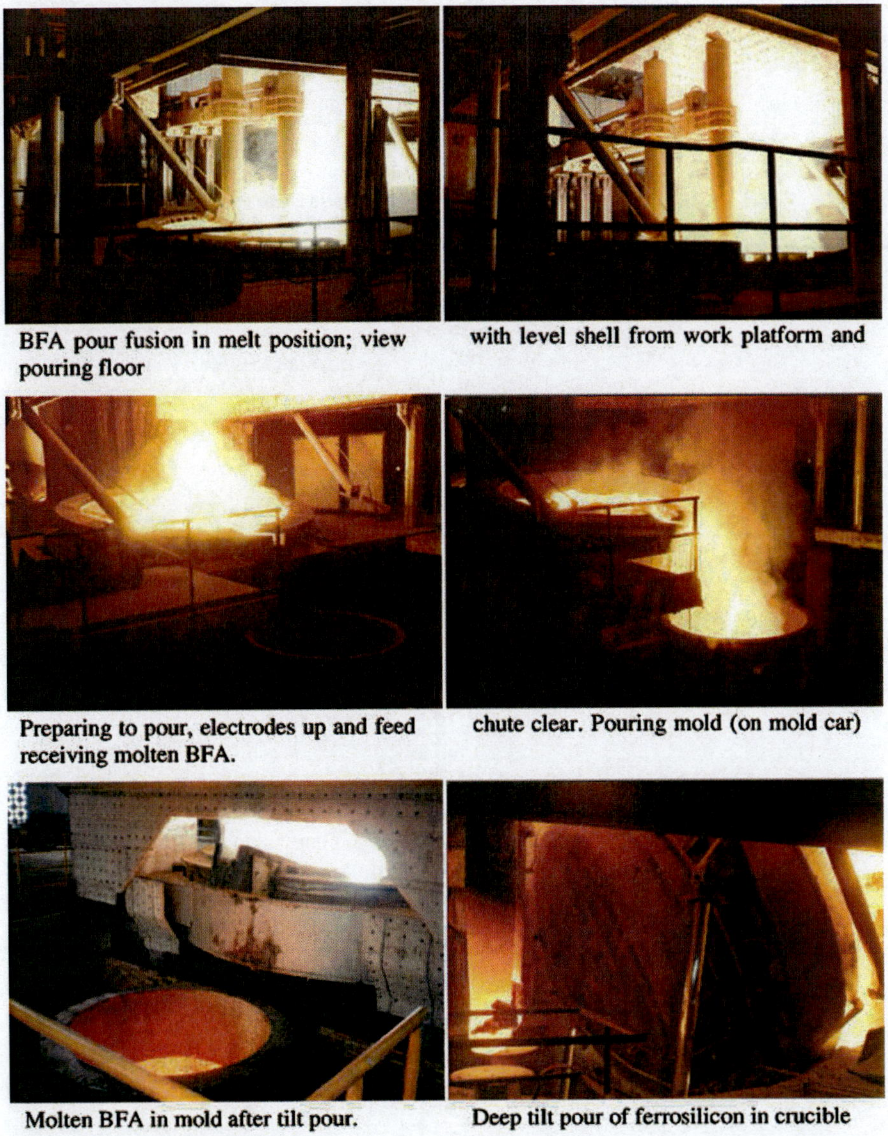

Fig. 1.23 Examples of electric arc tilt pour furnaces and operation (courtesy Whiting Equipment, Canada Inc.)

pour" into a rail cart with a bed of sand. The quality and consistency from large pour furnaces tends to be better because it is less expensive to monitor per tonne taking regular dip samples; in particular avoiding over-reduction of the titania. Pour molds tend to be low profile and the resulting grain a mixture of dendritic with finer equiaxial α-alumina together with some ferrosilicon inclusions. For further

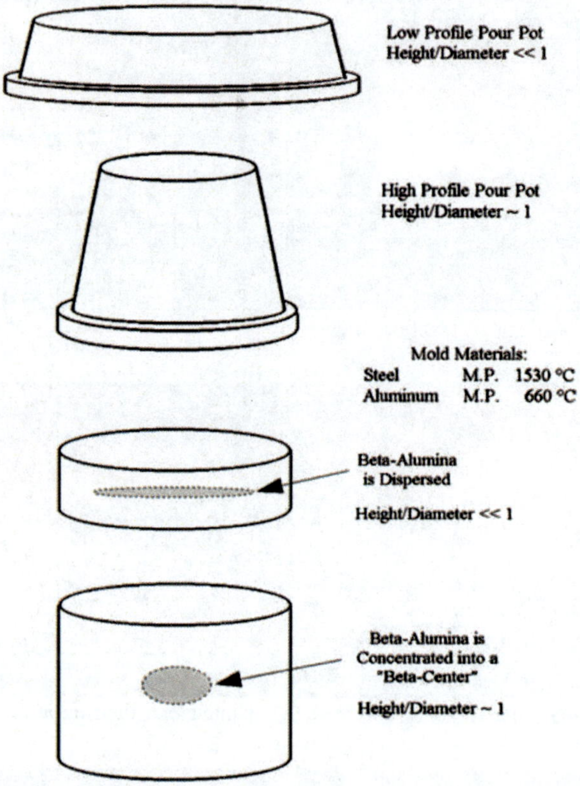

Fig. 1.24 Pour pot designs and resulting distribution of sodium β alumina content

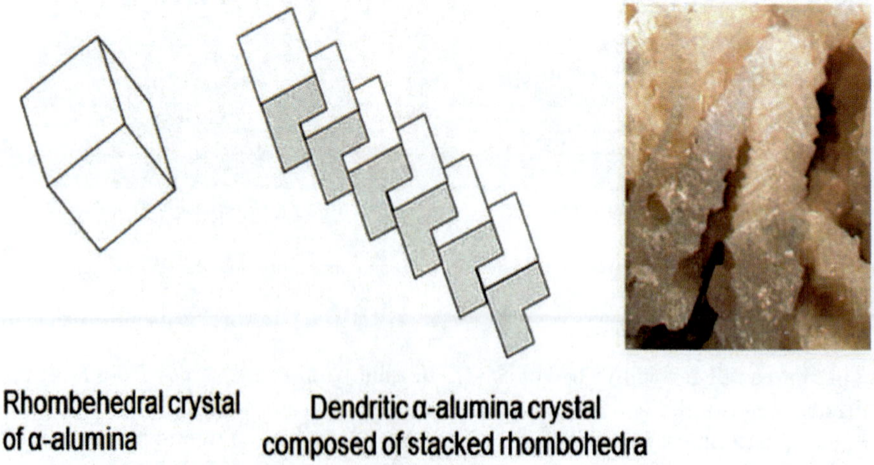

Fig. 1.25 Structural characteristic of WFA grain [30]

discussion on grain fusion and furnacing see Wolfe et al. [30], Lunghofer et al. [31], Whiting Equipment Canada (n.d.) [32].

The fusion of "Alundum" began over a 100 years ago on the shores of the Great Lakes using cheap hydro-electric power from Niagara Falls. Today, 2010, only three plants remain in N America with just 5% of global capacity although fused alumina has been substituted to a significant degree with more recent ceramic technology and manufacture as discussed below. In the last 10 years China has increased its capacity for fused alumina, especially BFA, to over 60% USGS [33, 34] with integrated manufacturing close to the bauxite mines and the largest capacity tilt pour furnaces. Eastern Europe, India, South Korea, and South America also continue to increase their prominence in world markets. Low cost electrical power availability, furnace capacity, quality control, and cost of raw material sourcing should be the dominant factors influencing the relative competitive stance of each.

1.3.1 Grain Types

Properties of the grain depend both on the fusion process and chemistry but also on the subsequent comminution process. The ingot is initially split and sorted and then pre-crushed in steel jawed Barmac and beater crushers. Further crushing is produced by passing the material through roll crushers. All these processes are high impact and will create major fractures resulting in a grain that is sharp edged, flawed and anisotropic, typically like a sliver in shape. Subsequent processing in steel or rubber lined ball mills will have a tendency reduce grain size by rounding of the grain edge. In this manner it is possible to control shape to a degree to have either angular or blocky forms from the same material. The grain availability can be divided into BFA and WFA based families.

Brown fused aluminum oxide: includes contains 2–4% titania which enhances toughness. This is still the most widely used abrasive in wheels to grind high-tensile-strength materials, and for rough grinding, deburring and snagging, as well as to cut low-alloy, ferrous materials and is generally viewed as the "workhorse" of the industry. Brown Fused Alumina is a tough, sharp but blocky abrasive. Depending on the processing regimes the grain is typically about 50% single crystal and can be provided in high, medium and low density based on shape packing characteristics. The grain may also be calcined after sizing to toughen it by annealing cracks generated in the crushing processes. The material is sometimes termed blue fired BFA as the grain changes color due to surface oxidation of impurities. Specialty coating such as silane (for resin bonded wheels to resist coolant interactions) or red iron oxide(for resin and rubber bonded wheels to increase surface area) may also be applied.

Low titania ("light" or "semi-friable") brown fused aluminum oxide: has 1–2% TiO_2 content, and is used in bonded or coated applications that require an abrasive that is slightly tougher than white aluminum oxide. Reducing the titania content reduces the abrasive's toughness, but increases its friability. Light BFA is commonly used in depressed center wheels, cut-off wheels, and for surface and cylindrical grinding of heat sensitive metals, alloys etc. where cool but fast cutting is required. The grain can be supplied with similar post and surface treatments as regular BFA.

White Fused Alumina: is standard multicrystalline WFA with sodium β-alumina contamination and is the most friable grain in the fused alumina family. It is considerably harder than BFA. Commonest applications include the grinding of tool, high-speed and stainless steels (Fig. 1.26).

Single crystal white fused alumina: is single crystal grain that has been produced in deep pour fusion pots and separated from any sodium β-alumina contamination. This is the hardest and most brittle of the alumina family of grains used most commonly for grinding tool and very high alloy steels that are very sensitive to heat.

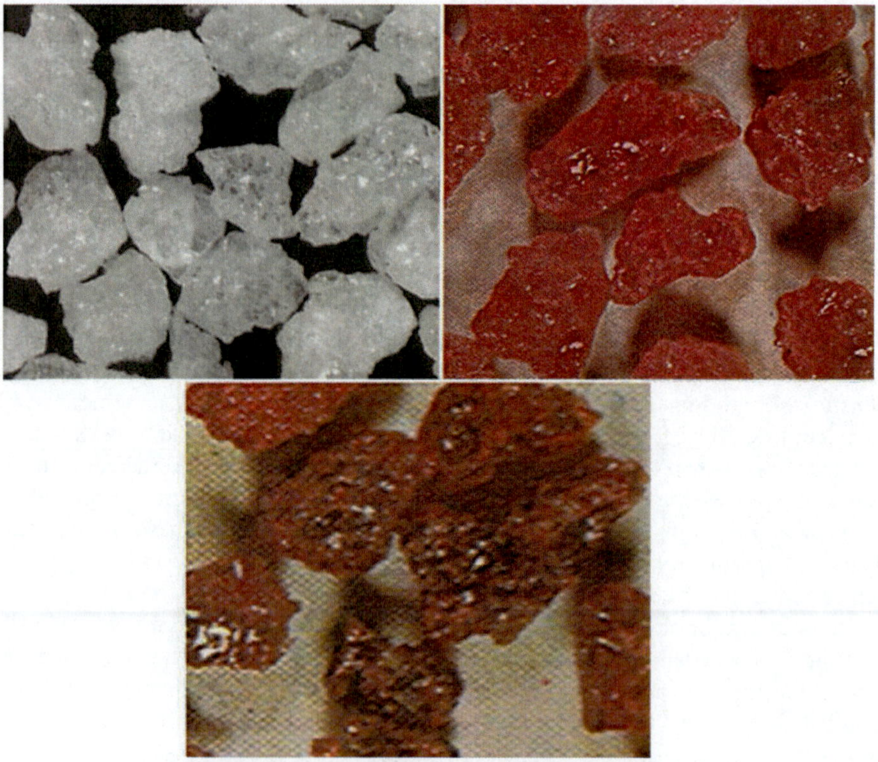

Fig. 1.26 Examples of *white*, *pink* and *red*-fused alumina grain

Pink alumina: is WFA to which <0.5% chrome oxide has been added in the fusion process to produce a grain that is slightly tougher than regular WFA used for grinding unhardened high alloy steels (Fig. 1.26).

Ruby alumina: is WFA to which 3% chrome oxide has been added to provide additional toughness over pink alumina (Fig. 1.26).

It can be inferred there is a steady increase in toughness but reduction in hardness in the following order:

Single crystal WFA → WFA → pink WFA → ruby WFA → light BFA → BFA → Blue fire BFA

In general the wheel maker will blend various grain types and sizes to combine the properties of each. In addition to chrome, other metal oxide additions have been investigated including vanadium and beryllium but not found to be commercially viable.

Sintered Alumina: is a family of grains developed in the 1950s produced from unfused alumina. Several processes exist based on both raw bauxite and Bayer processed aluminas. The most common is to use a feed material of raw bauxite milled to <5 μm. The mix with bonder is first extruded to produce rods which are cut into short cylinders or cones in the green state. They are then fired in rotary kilns at (1,350–1,500)°C using natural impurities in the bauxite as sintering agents, (Fig. 1.27), [35]. The resulting grain is extremely tough especially at the relatively large sizes the technology allowed to be produced (8#–20#) and the material found great success, until the advent of alumina-zirconia grain, in billet conditioning and other rough grinding operations. It is still used as a blend component with alumina-zirconias.

Fig. 1.27 Example of sintered extruded brown alumina grain (courtesy Saint-Gobain Abrasives)

1.4 Alumina Zirconia

Zirconia is a very high temperature refractory material, tougher but softer than alumina. It also has a higher melting point than alumina making fusion based on the Higgins furnace more demanding in terms of containment and control.

Fortunately, from a manufacturing viewpoint (if not a grinding perspective) the thermal conductivity of zirconia like alumina is very low making the Higgins fusion route still possible. However, as can be seen the liquidus curve for the zirconia – alumina system (Fig. 1.28) [36], compositions of these two materials combined, with the zirconia contents kept under about 65%, have comparable or lower melting points to alumina alone. This allows for relative ease of control of a combined alumina-zirconia fusion process including pouring from a tilt furnace.

Interest in zirconia as a potential abrasive grain or component of grain has been evident from the literature since at least the mid 1950s, e.g. [37, 38], in part because a relatively pure form known as baddeleyite began being produced in quantity at that time as a by-product of heavy metal mining especially of uranium ores in countries such as Russia, Brazil and South Africa. Another much commoner source of zirconia is as zircon (zirconium silicate $ZrSiO_4$) found as sands in USA, Australia and Brazil. Zircon sand can be refined by fusing it with coke, iron and lime until the silica is reduced and separates out as a denser, relatively low viscosity ferrosilicon liquid. Alumina – zirconias can be produced in a similar way by addition of Bayer process alumina to the fusion. Zirconia has the interesting

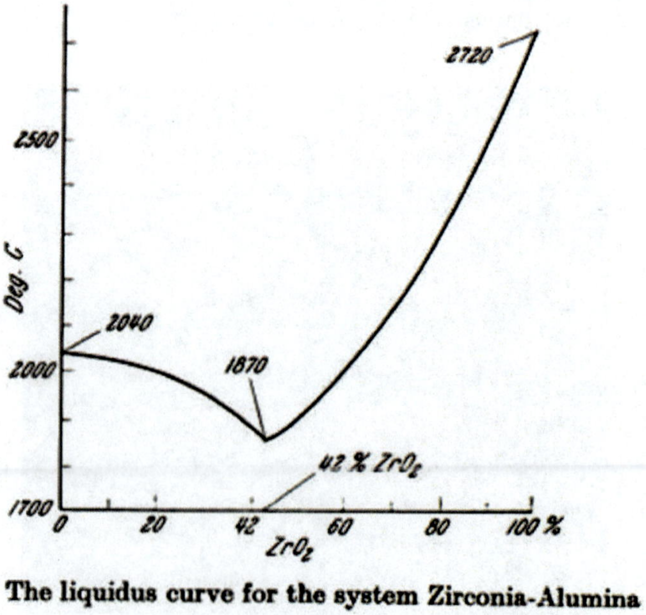

Fig. 1.28 Phase diagram for the zirconia-alumina system

property that it is metastable in the tetragonal state at certain crystal sizes when held under constraint. For pure zirconia the grain size upper limit is about 0.1–0.3 μm. With additions of small amounts of the alkali oxides CaO or MgO, or rare earths oxides such as Y_2O_3 or CeO_2, this limit can be raised into the micron range. Without constraint the tetragonal crystal converts to the monoclinic phase with a significant increase in volume of about 6%. If a crack from an active fracture intersects with a tetragonal crystal it releases the constraint but in the process the volume expansion to the monoclinic phase dissipates the ability of the crack tip to propagate. The result is an increase in the K_{1C} fracture toughness of the grain of an order of magnitude.

The technical superiority of a fused grain of zirconia over one of alumina especially at very coarse sizes for rough grinding was recognized by the mid 1960s but was cost prohibitive, although alumina-zirconia blends showed advantage, [39]. It was also recognized that an alumina-zirconia eutectic produced a strong structure due to a uniform dispersion of fine zirconia crystals in an alumina matrix. However, any excess alumina or zirconia from the eutectic would grow to a considerable crystal size depending on cooling rates from the melt. Rapid quench was therefore identified as a necessary pre-requisite of a processing route, [40]. This required quench rates of 100°C/s, two orders of magnitude faster than previously. Numerous attempts were made during the late 1960s and 1970s to develop a viable process [41–49], using various inert cooling media or hearth plates, but it was a process developed by Scott [50, 51], of the Norton company that was to prove commercially and technically effective. The process from the original patent is illustrated in Fig. 1.29.

Molten alumina-zirconia from an electric arc tilt furnace is poured into the relatively thin spaces between a plurality of relatively thick heat sink plates of graphite or iron as they pass underneath, before being emptied at a discharge station by the plates separating. The result is a very fine structure of α-alumina with high tetragonal zirconia content. The zirconia is in the form of rods (or platelets) which, on the average, are less than 0.3 μm in diameter. The solidified melt is made up of cells or colonies typically 40 microns or less across their width. Groups of cells having identical orientation of microstructure form grains which typically include from 2 to 100 or more cells or colonies [52]. Figure 1.30 shows a TEM micrograph that illustrates the fine, rod-like zirconia structures within the larger cell [53].

After solidification the material is comminuted by standard methods of crushing, milling and sizing to product grain. The processing will lead to some conversion of the tetragonal to the monoclinic phase depending on the processing energy and especially on the final grain size. The Smallest grain will lose much of its tetragonal toughening favoring this type of grain for use in coarse-sized, roughing operations.

The vast majority of alumina-zirconia grain for grinding wheels contains 25% zirconia, and sold under the trade names of ZF or ZS alundum, depending on the comminution method, and used in hot pressed resin bonds for rough steel, titanium and nickel alloy billet conditioning, or for foundry snagging. Grain size can be as coarse as 4# ("0.26" or 6.8 mm) and used either as a single grain type or blended to include extruded sintered brown alumina (for finish), SiC (grinding titanium) or

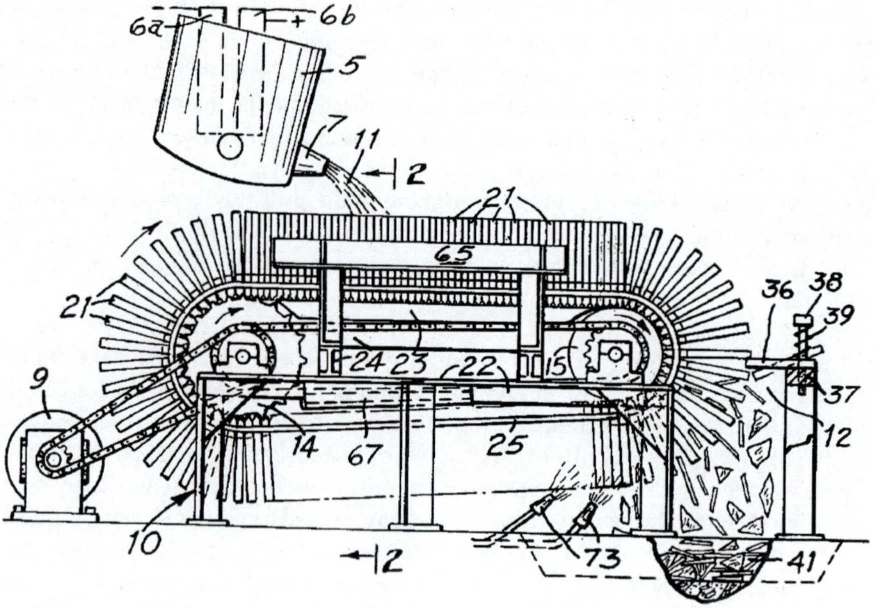

Fig. 1.29 Scott's patent detail for produces rapid quench alumina-zirconia

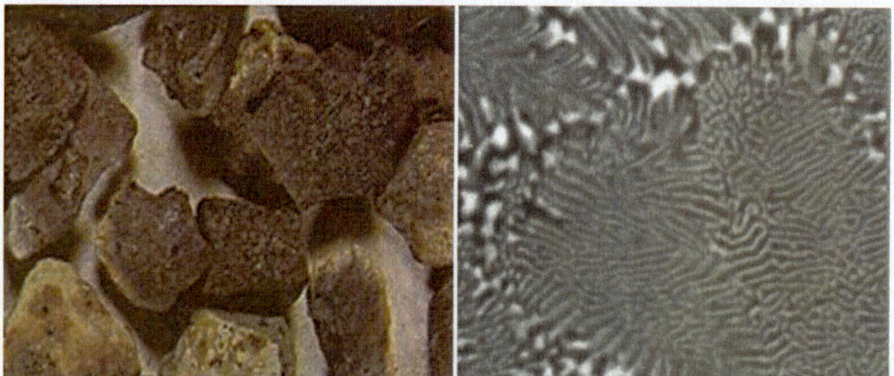

Fig. 1.30 Fused alumina-zirconia grain and a TEM micrograph of its rod-like zirconia structures[53]

regular alumina. Billet conditioning is very aggressive form of grinding. The operation is run dry with wheel speeds up to 80 m/s (16,500 sfpm) and spindle power as great as 500 Hp on the most modern equipment. The workpiece is still often red hot from the furnace. Metal removal rates are extraordinary and can exceed 2,500 lbs/h ($Q' \approx 40$ in.^{3}/in./min or 400 mm^3/mm/s) on steel or 400 lbs/h ($Q' \approx 12$

in.^{3}/in./min or 120 mm^3/mm/s) on titanium, far exceeding most other metal removal processes. Other applications include the re-grinding of rail track in situ to remove fatigue cracks using special trains travelling at speeds of up to 6 mph (10 km/h). Grinding dry, in all these processes the grains self-sharpen from cracks generated by thermal shock (Fig. 1.31).

Eutectic alumina-zirconia grain containing 40% zirconia is also produced, and sold under brand names of NZPlus, NZ®, NZP® and Norzon®. It is used primarily for coated applications and as such requires a different balance of hardness (−) and toughness (+) properties compared to grain for grinding wheels.

Fig. 1.31 Examples of (dry) rough grinding using grinding wheels containing coarse alumina-zirconia grain

1.5 "Ceramic" Sol Gel Alumina Abrasives

The development and commercial success of first the sintered extruded alumina family of grains and then the rapid chilled fused alumina-zirconia grain had a major impact on the research programs of abrasive manufacturer in regards to the importance of control of grain crystal size. Furthermore, for alumina grain it was known that reducing the crystal size from the macro scale equivalent to a single crystal per abrasive grain, common in fused material, to micron or ideally <0.5 micron crystalline structures significantly enhanced grain properties such as hardness (Fig. 1.32) [54].

The response was, rather than using traditional fusing or sintering processes with their general limitations on cooling and crystallization rates, to consolidate microstructures from finer building blocks by sintering well dispersed submicron precursors by the so called "sol-gel" route. This allowed the consolidation of an α-alumina based sub-micron, highly homogeneous and fully densified grain structure. The starting point of this new process is the manufacture of Boehmite, γ-aluminium oxide hydroxide γ-AlO(OH) from a modified version of the Ziegler process originally developed for the production of linear alcohols [55]. The material is produced as a sub-micron, narrowly sized powder which when mixed with water and a suitable acid dispersant forms an agglomerate-free sol-gel of aluminum hydrate ($Al_2O_3{\cdot}H_2O$), with a dispersant size of about 100 nm. The sol-gel is then dehydrated/shaped and sintered (Fig. 1.33).

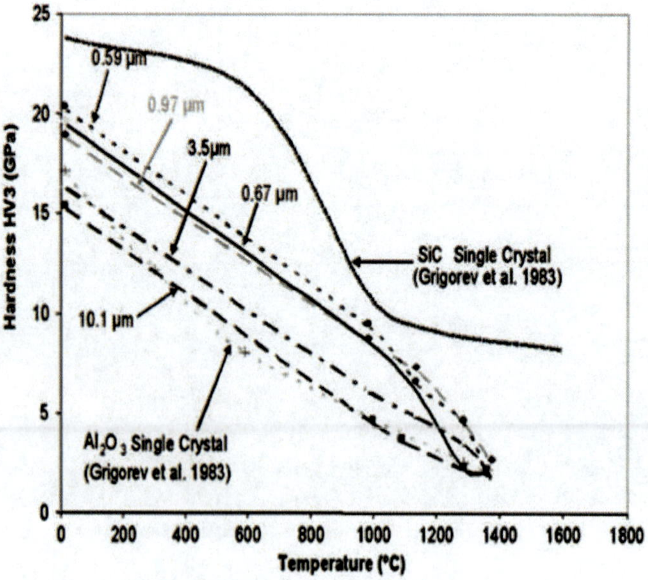

Fig. 1.32 Effect of crystal size on alumina grain hardness

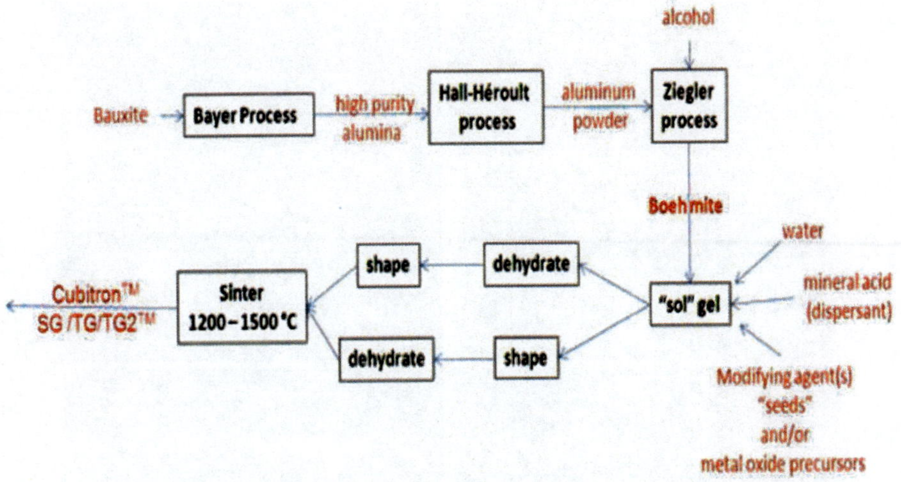

Fig. 1.33 Manufacturing route for production of "ceramic" alumina grain

The biggest hurdle to overcome was in the sintering process to maintain a uniform sub-micron crystal size and full densification. Firing of a sol-gel from a standard commercial boehmite at 1,400–1,500°C produces a large amount of porosity and relatively large grains of up >1 μm in size. This is believed due to a high activation energy to convert from a transitional τ to α-alumina phase resulting in infrequent nucleation with rapid uncontrollable growth rates. Attempting to control growth rates with lower temperatures, e.g., 1,200°C merely leads to larger crystals with higher porosity. There are two routes that have been developed to reduce the activation energy and control crystal size and densification. The first is the creation of a bi- or multi-composite structure through the use of modifying agents, the second is the controlled creation of a single α-alumina structure through the use of seeding agents (Fig. 1.34).

The earlier patents report the use of magnesia [56] which upon sintering forms a bi-composite structure of α-alumina plus a spinel structure of magnesium aluminate at about 25% by volume as shown in Fig. 1.34b. Note the fine needle like spinel structure and the still relatively coarse α-alumina phase. This particular grain was used primarily for low force coated abrasive applications. Later, numerous patents report various multi-phase systems using various modifying agents including zirconia, manganese oxide, chromia, nickel oxide and numerous rare earth oxides. One particularly effective material contains magnesia together with yttria and other rare earth oxides such as lanthana and neodymia to produce a dense and hard (19 GPa) grain. In Fig. 1.34c, the microstructure shows a finer α-alumina phase (although still relatively coarse compared to the feed material) but with a sub-micron "magnetoplumbite" type structure of needles/plates formed from the modifiers [57]. The structures created by the modifiers are believed to provide strength akin to rebar in reinforced concrete. This particular grain is the 3M™ Cubitron™ 321 grain [58].

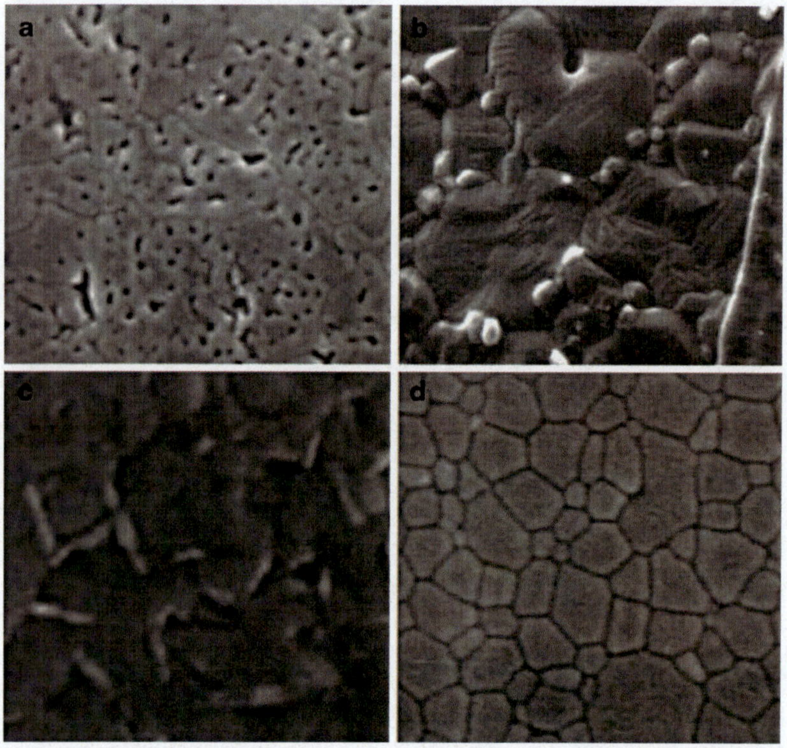

Fig. 1.34 (**a**) Sintered alumina microstructure from Boehmite with no modifying agent (image size 3 × 3 μm); (**b**) Sintered alumina microstructure from Boehmite, magnesia modifying agent (image size 3 × 3 μm); (**c**) Sintered alumina microstructure from Boehmite magnesia, yttria, lanthana and neodymia modifying agents (image size 1.5 × 1.5 μm); and (**d**) Sintered alumina microstructure from Boehmite with seeding agent (image size 1.5 × 1.5 μm)

The alternate route to controlling crystallization rates is by "seeding" the sol gel with nano-sized (<100 nm) α-alumina, or other materials with a crystallographic match to α-alumina such as α-ferric oxide or various titanates. Additions of 1–5% of seeding agent creates a heterogeneous nucleation condition by increasing the number of nucleation sites from 10^{11} to $10^{14}/cm^3$, and an average crystal size of about 400 nm (Fig. 1.34d) [59, 60]. This type of grain is sold commercially under the name Norton SG™. One limitation of such fine crystal size is surface reactivity with standard vitrified bonds for fabricating grinding wheels. Bonds had to be developed to be fired at <1,000°C rather than the 1,200°C of older bonds used for fused alumina abrasives [61].

A comparison of Fig. 1.35a–d indicate the single-phase seeded microstructure is finer than the multi-phase microstructure and would be expected to be slightly harder and tougher. It would also be expected to give a longer life but require higher force to micro-fracture when used as abrasive grain, or should be used at a lower

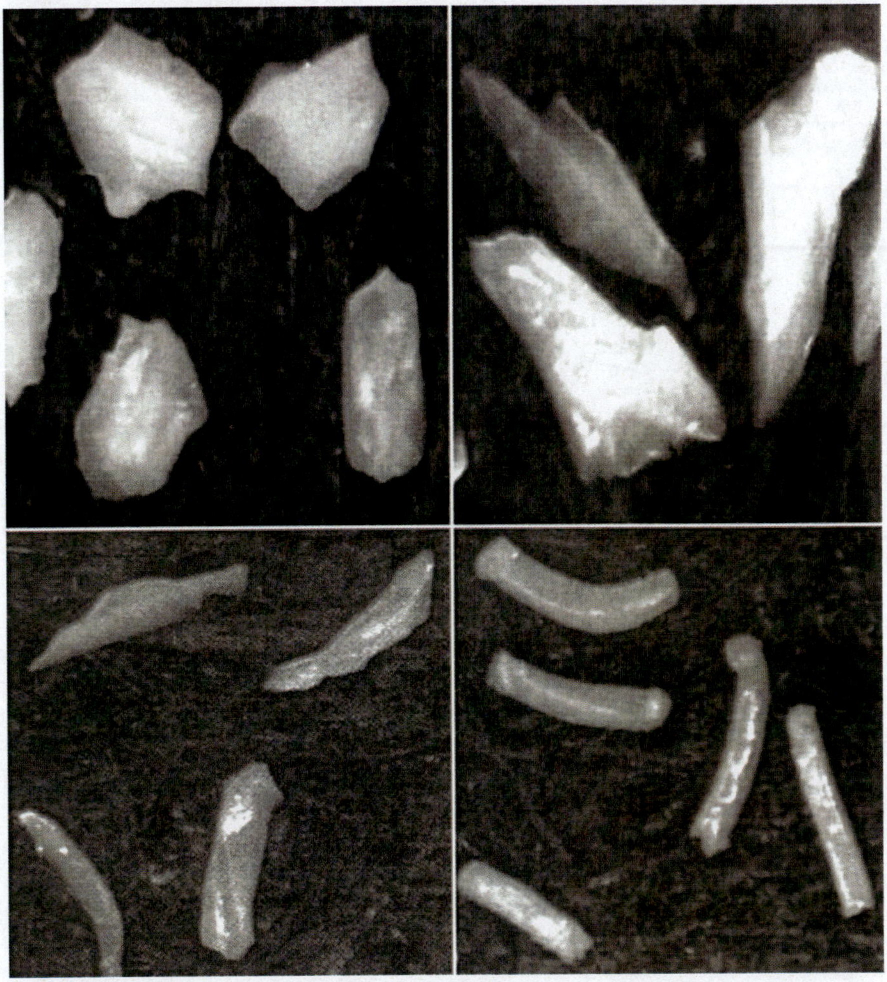

Fig. 1.35 (**a**) Tough blocky ceramic grain produced by milling; (**b**) Friable angular grain produced by crushing; (**c**) Weak extreme angular grain produced by crushing in green state; and (**d**) Extruded TG2™ ceramic grain (Courtesy Saint-Gobain Abrasives)

concentration in a blend. The multi-phase grain would be slightly more free-cutting, and also less reactive with high temperature vitrified bonds. These differences however are relatively minor compared to the difference in overall performance of this family of abrasives relative to fused alumina. Furthermore, further performance optimization can readily be obtained by wheel formulation and especially grain shape. Sol-gel manufacturing allows a much greater manipulation and control of grain shape. Standard crushing and milling methods can produce the typical strong blocky or weak angular shapes. The angularity can be further increased careful processing of soft, dried pre-sintered material (Fig. 1.36). As expected these

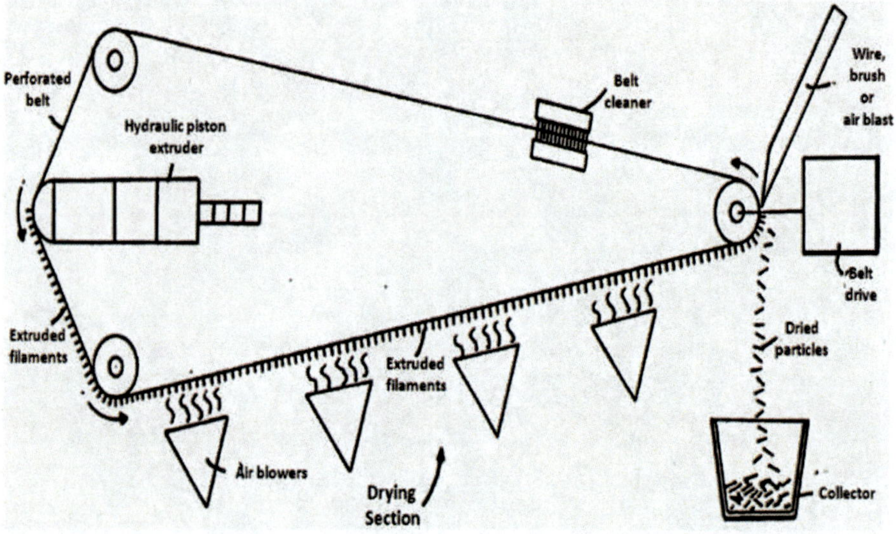

Fig. 1.36 Manufacturing process for producing extruded ceramic grain [62]

grains are also relatively weak but extremely successful if orientated on for example a coated application with relatively low grinding forces. More interesting however, are novel techniques [62], that have been developed to extrude as rectangular prisms with extraordinary aspect ratios, and having the appearance of smooth, surface defect free "worms" (Fig. 1.34d). Norton use TG™ grain with an aspect ratio of 5, and TG2™ with an aspect ratio of 8 [63]. Not only do these grains maintain a high toughness but they also have a very low packing density. Typical blocky grain may pack to about 50% by volume; an extruded grain with an aspect of 8 has a packing density closer to 30%. This provides for a very high level of permeability and excellent coolant access in the final fabricated wheel. Due to the toughness, shape and ability to provide coolant the stock removal capabilities on tough super-alloys such as Inconel or Rene alloys that exceed that of CBN grain for example by an order of magnitude.

1.6 Superabrasives

Superabrasive refers to natural and synthetic diamond, and Cubic Boron Nitride (CBN) materials characterized by having the most extreme hardness and thermal conductivity. Their manufacture, or in the case of natural diamond their extraction from the ground, is also extremely expensive resulting in material cost greater than 1,000 times that of conventional grain. The processing and use of superabrasive grain is in some ways very different to conventional abrasive but in other very similar if more exacting.

1.6.1 Diamond

Natural diamond – Natural diamond has been a familiar part of industry since its foundation and still represents a surprising portion (>50 M carats/annum) of the machining business primarily for cutting tools, dressing tools for wheels and wear parts with secondary use as actual abrasive for grinding wheels or coated belts. Demand continues to be strong [64], but the top quality product for industrial use is often under economic pressure from the jewelry trade and low labor costs for polishing small gem quality stones in countries such as India leading to scarcity [65]. Nevertheless natural diamond, generally of a color, shape or inclusion level unpopular for the gem business, remains the standard for single point dressing tools and stones in rotary dressing form rolls. Some crushed natural diamond is used in grinding wheels particularly in plated single products requiring grain with extreme sharpness and high angularity.

Natural diamond is formed at depths of 150–200 below the earth's surface under extreme temperature and pressure in the mantle. It may then be carried up in molten kimberlite and lamproite rock where it is found at the earth's surface within carrot shaped "pipes", or alluvial deposits produced from erosion of these pipes most commonly within old landmasses known as cratons. Large diamonds of the size used in dressing tools are believed to form and remain over great period of time in the mantle but microdiamonds <0.5 mm are believed to form in the kimberlite and lamproitic magma [66]. Not surprisingly each individual pipe can produce a very unique range of diamond sizes and shapes. Some fields may contain predominantly micro-diamonds that until recently were uneconomic due to the lack of traditional gem quality material. (Yields of diamond are of the order of <1 tonne per 13 million tons of ore processed). Furthermore most of the major diamond deposits are in politically unstable areas of the world especially South and Central Africa although mines in NW Australia and Canada have recently come on line while Russia has produced large quantities of both gem and industrial diamonds for many decades (Fig. 1.37).

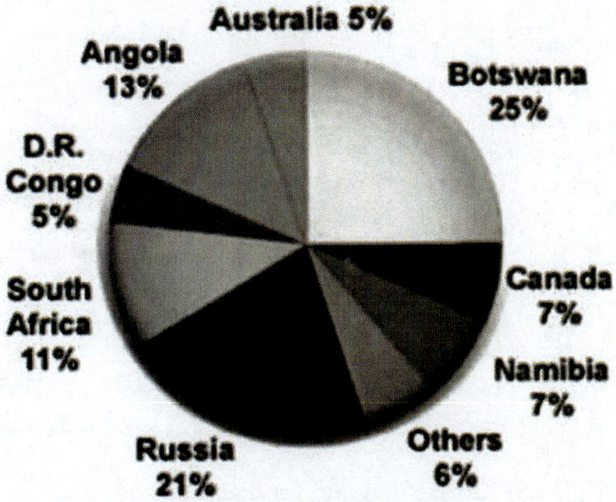

Fig. 1.37 Global supply of natural diamonds as of 2010

Synthetic HPHT diamond – The supply of natural diamond has for several decades lacked the consistency, security of supply and cost position necessary to meet the requirements of modern industry with its increased demands to machine carbide, ceramics and other advanced materials. Interest in synthesizing diamond was spurred after WW2 with the introduction of carbide tooling and the need for efficient means of fabricating them. The first to achieve synthesis of diamond was ASEA AB, Sweden under the Research Directorship of Erik Lundblad in 1953. Independently in 1954 GE's "Super-Pressure team" including Tracy Hall and Bob Wentorf produced its first synthetic diamond crystal which was the first to be repeatable and published. (See Finer points 2005 Superabrasive Resource Directory for further details [64, 67].)

Diamond is created by the application of extreme high temperatures and pressures to graphite. The stable form of carbon at room temperature and pressure is graphite with its familiar layered hexagonal lattice structure. Although bonding within the lattice is strong sp^3 covalent bonding between the layers is Van de Waals forces only, resulting in easy slippage and low friction. Diamond, which is meta-stable at room temperature and pressure, has a cubic arrangement of atoms with pure sp^3 covalent bonding with each carbon atom bonded to four others. The phase diagram for diamond/graphite is shown in Fig. 1.38.

The direct conversion of graphite to diamond requires temperatures of 2,500 K and pressures of >100 kbar. Diamonds produced by this route are termed high pressure, high temperature (HPHT). The severity of the growth conditions can be reduced significantly by the use of a metal solvent such as nickel or cobalt. Graphite has a higher solubility in these solvents than diamond; therefore at the high process temperatures and pressures the graphite dissolves in the molten solvent and

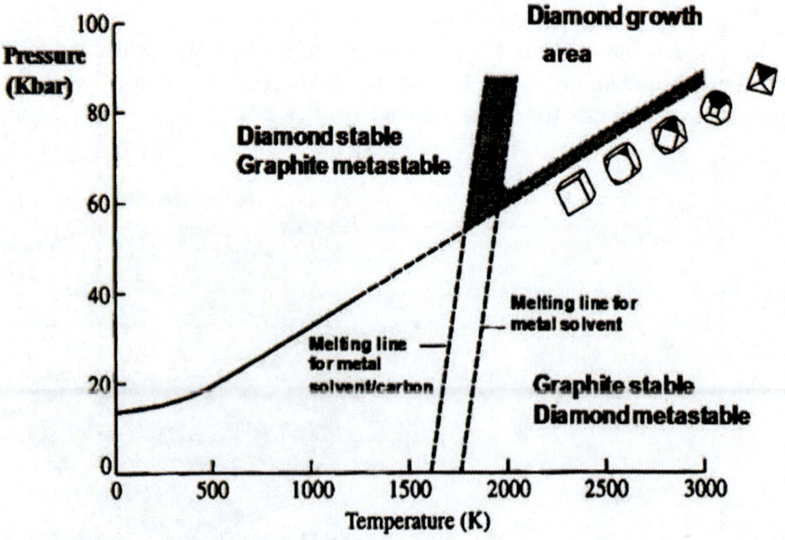

Fig. 1.38 Phase diagram for diamond and graphite

diamond then precipitates out. The higher the temperatures, the faster is the precipitation rate and the greater the number of nucleation sites.

The earliest diamonds were grown fast at high temperatures and had weak, angular shapes with a mosaic structure. Also, the principal crystallographic planes of diamond are the cubic (100), dodecahedron (011) and octahedron (111). The relative rates of growth on these planes are governed by the temperature and pressure conditions and the metal solvent present. In general at low temperatures the primary growth plane is cubic, while at the highest temperatures is it octahedron. Careful control of the growth conditions allows the shape to be engineered to specific applications. In particular the blockiest, strongest form is the intermediate cubo-octahedral used in the strongest metal bonds for cutting or grinding concrete and glass (Fig. 1.39).

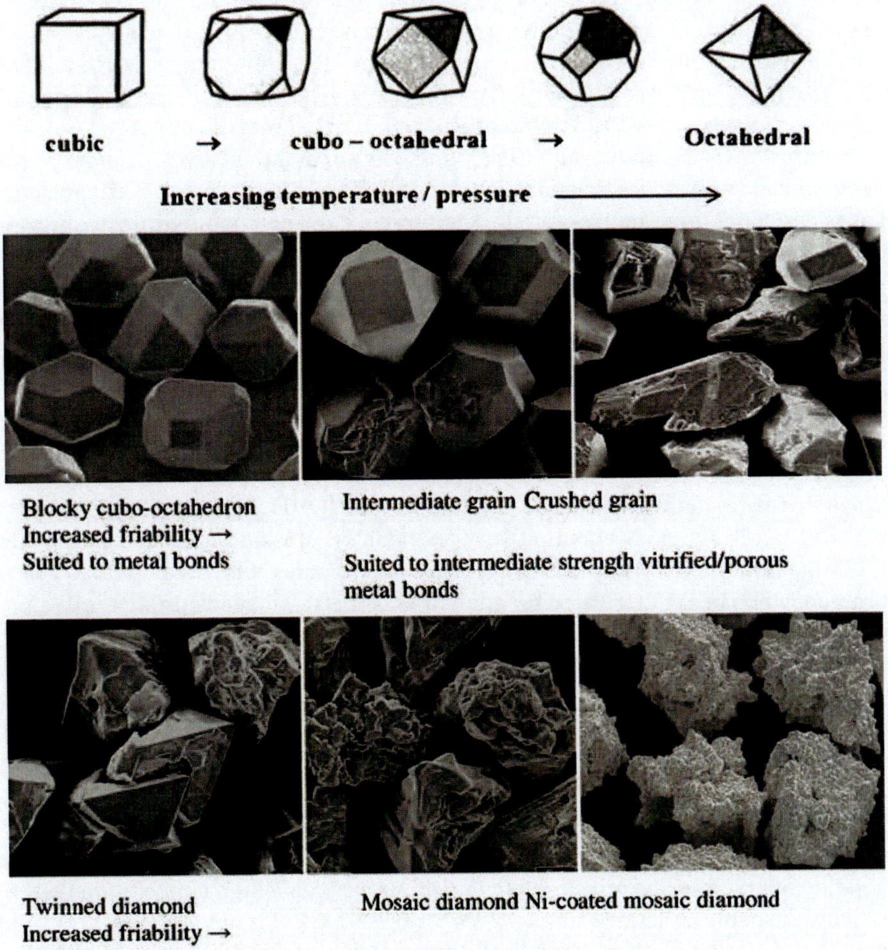

Fig. 1.39 Examples of synthetic diamond shapes and morphologies

High temperature and pressures are generated by three main press designs: the belt press, the cubic anvil press and the split-sphere (BARS) press. The belt press as developed for the first diamond synthesis by GE consists of an upper and lower anvil applying pressure to a cylindrical inner cell or bombe. The pressure is confined radially by a steel belt. Belt presses with substantial bomb volumes have been developed in recent years for the growth of large single crystals by companies such as Sumitomo and Ladd [68]. The bombe is doped with a seed crystal and a temperature gradient is created within such that diamond is gradually and steadily deposited over a prolonged period of time. The resulting diamond crystal is then laser cut along specific crystallographic directions to produce needles and blocks suitable for diamond dressing tools, rolls and wire drawing dies.

The cubic anvil press has six anvils that apply pressure simultaneously onto the faces of a cube-shaped bombe. This type of press has a relatively small bombe volume and is best suited to fast processing time of medium to high friability diamond. The labor input required is relatively high but is popular for the recent rapid increase in production of diamond in China. The split-sphere BARS apparatus, developed in Russia, is a method for growing primarily large high quality diamond for specialized and applications and the gem market.

Synthetic CVD diamond – The synthesis of diamond by Chemical vapor deposition (CVD) is a method first developed in Russia in the 1970s. Carbonaceous gas is reacted at high temperature in the presence of reducing hydrogen atoms in near vacuum to form the diamond phase on an appropriate substrate. Energy is provided by a hot filament or plasma to dissociate the carbon and hydrogen into atoms. Hydrogen is critical in that it interact with the carbon and prevents any possibility of graphite forming while promoting diamond growth on the substrate. The substrate composition, preparation and crystallographic orientation are all also critical. The resulting CVD diamond layer forms as a fine crystalline columnar structure with a thickness of up to 1–3 mm. There is only limited crystallographic orientation exhibited making wear characteristics much more uniform and less sensitive to orientation than single crystal diamond. CVD diamond is not used as an abrasive but is again is becoming very prevalent in dressing tools and form rolls. CVD diamond contains no metal solvent contaminates which can actually be a problem when being fabricated for applications such as shaped cutting tools, since it cannot be EDM wire cut. Diamond wetting also appears more difficult in brazing and must be compensated for by the use of an appropriate coating.

Cubic Boron Nitride (CBN) – Cubic boron nitride (CBN) is the most recent of major abrasive types, and the brain-child of Bob Wentorf of the same Super-Pressure Team at GE that developed synthetic diamond. Boron nitride occurs normally in the hexagaonal form, a white slippery substance with an hexagonal layered atomic structure called HBN (or α-BN) similar to graphite but with alternating nitrogen and boron atoms. Wentorf noted its similarities to the structure and bonding of graphite and proceeded to determine a suitable high temperature solvent to grow the cubic structured form – CBN (or β-BN). CBN is not found naturally but must be synthesized at pressures and temperatures comparable to those for diamond. The chemistry however is quite different; CBN shows no affinity for

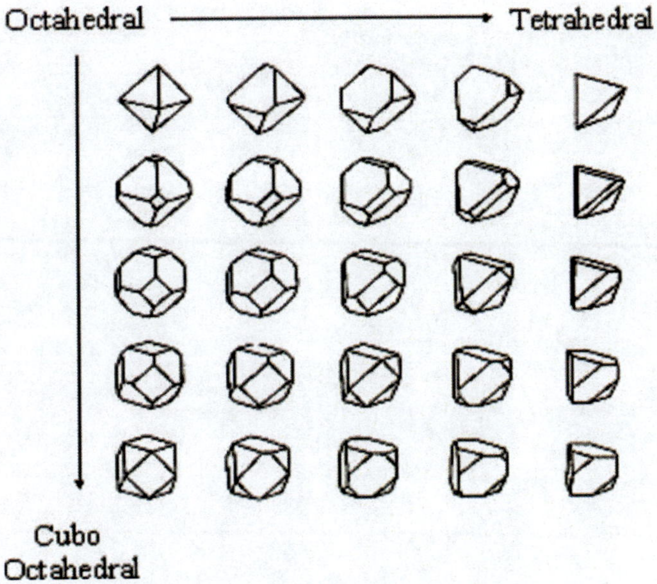

Fig. 1.40 CBN crystal growth morphologies [69]

transition metals. Instead the successful solvent/catalysts are metal nitrides, borides and oxide compounds of which the commonest is Li_3N. CBN was introduced commercially by GE in 1969 under the trade name Borazon®.

CBN grain morphology, like that of diamond, can be controlled in synthesis by the relative growth rates on the octahedral (111) and cubic (100) planes. This is controlled again with temperature and pressure but also be doping. Growth on the (111) planes dominate but because of the presence of both boron and nitrogen in the lattice, some (111) planes are terminated by boron atoms and some by N atoms. In general boron (111) plane growth dominates and the resulting crystal morphology is a truncated tetrahedron but twinned plates and octahedra are also common (Fig. 1.40). In addition shape can also be driven towards the octahedral or cubo-octahedral morphologies. The net result is there is a much wider potential availability of grain shape to choose from than for diamond in addition to twinned and multi-crystalline material. Pure, stoichiometrically balanced boron nitride in the cubic form is colorless but commercial abrasives are various shades of amber color thru brown to black depending on the level and type of dopants present. The black color in particular is believed to be due to an excess of boron (Fig. 1.41).

1.7 Vitrified Bonding Systems

This section focuses on vitrified bonding systems as they are predominantly used in production grinding wheels using the abrasive grains described in the previous sections of this chapter. The grinding process is accompanied by wear of the

Fig. 1.41 CBN commercial grain examples [69]

vitrified abrasive wheel, and the rate of this wear plays an important role in determining the efficiency of the grinding process and the quality of the workpiece [70–98]. The structure of a vitrified grinding wheel is composed of abrasive grains, a bonding system, and a large number of pores. Figure 1.42 shows a typical porous composite grinding wheel structure. Krabacher [95] stated that wear mechanisms in grinding wheels appear to be similar to that of single-point metal cutting tools, the only difference being in the size of swarf. The general form of the wheel-wear curve with volume of workpiece material removed is similar to that of single-point cutting tools [85, 99–101]. The wear behaviour observed is similar to that observed in other wear processes – high initial wear is followed by steady-state wear. A third accelerating wear regime usually indicates 'catastrophic' wear where the wheel requires re-dressing. Accelerating wear is usually accompanied by workpiece burn. The performance index usually used to characterize wheel-wear resistance is the "grinding ratio", or G-ratio, and is the ratio of the volume of workpiece removed to the volume of grinding wheel removed, thus,

$$G = V_w/V_s \qquad (1.1)$$

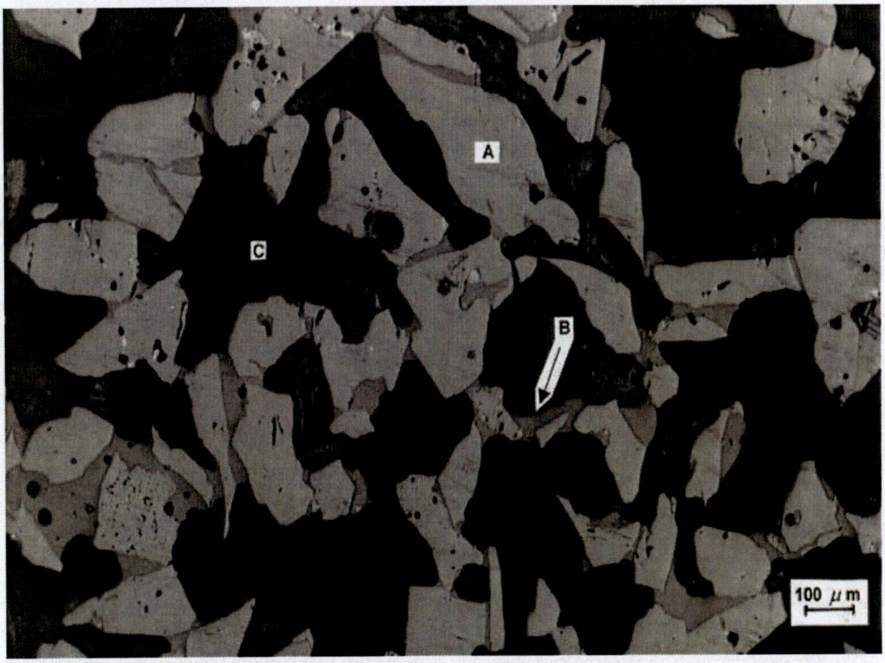

Fig. 1.42 Microstructure of a composite glass-ceramic grinding wheel. A – denotes abrasive grain, B – denotes vitrified bonding phase, and C represents distributed porosity

G-ratios cover a wide range of values ranging from less than 1 for vanadium-rich high-steels [93] to over 60,000 when internally grinding bearing races using cBN wheels [102]. Attempts have been made on vitrified wheels to address the problems related to the wear of abrasive grits in terms of the theory of brittle fracture [79, 98, 103]. The conclusions of various researchers lead us to believe that the variety of different and interacting wear mechanisms involved, namely plastic flow of abrasive, crumbling, chemical wear, etc., makes grinding wheel wear too complex to be explained using a single theoretical model.

1.7.1 Wear of Vitrified Grinding Wheels

The four different wheel wear mechanisms to which overall wheel wear can be attributed to the following are shown in Fig. 1.43 [79, 98, 99, 104–110], viz:

1. Attritious wear (grit dulling)
2. Fracture of bond bridges
3. Mechanical failure of grits and grit flaking
4. Fracture at the interface between grit and bond

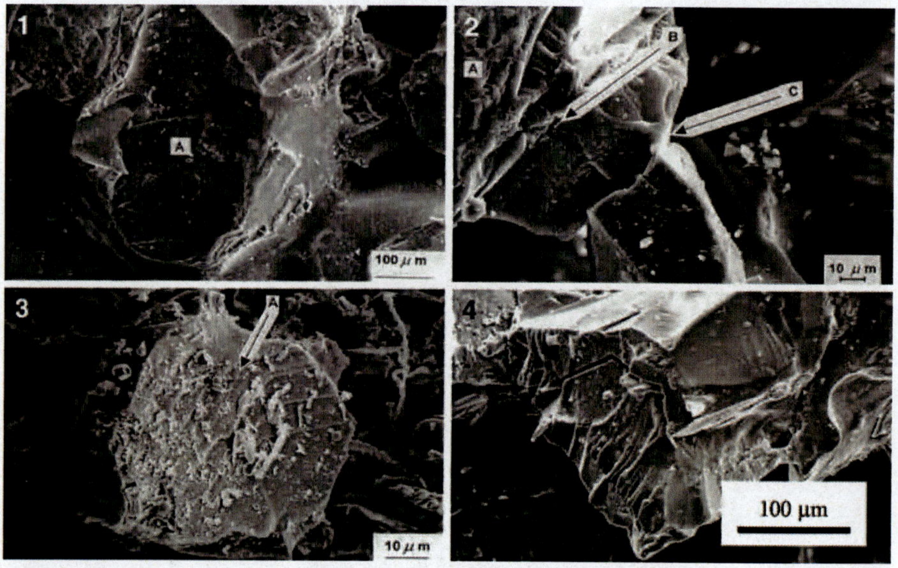

Fig. 1.43 Grinding wheel wear mechanisms: (**a**) abrasive wear – A denotes a wear flat generated by abrasion; (**b**) bond bridge fracture – A denotes the abrasive grain, B denotes the interfacial bond layer, and C denotes a crack passing through the bond bridge; (**c**) abrasive grain fracture – A denotes crystallographic grain fracture; and (**d**) interface fracture between abrasive grain and bond bridge

1.7.1.1 Attritious Wear

Grit dulling is the gradual deterioration of abrasive cutting edges leading to loss of sharpness. The sources of minute scale wear are:

1. Attritious wear due to mechanical friction [79, 98, 111, 112]
2. Plastic flow experienced by the abrasive at high temperatures and pressures [85, 103, 107, 108]
3. Crumbling due to thermal or mechanical shock [85, 103, 107, 108]
4. Chemical reaction between abrasive and workpiece material at elevated temperatures and in the presence of grinding fluids [99, 104, 110]

The last mechanism can induce lowering the resistance of the grit to other wear mechanisms. Grit dulling leads to the growth of wear flats on active grits, which increases the area of contact and the amount of rubbing between grit and workpiece. At the point of grit dulling very high temperatures existing in the area of contact greatly enhances adhesion and chemical reaction between the two surfaces. If grit or bond post fracture does not occur the plateau area on the grit widens, and hence the wear rate increases. If fracture is further delayed, as with hard grade wheels, the wheel becomes glazed and the workpiece tends to burn.

It has been shown experimentally [110] that chemical affinity between the abrasive and workpiece material can be used as a guide for the selection of grinding wheels. Their observations of solid diffusion of silicon carbide into ferrous materials explains the catastrophic wear rates exhibited by these "workpiece-wheel" combinations. The most common method for measuring wear flat area is by measuring wear flats at the grinding wheel surface using optical or electron microscopic techniques [79, 98, 111]. Hahn [83] observed and analysed theoretically the effect of the wear flat area through attritious wear during plunge grinding of various workpiece materials. Hahn concluded that grinding forces will gradually increase during wear-flat formation up to a point where the wheel will restore its sharpness due to grit fractures.

1.7.1.2 Fracture Wear

Grit and bond fracture are usually considered simultaneously for the following reasons:

1. They are of the same nature, i.e. fracture of brittle materials and hence the theory of brittle fracture is applicable to both bond and grit [79, 98, 103, 113]. The applied thermal and mechanical loads usually under cyclic conditions cause initiation and further development of cracks that leads to fracture and the formation of new irregular surfaces
2. They are related to the dressing methods used and occur simultaneously. The initial and final stages of wheel life between dressing exhibit exclusively fracture wear which is a combination of grit and bond fracture and
3. The relative amounts of bond and grit fracture cannot always be found. An investigation into precision grinding [79, 98], where light grinding conditions were involved, employed a soft wheel which gave a high percentage of bond fracture whereas a hard wheel gave mainly partial grit fracture – attritious wear occurring in both cases.

However, the combination of grinding parameters such as equivalent chip thickness and workpiece material determines the effective wheel hardness, and so no single feature of the grinding process can be used to predict the fracture pattern of the wheel in advance. The main difficulty in relating wheel wear due to fracture to the particular grinding condition arises from the lack of knowledge about the loads applied to both the grit and bond and their response to these applied loads.

Tarasov [102] suggests that grit fracture occurs as a result of mechanical forces due to chip formation or thermal shock induced by instantaneous high temperatures. Hahn [83] proposed a thermal stress hypothesis to explain the fracture of abrasive grits. Plunge grinding tests were conducted under fixed normal force conditions. Hahn asserted that as wear progresses, measurements of torque indicated that the tangential force actually decreases, therefore grit fracture due to mechanical loading will not occur. Mechanical stresses were also considered as an explanation for wear rates of wheels tested.

Bhattacharyya et al. [112] observed grit loss due to fracture using an electron microscope. They concluded that they could not differentiate between Peklenik's "crystal splintering", i.e., grit flaking due to thermal stress, and grit fragmentation. However, they did explain their results in terms of Hahn's thermal shock hypothesis. Hahn's experimental conditions suggested that attritious wear was expected to have a major contribution to the thermal shock hypothesis as Mohun [113] observed with abrasive discs. The wear measurements of Hahn [83] were based on the reduction in wheel diameter which Malkin and Cook [79, 98] attributed to attritious wear. Wear rates recorded were of the order of 50 μin./s on wheel diameter. For purely attritious wear, wheel wear rates of the order of 5 μin./s are normally observed. This indicated that the wear mechanism was not solely due to attritious wear. The amount of fracture wear present may consist of fragments of uniform average size particles due to partial mechanical grit fracture, or thermal flaking.

Malkin and Cook [79, 98] collected wheel wear particles for each grade of wheel tested when grinding with a fixed set of operating conditions then analysed their size distribution statistically. They found that with a soft grade wheel (G-grade), approximately 85% of the total wheel wear was due to bond fracture whilst with a harder K-grade wheel, this value reduces to around 55%. Attritious wear particles accounted for only 4% of the total wear in both cases. The strongest evidence in support of the idea of fracture due to mechanical loading is that fracture occurs at some distance away from the cutting tips [105, 114]. Yoshikawa [103] concluded that the heat generated by cutting has no effect on grit fracture since the peak temperature of the grip occurs at the surface of the grit in contact with the workpiece where fracture would be initiated upon cooling according to the thermal stress hypothesis. The hypothesis does not take into account any difference in coefficient of thermal expansion between grit and bond materials, and also of the effect of thermal shocks due to the quenching action of grinding fluids on the grit leaving the cutting zone. Saito and Kagiwada [115] analysed the latter case and reported that the thermal stress in a grit due to a pulsating heat source showed that the magnitude of the maximum tensile stress is not large enough to cause fracture of the grit.

Eiss [111] and Malkin and Cook [79, 98] both adopted the mechanical loading approach. Eiss applied a theoretical model of an idealised grit and compared it with grinding data. Malkin and Cook [79, 98] derived an expression, from first principles, for the probability of bond fracture against the bond stress factor, $(F_t - 20F_n)/V_B$. Yoshikawa and Sata [103] and Yoshikawa [101] developed expressions for the probability of grit and bond post fracture as functions of grit stress, σ_{grit}, and the product $(1/V_B \cdot f_{grit})$, where f_{grit} is the grit grinding force, V_B is the percentage of bond in the wheel by weight, and $1/V_B$ is the bond weakness factor.

Although bond and grit fracture are similar mechanisms they have a different effect on the economics of the grinding process. The first mechanism results in a rapid loss of wheel, and the second mechanism, on a comparable scale with the uncut-chip thickness, generates sharp cutting edges and is known as the self-dressing action. Both mechanical and thermal stresses seem to be responsible for fracture wear. The effect of heat at the grit interface is responsible for locally changing the mechanical properties of the abrasive material. However, fragments of larger sizes are likely to occur through mechanical loading which governs both grit pull-out wear and the self-sharpening action.

1.7.1.3 Wheel Wear Mechanisms

In view of the large numbers of independent variables involved in grinding it is self-evident that the more dominant wheel wear mechanisms depend greatly on the conditions applied to each grinding application. Grinding wheel wear consists of a large variety of physical and chemical mechanisms of very different nature.

The wear processes involved in grinding are classified as follows:

- Attritious wear, i.e. progressive wear leading to loss of grit form leading to a deterioration of cutting ability, and to excessive heat generation, and
- Fracture wear (of grit and bond) which restores the cutting ability of the wheel and removes worn grits and allows grinding to progress efficiently

A typical sequence of events for a grinding wheel according to Tarasov [102], involves dulling of the active grits by attritious wear and then to regain sharpness by fracture wear until bond post failure releases the grit. After the wheel has been dressed, stage I of the wheel wear diagram illustrates the removal of weakened grits by the fracture wear mechanism. Tsuwa and Yasui [114] reported the existence of a layer on the wheel surface after dressing which is progressively removed after grinding has started. High wear rates are also exhibited during the final stage of grinding where catastrophic breakdown of the wheel occurs due to mechanical overloading of the grinding grits.

The area exposed on the surface of active grits directly affects the magnitude of the grinding energy required for metal removal and, hence, the amount of power required before the onset of workpiece burn. The steady-state wear regime (stage II) occurs due to the combined effect of attritious wear and fracture wear. However, the ability of the grinding grits and bond posts to fracture, when the load exceeds some limit, determines the duration of this stage and prevents forces from becoming too excessive. Fracture also limits the amount of heat generated at the cutting zone and gives better workpiece quality. Grisbrook [99] found that the greatest amount of wheel wear results from diamond dressing rather than from wheel wear, which emphasises the need for longer steady-state periods and fewer dressings.

Stetiu and Lal [116] found that the mechanism of wheel wear can be changed from attritious wear to one of fracture wear by selecting the appropriate wheel hardness without changing the grinding conditions. The particle-size distributions gave evidence of a self-dressing action. In terms of volumetric wear Malkin and Cook [79, 98] found that in plunge-feed grinding experiments only 4% of the total wear volume was due to attritious wear, the rest being due to grit and bond fracture according to wheel grade when the grinding parameters are fixed. Tsuwa [104–106] recorded the changes occurring in wheel cutting edges during grinding. He stated that the section of grit that forms a cutting edge is always the worn surface of the grit, and that edges which fracture by a considerable amount cease to become active. It was concluded that self-dressing takes place only if the attritious wear is very great, or the operation is continued up to a very high degree of dulling of the active grits so that fracture edges become active.

It is certain that the rate of wear is dependent upon this fracturing tendency and that forces upon the grit will undoubtedly increase. There is a fundamental difference between the two forms of grinding wheel wear. Attritious wear is undesirable and in that case all practical measures aim to reduce it. This form of wear is very similar to the corresponding mechanism which causes deterioration of the cutting ability of turning tools were large scale failure is unlikely to occur [117]. Fracture wear in grinding is advantageous in that it can be a controlled form of wear. Although it is a form of volumetric wheel loss it opposes attritious wear in terms of grinding efficiency. Fracture wear gives the wheel its functional significance and considering very little is known about fracture behaviour in abrasive particles, it is clearly important to understand how grit and bond composition affects fracture behaviour of grinding wheels and their performance.

1.7.2 Wheel Wear and Grinding Forces

Grinding forces and their effect in wheel wear of grinding tools have been subjected to detailed investigations. The earliest dynamometer used for the measurement of grinding forces was probably one used by Marshall and Shaw [118]. For simplicity, two force components operate, namely: the normal force, F_n; and the tangential force, F_t. A notable feature of grinding force measurements is the high F_n/F_f ratio ranging from 1.5 to 3. This value seems high compared with conventional machining operations where a typical value lies between 0.5 and 1.

The difference is attributed to the different effective cutting geometries and the pattern of metal removal by the grits. Grisbrook et al. [99] and Grisbrook [119] recorded the magnitude of grinding force components during the whole period of wheel life between dressings, and for fixed downfeed conditions.

The typical change in force pattern can be divided into four phases.

1. An unstable phase where forces rise abruptly up to a peak, then fall to a steady-state value as the initial high wear rate (stage I), due to the effects of dressing, slows down. Davis and Rubenstein [120] showed that the rate of change in grinding force increases slightly after a transition point is reached then settles down to steady-state conditions. Pattison and Chisholm [96] showed that grinding forces are affected by the dressing conditions, which has been confirmed by Rowe et al. [121, 122]
2. A phase where forces are constant and heat flow into the grit and workpiece is in equilibrium. The region coincides with the self-dressing action of the wheel. Bond bridge fracture does not normally occur in this region although

bond strength does play an important part here as it determines the magnitude of the applied load the grit can sustain without fracture

3. A third phase where there is a progressive build-up of power and grinding forces. The mechanisms that proceeds from the second phase assumes that grits adopt stable geometries and shapes which are harder to fracture, and from then on the number of fractures decreases and grits become dull. At this point, grits are prone to overheating and grinding becomes inefficient. As rubbing and ploughing increases, and since the metal removal rate is constant, the normal force component increases at a much higher rate than the tangential force component. However, higher forces at this stage are not accompanied by higher wear rates which indicates that the rate of wheel wear is a function of the absolute values of the forces and their relative magnitude, i.e., F_t/F_n, which is referred to as the grinding coefficient. Marshall and Shaw [118] suggested that a higher grinding coefficient produced a more efficient grinding process. This is explained by considering that the normal force component induces compressive stresses into the abrasive grit whilst the tangential force component causes tensile stresses to be exerted at the rake face of the grit. Therefore, a higher F_t/F_n ratio means a higher probability of grit fracture since the grit material has a lower tensile than compressive strength which means that cutting edges retain their sharp facets for a greater period of time and
4. In this period, the rate of change of grinding forces becomes less and the effects of vibration become evident. This phase corresponds to stage III of the wheel-wear curve where wear rates are detrimental to economic grinding

The progress wear of grinding wheels operated under fixed normal force conditions has been studied by Hahn [83]. High force values lead to increases in depths of cut and rapid catastrophic wear due to fracture. For very low force intensities Hahn found that metal removal ceases quickly, and the wheel subsequently glazes and the wear rate becomes negligible.

Lindsay [123] and Lindsay and Hahn [124] found that radial wheel wear is linearly proportional to the normal force existing between wheel and workpiece, irrespective of whether the force is applied under fixed force conditions, or under fixed feed conditions. Volumetric wear is reported in Lindsay and Hahn's further work in precision grinding [125]. They related volumetric wear to be an approximate quadratic function of normal force intensity. Their conclusions showed that wheel wear could be related to any wheel-wear geometry, or conformity, in terms of interface force intensity and contact pressure.

1.7.3 Assessment of Grinding Forces and Wear

The dominant wear mechanisms in grinding wheels are attritious and fracture wear. In the author's opinion, fracture wear (bond and grit) should be considered the most important wear mechanisms for these reasons:

1. Fracture wear constitutes approximately 95% of the total volumetric wheel wear
2. Wear during the initial and final stages of the wheel-wear curve (I and III) is due to gross fracture
3. The dressing mechanism that affects wheel life and grinding forces is essentially fracture-sharpening of cutting edges
4. Fracture wear reduces grinding forces and heat generation which results in lower power consumption, and smaller workpiece distortion
5. Fracture wear, in the form of self-dressing, affects the duration of useful wheel life (region II of the wheel-wear curve)

It should be noted that the mechanism of grinding wheel fracture is an extremely complex process caused by the action of thermal and mechanical stresses induced into the grit. Thermal stresses are thought to be responsible for grit flaking whilst larger fragments are associated with mechanical stresses that are directly related to forces acting upon the grinding grit.

Fracture wear due to mechanical loading seems to be dependent on absolute force components and their relative magnitude. For fixed normal force operations the force is sufficiently large to allow fracture to occur at a controlled rate. Too high a force will cause catastrophic wear rates to dominate, whilst very small forces will impair metal cutting. For fixed-feed operations the grinding force components increase steadily during stage II of the wheel-wear curve. If the cutting conditions produce a friable wheel then metal cutting will take place efficiently. The rate of change of force depends upon the initial value of the force component ratio.

1.7.4 Effect of Workpiece Material on Wheel Wear

The suitability of abrasive materials to grind workpiece materials efficiently depends on their attritious wear resistance. The abrasive should be harder than the workpiece material being ground, however, hardness is not the dominant factor. In fact one would not use the two hardest known of natural abrasives, diamond and silicon carbide, to grind ferrous alloys.

Attritious wear of grits is both mechanical and chemical [110, 122–126]. Chemical effects are significant when the abrasive is appreciably harder than the workpiece material and its associated metallurgical phases. At higher temperatures during cutting, chemical reactions may occur between workpiece, grinding fluid, the surrounding atmosphere and the abrasive and bond. Diamond is not suitable for grinding ferrous metals despite its hardness. This is attributed to attritious wear caused by reversion from diamond to graphite [127]. Degradation of diamond appears to be aggravated in the presence of iron low in carbon. Loladze and Bockuchava [128] listed five types of diamond wheel wear based on adhesion, abrasion and diffusion wear. Cubic boron nitride is more stable than diamond in the presence of ferrous metals. However, the success of CBN on various steels is dependent on the complex carbide phases within the steel workpiece. The hardness of the carbides are quoted as "abrasive numbers" which are essentially weighted averages of their Vickers' hardness values [127].

When grinding ferrous metals with aluminium oxide abrasives, the most important chemical reaction usually involves the oxidation of iron and the reaction of the oxide with the abrasive to form spinel, $FeAl_2O_4$ [129], thus

$$2\ Fe + O_2 + 2Al_2O_3 \rightarrow 2\ FeAl_2O_4 \tag{1.2}$$

Spinel is an intermediate compound between the oxidized workpiece material and aluminium oxide. A high attritious wear rate on steel in humid air rather than dry air was found to be due to the catalytic effect of water on the oxidation of iron [129].

However, despite the role of oxygen and water to promote adhesion and attrition during grinding, their elimination by grinding in a vacuum has a dramatic effect on the process [130–133]. Chemical reaction between workpiece and abrasive was reported to be reduced. However, loading of the wheel surface increased. Surface oxidation and corrosion in normal grinding environments tends to reduce adhesion between metal particles and the workpiece. This same effect might explain the difficulties encountered when grinding high-temperature oxidation-resistant metals, including stainless steels, nickel-based alloys, and titanium [134].

Silicon carbide abrasives are harder than aluminium oxide abrasives but are inferior when grinding ferrous materials. The main chemical reaction tends to involve the dissociation of silicon carbide [135, 136], which promotes attritious wear when grinding titanium and other non-ferrous metals. In addition to chemical affinity, mechanical factors contribute significantly to attritious wear. When grinding carbon and alloy tool steels, the G-ratio is greatly reduced when grinding the material in its fully hardened state [137], which suggests that a mechanical effect is taking place. However, hardness is not indicative of grindability especially when the material's hardest phases are softer than the abrasive.

High-speed tool steels contain complex carbides that tend to reduce grindability of alumina abrasive wheels. The hardest carbides in high-speed steels are carbides of tungsten, molybdenum and vanadium. The volume fraction, C*, of these carbides in tool steels relative to that of tungsten carbide by itself can be approximated in terms of the weight percentages of tungsten (W), molybdenum (Mo), and vanadium (V) as [137];

$$C^* = W + 1.9Mo + 6.3V \tag{1.3}$$

The relative grinding ratio G* is the G-ratio for different tool steels expressed as a percentage relative to that of an M2 tool steel. The results tend to indicate that higher carbide content reduces grindability of alumina grinding wheels. This can be compared favourably with a similar correlation between G-ratio and the vanadium content for grinding high-speed tool steels with aluminium oxide wheels [93]. Higher G-ratios have been obtained for materials produced by powder metallurgical methods that results in a fine dispersion of small hard carbides which tends to be less abrasive than large hard carbides [138].

1.7.5 Effect of Abrasive and Bond Composition on Wheel Performance

Attempts were made to describe the process of wheel wear in terms of mechanical stresses applied to abrasive grits during stage II of the wheel-wear curve. Graham and Voutsadopoulos [137] presented data which supported the argument that fracture-type wear is the most important wear mechanism related to the loss of abrasive material from the grinding operation as a consequence of high stresses induced in the abrasive grit by grinding forces. Grit fracture was assumed to be caused by induced tensile stresses of relatively small magnitude. These workers used existing experimental data to correlate G-ratio to a number of process variables using a finite-element model of an idealised wedge. The wedge was assumed to be rigidly held in an infinitely strong bond. Graham and Voutsadopoulos [137] applied a tangential force of 5.4 lbs force and a normal force of 10.41 lbs force to produce a stress pattern showing lines of constant maximum shear stress (isochromatics). They applied Griffith's brittle fracture criterion to the post-processed finite element results in order to locate areas of compressive, tensile and neutral stresses within the model grit. The area of brittle failure was also located on this diagram in order to illustrate points at which the abrasive material is likely to fail, i.e. tensile fracture of the grit. A good correlation was found between the maximum tensile stress in the grit and the G-ratio using experimental data contained in the literature. However, Graham and Voutsadopoulos assumed that grit fracture was the predominant wear mechanism assuming that grit to be rigidly held. Wear during this period of grinding (stage II) is mixed, i.e., bond and grit fracture which explains why the correlation is close but not exact. The vitrified bond is not infinitely stronger than the grit, and its magnitude governs the duration of stage II wear. These workers postulated that wear during stages I, II and III was probably due to tensile stresses induced in the bond and the grit material, i.e., grit fracture and pull-out. The main criticism of the work is due mainly to the assumption of a rigidly held wedge finite element model, and the correlation of results of this model to wear data relating to stages I, II and III of the wheel-wear curve.

1.7.6 Vitrified Bonding Materials

The tensile strength of a ceramic material is determined by the most serious flaw in it. Such flaws are known as Griffith flaws and can appear as cracks, pores, or irregular-shaped grains that have a sharp notch acting as a stress raiser around the notch to a high level. Any inclusion which has a different elastic modulus from the matrix will produce a small stress concentration in its vicinity, e.g., a perfectly spherical pore will increase the average tensile stress by a factor of three at the pore surface on a plane perpendicular to the tensile stress direction. The low strengths observed in many ceramics must be caused by the presence of sharp notches. In ceramics of moderate strength (>70 MN/m^2), flaws are typically 100 μ and will be frequently found to be pores.

In ceramics showing high strength (350–700 MN/m^2), flaws are considerably smaller and are of grain size dimensions such as grain boundary cracks of fractured grains – grain sizes typically a few microns. The most serious flaw on a body is one situated at the surface orientated so that its maximum dimension is perpendicular to the applied tensile stress. Surface flaws are the most serious because the effective flaw length is the complete flaw length, whereas inside the volume of the stressed body, the effective flaw size is less than the flaw length. In addition, if the body is subject to bending, for example a bond post connected to two adhesive grains, it will experience the highest stress at its surface. Factors other than flaws that affect the strength of the body are related to the nature of the body, i.e., composition, grain size, and general porosity. These factors are important since they control the energy required to extend the flaw. The energy required for fracture initiation is higher than the energy required to form a new surface, i.e., the surface energy.

This is partly because it includes the energy absorbing process of plastic deformation that occurs in the highly stressed region of the crack tip. Generally for ceramics the fracture initiation energy $\gamma_{1,}$ is usually a few times 10 J/m^2. This value varies with fracture surface roughness so that smooth surfaces, such as glassy materials, have low values of, $\gamma_{1.}$ In the ideal case of a dense, homogeneous material containing a single flaw, the stress multiplication factor of a flaw can be evaluated for some simple flaw geometries. The expression relating the stress at failure, σ, to the size of the flaw causing failure, and the failure energy for fracture initiation for an elliptical crack is,

$$\sigma_{failure} = \sqrt{\frac{2.E.\gamma_1}{\pi.C}} \tag{1.4}$$

where E is Young's modulus, C is the crack length if it is a surface flaw or half a crack length of it is an internal flaw, and σ_{failure} is taken as the average stress calculated from the specimen's geometry and dimensions and the applied load. This is Griffith's equation but for any flaw geometry it is expected that,

$$\sigma_{failure} \alpha \sqrt{\frac{E.\gamma_1}{\pi C}} \tag{1.5}$$

This equation should be valid for bodies containing several well-separated serious flaws. The stress, σ_{failure}, includes any residual stresses in the body that are usually of unknown magnitude. Many experimental studies on ceramics have shown that strength depends on the total porosity, i.e., small and large pores have an effect. An empirical relationship between the failure stress and the porosity, p, of a body has been found to be:

$$\sigma_{\text{failure}} = \sigma_0 e^{-bp} \tag{1.6}$$

where, σ_0, is the strength found by extrapolating the date to zero porosity, and, b, is a constant determined from the experimental data plotted as in, σ_{failure}, versus porosity, p. The value of, b, has been found to vary considerably for the same ceramic material depending on the shape, size and distribution of the porosity [139, 140]. The relationship is known as the Ryshkewitch-Duckworth [139–142] equation and was shown by Knudsen [143, 144] to be based on the increased average stress caused by the reduction in load-bearing area resulting from the porosity.

However, Knudsen did not consider factors that control strength at zero porosity. Carniglia [138] considered that the flaw should be enclosed within a volume λ, i.e., the Saint Venant volume, such that at the periphery of this volume the stress re-distributing effect of the flaw was negligible. This volume was then considered in relation to the spacing, L, between general porosity. Carniglia showed that the Ryshkewitch-Duckworth equation was applicable only when $\lambda \gg L$. This is, when the Griffith flaw that initiates fracture is larger than the pores which form the general porosity, and when the spacing between pores is small compared to the size of the Griffith flaw.

Under these conditions, the average flaw stress acting on the Griffith flaw has increased by the reduction in load-bearing area and can be considered as uniform. When $\lambda \gg L$, a local stress model, such as the Griffith model for an elliptical crack, can be used. That is the stress magnification produced by a single flaw, with no interference from other flaws that cause failure. The case when $\lambda < L$ cannot be treated theoretically in a general manner, because each Griffith flaw is close to another stress re-distributing flaw and therefore the average stress around each flaw is highly variable. In practice, $\lambda \gg L$ would occur quite frequently for ceramics. Most ceramics contain a number of smaller pores and frequently more serious Griffith flaws. If the samples tested have constant Griffith fracture initiating flaws and the general porosity is variable, then the strength data should fit the Ryshkewitch-Duckworth equation. Carniglia [138] developed a more complex equation that showed the Ryshkewitch-Duckworth equation to be an approximation. If the general porosity remains constant but the size of the Griffith flaws is variable, then the fracture strength values, σ_{failure}, should yield a straight line when plotted against $\sqrt{(\gamma_1 / C)}$. This approach should work because, γ_1, and, E, are determined for the material containing the same general porosity. Flaws can be classified as gross macroscopic, microscopic and sub-microscopic [145]. This means this classification encompasses size only. A flaw is considered to be gross if it is readily visible to the unaided eye, so that the origin of failure can be viewed, i.e., large surface cracks, inclusions at the surface, etc.

A microscopic flaw is not easily identifiable and is usually a small crack, void or a small inclusion. A sub-microscopic flaw is identified using a scanning electron microscope. Flaws in ceramics result from several causes, i.e. pores might be caused by differential firing shrinkage on a small scale of size, burn-out of organic matter (dextrin and fillers), gaseous evolution caused by a reaction on firing, diffusion of gases or some other mechanism, etc. Differential shrinkage on a small scale of size, is caused by the non-uniformity of the characteristic properties on a related scale of size. The characteristic properties are porosity, particle size, composition, or particle alignment. Large-scale non-uniformities of any of these characteristics can cause the constituent materials to re-distribute during firing and may possibly lead to splitting of the abrasive wheel. The tensile stress may be relieved by the formation of one large fissure or possibly by the formation of numerous small fissures. A number of factors involved in pressing these wheels that could be responsible of the non-uniformity of porosity on a large scale include, segregation of fines, friction at the die wall, non-uniform powder packing on deposition, and variations in compaction ratio.

In addition to these, lamination problems might occur, depending on the state of the powder and the pressing technique used. It is caused by the elastic recovery of the compact that occurs when the compaction pressure is removed, as a result of the entrapment of air, particularly in fine powders. Spontaneous micro-cracking on cooling from the firing temperature is a common source of cracks around large inclusions of a different phase. This micro-cracking occurs either because of stresses arising from either a mismatch in thermal expansion of the inclusion and the matrix, or is caused by a phase transformation of the inclusions. In general, pores in ceramic bodies may be described by pores, cracks, fissures, inclusions, large grains, and surface defects. These flaws might be combined and produce complicated fracture-initiating combinations.

1.7.6.1 Effect of Particle Size of Constituent Materials

Parmalee and Morgan [139, 140] found that decreasing the particle size of quartz altered the vitrification behaviour and strength of a ceramic body. The bodies examined were classed as coarse, commercial, and fine, with average diameters of quartz of 68, 45, and 11 μm, respectively. The finer the quartz, the greater the reduction in porosity and the higher the strength. Koenig [143, 144] studied the effects of feldspar, quartz, and kaolin particle size in a vitreous china body. It was found that firing shrinkage and flexural strength were greater for the finer-ground feldspar body, and that there was less water absorption. The bodies containing finely ground quartz, as well as feldspar, had considerably greater strength than the regular china clay body, and they showed greater strength in which fine quartz and regular feldspar, or regular quartz, and fine feldspar were used. Koenig concluded that more finely divided quartz affects vitrification quite markedly and that finely ground feldspar increases vitrification behaviour. Sane and Cook [146] discovered that ball milling for 100 h reduced the final porosity of a clay-feldspar-quartz composition from 17.1 to 0.3% using the same firing conditions. The change is caused by intimate mixing of the constituents and the reduced distances fluxing ions are required to diffuse during firing to enhance chemical reactions. Increased densification of the fired body, due to smaller particle sizes, tended to produce a stronger body.

During vitrification of the body, a large mass of viscous liquid is formed. The liquid wets solid particles that are pulled together under the action of surface tension when the liquid flows into the pores. When the firing temperature is increased, more melt is formed which is less viscous. However, the dissolution of quartz opposes this reduction in viscosity which helps grinding wheel manufacturers, as well as manufacturers of clay-based materials, to fire products over a wide range of soaking temperatures. Fine grinding of quartz produces more surface area of quartz per volume, which promotes its dissolution in the liquid phase, and consequently aids vitrification, which increases strength. Finer quartz particles have also been reported to inhibit inversion cracking at 573°C [147].

1.7.6.2 Effect of Mullite and Glass Content

It is generally accepted that the development of interlocking fine mullite needles in the body increases the strength of the clay-based material. However, this hypothesis is not free from controversy. According to Zoellner [145] and Budnikov [148], mullite provides porcelain its strength. Zoellner dissolved pieces of porcelain in a cold 25–33% solution of hydrofluoric acid for several days. The vitreous matrix and quartz were dissolved and the crystalline phase remained. Zoellner assumed that these crystals were sillimanite, since mullite at that time was unknown. He suggested that increasing the firing time and temperature to increase the formation of these crystals would increase the strength of porcelain. Budnikov [148] published data concerning the strength of electrical porcelain and mullite, from which the strength of mullite is greater than porcelain. Geller [149] showed that the effect of firing increased the size of mullite crystals to such an extent that the strength of porcelain decreased. This was reported also by Krause and Keetman [150] and Eitel [147]. Grofcsik [151] reported that the total Al_2O_3 content of kaolinite transforms by exothermic reaction into mullite at about 960°C. Therefore, repeated or prolonged firing will not change the amount of mullite but may change its size. According to Krause and Keetman [150], the size of mullite crystals will increase with the logarithm of firing time at a suitable soaking temperature. They found that by maintaining the samples at 1,400°C for 6,60,600, and 6,000 min, the average length of acicular crystals of mullite increased in size to 5, 7.2, 11, and 14.2 μm, respectively.

Sane and Cook [146] milled a clay-based body (20% feldspar, 30% quartz, 42.5% kaolin and 7.5% ball clay) for different periods of time which were fired at three different temperatures. They discovered that increasing milling time would increase the wt.% glass in the body and reduced the amount of quartz. However, it was discovered that increasing the mullite content of the body, increased strength of the body. The glassy plane was considered the major component of porcelain, which was considered the weakest part. This aspect of the body has attracted much attention to the strength of clay-based materials. Mattyasovsky-Zsolnay [152] assumed that the tensile strength of porcelain is influenced by stresses set-up in the glassy phase rather than by the amount and size of mullite crystals. He suggested that if mullite controls the strength of porcelain then an increase in the Al_2O_3 content (by increasing kaolinite fraction) to form more mullite would increase the strength of porcelain. Mattyasovsky-Zsolnay demonstrated that mechanical strength increased when quartz content was increased and reduced when kaolinite content was increased. This statement is in contradiction to the experimental data published by Weidman [153]. Weidman increased the kaolin content and the result was an increase in strength.

Experimental samples containing 50, 60, 65 and 70% weight kaolin (corresponding to 30, 16, 10, 1% weight quartz and 20, 24, 25 and 29% weight feldspar) showed bending strengths of 71, 91, 94 and 130 MPa, respectively. Kalnin et al. [154] published results on the strength and elasticity of quartz-free clay-based bodies using compositions of kaolin and nepheline syenite, focusing on mullite content and porosity. They concluded that elastic moduli and flexural strength of the bodies increased with the proportion of mullite present over the range 11–36% weight. In kaolin-rich preparations, i.e., 2:1 and 1:1 kaolin: nepheline syenite ratios, the amount of mullite obtained by powder X-ray methods agrees with the calculated values assuming complete decomposition of kaolin into mullite and silica. However, in the nepheline syenite system (1:2 ratio), the amount of mullite present is much less than expected, and it appears that in this case silica and alumina dissolved in vitrified nepheline syenite. Koch [155] noted that porcelain bodies should be fired such that the microstructure contains plate-like primary mullite, and a high proportion of fine acicular secondary mullite concentrated in the glassy phase and also at grain boundaries to give a felted structure. Lach [156] explained that the presence of a viscous, glassy phase aids the diffusion of cations so that mullite forms more regular-shaped crystals. Therefore, in sintering wheel bonds, high temperatures are needed to obtain well-developed mullite crystals. It has been shown that using mineralizers such TiO_2 [157] and MgO [158], the mullite content can be increased. Primary mullite is formed by the decomposition of clays and secondary mullite is formed by re-crystallization. It is possible that both forms of mullite affect mechanical strength in many ways [159].

1.7.6.3 Effect of Quartz Content

The study of the effect of quartz on the strength of clay-based materials has concentrated on finding an optimum size of quartz particle that inhibits "de-bonding" from the matrix. The "pre-stress theory" espoused by many researchers, was proposed to account for observed increases in strength that occurred with an increase in fine quartz content. It was assumed that fine quartz generates a compressive stress in the matrix and that any applied tensile force would be effectively reduced by the "pre-stress". Therefore, higher tensile stresses would be required to cause failure [160–163]. However, cracks were found around small particles after loading that were present before loading. This indicated that small quartz particles should have a weakening effect on the matrix rather than a strengthening effect [164]. The weakening effect of quartz was demonstrated by comparing the strength of the matrix material, composed of equal weight fractions of kaolin and nepheline syenite, fired to 1,265°C and subsequently ground; with and without quartz. The ground matrix was isostatically pressed with an equal amount of quartz of different sizes, then fired to a pre-determined bulk density. The maximum strength was found to occur at a quartz particle size of 25 μm having a sharp fall above and a slight decrease below this size. But the matrix material without the quartz was much stronger, which contradicts the "pre-stress theory". Weyland [159] showed that the strength of clay-feldspar-alumina porcelain bodies greatly decreased when a small percentage of quartz was added. Another problem is that when Al_2O_3 replaced SiO_2, in the composition of clay-based materials, the alumina-containing body showed higher strength [165–167]. Pass and German [164] suggested that the development of high strength with alumina was probably due to a reduction in the number of Griffith cracks that can be caused by silica inversion. Smothers [168] opposed the "pre-stress theory". He performed hot-strength tests on both quartz and alumina-containing bodies. Above the transition temperature of quartz the strength of the quartz body was as high as that for the alumina body. Weyl [165] and Dunsmore et al. [166] have also shown that the strength of a quartz-containing body is higher above the displacive polymorphic change of quartz.

On cooling a clay-based material from the soaking temperature, the quartz transforms at 573°C. This transformation, which involves a contraction of the quartz particles, produces stresses both in the quartz particles and in the matrix, and can cause circumferential cracks surrounding the quartz grains in a quasi-pore within the matrix. However, it has been reported in studies both porcelain bodies [169] and on model systems [170] that it is only above a critical size that circumferential cracking tends to occur during cooling. Investigations have led to relationships being proposed which relate the effect of quartz particle size on strength and the existence of an optimum grain size. However, the reported optimum value of the quartz size varies significantly between researchers. Krause [167, 171, 172] carried out experiments with finely milled and graded quartz and established that the maximum bend strength is obtained with quartz of particle sizes in the range 15–20 μm. The data showed that up to 45 μm particle size, the body with the highest % weight gave the strongest matrix. Ludas [169] reported that the highest mechanical strength of porcelain occurred when quartz was sieved to the size of 30–35 μm. Beech and Norris [173] investigated a porcelain made using 40% weight quartz of a particle size 10–30 μm which produced the highest strength. Beech and Norris also varied the firing conditions. Their results indicated that the maximum strength occurred at a temperature below the required for maximum bulk density. Grofcsik [151] used the disc compression test on circular discs to investigate the effect of quartz particle size on mechanical properties. Experimenting with a series of bodies of different composition, particle size and firing schedule, Grofcsik concluded that for fine quartz, increasing the quartz content increases strength. He also pointed out that having optimum grains of 20–60 μm may be causing the grains to dissolve in the glassy phase during firing, whilst coarser grains allow harmful stresses to accumulate. Many researchers have published data on optimum particle sizes for quartz. However, it must be borne in mind that the strength of clay-based materials depends not only on quartz particle size, but also on other characteristics of the fired body. Therefore, it is difficult to relate the strength of clay-based materials on quartz particle size alone.

Dinsdale and Wilkinson [174] related the properties of whiteware bodies to the size of constituent particles. By assuming, ε, to represent the mean size of the crystalline particles present in the fired system, they suggested the modulus of rupture, S, to be,

$$S = K \cdot \varepsilon^{-a} \tag{1.7}$$

where, K, is a constant and the index, a, would be expected from Griffith's crack theory to be about 0.5. Work on single-phase crystalline materials has shown that there is an exponential relationship between the strength, S, and the true porosity, P, of the form.

$$S = S_o e^{-bp} \tag{1.8}$$

where, $S_{o,}$ is the strength at zero porosity. Knudsen [143, 144] presented the above equations as,

$$S = K \cdot \varepsilon^{-a} \cdot e^{-bp} \tag{1.9}$$

where K, a, and, b, are empirical constants. Dinsdale and Wilkinson [174] agreed that the grain size distribution of the starting materials will strongly influence both the porosity and the grain size in the fired body. If the size of the filler material such as quartz is increased, packing may be increased, and unfired porosity reduced. However, fired strength is adversely affected. Improvements in fired strength can be attained by reducing filler size that is obtained at the expense of firing contraction. In a paper by Evans and Linzner [170], an acoustic emission study was carried out on a clay-based material as it was loaded to failure.

This showed that the acoustic emission rate increased rapidly as the stress in the sample approached the failure value. The sound pulses were expected to arise from cracks that de-bond quartz particles, cracking of quartz particles, themselves and cracks linking up or starting to run after being arrested. The results of this study indicate the quartz particles that remain attached to the matrix after cooling, and are therefore residually stressed, can be detached when the applied stress reaches a sufficiently high value. As the highest stresses in the material, resulting from the applied stress, occur adjacent to the tips of many cracks in the sample, the formation of a process zone around a crack which starts to grow can be expected if bonded quartz particles are in its vicinity.

In another acoustic emission study of porcelain [175], it was found that a maximum acoustic emission rate occurred at a temperature below the β- to α-quartz transition point. This suggests that the residual stress, resulting from the transition, increased as a result of the thermal expansion mismatch between the quartz particles and the matrix on cooling below 573°C. It was found that the temperature at which the maximum emission rate occurred was reduced as the quartz particles were made smaller. This is consistent with the assumption that smaller inclusions require higher stresses to become de-bonded. A further interesting finding was that a second maximum emission rate occurred at 200°C, the phase transition from β- to α-cristobalite. The presence of cristobalite arises from the conversion of quartz during densification heat treatment. Results published by Oral et al. [176], showed that for a variety of clay-based material compositions, the strength of ring specimens showed considerable scatter in test pieces that had optimum heat treatments. Failure was reported to occur by cracks around quartz particles or by quasi-spherical pores.

1.7.6.4 Ceramic Bonding Materials and Bond Strength

During grinding, the action of tangential and normal grinding forces create stresses within the abrasive grit and adjacent bond posts which inevitably causes bond-post failure. When the bond strength is optimized, this process takes place gradually, i.e., when abrasive grits have lost their ability to cut the workpiece. The most frequently used bonding materials for vitreous-bonded grinding wheels containing clay minerals, quartz, and feldspar, which contain crystalline phases that reduce their melting points. As previously mentioned, bonding materials used for alumina wheels resemble high-strength tough enamels (vitreous), and those for silicon carbide wheels are referred to as stoneware or soft porcelain bonds.

In alumina grinding wheels, the bond not only dissolves other crystalline phases in the bond, but also dissolves the surface of the alumina grain. The bond must not produce any "rounding" of the grains, therefore, the bond must produce the same "hardness" at the interface of the bond/grit couple. This explains why it is possible to fire grinding wheels at temperatures 200 or 300°C higher than the melting point of the bonds without discharging the bonding material or deforming the wheel itself. Guilleaume [176, 177] characterised vitreous bonds using the Seger formulae. Guilleaume tested 60 types of clay bond which were described by the formula:

$$RO \cdot (1.25 - 3.0Al_2O_3) \cdot (4.5 - 10SiO_2) \quad (1.10)$$

where RO is the sum of alkali oxides contained in the bond, i.e., CaO, MgO, MnO etc. These bond compositions were composed of a 'clay substance', feldspar, and quartz with powdered marble and magnesite as mineralizers. Guilleaume [178, 179] first examined fired bonds without the admixture of alumina grains, then determined the strength of bar specimens containing the alumina grains. Specimens were fired at different temperatures and specimens with variations in bond composition fired at constant temperature.

The effects on strength and hardness were examined. Guilleaume's results gave only a limited understanding since no information was provided on mineral composition and fired bond composition. Furthermore, firing conditions characterised by Seger cones (Sg) do not represent well-defined conditions. Firing conditions characterised by the same Sg temperature can produce grinding wheels with very different properties. Franz [180] reported that a deeper analysis on the properties of bonding materials on the performance of grinding wheels is required. He examined the bonding materials used by Guilleaume and reported that on firing grinding wheels containing bonds of the same composition in a small temperature range (between Sg8 and Sg12 – a temperature difference of 100°C), deviations in bond hardness were as high as 7° [180]. Rieke and Haeberle [181] noted that the strength of specimens prepared with bond materials that have low melting points are higher than those wheels prepared with classic bond materials.

Filonenko and Lavrov [177] pointed out that during the cooling period of firing, when grinding wheels are quickly cooled from peak temperature to 800°C, devitrification is inhibited. Thereafter, slow cooling is required in order to relieve internal stresses. The mechanical strength of the bond is dependent on the amount of its vitreous phase. According to Filonenko and Lavrov [177] of all the crystalline compounds, only spinel ($MgO{\cdot}Al_2O_3$) is capable of raising the strength of vitreous bonds. This compound surrounds alumina grains, as a quasi-embedding material, with octahedral spinel crystals smaller than 8 μm diameter formed in the alumina-rich melt zone developed at the interface between bond and grit. The high strength bonds used were located in the proximity of the SiO_2 peak in the Na_2O-SiO_2-Al_2O_3 ternary diagram. As a consequence of the alumina-rich melting zone, the following compounds may be formed: anorthite, cordierite, mullite, spinel, plagioclases, anastase, rutile, hematite and magnetite [177]. The composition of ceramic bonds is important for maintaining strength.

In order to achieve complete penetration of the abrasive grain surface, the viscosity of the melt plays an extremely important part in complete adhesion between grit and bond. The flow characteristics of ceramic bonds have been investigated by a number of researchers. Moser [182] used a heating microscope to observe, qualitatively, the changes in contact angle of various bonding materials up to 1,460°C. Bond materials and bond containing alumina grains were examined using three crude bond compositions, viz, (1) illite bond, (2) modified illite bond, and (3) fritted borosilicate bond. It was found that the fritted borosilicate bond melted at a lower temperature and produced better wetting characteristics. Moser reasoned that increased wetting of bond to grit would increase wheel strength.

Hartline [183] conducted work on the strength and fracture of alumina abrasive wheels and concluded that the fracture process involved in grinding wheels proceeds through the bond, and fractography conducted on bond posts showed a signs of de-vitrification and porosity. Hartline stated that increases in strength were achieved by developing uniform bond post strengths and/or higher strength wheel structure. Experimental grinding wheel bonds were examined by Barry, Lay, and Morrell [184] using novel glass-ceramics and conventional feldspar bonds. It was concluded that the strongest bonds were those based on traditional clay-feldspar-quartz composition, and that strength was dependent on wetting and flow properties of the bond on the surface of the grit.

Ogawa and Okamoto [185] found that increasing the feldspar content of bonds increased the adhesion of bond to grit in alumina wheels by producing a less viscous mixture at the soaking temperature. The development of superabrasive grinding wheels has resulted in a number of papers published on ceramic bonding of diamond and cBN [186–195]. Yang et al. [187] have published results concerning the strength of vitreous-bonded cBN wheels. They concluded that the strength of their specimens was dependent on glass composition and, to a much lesser extent, on porosity. The bonds examined were full of pores with an optimum composition of 51% SiO_2, 15% Al_2O_3, 26% B_2O_3, 3% Na_2O and 5% CaO.

Jackson, Barlow and Mills [189] concluded that the strength of vitreous-bonded alumina wheels depended on the K_2O-CaO ratio. They reasoned that the increased mass of glass-network modifiers would release liberated gases within the bonds by reducing the bond's viscosity. It appears that a reduction in bond viscosity not only releases gases liberated during the breakdown of clays and fluxes, but also improves interface strength by improving wetting and flow characteristics of the glassy bond.

1.7.6.5 Effect of Interfacial Cohesion on Bond Strength and Wheel Wear

Interfacial cohesion in vitreous-bonded grinding wheels is as important, if not more important, as the effect of bond composition on their strength and performance. In grinding wheels, interfacial strength dictates whether a grinding grain cuts efficiently or not at all. According to Hondros [190], a knowledge of interfacial properties holds the key to the design of bulk properties. Moseley et al. [191] compared the cohesion of various borosilicate bonding materials in contact with white-fused alumina (containing 99% wt Al_2O_3) and impurity phases 3% wt TiO_2, 1% wt SiO_2 and MgO and CaO, Fe_2O_3 and ZrO_2 in smaller quantities. They compared the properties of each grit/bond composition by measuring their fracture toughness and related these values qualitatively to observations of fracture surfaces and interfaces. Fired test specimens were inspected which showed preferential etching along crystallographically controlled directions in white alumina grit. This was observed to be dissolution of planar blocks of sodium aluminate or β-alumina ($Na_2O{\cdot}11Al_2O_3$) present in α-alumina (essentially pure Al_2O_3), established by X-ray diffraction of samples. β-alumina is thought to be detrimental to grit strength which, when in small well-dispersed amounts, can control the self-sharpening effect during grinding. However, in large amounts leads to loss of strength of the grit. Bragg, Gottfried and West were the first to determine the crystal structure of β-alumina in 1931. Beevers and Ross [192] came to the same conclusion as Bragg et al. when they discovered that the crystal structure and chemical composition do not readily agree with each other. Deviations in the chemical composition, noted by various workers, report: $Na_2O{\cdot}9Al_2O_3$; $Na_{1.5}Al_{10.83}O_{17}$; and $Na_2O{\cdot}6Al_2O_3$. According to Harata [193] the non-stoichiometric composition alters the measured lattice parameters. Based on his own measurements, β-alumina is represented by the formula,

$$(1.16 + X)Na_2O\,11Al_2O_3 \tag{1.11}$$

where x denotes the molar fraction of Na_2O, which varies from 0.19 to 0.59. Below this limit, α-Al_2O_3 separates from β-alumina, above it $NaAlO_2$ dissociates. Moser [194–197] performed electron microscopic studies on white-fused alumina grits. He observed that β-alumina was present as spots and bands on the surface of the grits, and noted that heating uncoated grits up to a temperature of 1,000°C led to a lower amount of β-alumina due to the evaporation of Na_2O above 900°C, thus leaving α-alumina. He further observed that grits coated with borosilicate bonding material contained a higher level of Na_2O at the interface. The most likely cause of increasing the strength of white alumina grinding wheels is the dissolution of β-alumina that would locally enrich the melted bond with Na_2O which promotes fluidity in alumino-borosilicate and alumino-alkalisilicate glasses. The glassy phase would then fill dissolution bands created in the surface of the grit thus promoting a better mechanical bond between grit and glass phase, i.e. enhanced shear resistance when subjected to grinding forces. In their samples, Moseley et al. [198] also identified small (50 μm dia × 10 μm thick) Ca-rich platelets, of which two types were identified; alite (Ca_3SiO_5) and an unnamed oxide, $NaCaAlO_3$, known to have several polymorphic forms. S.E.M. analysis performed by Moseley et al. on brown-fused alumina grit, showed that the grit contained 96–98 wt.%, Al_2O_3, 1–2 wt.% TiO_2 and up to 0.5 wt.% SiO_2. The amount of TiO_2 in solution is inconsistent with earlier work which had determined that the maximum solubility of TiO_2 in Al_2O_3 is less than 0.3 mol.%. at 1,300°C [199]. However, it was stated that MgO may increase solubility of TiO_2 in alumina. Examination of this section of brown-fused grit showed the appearance of blade-like inclusions at the grit-bond interface. This morphology is consistent with rutile needles observed in some synthetic sapphires [200]. These inclusions accounted for the variability in TiO_2 content and hence its solubility in alumina. Further examination of the fired specimens in both white and brown-grit samples, showed evidence of crystal formation in the glassy bond. These crystals were lath-shaped with square sections distributed evenly throughout the glass. X-ray diffraction indicated an alumino-borate with high alumina content. The best match was with $Al_{18}B_4O_{33}$. Titanium was considered deleterious to fracture toughness due to the presence of rutile needles on the surface of the grit and titania in the bond itself. Moseley et al. suggested using oxides to allow the formation of titanates insteadof rutile needles. Unfortunately, Mn-doped glasses did not improve strength appreciably. A second attempt was made to form a protective coat around the grit. Oxides such as MgO, ZnO, and CoO were used to form spinel layers between grit and bond. Unfortunately, the test specimens were lower in strength. They concluded that white-fused alumina grits were stronger due to planar dissolution of β-alumina, whilst brown-fused alumina grits were considered weaker due to rutile needle formation a the interface between bond and grit.

Similar interfacial studies were conducted on alumina abrasives [201] and silicon carbide vitreous-bonded wheels [202]. Comparison of wheel performance using wheels with brown-fused and white-fused alumina grits was examined by Reichenbach [93]. Interface studies of metals in contact with super-abrasive grits have recently been conducted by Scott et al. [203]. These workers provided substantial quantitative information on the wetting behaviour of copper-based alloys on diamond. This work was further extended by Evens et al. [204], whom correlated G-ratio of abrasive buttons with interfacial bond strength. Evens et al. concluded that increases in titanium and tin concentrations enhance wetting to the (III) plane of diamond, and that titanium segregated to the interface where alloy compositions promoted wetting on the diamond surface and to the free surface where alloy compositions did not promote wetting. The interfacial layer was characterized by electron probe analysis to be composed of bronze ($\alpha + \delta$) eutectoid and $CuSnTi_2$. Good interfacial bonding was found not to be associated with good wetting. The interfacial bonding layer with promoted wetting was identified as a 100 nm reaction product (TiC), which was associated with low temperatures, i.e., 900–950°C. The best composition was found to be a mixture of 20 wt.% copper–10 wt.% tin-titanium alloy which wets and bonds well to diamond at 900–950°C. Evens et al., [205] stated that better bonding and wetting was achieved with nickel-coated diamond grinding buttons.

Shilo et al. [199] conducted a comprehensive study on the wettability of glass on substrates of cBN They found that wettability was dependent on the reaction product at the grit/bond interface, which was identified as B_2O_3. In the case of sodium-borosilicate glasses, with Na_2O/B_2O_3 ratios less than or equal to 0.5, boron atoms tend to co-ordinate from ternary to tetrahedral form owing to the introduction of oxygen atoms by alkali oxides. The best wetting conditions were achieved at 900°C using glasses of the system Na_2O-B_2O_3-PbO-SiO_2. The addition of 4–7 wt.% Li_2O was responsible for the best wetting condition that was due to the network modifying effect of Li_2O.

1.7.7 *Reactions in Ceramic Bonds*

Reactions in bond compositions used in this study are similar to those that occur in clay-based materials used in whiteware bodies. The reactions outlined in this chapter form the basis for studies on the effect of bond composition on wheel performance. Clays provide the wheel bond with plasticity for forming and strength in the green state. The two principal clays used for wheel bonds are china clay and ball clay. Both clays are primary clays formed from decomposed feldspar.

The main mineral constituents in both china and ball clay is kaolinite ($Al_2O_3 \cdot 2SiO_2 \cdot 2H_2O$). However, clays differ in purity and plasticity. Kaolinite has a layered structure and the particles reflect this by having a plate-like morphology. Fluxes are used in bonds in order to lower the firing temperature by reacting primarily with the clay to form a viscous liquid phase that promotes densification.

The flux is usually a feldspathic mineral such as feldspar, nepheline syenite, Cornish stone, talc and sometimes lithia. The complete range of fluxes has been reviewed by Royle [200]. The fluxing effect of feldspar in clay-based materials was studied by Schramm and Hall [201]. Orthoclase ($K_2O{\cdot}Al_2O_3{\cdot}6SiO_2$) and albite ($Na_2O{\cdot}Al_2O_3{\cdot}6SiO_2$) and a combination of these fluxes in the formation of clay-based materials. In high strength bonds, used for conventional and superabrasives vitreous products, the fluxing effect is provided by powdered glass frits and borax in addition to feldspar. These additions provide the bond with fluxes that melt over a wider temperature range than those using traditional bond materials. This reduces the firing temperature that tends to promote increased adhesion between grit and bond.

1.7.7.1 Densification and Phase Analysis

When a grinding wheel has been formed, it is fired at an appropriate soaking temperature to be densified in order to mature to the optimum state. During firing the bond materials, abrasive, and fillers initially behave independently of one another. Water occurring in the pores and in the clay is driven off below 250°C, whilst at 500°C, the organic matter has been burnt off.

Theoretical Phase Analysis: Use of Equilibrium Diagrams

The overall final composition of clay-based grinding wheel bonds is composed of unreacted quartz, mullite, glass and sometimes cristobalite depending on the reaction conditions. The formation and growth of mullite crystals is thought to occur in the following manner:

1. Formation in the kaolinite platelets
2. Formation in the feldspar platelets
3. Formation in the mica platelets
4. Growth by recrystallisation
5. Formation during cooling

Reaction rates and mechanisms of mullite formation may be different when the mullite is formed in the mica, kaolinite and feldspar platelets. The formation rate in one region is influenced by formations in other regions unless concentration gradients of atoms are distributed in such a way as to promote equal reaction rates throughout the total reaction volume. It is clear that the formation of mullite in clay-based bonds is very complicated. The formation of mullite can be explained qualitatively using equilibrium diagrams, and quantitatively by comparing equilibrium phases with experimental results.

The application of equilibrium diagrams to ceramic manufacturing processes has been discussed by many authors. However, when applied to clay-based materials, Bowen [202], Hall and Insley [206] and Foster [207] express reservations in using them due to the coarseness of the clay particles, the high viscosity of molten feldspar [208, 209], and the slow diffusion of alkali ions which tend to prevent equilibrium and homogeneity of the mass. Shelton [210] used equilibrium diagrams with success to correlate the properties of fired whiteware bodies to the amount of eutectic melt formed. Dietzel and Padurow [211] used phase diagrams in their discussion of quartz dissolution in porcelain.

Although equilibrium is not obtained for clay-based bonds, equilibrium diagrams are useful for describing reaction rates towards equilibrium. The equilibrium diagram used in this work is the K_2O-Al_2O_3-SiO_2 ternary system derived by Schairer and Bowen [212]. When heating a body consisting of quartz, kaolinite and potash feldspar, there are no appreciable reactions that take place between quartz, feldspar and kaolinite relicts until the temperature approaches the eutectic temperature of the total system, 985°C. In each isolated sub-system below 985°C, quartz has transformed to its high temperature form β-quartz and the feldspar grains have transformed to sanidine. Kaolinite has lost its chemically bonded water and the decomposition products have been transformed to γ-Al_2O_3 and an amorphous phase with a high SiO_2 content. This is the primary composition of the reaction system.

Considering complete equilibrium conditions within the total system immediately under the eutectic temperature there are three solid phases, feldspar, tridymite, and mullite. When the system is heated to the eutectic temperature a fourth phase is formed, a melt phase with a composition of the eutectic. With continued heating at constant temperature the amount of the melt phase is increased until one of the phases is consumed. If feldspar is consumed the composition point of the melt moves with rising temperature along the boundary line between the primary phase regions of tridymite and mullite until one of these two phases is consumed. The composition point then enters the remaining solid phase in the direction of the composition of the total system. If tridymite is first consumed at the phase reaction in the eutectic, the composition point of the melt moves along the boundary line between the feldspar and mullite regions. If neither of these two phases is consumed before 1,140°C, the ternary invariant point is reached where feldspar is consumed and leucite occurs as a new phase.

After this phase reaction is completed the composition follows the boundary line until one of the solid phases has been consumed. The consumption of phases is dependent on the rate of reaction of the body until equilibrium conditions have been reached. These conditions are applicable under conditions of equilibrium. For clay-based bonds equilibrium is rarely achieved and as a consequence, the grain size of the raw material is the most important reaction rate variable. Quartz shows very low reactivity up to the eutectic temperature. At that temperature, the quartz phase present is not affected by the other phases.

Formation of Mullite in Kaolinite Clays

The effect of heat on ceramic raw materials has been studied by many workers as far back as 1887 when le Chatelier charted discontinuities in the thermal analysis of kaolinite. Ford and Todor presented differential thermal analysis curves for china clay and ball clay. The major phase transformations for potash feldspar, as measured using a DuPont 1600 D.T.A. cell, of kaolinite → metakaolinite → spinel-type phase → mullite occurs as heating continues. There appears to be a controversy regarding the products formed in heat treatment. One theory suggests that, on dehydroxylation, kaolinite forms a mixture of alumina and amorphous silica. The other theory considers the dehydrated product as an aluminosilicate. To explain the exothermic reaction it is believed that γ-Al_2O_3 is formed and in the other case, an Al-Si spinel is formed. Comeforo et al. [213], show the appearance of hexagonal particles far above the temperature for dehydroxylation, illustrating a residual structure still present in the non-crystalline compound of metakaolin.

Many researchers have investigated reactions that occur in clay-based materials containing these compounds. Weiss et al. [214], announced that the cubic phase that appeared at 900°C from kaolinite was an Al-Si spinel. These workers isolated the spinel phase by leaching amorphous SiO_2 from the fired kaolinite and found that its chemical formula agreed with the theoretical formula. The Al-Si spinel analyzed by Weiss et al. is different from the γ-Al_2O_3 spinel because its lattice constant is 0.002 nm lower owing to the replacement of Al-ions by smaller Si-ions present in the cubic γ-Al_2O_3 spinel structure [215]. However, other researchers have supported the crystallization of γ-Al_2O_3 as the cause of the exothermic peak near 100°C [216–221]. There are two other exothermic peaks due to the crystallization of primary mullite and formation of cristobalite from amorphous silica that occur at 1,150 and 1,250°C respectively. The transformation of pure kaolinite to mullite has been reviewed by Chaudhuri [221].

Effect of Heat on Feldspar and Quartz

α-quartz is known to invert to β-quartz at 573.3°C [222]. The transition is accomplished by a 0.8% volume expansion and is reversible and rapid. Any cristobalite present in the matrix is transformed ($\alpha \rightarrow \beta$ structure) between 200 and 270°C [223]. The breakdown and melting of potash feldspar has been studied by Morey and Bowen [224] who reported incongruent melting at 1,170°C. However, the thermal behaviour of feldspar has had little attention paid to it. The solubility of quartz in feldspar [225] and the solubility of quartz in clay and feldspar [226] has been investigated extensively. These workers report that mullite ($3Al_2O_3{\cdot}2SiO_2$) crystallizes in molten feldspar.

Effect of Heat on Clay-Based Materials

Clay-based materials principally contain the oxides SiO_2, Al_2O_3, K_2O and Na_2O. Equilibrium diagrams were used by Lundin [226] to study the formation of mullite in detail. For mullite content, the experimental values were 3.4% lower than those calculated from the phase diagram. For glass content, the difference was 5% lower. Lundin concluded that the quartz had only partly reacted. The difficulty in attaining equilibrium in clay-based materials is thought to be due to the high viscosity of molten feldspar, retarded dissolution of quartz, and the slow diffusion of partially mixed and partially melted constituents. When the temperature increases mullite dissolves partially in the melt if equilibrium is achieved. At around 1,000°C the surface tension of the liquid draws unreacted/partially reacted particles together that reduces porosity and increases bulk density. A loss in surface area causes shrinkage in the body. The porosity is initially interconnected and is referred to as open porosity. However, the reduction in the volume of the body produces closed porosity. When the open porosity is removed, the body is vitrified.

The densification of the body is slow because of the high viscosity of the liquid phase. The decrease in the viscosity is expected from the increase in temperature that is partially offset by the enrichment of the liquid with silica by partial dissolution of quartz. In practice, the slow dissolution of quartz results in bonds having wide range of firing. The initial densification of clay-based bonds can be modelled simply using two contacting spherical particles. The rate of initial neck growth is,

$$\frac{x}{r} = \sqrt{\frac{3 \cdot \gamma_t \cdot t}{2 \cdot \eta \cdot \rho}} \tag{1.12}$$

The increase in contact diameter is proportional to t, and the increase in area between particles should be directly proportional to time. The factors affecting the rate of densification are surface tension, γ_t, viscosity, η, and particle size, r. The volume shrinkage, or linear shrinkage which takes place is determined by the approach between particle centers is given by the model as,

$$\frac{\Delta V}{V_o} = 3\frac{\Delta L}{L_o} = \frac{9 \cdot \gamma_t \cdot t}{4 \cdot \eta \cdot r} \tag{1.13}$$

This equation shows that the initial rate of shrinkage is directly proportional to the viscosity and particle size. As firing develops, more liquid is formed and mullite crystals appear. Primary mullite is formed from clay relicts whilst secondary mullite is formed from the melt. At 1,200°C, a considerable amount of mullite is formed in quartz-rich bonds. Shelton and Meyer [227] reported that increased rates of heating cause less liquid and mullite formation, less quartz corrosion, and more pores. An optimum heating rate of 50–90°C/h was preferred. The crystallization of glasses in the K_2O-SiO_2-Al_2O_3 system was studied extensively by Hermansson and Carlesson [228]. They concluded that crystallization from this high viscosity ternary system is possible. Tuttle and Cook [229] claimed to have found quartz, mullite, cristobalite and wollastonite as the crystalline constituents of kaolin-flint-feldspar blends. However, in many cases only quartz and mullite, and occasionally cristobalite are the phases present in clay-based kaolin-quartz-feldspar composites.

Effects of Cooling

When densification occurs, the cooling rate is reduced in order to prevent thermal stress cracking of the body. It is better to reduce the cooling rate when crystalline inversions occur that involve volume changes. The inversion ranges for quartz and cristobalite are 550–580°C and 200–300°C, respectively. When quartz-containing bonds begin to cool from the soaking temperature, it is considered that the liquid phase relieves stresses resulting from thermal expansion mismatch between itself and the phases β-quartz, β-cristobalite and mullite to at least 800°C. At 800°C, stresses will develop in quartz particles and the matrix which causes micro-cracking. The shrinkage behaviour of quartz and the glass phase has been described by Storch et al. [230].

Between the temperature range $573 < T < 800$°C, the glass phase shrinks more than quartz phase which causes tangential tensile stresses to form cracks in the matrix. At 573°C, β-quartz transforms to α-quartz that causes tensile residual stresses to produce circumferential cracking around quartz particles. Some of these cracks have been observed to propagate into the glass phase [231]. Similar observations occur in the cristobalite phase. Spontaneous cracking of quartz has been found to occur over a temperature range that depends on the size of the quartz particles [232]. Particles larger than 600 μm diameter cracked spontaneously at 640°C whereas smaller particles (<40 μm diameter) cracked at 573°C. This observation agrees with temperature-dependent microcracking reported elsewhere [155].

1.8 Conclusions

A review of the literature has provided a mathematical foundation that quantifies grinding wheel performance in terms of grinding parameters and wheel wear up to the burn boundary of a particular workpiece tested. The transformation of these fundamental factors presented in the form of performance diagrams show optimum bond compositions and wheel grades for a particular workpiece-wheel combination under pre-determined operating conditions.

A review of the literature has provided an insight into the causes and effects of wheel wear in terms of applied grinding loads, abrasive grain type, and abrasive strength. Relationships have also been formulated and tested that relate the wheel wear parameter, G-ratio, and the maximum tensile stress acting within the abrasive material. However, the effect of bond composition on wheel wear and performance has not been examined.

A review of the strength of clay-based materials thus provides a basis for studying the effect of bond composition on bond strength and wheel performance. Studies of the influence of workpiece material composition on grinding wheel wear also ensures that any study on grinding wheel performance should include factors that affect workpiece hardness and vitreous-bond strength.

The fired microstructure of all clay-based materials will depend on the structure of the raw materials used, the processing sequence and the heat treatment schedule. The fired microstructure can contain:

1. Feldspar
2. Quartz and possibly cristobalite
3. A glass phase of variable composition
4. Cracks around quartz and cristobalite particles
5. Mullite-primary and secondary
6. Pores-open and/or closed, depending on the heat treatment, and fissure-like pores resulting from differential shrinkage

The complexity of the grinding wheel bonding system can be appreciated given the shape, size, amount, distribution, and orientation of constituent particles and how these variables affect the bond's tribological properties such as wheel wear and grinding friction.

References

1. De Pellegrin D.V., Corbin N.D., Baldoni G., Torrance A.A., Wear 253 (2002) 1016–1025.
2. De Pellegrin D.V., Corbin N.D., Baldoni G., Torrance A.A., Tribol. Int. 42(1) (2008) 160–168.
3. De Pellegrin D.V., Stachowiak G.W., Tribol. Int. 37 (2004) 255–270.
4. Diashape – World Wide Web, (2006), "Diashape Parameters, Innovative Sintering Technologies" http://www.istag.ch/diamonds/parameters.html.
5. Sysmex – World Wide Web, "FPIA-3000 Particle Analyzer – Parameters," Sysmex (2006), http://www.particle.sysmex.co.jp/en/fpia/parameter.html.
6. Verspui M.A., van der Velden P., de With G., Slikkerveer P.J., Wear 199 (1996) 122–126.
7. Hamblin M.G., Stachowiak G.W., Tribol. Trans. 39(4) (1996) 803–810.
8. Kaye B.H., "Fractal description of fine particle systems," in: J.C. Beddow (Ed.), Particle Characterization in Technology, Applications and Microanalysis, vol. 1, CRC Press, Boca Raton, FL (1984).
9. Swanson P.A., Vetter A.F., ASLE Trans. 28(2) (1984) 225–230.
10. Leavers V.F., IEEE Trans. Pattern Anal. Mach. Intell. 22(12) (2000) 1411–1423.
11. Roquefeuil F., Abras. Mag. December (2003) 24–29.
12. Komanduri R., Shaw M.C., "Attritious Wear of Silicon Carbide," ASME Paper #75-WA/Prod-36 (1975).

13. Kugemai N., Kiyoshi K., Titanium Science & Technology, Proc. Fifth Intl. conf. on Titanium, September 10–14 (1984), Munich.
14. Hirosaki K., Shintani K., Kato H., Asakura F., Matsuo K., JSME Int. J. Ser. C 47(1) (2004) 14–20.
15. Kumar K.V., Fourth Intl Grinding Conf. SME, October 9–11 (1990), Detroit.
16. Hagiwara S., Obikawa T., Usui E., Trans. ASME 120 (1998) 708–714, J. Manufact. Sci. Eng.
17. Ishikawa T., Kumar K., "Conditioning of Vitrified Bond Superabrasive Wheels" Superabrasives' 91 SME MR91–172 (1991).
18. Takagi J., Liu M., J. Mater. Process. Technol. 62 (1996) 397–402.
19. Jakobuss M., Webster J. Abras. Mag. August/September (1996) 23.
20. Fujimoto M., Ichida Y., Sato R., Morimoto Y., JSME Int. J. Ser. C 19(1) (2006) 106–113.
21. Yonekura M., Yokogawa K., Bull. JSPE 17(2) (1983) 113–118.
22. Mindek R., MSc Thesis (1992), Univ. Connecticut.
23. Acheson E.G., US Patent 492,767 (1893).
24. Cowles A.H., Cowles E.H., US Patent 319945 (1885).
25. Kern, E.L., Hamill, D.W., Deem, H.W., Sheets, H.D., Mater. Res. Bull. 4 (1969) S25–S32.
26. Kern, E.L., Hamil D.W., Deam H.W., Sheets, Mater. Res. Bull., Special Issue 4, 1969, S25–S32, Proceedings of the International Conference on Silicon Carbide, University Park, Pennsylvania, USA, October 20–23 (1968).
27. Nilsson O., Mehling H., Horn R., Fricke J., Hofmann R., Muller S.G., Eckstein R., Hofmann D., High Temp. High Press. 29 (1997), 73–79.
28. Tymeson M.M., "The Norton Story," Norton Co., USA (1953).
29. Wikipedia "Bayer Process" Online Encyclopedia Reading, December 12 (2009).
30. Wolfe L.A., Lunghofer E.P., 13th Industrial Minerals International Congress, Kuala Lumpur, Malaysia, April 26–29 (1998).
31. Lunghofer E.P., Wolfe L.A., "Fused brown alumina production in China," Posted: August 6 (2000), http://www.ceramicindustry.com/copyright/77d58fabca9c7010VgnVCM100000-f932a8c0_...9/24/2009.
32. Whiting Equipment Canada Inc., "Metallurgical Equipment," Commercial brochure (n.d.).
33. U.S.G.S. Geological Survey, Mineral Commodity Summaries, January (2009).
34. U.S.G.S. Geological Survey, Minerals Yearbook Abrasives, Manufactured (advance release) (2008).
35. Wellborn W.W., AES Mag. 31(1) (1991) 6–13.
36. Coes L., "Abrasives," Springer-Verlag, Berlin (1971) p. 65.
37. Polch, US Patent 2769699, November (1956).
38. Robie N.P., US Patent 2877104, March 10 (1959).
39. Foot D.G., US Patent 3,175,8943/30 (1965).
40. Marshall D.W., US Patent 3181939 May 4 (1965).
41. Cichy P., US Patent 3,726,6214/10 (1973).
42. Richmond W.Q., Cichy P. US Patent 3,861,849, January 21 (1975).
43. Richmond W.Q., Cichy P. US Patent 3,928,515, December 23 (1975).
44. Sekigawa H., US Patent 3,977,132, August 31 (1976).
45. Ilmaier B., Zeiringer H., US Patent 4,059,417, November 22 (1977).
46. Cichy P., US Patent 4,061,699, December 6 (1977).
47. Ueltz H.F.G., US Patent 4,194,887, March 25 (1980).
48. Richmond W.Q., US Patent 4,415,510, November 15 (1983).
49. Richmond W.Q., US Patent 4,439,895, April 3 (1984).
50. Scott J.J., US Patent 3,993,119, November 23 (1976).
51. Scott J.J., US Patent 4,070,796, January 31 (1978).
52. Rowse R.A., Watson G.R., US Patent 3,891,408, June 24 (1975).
53. Bange D., Wood B., Erickson D., Abras. Mag. June/July (2001) 24–30.
54. Webster J.A., Tricard M., "Innovations in abrasive products for precision grinding," CIRP Innovations in Abrasive Products for Precision Grinding Keynote STC G, August 23 (2004).

55. Bauer R. "Process for production of alpha alumina bodies by sintering seeded boehmite 2039 made from alumina hydrates," US Statutory Invention Disclosure H000189, January 6 (1987).
56. Leitheiser M.A., Sowman H.G., US Patent 4,314,827, February 9 (1982).
57. Wood W.P., Monroe L.D., Conwell S.L., US Patent 4,881,951, November 21 (1989).
58. Bange D., Wood B., Erickson D., Abras. Mag. June/July (2001) 24–30.
59. Cottringer T.E., van de Merwe R.H., Bauer R., US Patent 4,623,364, November 18 (1986).
60. Schwebel M.G., US Patent 4,744,802, May 17 (1988).
61. Rue C.V., US Patent 4,543,107, September 24 (1985).
62. Pellow S.W., US Patent 5,090,968, February 25 (1992).
63. DiCorletto J., Conf. Precision Grinding & Finishing in the Global Economy, Oak Brook, IL, October 1–3 (2001).
64. Roffman R., Finer Point Mag. (2005) 26–30.
65. Cockburn C., National Geographic, March (2002) 2–3.
66. Trautman R., Griffin B.J., Scharf D., Sci. Am. 279(2) (1998) 82–87.
67. U.S. Geological Survey, Data Series 140 (2005).
68. Ladd R. "Manufacutured large single crystals" Finer Points – Wire Die Products and Applications (n.d.) 23–28.
69. Bailey M.W., Juchem H.O., "The advantages of CBN grinding: low cutting forces and 2064 improved workpiece integrity," IDR Pt. 3 (1993) 83–89.
70. Woodbury R.S., "History of the Grinding Machine – A Historical Study in Tools and 2066 Precision Production," Technology Press, Massachusetts Institute of Technology, USA (1959).
71. Bateman A.H., J. Soc. Arts (London), March 22 (1878) 362.
72. Grinding Data Book. Unicorn Abrasives, Stafford, England (1995).
73. Black J.T., Trans. ASME J. Eng. Ind. 101 (1979) 403.
74. Rowe G.W., Wetton A.G., J. Inst. Met. 97 (1969) 193.
75. Marshall E.R., Shaw M.C., Trans. ASME 74 (1952) 51.
76. Backer W.R., Marshall E.R., Shaw M.C., Trans. ASME 74 (1952) 61.
77. Merchant M.E., J. Appl. Phys. 16 (1945) 207.
78. Doyle E.D., Agahn R.L., Metall. Trans. B 6B (1975) 143.
79. Malkin S., Cook N.H., Trans. ASME J. Eng. Ind. 93 (1971) 1120.
80. Shonazaki T., Shigematu H., Bull. Jpn. Soc. Precision Eng. 2 (1966) 8.
81. Takenaka N., Proc. Int. Conf. on Manufact. Tech., Ann Arbor, MI, USA (1967) 617.
82. Okamura K., Nakajima K., Proc. Int. Conf. Grinding, Pittsburgh, USA (1972) 305.
83. Hahn R.S., Proc. Third. Int. Mach. Tool Des. and Res. Conf., UK, (1962) 129.
84. Doyle E.D., Dean S.K., Ann CIRP 29 (1980) 571.
85. Mulhearn T.O., Samuels T.O., Wear 6 (1963) 457.
86. Malkin S., Proc. Ninth North American Metalworking Research Conference, USA, Carnegie Press (1981) 235.
87. Malkin S., Proc. Third North American Metal Working Research Conference, USA, Carnegie Press (1975) 453.
88. Malkin S., Joseph S., Wear 32 (1975) 15.
89. Werner G., Doctoral Dissertation, Technische Hochschule Aachen (1971).
90. Malkin S., Ann. CIRP 29 (1979) 209.
91. Tarasov L., Am Soc. Met. 43 (1951) 1144.
92. Opitz H., Ernst W., Meyer H., Proc. Sixth Int. MTDR Conf., Manchester, England (1962) 155.
93. Reichenbach G.S., ASTME Paper no. MR67-594, USA (1967) 1.
94. Malkin S., Grinding Technology. Ellis Horwood, England (1989).
95. Krabacher E.J., Trans. ASME J. Eng. Ind. 81 (1959) 187.
96. Pattinson E., Chisholm A., Int. Conf. On Manufacturing Technology, ASTME (1967) 601.

97. Malkin S., Murray T., Proc. Fifth North American Metalworking Research Conference, S.M.E., USA (1977), p. 278.
98. Kannapan S., Malkin S., Trans. ASME J. Eng. Ind. 94 (1972) 833.
99. Grisbrook H., Hollier R.H., Varley P.G., Int. J. Prod. Res. 1 (1961–1962) 57.
100. Tonshoff H.K., Grabner T., Proc. Fifth Int. Conf. On Prod. Eng., Jap. Soc. of Precision Engrs., Tokyo (1984) 326.
101. Yoshikawa H., Int. Res. Prod. Eng. ASTME 23 (1963) 209.
102. Tarasov L.P., Int. Res. Prod. Eng. ASTME 21 (1963) 196.
103. Yoshikawa H., Sata T., Trans. ASME J. Eng. Ind. 85 (1963) 39.
104. Tsuwa H., Tech. Rep. of Osaka Univ., Japan 10 (1960) 733.
105. Tsuwa H., Tech. Rep. of Osaka Univ., Japan 11 (1961) 287.
106. Tsuwa H., Tech. Rep. of Osaka Univ., Japan 11 (1961) 299.
107. Tanaka Y., Ikawa N., Tech. Rep. of Osaka Univ., Japan 12 (1962) 345.
108. Geopert G.J., Williams J.L., Mech. Eng. 81 (1959) 69.
109. Tsuwa H., Trans. ASME J. Eng. Ind. 86 (1964) 371.
110. Lal G.K., Shaw M.C., Proc. Int. Grinding Conf., Pittsburgh, Pennsylvania, USA (1972) 107.
111. Eiss N.S., Trans. ASME J. Eng. Ind. 89 (1967) 463.
112. Bhattacharyya S.K., Grisbrook H., Moran H., Microtechnic 22 (1965) 114.
113. Mohun W., Trans. ASME J. Eng. Ind. 84 (1962) 431.
114. Tsuwa H., Yasui H., Proc. of the Int. Grinding Conf., Pittsburgh, Pennsylvania, USA (1972) p. 142.
115. Saito K., Kagiwada T., Bull. Jpn. Soc. of Precision Eng. 8 (1974) 125.
116. Stetiu G., Lal G.K., Wear 30 (1974) 229.
117. Trent E.M., Metal Cutting. 3rd Edition, Butterworth-Heinemann, London (1991), p. 113.
118. Marshall E.R., Shaw M.C., Trans. ASME 72 (1952) 51.
119. Grisbrook H., Prod. Eng., 37 (1960) 251–269 and 341–346.
120. Davis C.E., Rubenstein R., Int. J. Mach. Tool. Des. Res. 12 (1972) 165.
121. Rowe W.B., Chen X., Morgan M.N., Proc. 30th Int. Mach. Tool Des. and Res. Conf., Manchester, UK (1993) 195.
122. Rowe W.B., Chen X., Morgan M.N., Proc. 31st Int. Mach. Tool Des. And Res. Conf., Manchester, UK (1995) 415.
123. Lindsay R.P., Trans. ASME J. Eng. Ind. 92 (1970) 683.
124. Lindsay R.P., Hahn R.S., Proc. Fourth All India Mach. Tool Des. Res. Conf., Indian Institute of Technology, Madras, India, p. 537.
125. Lindsay R.P., Hahn R.S., Society of Manufacturing Engineers' Technical Report no. MR71, USA (1971).
126. Coes L. Abrasives. Springer-Verlag, New York (1971), Chapter 14.
127. Graham W., Nee A.Y.C., Prod. Eng. (1974) 186.
128. Loladze T.N., Bockuchava G.V., The Wear of Diamonds and Diamond Wheels. Mashinostroneniye, Moscow, USSR (1967).
129. Tanaka Y., Ueguchi T., Ann. CIRP 19 (1971) 449.
130. Outwater J.O., Shaw M.C., Trans. ASME 74 (1972) 73.
131. Duwell E.J., Hong I.S., McDonald W.J., ASLE Trans. 12 (1969) 86.
132. Yossifon S., Rubenstein C., Trans. ASME J. Eng. Ind. 103 (1981) 144.
133. Komanduri R., Shaw M.C., Trans. ASME J. Eng. Ind. 98 (1976) 1125.
134. Konig W., Messer J., Ann. CIRP 30 (1981) 457.
135. Komanduri R., Ann. CIRP 25(1) (1976) 161.
136. Komanduri R., Shaw M.C., Proc. Third North American Metalworking Research Conference, USA (1975), p. 481.
137. Graham W., Voutsadopolous C.M., Int. J. Mach. Tool Des. Res. 18 (1978) 95.
138. Carniglia S.C., J. Am. Ceram. Soc. 55 (1972) 610.
139. Duckworth W., J. Am. Ceram. Soc. 36 (1953) 68.
140. Parmalee C.W., Morgan W.R., Ceram. Ind. 22 (1934) 146.

141. Ryskewitch E., J. Am. Ceram. Soc. 55 (1953) 610.
142. Fiero D.C., et al., IEEE Trans. Power App. Syst. 89 (1970) 680.
143. Knudsen F.P., J. Am. Ceram. Soc. 42 (1959) 376.
144. Koenig J.H., Bull. Am. Ceram. Soc. 19 (1940) 424.
145. Zoellner H., Sprechsaal 41 (1908) 71.
146. Sane S.C., Cook R.L., J. Am. Ceram. Soc. 34 (1951) 145.
147. Eitel W., Phys. Chem. Der Silic. Barth 2 (1941) 649.
148. Budnikov P.P., Technologie der Keramishen Erzeugnisse. V.E.B. Verlag, Berlin, (1958) p. 532.
149. Geller R.F., Am. Ceram. Soc. Bull. 12 (1933) 18.
150. Krause O., Keetman E., Sprechsaal 68 (1935) 177.
151. Grofcsik J., Proc. Eighth Conf. Silicate Ind., Budapest, Hungary (1965) 367.
152. Mattyasovsky-Zsolnay L., J. Amer. Ceram. Soc. 40 (1957) 299.
153. Weidman T., Sprechsaal 92 (1959) 2.
154. Kalnin I.C., et al., Bull. Am. Ceram. Soc. 47 (1968) 498.
155. Koch H., Ber. Deut. Keram. Ges. 47 (1970) 538.
156. Lach V., Intern. Ceram. 23 (1974) 27.
157. Winterling A., Ber. Deut. Keram. Ges. 38 (1961) 9.
158. Washaw S.I., Seider R., J. Am. Ceram. Soc. 50 (1967) 337.
159. Weyland J.D., MSc Thesis, University of Missouri, USA (1964).
160. Avgustinik A.I., Adamashvili I.P., Ceram. Abst. (1971) 98h.
161. Austin C.R., et al., J. Am. Ceram. Soc. 29 (1946) 341.
162. Batchelor R.W., Dinsdale A., Seventh International Ceramic Congress (1961), p. 31.
163. Allen C.J., Trans. Br. Ceram. Soc. 63 (1964) 13.
164. Pass D., German W.L., Ceramics 20 (1969) 15.
165. Weyl D., Ber. Deut. Keram. Ges. 36 (1959) 319.
166. Davidge R.W., Green T.J., J. Mater. Sci. 3 (1968) 629.
167. Krause O., Sprechsaal 70 (1937) 611.
168. Washaw S.I., Seider R., Ceramic Microstructures, in: R.M. Fulrath and J.A. Pask (Eds.), John Wiley and Sons, New York (1966), p. 559.
169. Ludas M., Proc. Sixth Conf. Silicate Ind., Budapest, Hungary (1961) p. 297.
170. Evans A.G., Linsner M., J. Am. Ceram. Soc. 56 (1973) 575.
171. Krause O., Sprechsaal 72 (1939) 611.
172. Krause O., Sprechsaal 75 (1942) 229.
173. Beech D., Norris A.W., Trans. J. Br. Ceram. Soc. 60 (1961) 556.
174. Dinsdale A., Wilkinson W.T., Trans. J. Br. Ceram. Soc. 65 (1966) 391.
175. Kirchoff G., et al., J. Mater. Sci. 17 (1982) 2809.
176. Oral M.S., et al., Trans. J. Br. Ceram. Soc. 82 (1983) 31.
177. Filonenko N.E., Lavrov J.V., Petrography of Artificial Abrasives. Mashgiz, Moscow, USSR (1958).
178. Guilleaume A., Ber. Deut. Keram. Ges. 11 (1930) 284.
179. Guilleaume A., Doctoral Dissertation, Technische Hochschule Berlin, Germany (1927).
180. Franz W., Doctoral Dissertation, Technische Hochschule Braunschweig, Germany (1936).
181. Reike R., Haeberle C., Ber. Deut. Keram. Ges. 24 (1934) 117.
182. Moser M., Microstructure of Ceramics. Akademiai Kiado, Budapest, Hungary (1980).
183. Hartline S.D., Ph.D. Thesis, Pennsylvania State University, USA (1975).
184. Barry T.I., et al., Trans. J. Br. Ceram. Soc. 79 (1980) 139.
185. Ogawa S., Okamoto T., Bull. Jpn. Soc. Precision Eng. 20 (1986) 264.
186. Tanaka T., et al., Bull. Jpn. Soc. Precision Eng. 19 (1985) 221.
187. Yang J., et al., Ceram. Int. 19 (1993) 87.
188. Kuznichenko A.N., et al., Steklo i Keramika 8 (1983) 22.
189. Jackson M.J., et al., J. Mater. Sci. Lett. 13 (1994) 1287.
190. Hondros E.D., Br. Ceram. Trans. J. 90 (1991) 139.

191. Moseley D., et al., Br. Ceram. Trans. J. 88 (1989) 41.
192. Beevers C.A., Ross M.A., Z. Kristallogr. 97 (1937) 59.
193. Harata M., Mater. Res. Bull. 6 (1971) 461.
194. Moser M., Microstructure of Ceramics. Akademiai Kiado, Budapest, Hungary (1971).
195. Winkler E.R., et al., J. Am. Ceram. Soc. 49 (1966) 634.
196. Deer W.A., et al., Rock-Forming Minerals, Vol. 5: Non-silicates, Longman, London (1962).
197. Moser M., Proc. Int. Conf. Grinding, Pittsburgh, Pennsylvania (1972), p. 91.
198. Moser M., Period Polytech. Chem. Eng. 21 (1977) 27.
199. Shilo A.E., et al., Sverkhtverdye Mater. 5 (1983) 5.
200. Royle J.D., Trans. J. Br. Ceram. Soc. 73 (1974) 291.
201. Schramm E., Hall F.P., J. Am. Ceram. Soc. 19 (1936) 159.
202. Bowen N.L., J. Am. Ceram. Soc. 26 (1943) 285.
203. Scott P.M., et al., J. Mater. Sci. 1 (1975) 1833.
204. Evens D., et al., India Rev. (1977) 306.
205. Shilo A.E., Malogolovets V.G., Khizhnyak G.I., Sverkhtverdye Materialy, (1983), Vol. 5, p. 5.
206. Hall F.P., Insley H., Phase Diagrams for Ceramists. Columbus Press, USA (1947).
207. Foster W.R., J. Am. Ceram. Soc. 34 (1951) 151.
208. N'Dala I., et al., Br. Ceram. Trans. J. 83 (1984) 105.
209. Deletta M., et al., Br. Ceram. Trans. J. 83 (1984) 108.
210. Shelton G.R., J. Ceram. Soc. 31 (1948) 39.
211. Dietzel A., Padurow N.N., Ber. Deut. Keram. Ges. 31 (1954) 7.
212. Schairer J.F., Bowen N.L., Am. J. Sci. 245 (1947) 193.
213. Comeforo J.E., et al., J. Am. Ceram. Soc. 31 (1948) 254.
214. Weiss A., et al., Proc. Int. Clay Conf., Tokyo, in: L. Heller (Ed.), Israel Universities Press, Jerusalem, Vol. 2, (1969) p.34.
215. Brindley G.W., Nakahira M., J. Am. Ceram. Soc. 42 (1959) 319.
216. Hyslop J.F., Rokksby H.P., Trans. Ceram. Soc. 42 (1928) 299.
217. Jay A.H., Trans. Br. Ceram. Soc. 38 (1939) 455.
218. Colegrave E.B., Rigby G.R., Trans. Br. Ceram. Soc., 51 (1952) 355.
219. Richardson H.M., Wilde F.G., Trans. Br. Ceram. Soc. 51 (1952) 387.
220. Mazumdar S., Mukherjee B., J. Am. Ceram. Soc. 66 (1983) 610.
221. Chauduri S.P., Trans. Indian Ceram. Soc. 36 (1977) 71.
222. McVay T.N., J. Am. Ceram. Soc. 19 (1936) 195.
223. Morey G.W., Bowen N.L., Am. J. Sci. 4(5) (1922) 1.
224. Klein A.A., Trans. Am. Ceram. Soc. 18 (1916) 377.
225. Parmalee C.W., Amberg C.R., J. Am. Ceram. Soc. 12 (1929) 699.
226. Lundin S.T., Studies on Triaxial Whiteware Bodies. Almqvist and Wiksell Publishers, Stockholm, Sweden (1959).
227. Shelton G.R., Meyer W.W., J. Am. Ceram. Soc. 21 (1938) 371.
228. Hermannson L., Carlsson R., Trans. J. Br. Ceram. Soc. (1978) 32.
229. Tuttle M.A., Cook R.L., J. Am. Ceram. Soc. 32 (1949) 279.
230. Storch W., et al., Ber. Deut. Keram. Ges. 61 (1984) 335.
231. Binns E., Sci. Ceram. 1 (1962) 315.
232. Ford W.F., White J., Trans. J. Br. Ceram. Soc. 50 (1951) 46.

Chapter 2
Heat Treatment and Performance of Vitrified Grinding Wheels

Mark J. Jackson

Abstract The heat treatment of vitrified grinding wheels is presented in this article. The article describes the minerals formed during the heat treatment of vitrified grinding wheels and presents two case studies where the effects of interfacial compounds and the dissolution of quartz are documented in terms of the wear of grinding wheels. The article should prove useful to industrial engineers who are responsible for selecting firing conditions and the raw materials used for high performance vitrified grinding wheels.

Keywords Heat treatment · Vitrification · Grinding wheels

2.1 Introduction

Grinding wheels of various types, sizes and shapes, are geometrically defined bodies consisting of abrasive grains bonded with various types of ceramic bond. During heat treatment, or firing, the alumina and bonding minerals react with the bond, and this has an important effect on the properties of the resultant tool. For this reason, studies of the structure of abrasive tools made of alumina, and the determination of the phase composition and structure of their bonds, are important research tasks. Generally speaking, grinding wheels are made in the following way. The bond materials are ground until they can pass through a 120 mesh screen without leaving a residue, weighed and mixed. The firing resistance of the bond is checked and it is then passed to the press shop. Here, alumina of the required size

M.J. Jackson (✉)
MET, College of Technology, Purdue University, 401 North Grant Street, West Lafayette, IN 47907, USA
e-mail: jacksomj@purdue.edu

M.J. Jackson and J. P. Davim (eds.), *Machining with Abrasives*,
DOI 10.1007/978-1-4419-7302-3_2,

and the bond materials are weighed in the quantities specified by the formulation, moistened with a dextrin or sodium silicate solution and thoroughly mixed in a planetary mixer. The finished molding material is carefully weighed into exact portions that are then poured into moulds and compressed to the required volume. The molded product is then dried at 90–100° and fired. Heat treatment, or firing, takes place at temperatures of 1,250–1,300°, in continuous or batch furnaces. The relatively long firing times (up to 80–120 h) are due firstly to the non-uniformity of abrasive tools, composed of abrasive grain and ceramic bond, and secondly to the need for the bond to become sufficiently mobile to coat the abrasive grains without disturbing the structure of the tool, or deforming it. Firing time, and more importantly, the length of time spent in the pre-heating and cooling zones of the kiln also depend on the dimensions of the abrasive tools, and increase with their size. Fired grinding wheels are machined on lathes to the required shape and size. Finished abrasive tools undergo hardness tests, and grinding wheels are additionally rupture tested at 1.5–2 times their peripheral velocity.

2.2 Grinding Wheel Structure Formation During Heat Treatment

Ceramic bond materials are made of refractory and fusible clays, feldspar, quartz, talc and sodium silicate. The chemical composition of ceramic bonds is typically 65–70% SiO_2, 20–25% Al_2O_3 and approximately 10% of alkali and alkaline earth metal oxides. Figures 2.1 and 2.2 give some idea of the structure of an abrasive tool. They are photographs of polished sections prepared from regular alumina grinding wheels (grit 46, K-hardness and medium structure (6)). The general appearance of a section is shown in Fig. 2.1. The micrograph demonstrates the non-uniform structure of the grinding wheel. In places, the alumina grains are surrounded with a light film of the bond material, with fine bridges linking the grains and large open pores, while elsewhere the grains seem to be immersed in the bond, which contains fine closed pores.

Detailed examination of individual areas of the section clearly demonstrates that the non-uniformity is much greater than can be seen at low magnifications. Most of the section consists of grains or fragments of monocrystalline alumina (corundum). The grains are usually surrounded by fine rims of the bond material, which form bridges between the corundum grains. Adjoining bridges are separated by large round pores with smooth internal surfaces (Fig. 2.3). Aggregate grains, consisting of several corundum crystals cemented by slag interlayers, are broken down by the bond during firing, and individual corundum crystals become immersed in the glass, which frequently contains crystals produced by the interaction between the accessory minerals of alumina and the bond (Fig. 2.4).

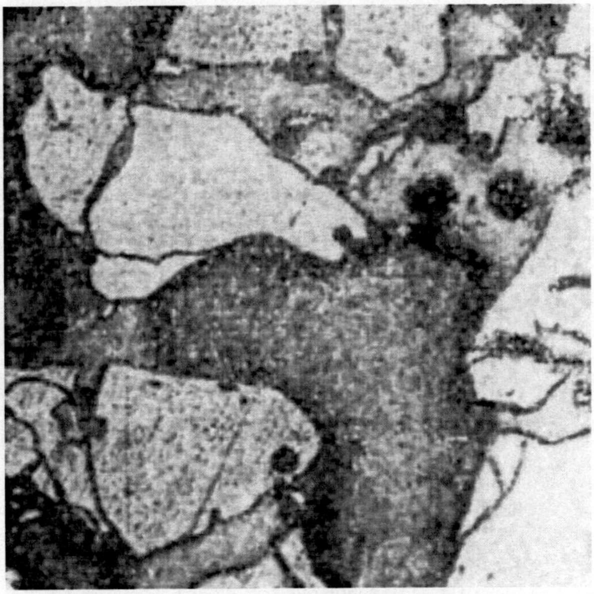

Fig. 2.1 Abrasive tool body: dense alumina aggregates cemented with a vitreous bond. Reflected light, ×60 magnification

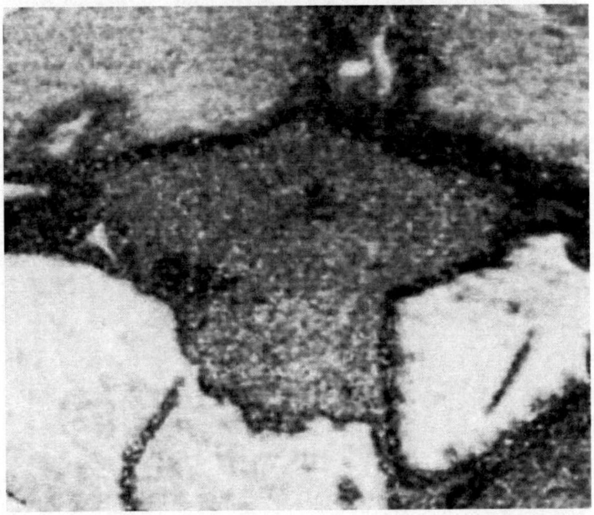

Fig. 2.2 A meniscus of vitreous bond material between grains. Reflected light, ×250 magnification

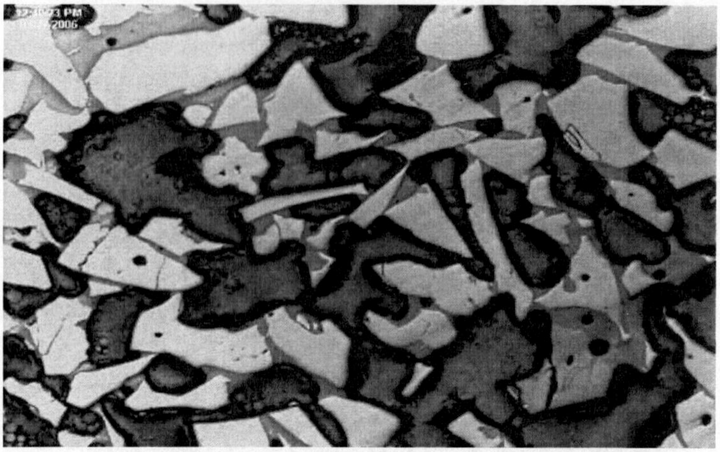

Fig. 2.3 A section of an abrasive tool showing adjoining bridges are separated by large round pores with smooth internal surfaces: an aggregate alumina grain broken down by the bond. The bond contains rutile aggregates (*white*), pores and microcracks (*black*). Reflected light, ×250 magnification

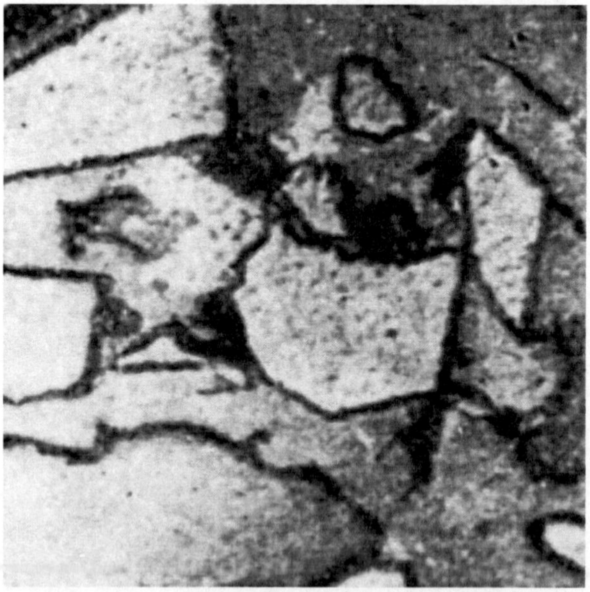

Fig. 2.4 A section of an abrasive tool: an aggregate alumina grain broken down by the bond. The bond contains rutile aggregates (*white*), pores and microcracks (*black*). Reflected light, ×250 magnification

Therefore, abrasive tools appear to be the most complex of silicate products, and it is quite evident that their properties, like those of any other silicate product, will be determined by their phase composition, the structure of the bond, the strength of its adhesion to the alumina grains.

2.2.1 *Physico-Chemical Processes that Occur During Firing*

Studies of ceramic bonds used in industry, both in their pure state and as part of alumina abrasive tools, have shown that the firing process converts bonds into non-uniform glass of complex composition, which on devitrification produces mullite, cordierite and spinel if the bond contains MgO, and anorthite and mullite if it contains CaO. The physico-chemical processes occurring during the heat treatment of abrasive tools can be described as follows. Firing causes the bond to form a liquid phase, and as the temperature rises, this phase dissolves the other components of the bond, as well as the surface of alumina grains. The temperature at which the liquid phase first appears depends on the melting point of the most fusible component of the ceramic bond. In the industry, this is soluble glass (sodium silicate), which is either introduced into the bond in its solid state, or forms part of the solution used to wet the abrasive material during molding. Undoubtedly, if other oxides were also introduced into the bond in the form of more fusible compounds, they too would act as solvents and would react with the alumina.

However, because alumina is relatively soluble in alkali oxide components, the most active solvent is Na_2O. Since sodium silicate becomes quite mobile even at 650°C, it can be assumed that the dissolution of alumina (and of bond components) begins at this temperature. Initially, the concentration of Na_2O in the newly formed liquid phase is relatively high, up to 25%. As the other bond components and alumina dissolve (as the temperature increases), the concentration of Na_2O in the melt falls, its viscosity rises and at the holding temperature dissolution is virtually complete.

The interaction between alumina and the bond is illustrated by the figures contained in Tables 2.1 and 2.2. Table 2.1 gives the fired chemical composition of several bonds and their heat resistance, or temperature at which melting first

Table 2.1 Fired chemical composition of several bonds and their firing resistance

	Chemical composition (wt.%)									Firing resistance
Bond no.	SiO_2	TiO_2	Al_2O_3	Fe_2O_3	CaO	MgO	K_2O	Na_2O	Total	(°C)
1	67.68	1.34	16.58	2.40	1.45	0.54	Nil	9.91	99.88	1,150
2	67.73	0.80	12.50	1.70	1.40	0.60	–	14.80	99.53	970
3	67.48	Nil	3.04	1.27	1.40	0.82	–	25.74	99.75	650
4	69.14	1.14	11.31	1.91	5.40	0.62	–	10.14	99.66	1,060
5	67.92	1.20	11.56	1.94	1.25	5.10	–	10.66	99.63	1,090
6	67.72	0.45	13.02	2.36	0.40	0.56	5.43	10.08	100.02	1,080

Table 2.2 Determination of Al_2O_3 in these bonds, extracted from the samples after heat treatment at various temperatures

	Al_2O_3 content (%) of bonds heat treated at various firing temperatures (°C)						
Bond no.	450	650	850	1,050	1,250	1,250[a]	1,250[b]
1	16.62	17.76	20.25	23.18	22.45	25.47	26.50
2	12.58	16.73	21.08	23.16	24.28	26.13	28.07
3	3.06	4.67	10.52	23.57	25.98	28.34	28.98
4	11.34	16.25	19.53	22.73	25.52	26.21	26.48
5	11.57	13.44	18.94	23.53	26.08	29.98	30.61
6	13.12	15.58	18.76	23.08	24.33	26.46	27.44

[a]Holding at the final firing temperature (8 h)
[b]Holding at the final firing temperature (16 h)

occurs (also known as the bond softening temperature). Table 2.2 contains the results of determination of Al_2O_3 in these bonds, extracted from the samples after heat treatment at various temperatures. The tables demonstrate that there is a direct relationship between the amount of dissolved alumina, the final firing temperature, and the alkali content of the bond.

Irrespective of its initial value, the heat resistance reached by the bond material during the firing process is close to the abrasive tool holding temperature. This explains, why abrasive products can be fired at temperatures 200°C and even 300°C higher than the bond softening temperatures, without the bond leaking and the products being deformed. It is interesting to note that bond 5, which contains magnesium oxide and has high initial heat resistance, dissolved alumina faster than all the others. This is explained by the formation of spinel at the contact interface with alumina, which prevents rapid increase of the alumina content of the liquid phase and thus allows it to retain greater mobility.

As the material dissolves, new mullite, anorthite, cordierite and spinel crystals appear in the liquid phase at 1,200°C. Thus, the first heating stage of the firing process is the phase of chemical reactions and physico-chemical transformations through which the bond reaches a state of equilibrium and partial crystallization. During the holding period, the composition of the bond equalizes and new crystals appear. This is conditional on the local dissolution of alumina during the heating period causing its composition to move into the area of molten bond that is capable of devitrification. The holding period is therefore chiefly a period when new compounds formed in the bond crystallize out of the liquid phase.

During the cooling process, the viscosity of the bond increases so much that the crystallization of new compounds virtually ends. This is why the cooling pattern seems to play an important part, not in the sense of giving the bond a specific structure, but in the sense of preventing mechanical stresses arising in grinding wheels due to their heterogeneous composition, and, more importantly, due to the simultaneous presence of a crystalline component, i.e. alumina grains and the vitreous bond. The cooling period is essentially a period when physical changes take place in the grinding wheel. Studies have shown that the vitreous content of the bond determines the mechanical properties of the tool. The vitreous content of the bond increases from 75 to 100%, with resulting breaking strength increases from 100 to 200 kg cm^{-2}.

Of the minerals studied, the only one that appears to improve the mechanical properties of grinding wheels is spinel $MgO{\cdot}Al_2O_3$, which is formed at the contact of the bond with alumina and encloses its grains in a casing of fine octahedra, no larger than 8 μm in diameter. The Na_2O contained in the tool dissolves the alumina grain, forming a small area of melt enriched with Al_2O_3, thus aiding the formation of spinel. The dissolution of up to 4% of alumina in the bond increases the mechanical strength of the grinding wheel, provided that the bond retains its vitreous structure, or small quantities of minerals form at the contact interface.

Ceramic bond materials producing abrasive tools with good mechanical properties are those located near the SiO_2 apices of the tetrahedra of two systems: Na_2O-K_2O-Al_2O_3-SiO_2 and Na_2O-MgO-Al_2O_3-SiO_2, containing these materials in the following proportions: $SiO_2 = 70–75\%$; $(K_2O)MgO = 5\%$; $Al_2O_3 = 15–10\%$, and $Na_2O = 10\%$. In modern grinding wheel firing conditions, these compounds react vigorously with alumina. In this process, the bond is enriched with Al_2O_3, whose content (in the four-part system) rises to 30–35%. These compounds form glass that does not devitrify during firing and produces grinding wheels with breaking strengths of 170–200 kg cm^{-2}.

2.2.2 *Ceramic Bond Minerals that Form During Firing*

Alongside anorthite, mullite, cordierite ($2MgO{\cdot}2Al_2O_3$-$5SiO_2$) and spinel, which form in the bond when it is enriched with alumina, firing gives rise to other new formations, produced by the interaction between the bond and the accessory minerals present in alumina. They include plagioclases, anatase, hematite, magnetite and rutile. Let us now consider the process of formation of each of these minerals.

Anorthite glass (slag) contained in regular alumina grains in the form of streaks or interlayers, is dissolved by the bond. The composition if the bond is significantly changed as a result, and on crystallization it forms plagioclase and anatase (Fig. 2.5). Silica-rich slag is also absorbed by the bond, but does not disturb its vitreous structure. The resultant sections of the grinding wheel usually consist of glass containing small quantities of mullite prisms, acicular rutile crystals and magnetite dendrites (Fig. 2.6).

Titanium oxides occurring in glass present as accessory minerals in alumina convert into disperse rutile grains on firing. The grains are then recrystallized in the bond, forming acicular aggregates. Anosovite behaves differently, converting into a pseudomorph after anatase and appearing in the bond in the form of brown colored crystals. Titanium carbide and nitride oxidize during firing, forming granular rutile aggregates. Polished sections distinctly show the explosive nature of their oxidation, and its adverse effects – crystallization of the rutile present in the bond and the formation of gas bubbles (Fig. 2.7). In addition, during the firing process, the action

Fig. 2.5 A bond interlayer separating alumina grains and consisting of plagioclase (*grey*) with anatase interpenetrations (*white*). Reflected light, ×300 magnification

Fig. 2.6 Mullite (*pale grey*), anatase (*white elongated sections*), magnetite (*white dendrites*) and rutile (*round white formations*) in glass separating fractured alumina grains. Reflected light, ×250 magnification

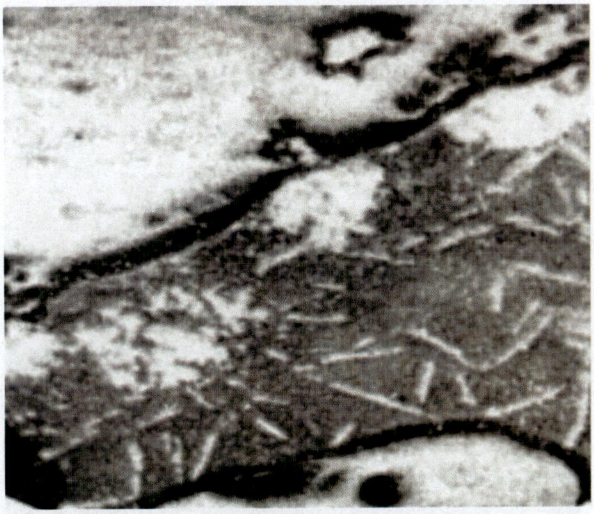

Fig. 2.7 A portion of the bond between alumina grains: the formation of granular rutile aggregates (*white*) through the oxidation of titanium carbide (*white*) embedded in the alumina. The bond contains large pores (*white*). Reflected light, ×250 magnification

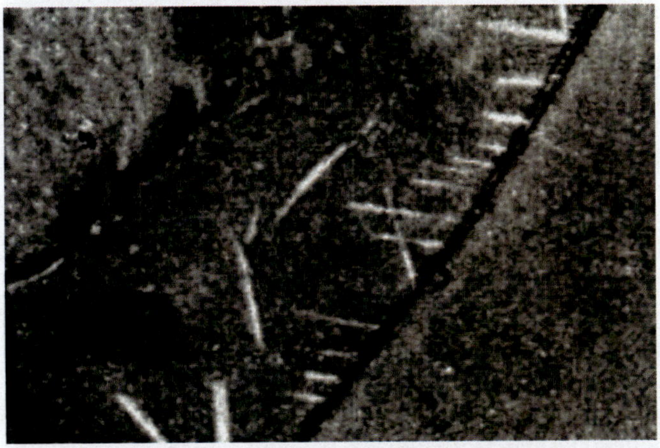

Fig. 2.8 A fringe of rutile needles at the alumina-bond contact. Reflected light, ×250 magnification

of Na_2O present in the bond causes the solid solution of Ti_2O_3 in alumina to break down, and a fringe of rutile needles form on the surfaces of the alumina crystals (Fig. 2.8).

Measurements have shown that the solid solution of Ti_2O_3 in alumina crystals break down to a depth of 30–40 μm. The presence of ferroalloy has a dramatic

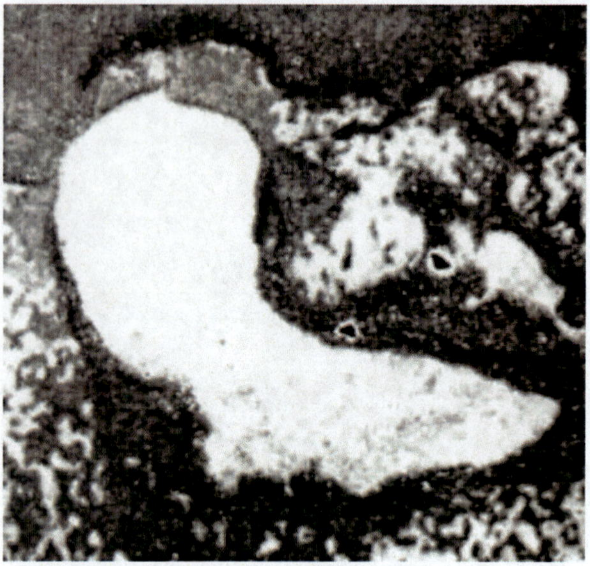

Fig. 2.9 Ferroalloy and iron oxides in bond material. Reflected light, ×250 magnifiction

effect on bond composition. The bond material surrounding the ferroalloy bead is saturated with hematite and magnetite formed through ferroalloy oxidation (Fig. 2.9). Titanium sulphide and carbide present in industrial monocrystalline alumina grains also convert to rutile during firing, forming granular aggregates distributed in the bond. A grinding wheel bond usually contains up to 20–25% of neocrystalline formations, with crystal sizes not exceeding 40–50 μm. If the bond is overfired, these crystals reach 80–120 μm and develop microcracks in the bond (Fig. 2.10). The features of minerals formed in grinding wheels after firing, which form the basis of microscopic analysis, are described below.

Anorthite: ($CaO{\cdot}Al_2O_3{\cdot}2SiO_2$) appears in the bond in the form of randomly arranged colorless lamellae and columnar (elongated tabular) crystals displaying negative elongation and polysynthetic twinning. Moderate refractive index and birefringence: $N_g = 1.588 \pm 0.002$, $N_p = 1.575 \pm 0.003$, $N_g - N_p = 0.013$.

Mullite: ($3Al_2O_3{\cdot}2SiO_2$) is always colorless in the bond (Fig. 2.11). It crystallizes in the form of fine needles, mostly gathered into felted and radiated aggregates. Moderate refractive index and birefringence: $N_g = 1.654 \pm 0.002$, $N_p = 1.642 \pm 0.003$, $N_g - N_p = 0.012$.

Cordierite ($2MgO{\cdot}2Al_2O_3{\cdot}5SiO_2$ (Fig. 2.12)) crystallizes in the form of colorless short prismatic pseudohexagonal crystals (in the rhombic system), which have a moderate refractive index and low birefringence: $N_g = 1.525 \pm 0.003$, $N_p = 1.521 \pm 0.002$, $N_g - N_p = 0.004$. In polished sections, it is dark grey, with low reflectivity and a relief similar to that of glass.

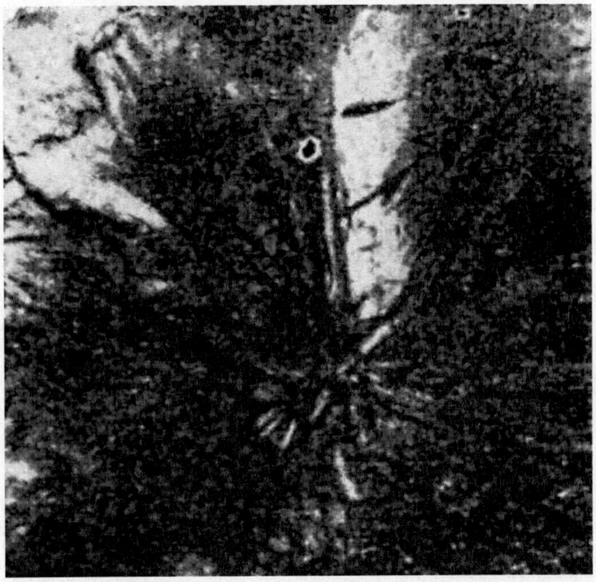

Fig. 2.10 An anorthite spherolite in bond material showing microcrack formation. Analyser out, ×200 magnification

Fig. 2.11 Mullite in bond material. Analyser out, ×80 magnification

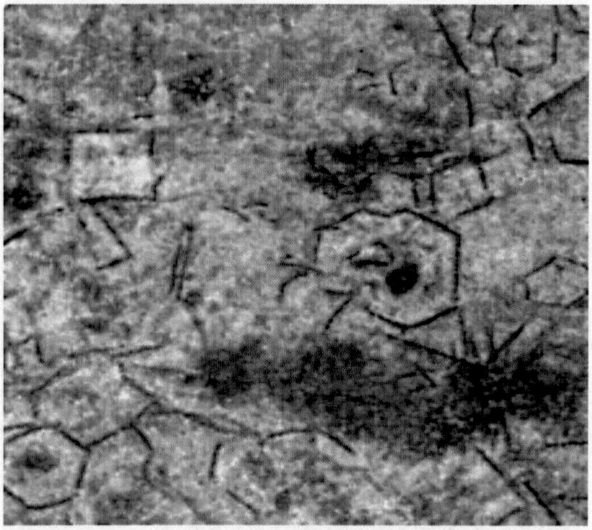

Fig. 2.12 Cordierite in bond material. Analyser out, ×160 magnification

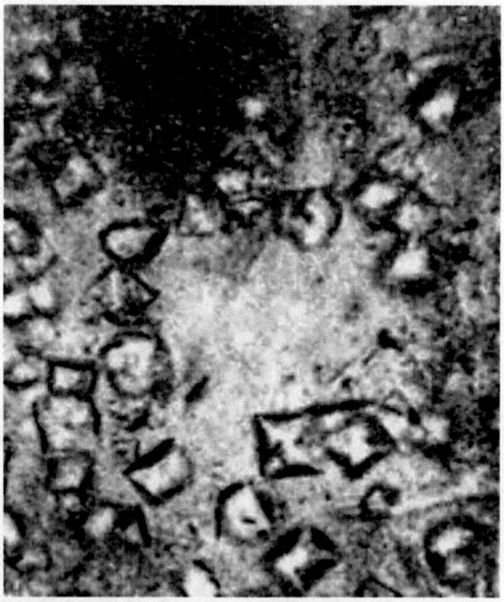

Fig. 2.13 Spinel in bond material. Analyser out, ×250 magnification

Spinel: ($MgO{\cdot}Al_2O_3$) is yellow or colorless in transmitted light (Fig. 2.13). It appears in the form of octahedra and grains at the alumina contact. It has a high refractive index: $N = 1.722 \pm 0.003$.

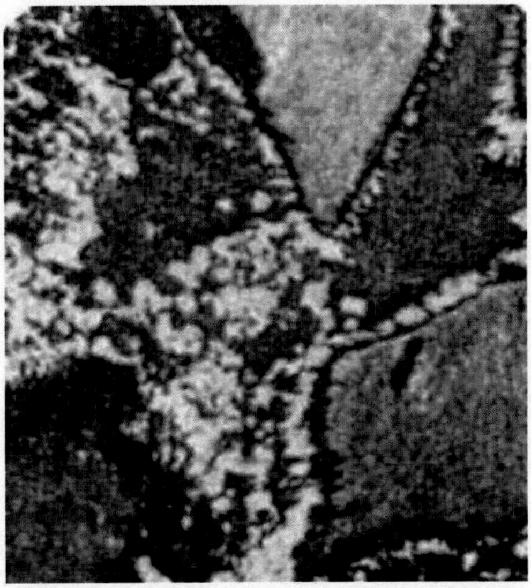

Fig. 2.14 Hematite on the alumina-bond contact. Reflected light, ×150 magnification

Plagioclases: crystallize in the triclinic system. In transmitted light, they appear in the form of elongated tabular crystals, frequently displaying polysynthetic twinning. Moderate refractive index and birefringence: $N_g = 1.553–1.558$, $N_p = 1.547–1.552$. In polished sections they are grey, with low reflectivity and relief equal to that of as glass.

Hematite: Fe_2O_3 crystallizes in the trigonal system. In transmitted light, it appears in the form of orange-red irregular accumulations, less often in the form of hexagonal or triangular lamellae. High refractive index and birefringence: $N_g = 3.01$ Li, $N_p = 2.78$ Li. The irregular accumulations result from ferroalloy oxidation, while the regular lamellae, frequently with regular orientation, appear on the surfaces of alumina crystals (Fig. 2.14). They are formed by the recrystallization of ferrous oxide film on alumina grains (in monocrystalline alumina). In reflected light, hematite is white, with above-average reflectivity and a higher relief than glass.

Magnetite: Fe_2O_4 crystallizes in the cubic system. It is opaque. In transmitted light it appears in the form of irregular black accumulations associated with ferroalloy, less often in the form of fine cubic crystals and skeletal cruciform dendrites. In reflected light, it is white with high reflectivity and a higher relief than glass.

Rutile: TiO_2 crystallizes in the tetragonal system. It is colorless, and forms either granular formations or acicular or prismatic (frequently hollow) crystals. Its refractive index and birefringence are extremely high: $N_g = 2.903$, $N_p = 2.616$. In reflected light, it appears as either sinuous lacy aggregates and accretions formed by the oxidation of titanium carbide and nitride, or in the form of white rectangular and rhomboid sections, formed by the alteration of anosovite. It has high reflectivity and a relief higher than glass.

Anatase: TiO_2 crystallizes in the tetragonal system, forming prismatic and rod-like crystals, which can be either colorless or colored yellow or brown. Its refractive index and birefringence are very high, $N_g = 2.56 \pm 0.02$, $N_p = 2.48 \pm 0.02$. In reflected light, it appears in the form of white, rectangular and rhomboid sections. It has above-average reflectivity and a relief higher than glass.

2.3 Case Study I: Interfacial Compounds and Their Effect on Grinding Wheel Wear

The type of grinding wheel considered in this case study is made using aluminum oxide (α-Al_2O_3), a hard material with a Knoop hardness of up to 2,000 kg mm^{-2}, is used in the grinding industry in two principal forms: a high purity, fused form of alumina containing over 99.9 wt.% Al_2O_3 that is white in appearance; and a fused, brown coloured, alumina of 95 wt.% purity. The main impurity in this latter form is TiO_2 at a level no greater than 3 wt.%. This tends to increase the toughness of the grain and is accompanied by other impurities such as MgO, CaO, Fe_2O_3, and ZrO_2. Other grinding wheels described in this case study use cubic boron nitride (cBN) that has a Knoop hardness in excess of 4,500 kg mm^{-2}.

The range of vitreous bonding systems and abrasive types is very large, though only alumino-alkalisilicate and alumino-borosilicate bonding systems are used by the abrasive wheel industry. The normal practice is to adjust the proportions of Al_2O_3, B_2O_3, SiO_2, and alkali oxides to achieve the desired fluidity. Other chemical and physical properties can be modified by the addition of alkaline-earth oxides. Vitreous bonds are composed of mixtures of quartz, feldspar, clay, borate minerals, and ground frits. In practice, the bonds are mixed with a variety of abrasive grains. However, this case study considers high purity and titanium-doped varieties (using a typical mesh size of 220, which is approximately 62 μm diameter abrasive grain size), and cBN with B64 grain size (approximately 63 μm in diameter).

The grinding process is accompanied by wear of the abrasive wheel, and the rate of this wear plays an important role in determining the efficiency of the grinding process and the quality of the workpiece. The structure of a vitrified grinding wheel is composed of abrasive grains, a bonding system, and a large number of pores. Figure 2.15 shows a typical porous grinding wheel structure [1]. Krabacher [2] pointed out that wear mechanisms in grinding wheels appear to be similar to that of single-point cutting tools, the only difference being the size of swarf particles generated. The wear behaviour observed is similar to that found in other wear processes; high initial wear followed by steady-state wear. A third accelerating wear regime usually indicates catastrophic wear of the grinding wheel, which usually means that the wheel will need to be dressed. This type of wear is usually accompanied by thermal damage to the surface of the ground workpiece. The performance index used to characterize wheel wear resistance is the grinding

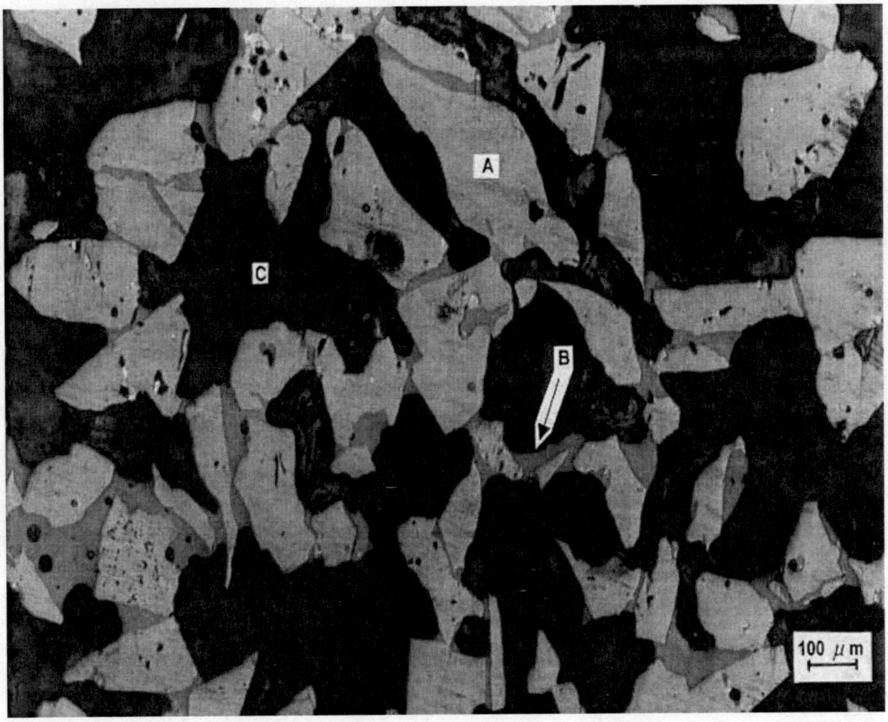

Fig. 2.15 Microstructure of a vitrified grinding wheel. A – denotes abrasive grain, B – denotes vitrified bonding phase, and C represents distributed porosity

ratio, or G-ratio, and is expressed as the ratio of the change in volume of the workpiece ground to the change in the volume of the grinding wheel removed, thus,

$$G = \frac{\Delta v_w}{\Delta v_s} \tag{2.1}$$

Grinding ratios cover a wide range of values ranging from less than 1 $mm^3\,mm^{-3}$ for vanadium-rich high speed steels to over 60,000 $mm^3\,mm^{-3}$ when internally grinding bearing races using cubic boron nitride abrasive wheels. Attempts have been made on how to address the problems related to the wear of abrasive grains in terms of the theory of brittle fracture. The conclusions of various researchers lead us to believe that the variety of different and interacting wear mechanisms involved, namely, plastic flow of abrasive, crumbling of the abrasive, chemical wear etc., makes grinding wheel wear too complicated to be explained using a single theoretical model. High efficiency precision grinding processes place extreme loads onto the grain and the vitrified bonding bridges.

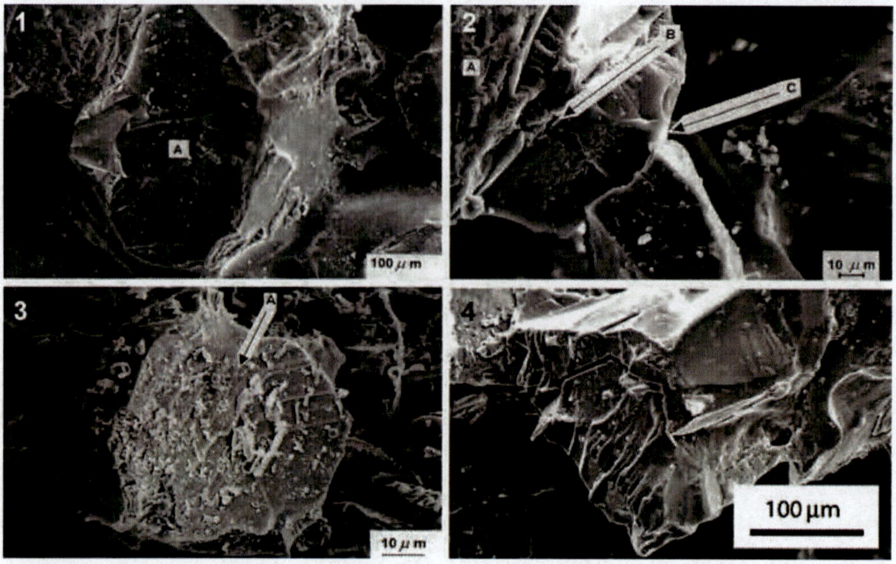

Fig. 2.16 Grinding wheel wear mechanisms: (1) abrasive wear – A denotes a wear flat generated by abrasion; (2) bond bridge fracture – A denotes the abrasive grain, B denotes the interfacial bond layer, and C denotes a crack passing through the bond bridge; (3) abrasive grain fracture – A denotes crystallographic grain fracture; and (4) interface fracture between abrasive grain and bond bridge

2.3.1 Wear Mechanisms

Four distinct wheel wear mechanisms that contribute to the wear of grinding wheels are identified as (Fig. 2.16):

(a) Abrasive wear (formation of wear flats on the surface of abrasive grains);
(b) Fracture of bond bridges;
(c) Fracture of abrasive grains due to mechanical and thermal shock loads; and
(d) Fracture at the interface between abrasive grain and bond-bridge.

2.3.1.1 Abrasive Wear

The formation of wear flats on abrasive grains leads to a loss of grain sharpness. The sources of minute scale wear are:

(a) Wear due to frictional interaction between workpiece and abrasive grain;
(b) Plastic flow of the abrasive grain at high temperature and pressure;
(c) Crumbling of the abrasive grain due to thermal diffusion and microscale mechanical impact; and

(d) Chemical reaction between abrasive and workpiece material at elevated temperatures and in the presence of grinding fluids.

The latter mechanism can reduce the resistance of the abrasive grain to other wear mechanisms. Dull abrasive grains are caused by the generation of wear flats on active grains that leads to an increase in the area of contact and frictional interactions between abrasive grain and the workpiece. At the point of dulling of the abrasive grain, very high temperatures exist in the area of contact that greatly enhances adhesion and chemical reaction between two surfaces. If grain and bond bridge fracture does not occur during grinding then the plateau area on the grain widens and the rate of wear increases. If fracture is delayed further, as with hard grinding wheels, then the wheel becomes glazed and the workpiece is thermally damaged.

It has been shown experimentally [3] that chemical affinity between the abrasive and the workpiece material can be used as a guide for the selection of grinding wheels. Their observations of solid diffusion of silicon carbide into ferrous materials explain the catastrophic wear rates exhibited by these "wheel-workpiece" combinations. The most common method used for measuring wear flat area employs an optical, or an electron microscope. Hahn [4] observed and analysed the effect of the increasing wear flat area during the plunge grinding of various workpiece materials. Hahn concluded that grinding forces gradually increase during wear-flat formation up to a point where the grinding wheel will restore its sharpness due to abrasive grain fracture.

2.3.1.2 Fracture Wear

The occurrence of abrasive grain and bond fracture are considered simultaneously for the following reasons:

(a) They are of the same nature, i.e. fracture of brittle materials and hence the theory of brittle fracture is applicable to both bonding bridge and abrasive grain. The applied thermal and mechanical loads, usually under cyclic conditions, cause initiation and further development of cracks that leads to fracture and the formation of new irregular surfaces;
(b) They are related to dressing methods used and occur simultaneously. The initial and final stages of wheel life between dressings exhibit fracture wear that is a combination of abrasive grain and bonding bridge fracture;
(c) The relative amounts of bond bridge and abrasive grain wear cannot always be calculated. An investigation into precision grinding employed a soft wheel that gave a high percentage of bond fracture, whereas a harder wheel gave partial abrasive grain fracture. Wear by attrition occurred in both cases.

However, the combination of grinding parameters such as equivalent chip thickness and the grindability of the workpiece material determines the effective wheel hardness, and so no single feature of the grinding process can be used to predict the fracture pattern of the wheel in advance. The difficulty when relating grinding wheel wear due to fracture to a particular grinding condition arises from the lack of knowledge about the loads applied to both abrasive grains and their bonding bridges and their response to these applied loads.

Tarasov [5] suggests that abrasive grain fracture occurs as a result of mechanical forces due to chip formation, or thermal shock, induced by instantaneously high temperatures. Hahn [4] proposed a thermal stress hypothesis to explain the fracture of abrasive grains. Plunge grinding experiments were conducted under fixed normal load conditions. Hahn asserted that as wear progresses measurements of torque indicated that the tangential force decreases. This led to the conclusion that abrasive grain fracture due to mechanical loading will not occur. Mechanical stresses wear also considered as an explanation for the different rates of wear of the grinding wheels used in the experiments.

Bhattacharyya et al. [6] observed abrasive grain loss due to fracture using an electron microscope. They concluded that they could not differentiate between Peklenik's crystal splintering, i.e. grit flaking due to thermal stress, and abrasive grain fragmentation. However, they did explain their results in terms of Hahn's thermal shock hypothesis. Hahn's experimental conditions suggested that attrition of the abrasive was expected to occur through abrasive wear. Wear measurements by Hahn [4] were based on the reduction in grinding wheel diameter, which Malkin and Cook [7] attributed to abrasive wear. Wear rates recorded were of the order of 50 microinches per second. It was expected that abrasive wear rates were in the region of 5 microinches per second. This rate was observed under light grinding conditions. Under heavy grinding conditions, the conditions of wear appeared to be more complex.

Malkin and Cook [7] collected wheel wear particles for each grade of grinding wheel when grinding using a fixed set of operating conditions. They analysed their size distributions statistically and discovered that a soft-grade grinding wheel (G-grade) produces 85% of grinding debris associated with bonding bridge fracture, whilst a harder-grade grinding wheel (K-grade) produces 55% of grinding debris associated with fractures of bonding bridges. Abrasive wear accounted for 4% of the total wear in both cases.

The strongest evidence in support of the idea of fracture due to mechanical loading is that fracture occurs some distance away from the cutting edge [8]. It was concluded that the heat generated by cutting has no effect on abrasive grain fracture since the peak temperature of the abrasive grain occurs at the surface of the grain in contact with the workpiece where fracture is initiated on cooling according to the thermal stress hypothesis. The hypothesis does not take account of any difference in the coefficient of thermal expansion between abrasive grain and bond bridges, and also the effect of thermal shocks on the quenching action of grinding fluids on the abrasive grain leaving the cutting zone. The latter case was analysed and it was reported that the thermal stress in an abrasive grain due to a pulsating heat source showed that the magnitude of the maximum tensile stress is not large enough to cause fracture of the grain. Malkin and Cook [7] adopted the mechanical loading approach. Malkin and Cook [7] derived an expression from first principles for the probability of bond fracture in terms of a bond stress factor.

Although bond and grain fracture are similar mechanisms, they have a different effect on the economics of the grinding process. The first mechanism results in a rapid loss of the grinding wheel, while the second mechanism, on a comparable scale with the un-cut chip thickness, generates sharp cutting edges and is known as the "self-dressing action". Both mechanical and thermal stresses appear to be responsible for fracture wear. The effect of heat at the abrasive grain and workpiece interface is responsible for locally affecting the mechanical properties of the abrasive grain. However, fragments of larger sizes of abrasive grain are more likely to occur through mechanical loading that governs bond fracture and the self-sharpening action. A method of alleviating the onset of bond fracture due to unusually large mechanical loads is to dissolve deleterious particles in the bonding system that weakens the structure of the bonding bridge.

In vitrified bonds, these particles are quartz particles that naturally occur in ceramic raw materials. These particles reduce the load-bearing strength of the bonding bridges during vitrification heat treatment. The study of the effect of the elastic modulus on the fracture behaviour of vitrified abrasive grinding wheels was conducted by Decneut, Snoeys, and Peters [9]. They discovered that vitrified grinding wheels with a high modulus of elasticity wear by a mechanism of abrasive grain fracture rather than fracture of the glass bond bridges that hold the abrasive grains in place. As the modulus of elasticity increases the "self-sharpening effect" is lost because abrasive grains cannot be released from the bonding matrix. This leads to a condition where the temperature of the workpiece material begins to increase and is associated with phase transformations and thermal cracking of the surface layers that results in a reduction in fatigue strength.

In this case, the performance of the abrasive grinding wheel for a specific metal removal rate and workpiece material depends on the selection of the appropriate grade of abrasive grinding wheel that is a function of its modulus of elasticity and strength. In the present study, the elastic modulus, bending strength, and nature of fracture was found to be dependent on the vitrification behaviour of the glass bonding system, the amount of bond, and the type of abrasive grain used in the vitrified grinding wheel. It was found that the wear of vitrified grinding wheels is highly dependent on the way the grinding wheel "vitrifies" during heat treatment.

2.3.2 *Microstructure of Abrasive Grains*

2.3.2.1 High Purity Aluminum Oxide

Examination of high purity aluminum oxide in a scanning electron microscope using an electron probe micro analyzer showed that 99.5 wt.% of the grain was Al_2O_3 with the balance consisting of Na_2O and SiO_2 in equal proportions.

However, local Na_2O-enriched areas were observed within parts of the grain. Figure 2.17 shows the areas of Na_2O local enrichments within the grain as white reflections when viewed under an optical microscope. Under close examination,

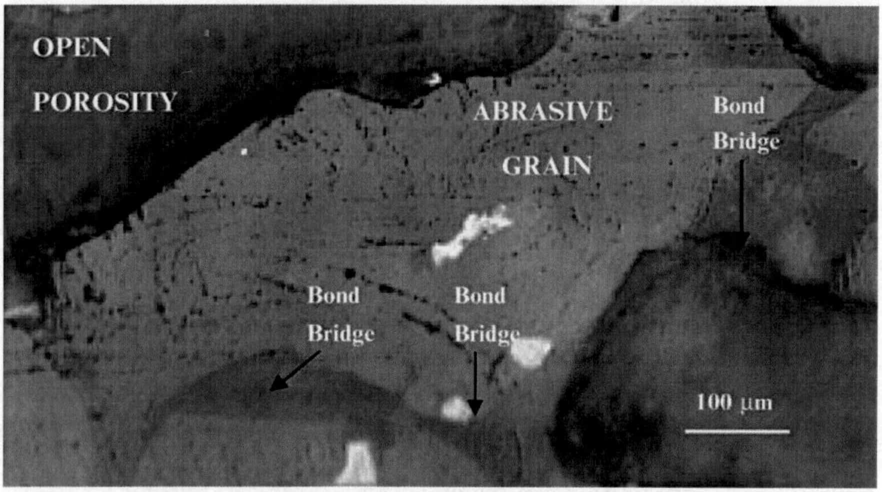

Fig. 2.17 High purity aluminum oxide grinding wheel showing enriched regions of Na_2O (denoted by *black arrows*) determined using an electron probe micro analyzer

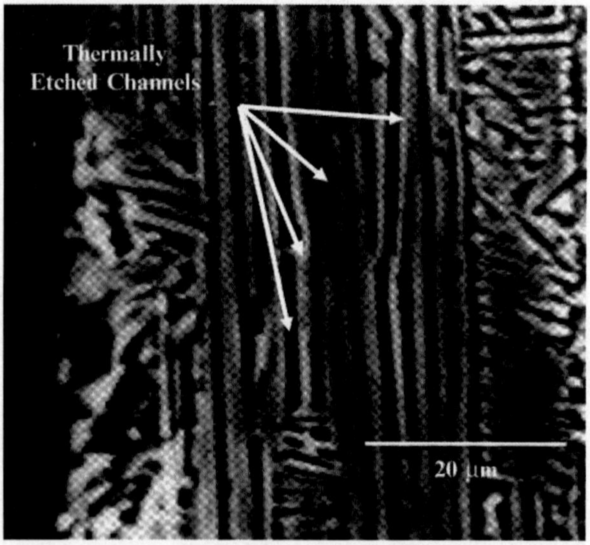

Fig. 2.18 High purity aluminum oxide grain showing thermally etched channels of β-Al_2O_3 layers that are present in the α-Al_2O_3 structure

high purity aluminum oxide contains aluminum oxide, sodium aluminate, carnegieite, sodium monoaluminate, nepheline, and glass of variable composition. In heat-treated high purity abrasive grains, preferential etching at the surface of the grain appears to occur along crystallographically controlled directions (Fig. 2.18). This is assumed to be due to the dissolution of planar blocks of β-Al_2O_3

($Na_2O \cdot 11Al_2O_3$) that is present in the α-Al_2O_3 host material. X-ray diffraction of high purity alumina established the existence of β-Al_2O_3 prior to the optical examination of the abrasive grains. Other impurities found include rarely seen calcium rich platelets in the form of alite (Ca_3SiO_5), and an un-named oxide, $NaCaAlO_3$, which is known to have several polymorphic forms.

2.3.2.2 Titanium-Doped Aluminum Oxide

The amount of TiO_2 in titanium-doped aluminum oxide was measured using an electron probe and was found to be in the range of 1–2 wt.%. The amount of titania present is inconsistent with earlier work that had determined that the maximum solubility of TiO_2 in Al_2O_3 is less than 0.3 mol.% at 1,300°C [10]. Although some of the excess can be accounted for in the formation of Ti_2O_3, It is possible that not all titania is in solid solution. This was confirmed by the occurrence of blade-like inclusions that is consistent with rutile (TiO_2) needle morphology. This would account for the variability in measured titania and its presence in amounts greater than its solubility in Al_2O_3. In heat-treated and titanium-doped aluminum oxide, calcium hexaluminate, anorthite, and spinel are not affected by the heat treatment process. However, glass is devitrified forming anorthite spores. Titanium minerals are oxidised to higher oxides such as anatase and rutile. These changes are accompanied by large changes in volume that may affect the performance of any abrasive tool. As a precaution, Ti-doped aluminum oxide must be heated to 1,000°C before it can be used for making abrasive cutting tools.

2.3.2.3 Cubic Boron Nitride

Cubic boron nitride (cBN) abrasive grains are made by compacting grains of cBN in the presence of aluminum. Aluminum reacts with BN to form a mixture of AlN and AlB_2 that forms a stable and catalytically inactive binder. Interaction between aluminum and BN is intimate and can be observed directly using scanning and transmission electron microscopes. There is very little interaction between cBN grains. The edges of cBN grains not in contact with each other form rinds of AlN in thin, continuous lines with several nodules along its length.

The rind that encloses the exposed cBN grain is always orientated so that it has crystallographic directions parallel to particular directions in the cBN lattice. The selected area diffraction pattern shown in Fig. 2.19 shows a [110] cBN pattern with a rectangular AlN[$11\bar{2}0$] pattern superimposed. The AlN has grown with its basal planes parallel to the cBN facet plane. This orientation with cBN (110) // AlN (0001) and cBN [110] // AlN [$11\bar{2}0$] is the most common orientation observed even when facet planes deviate away from being octahedral. At cube surfaces the orientation the orientation cBN (001) // AlN (0001) and cBN [110] // AlN [$11\bar{2}0$] occurs. While most of the AlN can be located at cBN grain surfaces, AlB_2 nucleates

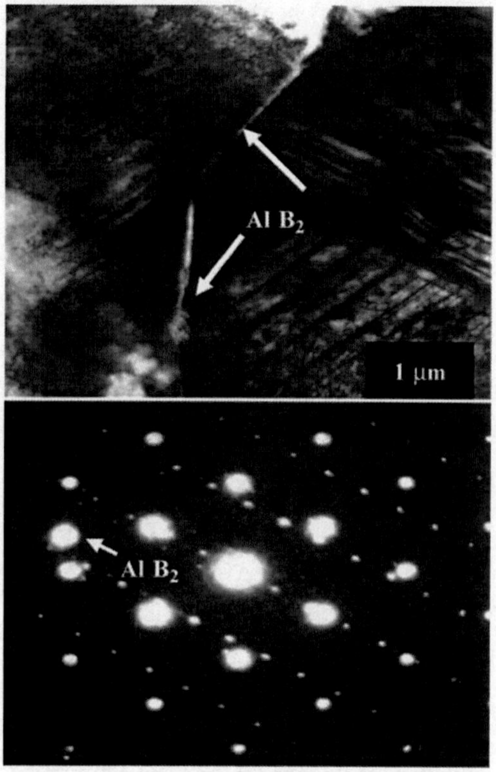

Fig. 2.19 (**a**) Two contacting cBN grains separated at intervals by an AlN rind which is parallel to the cBN [110] planes. The outer edges of the grains are in contact with a featureless AlB_2 layer, (**b**) Selected area diffraction pattern from part of the field of contact showing the relative orientation of phases present. The smaller spots are AlN, and the larger spots are cBN. The *arrow* indicates the single spot generated by AlB_2 phase

independently in liquid aluminum at the later stages of consolidation. A single crystal of AlB_2 produces the reflection to the left of the SAED pattern that produces a single bright spot [11].

2.3.3 Experimental Procedure

2.3.3.1 Measurement of Mechanical Properties

The experimental procedure involved making experimental samples of abrasive grain and glass bond as a vitrified product using high purity aluminum oxide, titanium-doped aluminum oxide, and cBN bonded with an alumino-borosilicate bond containing 61.4 wt.% SiO_2, 17 wt.% Al_2O_3, 0.4 wt.% Fe_2O_3, 3.2 wt.% CaO, 0.1 wt.% MgO, 2.7 wt.% Na_2O, 3.1 wt.% K_2O, and 10.1 wt.% B_2O_3. Experimental samples were made by pressing abrasive grains and glass bond ingredients to a known density. The samples were moulded in the form of bars. The dimensions of the bars were 60 mm length, 12 mm height, and 12 mm depth. The samples were fired at the vitrification temperature (between 1,000 and 1,300°C) for 6 h in an electric furnace. The samples were prepared for four-point loading and for measuring their elastic modulus using the sonic method developed by [9]. A total of 20 experimental test samples were loaded in uniaxial tension. The Weibull modulus for the fractured samples was calculated to be 18.3 for aluminum oxide samples, and 18.8 for cBN samples. A section of one of the bar samples was cut, mounted in resin, and polished to reveal the nature of bonding between glass and aluminum oxide. Figure 2.20 shows the section revealing abrasive grains bonded together by the vitrified glass bonding system. The black areas represent the pores between abrasive grains that are essential to provide free space for chips of metal and for coolant access. Figure 2.20 also shows the characteristic x-ray spectra for abrasive grains and glass bond. The abrasive grain spectrum shows aluminum and titanium (indicative of titanium-doped aluminum oxide), and the glass bond spectrum shows elements such as potassium, calcium, and sodium that are glass

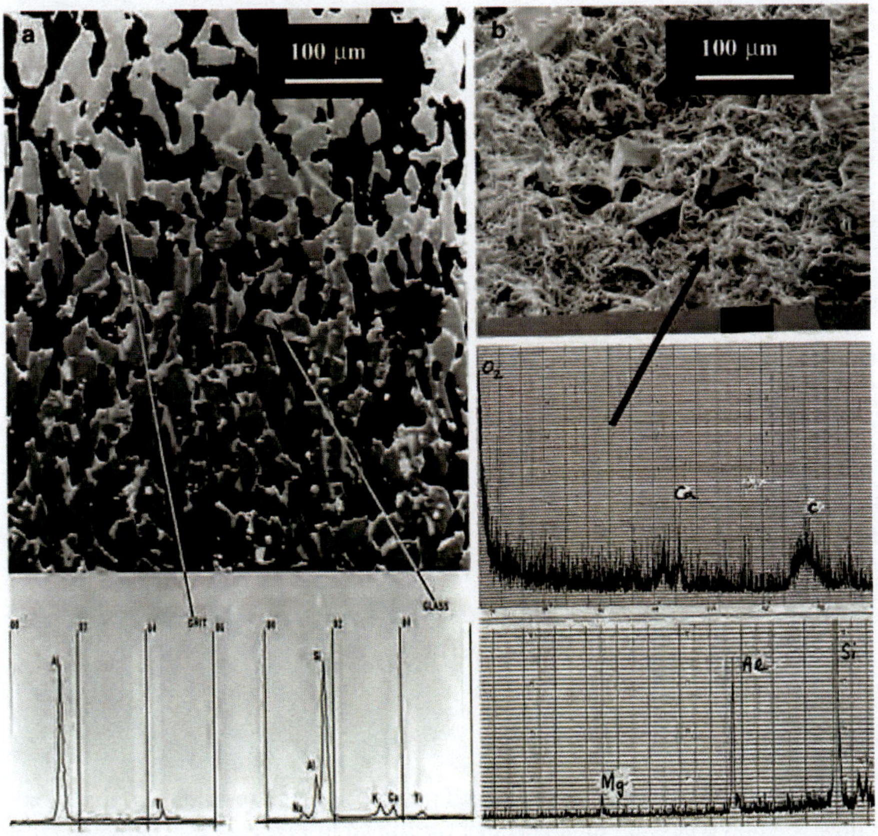

Fig. 2.20 Electron probe microanalysis of (**a**) titanium-doped aluminum oxide and vitrified glass bonding system, and (**b**) cBN and vitrified glass bonding system

network-modifying elements, and aluminum and silicon that are network-forming elements. For the vitrified cBN grinding wheel, the bonding system contains magnesium, aluminum, silicon, calcium, and oxygen.

2.3.3.2 Manufacture of Grinding Wheels

Grinding wheel segments were made by pressing abrasive grains and glass bond ingredients to a known density. The samples were moulded in the form of segments to be attached to a pre-balanced grinding wheel body. The dimensions of the Segments were 60 mm length, 15 mm height, and 20 mm depth. The samples were fired at the vitrification temperature (between 1,000 and 1,300°C) for 6 h in an electric furnace. Once fired, the segments were measured in terms of their hardness and grade and were bonded onto a steel backing using a high strength adhesive. The steel backings were then bolted onto a steel body containing the rest of the abrasive segments.

2.3.3.3 Measurement of Wear

The method of grinding wheel wear measurement adopted was the "razor-blade" technique. The method involves grinding a workpiece that is less wide than the grinding wheel. A groove is worn into the wheel profile, which was measured with reference to the non-grinding portion of the grinding wheel using a razor blade. The grinding wheel was initially dressed using a single point diamond and the wheel conditioned until steady-state grinding wheel wear was achieved. In order to achieve the conditions of bond fracture, the depth of cut for all experiments was set at 10 μm per pass with a table speed of 0.2 $m\,s^{-1}$, and a grinding wheel speed of 60 $m\,s^{-1}$.

Immediately after the grinding experiments were performed, the razor blade was then lowered into the grinding position with the grinding wheel touching the blade. After grinding the blade, the wear of the grinding wheel was measured using a surface profilometer. The grinding ratio was calculated by measuring the volume of the grinding wheel removed, and the volume of the workpiece removed.

2.3.4 Experimental Results

2.3.4.1 Mechanical Properties

The relationship between the elastic modulus and firing temperature as a function of abrasive grain type and bonding content is shown in Fig. 2.21 for both high purity and titanium-doped aluminum oxide structures. It is shown that the elastic modulus is developed as the vitrification temperature is increased, and is highly dependent

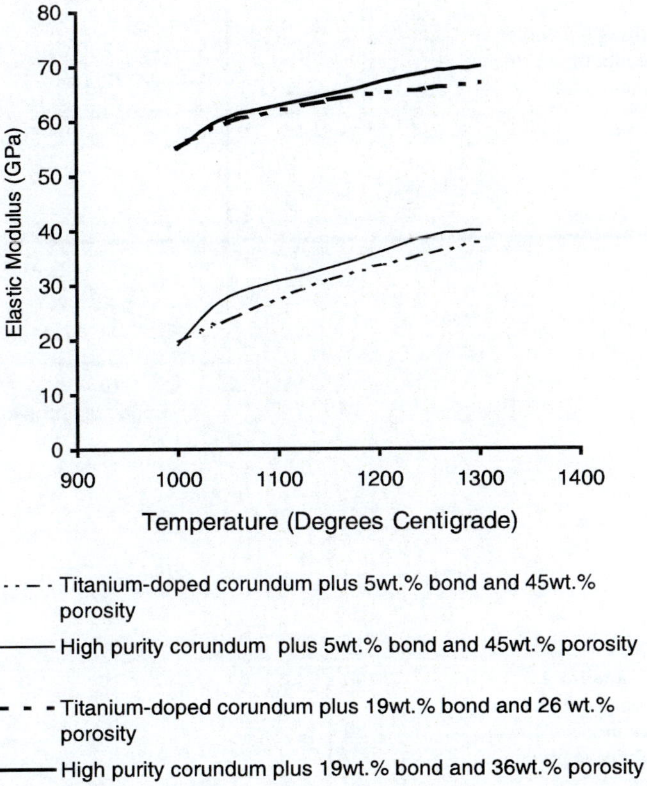

Fig. 2.21 Elastic modulus as a function of firing temperature for a number of abrasive grain types and bond contents

on the amount of bonding material that surrounds the abrasive grain. This is confirmed in Fig. 2.22, which shows the effect of the increase in bonding content on the elastic modulus at three different vitrification temperatures for high purity aluminum oxide structures. An interesting observation is that up to the softening point of the glass bond, high purity and titanium-doped aluminum oxide vitrified structures developed strength in the same way then declines for titanium-doped aluminum oxide structures depending on the amount of bonding material. The relationship is shown in Fig. 2.23. The same general trends are observation with vitrified cBN grinding wheels.

2.3.4.2 Wear of Grinding Wheels

The relationship between the wheel wear parameter, grinding ratio (G), and the firing temperature is shown in Fig. 2.24 for both high purity and titanium-doped aluminum oxide grinding wheel structures containing a different amount of vitrified

Fig. 2.22 Effect of bond content and firing temperature on the elastic modulus of high purity aluminum oxide structures

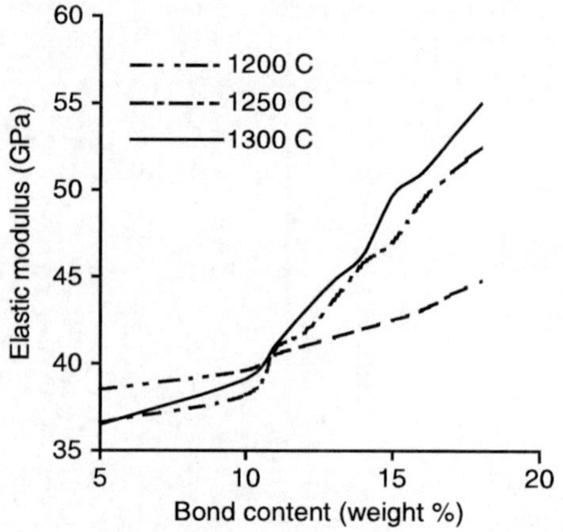

Fig. 2.23 Relationship between bending strength and firing temperature as a function of abrasive grain type and bond content

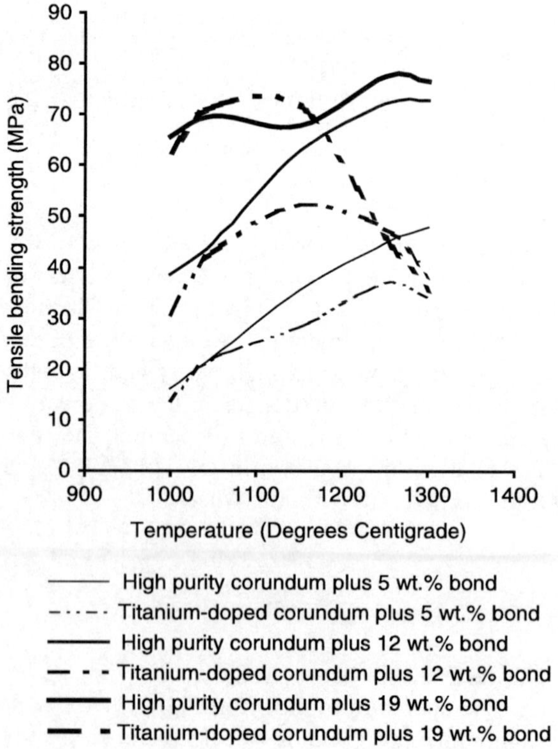

Fig. 2.24 Relationship between grinding ratio and firing temperature as a function of abrasive grain type and bond content

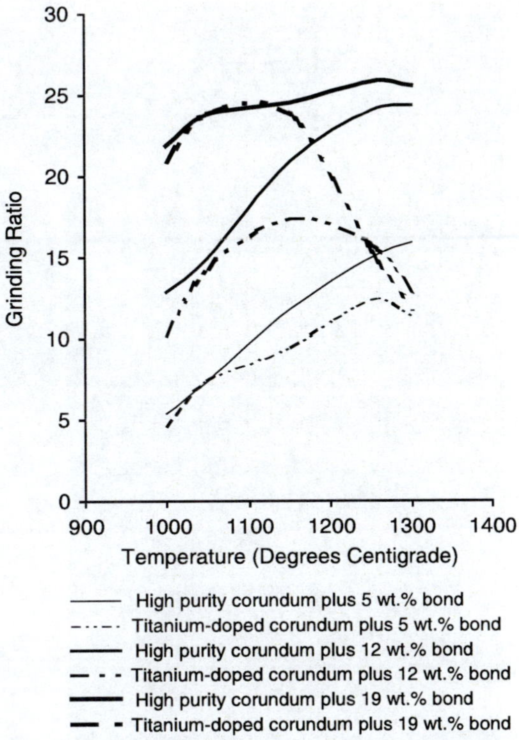

bonding material. Again, the observation that up to the softening point of the glass bond, high purity and titanium-doped aluminum oxide structures develop wear resistance in the same way is noteworthy. Figure 2.24 shows that the grinding ratio is a function of vitrification temperature, but at a certain temperature, it is highly dependent on the type abrasive grain used in the grinding wheel and the amount of bonding material used.

Examination in a scanning electron microscope showed that certain parts of the glass bond had de-vitrified in both high purity and titanium-doped aluminum oxide structures. The crystals are elongated with square sections and have a high Al_2O_3 content. An x-ray diffraction spectrum indicated that the phase is an aluminoborate solid solution. The best match was with Al_{18} B_4 O_{33}. In addition to this phase, a second crystalline phase was observed in titanium-doped aluminum oxide structures. The phase consists of needles of rutile (TiO_2) orientated on the faces of titanium-doped aluminum oxide grains that penetrate into the glass bond. Figure 2.25a, b shows orientated rutile needle formation in the glass bond emanating from the aluminum oxide crystals. The structure in Fig. 2.25b was etched with a solution of 40% hydrofluoric acid in water. Figure 2.25c shows the growth of rutile needles from the interface between aluminum oxide and the glass bond using the electron backscatter mode. Figure 2.25d shows the de-vitrification of glass in the form of Al_{18} B_4 O_{33} crystals.

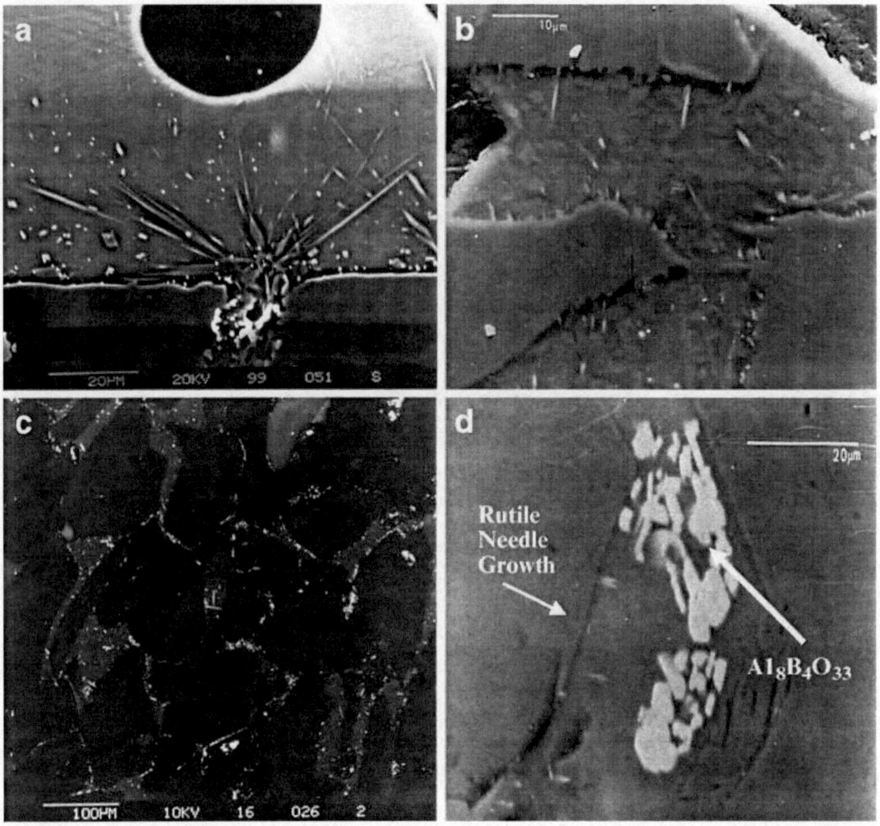

Fig. 2.25 (**a**) Titania (TiO_2), in the form of rutile needles, on the surface of the vitrified glass bond; (**b**) vitrified glass bond etched with 40% HF in water to show rutile formation within the glass bonding system; (**c**) electron backscattered image showing needle growth into the glass bond from the abrasive; (**d**) de-vitrified glass bond containing crystals of Al_{18} B_4 O_{33} bounded by two abrasive grains

Fractured samples revealed a higher proportion of intergranular fracture than cut and polished samples. High purity aluminum oxide did not exhibit intergranular fracture at the interface between abrasive and bond but did exhibit the bond fracture mode. It appears that titania is an undesirable constituent in bonding systems that tends to promote interfacial fracture at the abrasive grain-bond bridge interface. Even if its presence does not cause a reduction in cohesive strength, one method of reducing its effect is for it to form a titanate compound that does not reduce interfacial strength. Examination of fractured high purity aluminum oxide samples revealed preferential etching of the abrasive grain by the glass bond. This is assumed to be dissolution of blocks of β-aluminum oxide ($Na_2O{\cdot}11Al_2O_3$) present in α-aluminum oxide (pure aluminum oxide). The relationship between the wheel

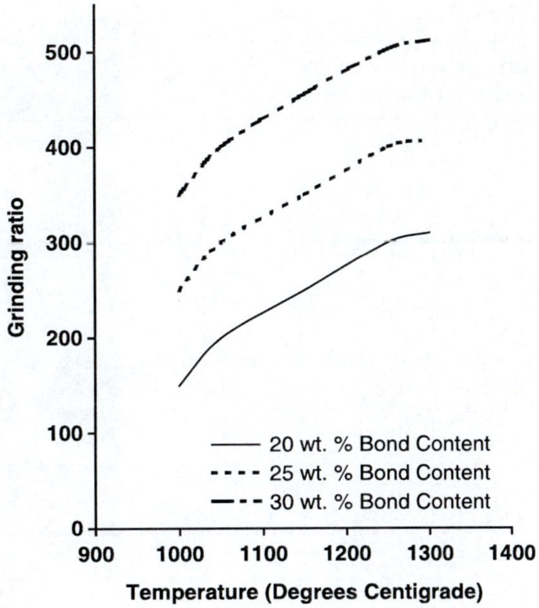

Fig. 2.26 Relationship between grinding ratio and firing temperature as a function of bond content for vitrified cBN grinding wheel structures

wear parameter, grinding ratio, and the firing temperature for vitrified cBN grinding wheel structures containing different amounts of bonding content is shown in Fig. 2.26. An interesting observation one can observe is that the retention of the abrasive grains in the vitrified bonding matrix can be improved by increasing the sintering temperature. In order to investigate the mechanism of cBN retention, samples of the post-fired abrasive structures were polished and etched. Figure 2.27 shows the unpolished fracture surfaces of the vitrified cBN grinding wheels. A magnified image of the interface between abrasive grain and bonding bridge is shown in Fig. 2.27b. Interfacial cohesion appears to be quite apparent in this image. Figure 2.28 shows a polished and etched fracture surface in the vicinity of the abrasive grain and bonding bridge. The associated electron probe microanalysis of the image clearly shows a concentration of oxygen at the interface between cBN and glass bonding bridge. The concentration of oxygen appears to be associated with boron and the formation of a boron-containing oxygen layer that separates the alumino-borosilicate bonding system and the cubic boron nitride abrasive grain. This is thought to be a relatively thin layer of B_2O_3 (boric oxide). As the sintering temperature is increased, the thickness of this layer is also increased with a subsequent loss of boron from the abrasive grain. Figure 2.29 illustrates the relationship between the interfacial layer thickness and sintering temperature. As the temperature is increased further, the width of the interfacial layer tends to stabilize and reaches an equilibrium thickness.

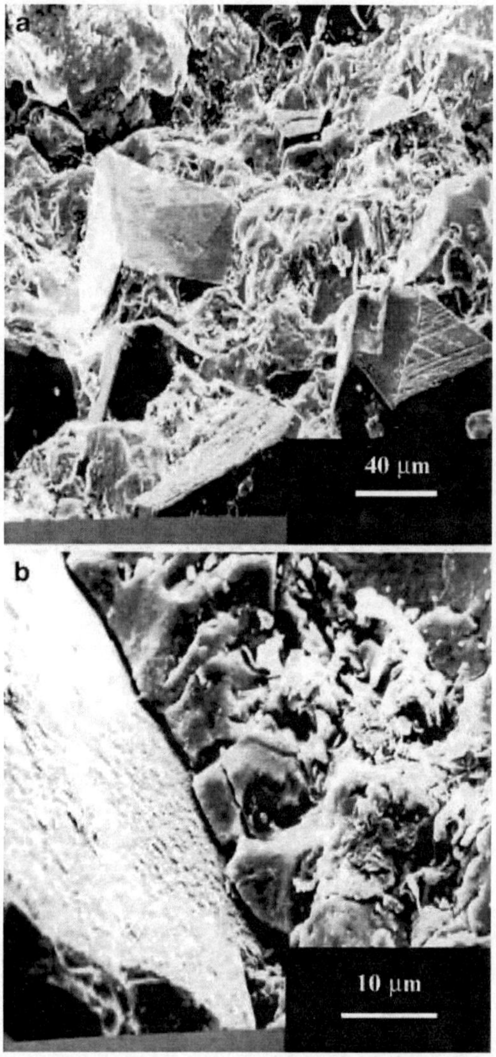

Fig. 2.27 (**a**) Vitrified cBN grinding wheel structure, (**b**) interface between cBN abrasive grain and vitrified bonding

2.3.5 Discussion of Interfacial Compounds on Grinding Wheel Wear

The existence of β-aluminum oxide was established by x-ray methods. When the bond content is low in samples made with high purity aluminum oxide, failure occurs by fracture of bonding bridges. At higher bond contents the mode of failure is one of abrasive grain fracture. Fracture at the abrasive grain-bond bridge interface was not observed. This is because the β-aluminum oxide phase is etched away preferentially due to the dissolution of Na_2O into the glass bond that

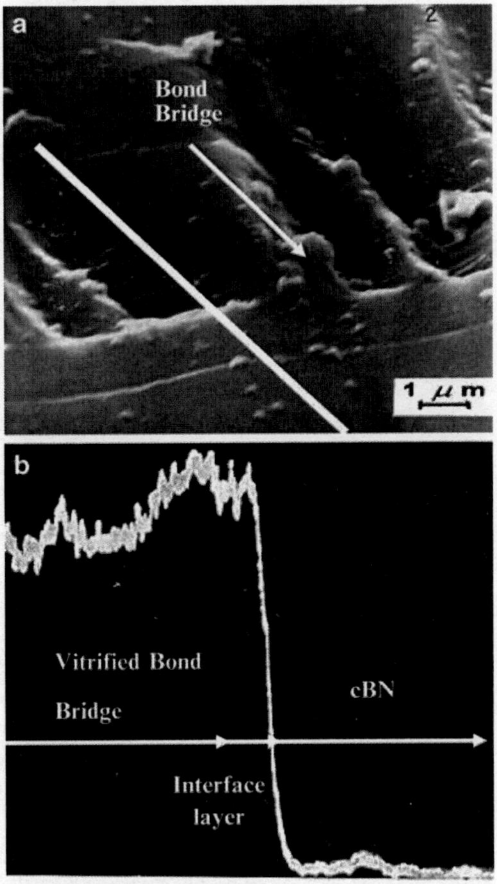

Fig. 2.28 (**a**) Polished cross section of cBN abrasive and bond bridge clearly showing the interface layer, (**b**) electron probe microanalysis of oxygen across the line scan shown in (**a**), *left-to-right*

locally increases the fluidity of the bond. This allows the bond to penetrate the surface of the abrasive grain and provides it with enhanced shear resistance.

This effect does not happen with titanium-doped aluminum oxide, in fact, the strength decreases at the softening point of the glass because of enhanced dissolution of aluminum oxide that releases more TiO_2 into the glass bond for rutile needle growth. Therefore, in contrast to Decneut et al. [9] the mode of fracture in titanium-doped structures is interfacial between abrasive grain and glass bond and is not completely dependent on bond content.

Even in the case where bond bridges have preferentially fractured, the mode of fracture is always associated with rutile needle weakening. The vitrification temperature and glass bond content has a significant effect on the elastic modulus of high purity and titanium-doped aluminum oxide structures. The differences in strength between these structures when fired at temperatures above the softening point of the glass bond is due to differences in the crystal structures of the two types of abrasive grain. The presence of β-aluminum oxide in high purity aluminum

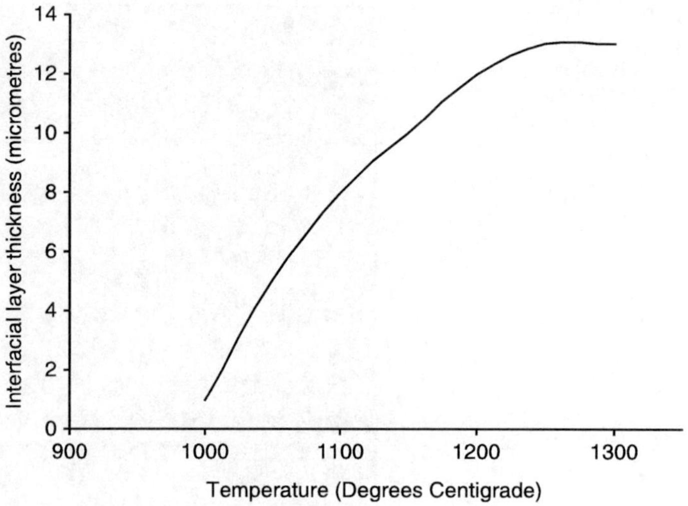

Fig. 2.29 Interfacial layer thickness between cBN and vitrified bonding bridge as a function of sintering temperature

oxide allows selective dissolution of aluminum oxide to occur that enables stronger bonding to take place between aluminum oxide and glass. This effect does not happen with titanium-doped aluminum oxide where dissolution allows the precipitation of TiO_2 into the glass bond in the form of rutile needles that reduces the cohesive strength between aluminum oxide and glass.

The existence of an interfacial layer between cBN and glass was thought to be that of the formation of boric oxide (B_2O_3). As sintering continued, the layer became thicker and tended to strengthen the interfacial layer. This is assumed to be the reason why the grinding ratio of the abrasive tool increased as a function of sintering temperature. It was also noted that the size of the cBN grains decreased as sintering temperature increased until an equilibrium interfacial layer thickness was reached. It would also be right to assume that at this point, that diffusion of oxygen into the cBN abrasive grain ceases to occur. The fracture surface of the vitrified cBN structure shows that fracture is associated with fracture within the bonding bridge rather than fracture at the cBN-bond bridge interface. This tends to imply that the interfacial bonding layer is stronger than bonding bridge.

2.4 Case Study II: Dissolution of Quartz and Its Effect on Grinding Wheel Wear

When considering individual bond constituents, mineral fluxes and ground glass frits have little direct effect on the ability to manufacture grinding wheels. However, most clay minerals develop some plasticity in the presence of water, which improves the ability to mould the mixture so that the wheel, in its green state, can be mechanically handled [12]. Clays and clay-based fluxes contain an amount of free quartz that has a detrimental effect on the development of strength during vitrification heat treatment. Clays are used to provide vitrified grinding wheels with green strength during the heat treatment process. However, when the glass material solidifies around the particles of clay and quartz, the displacive transformation of quartz during the cooling stage of vitrification leads to the formation of cracks in the glass around the quartz particle (Fig. 2.30). The strength of the bonding bridge is reduced and leads to the early release of the abrasive particle during the cutting process (Fig. 2.31).

The basic wear mechanisms that affect vitrified grinding wheels are concerned with grain fracture during metal cutting, fracture of bond bridges, mechanical fracture of abrasive grains due to spalling, and fracture at the interface between abrasive grain and bond bridge [13–27]. Failure in vitrified silicon carbide grinding wheels is more probable due to the lack of a well-developed bonding layer between abrasive grain and the glass bond-bridge, which is typically only a few micrometres. The lack of adequate bonding is due to the use of a high clay content bonding system with very little opportunity for a glass to form at the interface. High glass content bonding systems tend to aggressively decompose the surface of silicon carbide abrasive grains. In vitrified aluminum oxide grinding wheels, high glass content bonding systems are used extensively and lead to bonding layers in excess of 100 μm in thickness.

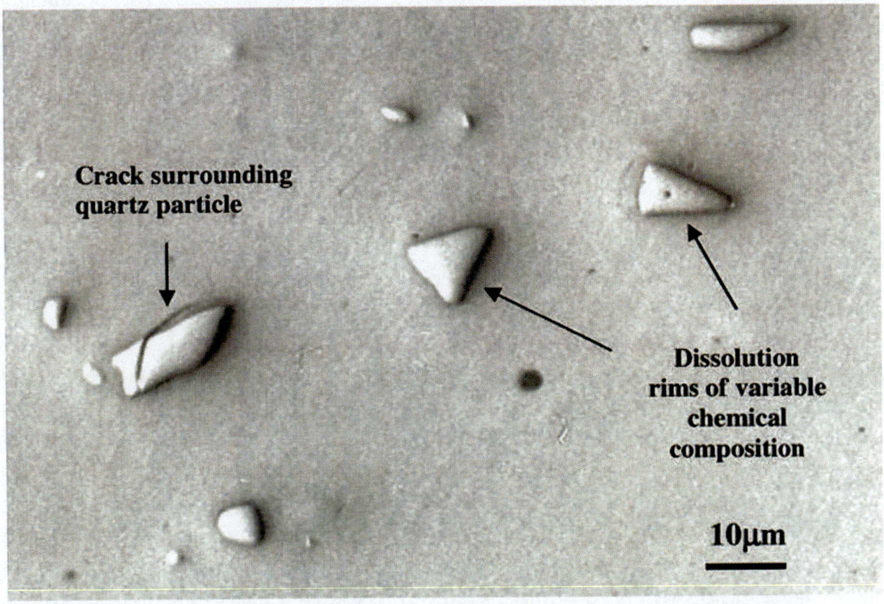

Fig. 2.30 A collection of quartz particles in a vitrified bonding system. The quartz particle on the *left* has a circumferential crack extending into the dissolution rim and abrasive grain

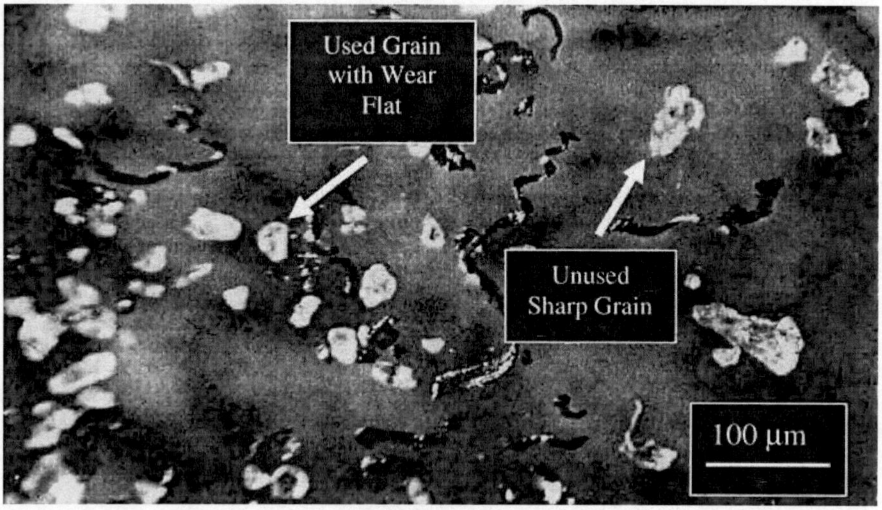

Fig. 2.31 Grinding swarf and a collection of used and unused abrasive cutting grains

In addition to the formation of very thin bonding layers in vitrified silicon carbide grinding wheels, the use of high clay content bonding systems means that there is an increase in the amount of quartz contained in the bond bridges between the abrasive grains. Although the likelihood of decomposition of silicon carbide surfaces is reduced, the probability of bond bridge failure is increased due to the increased quartz content. Therefore, the dissolution of quartz in these bonds is highly desired in order to compensate for a much thinner interfacial bonding layer. Difficulties encountered when developing a dissolution model arise from the fact that the phase boundary between quartz particle and molten glass moves during the diffusion process. The problem of a fixed boundary can be solved without difficulty although this is not equivalent to the conditions associated with a moving boundary between quartz particle and a highly viscous glass melt. The development of dissolution models are required to determine the magnitude of quartz remaining in the bonding system after a period of heat treatment. The models are then compared with experimentally determined quartz content of the bonding systems using x-ray diffraction techniques. Subsequently, dissolution models are used to specify the appropriate heat treatment schedule for a particular bonding system that is used in grinding wheels that grind automotive camshafts and crankshafts depending on the material removal rate and the nature of the material to be ground. The use of x-ray techniques is also applied to measuring phase transformations in grinding wheels that have been subjected to laser irradiation. When using a laser beam to dress the wheel, it is possible to form localized texture in the abrasive grains that allow the grinding wheel to remove material in the superfinishing regime. For the first time, it is reported that grinding wheels are able to provide roughing, finishing, and superfinishing operations in one grinding stroke.

2.4.1 Dissolution Models for Vitrified Grinding Wheel Bonds

When densification occurs in a vitrified grinding wheel after the peak soaking temperature has been reached, the cooling rate is reduced to prevent thermal stress cracking in the bonding layers between abrasive grains. Cooling rates are reduced when crystalline inversions occur that involve volume changes. The inversion range for quartz and cristobalite are 550–580°C and 200–300°C, respectively. Since the formation of cristobalite is rare in vitrified bonding systems used for grinding wheels, the rapid displacive transformation of quartz tends to promote the formation of cracks in bonding bridges.

When quartz-containing bonds begin to cool from the soaking, or vitrification, temperature it is thought that the liquid phase relieves stresses resulting from the thermal expansion mismatch between itself and the phases, β-quartz, β-cristobalite, and mullite, to at least 800°C. At 800°C, stresses will develop in quartz particles and in the matrix that causes cracking to occur around quartz particles. The shrinkage behaviour of quartz and the glass phase has been described by Storch et alia [28]. Between the temperature range, 573 and 800°C, the glass phase shrinks more than the quartz phase that causes tangential tensile stresses to form cracks in the matrix. At 573°C, β-quartz transforms to α-quartz that causes residual stresses around quartz particles to produce circumferential cracking around those same quartz particles (Fig. 2.30). Some of these cracks have been seen to propagate into the glass phase [29]. Similar observations occur in the cristobalite phase. Spontaneous cracking of quartz has been found to occur over a temperature range that depends on the size of the quartz particles [30]. Particles larger than 600 μm diameter cracked spontaneously at 640°C, whereas smaller particles of less than 40μm diameter cracked at 573°C. This observation agrees with temperature–dependent cracking reported by Kirchoff et alia [31]. To maintain the integrity of the bond bridges containing coarse quartz particles, the grinding wheel must remain at the vitrification temperature until the quartz particles have dissolved.

The dissolution model assumes that at a constant absolute temperature, T, a particle of quartz melts in the surrounding viscous glass melt, and that the rate of change of the volume of quartz present in the melt at a particular instant in time is proportional to the residual volume of quartz. The above assumption is based on the fact that alkali ions diffuse from the viscous glass melt to the boundary of the quartz particle thus producing a dissolution rim around each quartz particle. Diffusion rims around quartz particles are shown in Fig. 2.30.

A high reaction rate will initially occur which continuously decreases as the quartz particle is converted to a viscous melt. Previous models have provided an insight into how various factors contribute to the dissolution of quartz in vitreous bodies. However, Jackson and Mills [32] derived a mathematical relationship that accounts for the change in density when β-quartz transforms to α-quartz on cooling from the vitrification temperature, thus,

$$m_{T,t} = M\gamma \exp\left(-At^{1/2} \exp\left[\frac{-B}{T}\right]\right) \tag{2.2}$$

Where, $m_{T,t}$, is the residual mass fraction of quartz at a constant time and temperature couple, M is the original mass fraction of quartz prior to heat treatment, γ is the ratio of densities of β-quartz and α-quartz, A and B are constants, t is time, and T is absolute temperature. The model was compared with experimental data determined using the powder x-ray diffraction method. The experimental work was divided into two parts. The first part concentrates on comparing the dissolution model with x-ray diffraction data using "sintering" bond compositions that are used in vitrified silicon carbide grinding wheels, whilst the second part focuses on comparing the model with "fusible bond" compositions that are used in high-performance vitrified aluminum oxide grinding wheels.

2.4.2 *Experimental Procedures*

2.4.2.1 Raw Materials and Preparation

The raw materials used in the experimental study (case study 2) were Hymod Prima ball clay, standard porcelain China clay, potash feldspar, and synthetic quartz (supplied as silica flour). The chemical analysis of the raw materials is shown in Table 2.3. Rational analysis of the raw materials was performed to reveal the mineralogical composition of the raw materials. The rational analysis appears in Table 2.4. The bond mixture described is one typically used in vitrified silicon carbide grinding wheels where the erosion of the abrasive grain is reduced by using high clay content bonding systems. This bonding system is used where silicon carbide is predominantly used in grinding cast iron camshafts and crankshafts.

Fusible bonding systems using a mixture of ball clay and potassium-rich feldspar were made to test the model developed by Jackson and Mills [32]. The ball clay used contained 12.77 wt.% quartz, and the feldspar contained 4.93 wt.% quartz. The bonding system was composed of 66 wt.% ball clay, and 34% feldspar. The initial quartz content, M, of the bond mixture was 10.1 wt.%. The bond mixture described is one typically used in high-performance vitrified aluminum oxide grinding wheels, and is used when grinding steel camshafts and crankshafts.

The raw materials were mixed in a mortar, pressed in a mould, and fired at various temperatures. A heating rate of 3°C min^{-1} was employed until the vitrification temperature was reached. The typical soaking temperature was varied between 1,200 and 1,400°C for "sintering" bond compositions, and 950 and 1,050°C for "fusible" bond compositions in order to simulate industrial firing conditions. The samples were cooled at a rate of 2°C min^{-1} to avoid thermal stress fracture in the bonding bridges between abrasive grains. The fired samples were crushed to form a fine powder in preparation for x-ray diffraction.

Table 2.3 Chemical analyses of raw materials

Oxide (wt.%)	China clay	Ball clay	Potash feldspar	Quartz
Al_2O_3	37	31	18.01	0.65
SiO_2	48	52	66.6	98.4
K_2O	1.65	1.8	11.01	0.35
Na_2O	0.1	0.2	3.2	0.04
CaO	0.07	0.2	0.09	0.00
MgO	0.03	0.3	0.09	0.00
TiO_2	0.02	0.9	0.00	0.07
Fe_2O_3	0.68	1.1	0.11	0.03
Loss on ignition	12.5	16.5	0.89	0.20

Table 2.4 Mineralogical analyses of raw materials

Compound (wt.%)	China clay	Ball clay	Potash feldspar	Quartz
Quartz	4.05	12.77	4.93	98.40
Orthoclase	0.00	15.23	64.96	0.00
Kaolinite	79.70	62.71	2.17	0.00
Mica	13.94	0.00	0.00	0.00
Soda feldspar	0.8	1.69	27.07	0.00
Miscellaneous oxides/losses	1.51	7.60	0.87	1.60

2.4.2.2 X-Ray Diffraction of Vitrified Bonding Systems

The dissolution model was compared with experimental data using the x-ray powder diffraction method. X-ray diffraction of the raw materials was performed on a Phillips 1710 x-ray generator with a 40 kV tube voltage and a 30 mA current. Monochromatic Cu kα radiation, $\lambda = 0.154060$ nm, was employed. A scanning speed of 2° per minute for diffraction angles of 2θ was used between 2θ angles of 15° and 60°, and the x-ray intensity was recorded using a computer. The spectrum was then analysed and compared with known spectra. Powder specimens were prepared by crushing in a mortar and pestle in preparation for quantitative x-ray diffraction. To eliminate the requirement of knowing mass absorption coefficients of ceramic samples for quantitative x-ray diffraction, Alexander and Klug [33] introduced the use of an internal standard. Firstly, the ceramic sample is crushed to form a powder – the sizes of particles should be small enough to make extinction and absorption effects negligible. Secondly, the internal standard to be added should have a mass absorption coefficient at a radiation wavelength such that intensity peaks from the phase(s) being measured are not diminished or amplified. It should be noted that the powder diffraction mixture should be homogeneous on a scale of size smaller than the amount of material exposed to the x-ray beam, and was free from preferred orientation. The powder bed that is subjected to “x-rays” should be deep enough to give the maximum diffracted intensity. The expected equilibrium phases from the fired mixtures are quartz (unreacted and partially dissolved), mullite, cristobalite and glass. However, from the samples tested, the compounds quartz, mullite and glass were successfully detected. A calibration curve was constructed using a suitable internal standard (calcium fluoride), a diluent, and a synthetic form of the phase(s) to be measured. Synthetic mullite had a purity greater than 99.8%, whilst powdered quartz had a purity greater than 99.84% SiO_2. The method used for quantitative analysis of ceramic powders was developed by Khandelwal and Cook [34].

The internal standard provides an intense (111) reflection (d = 0.1354 nm) lying between the (100) reflection for quartz (d = 0.4257 nm) and the (200) reflection for mullite (d = 0.3773 nm). Using copper kα radiation ($\lambda = 0.15405$ nm), the corresponding values of diffraction angle 2θ are: (100) quartz = 20.82°; (111) calcium fluoride = 28.3°; and (200) mullite = 32.26°. Figure 2.32 shows the calibration curve generated by varying proportions of calcium fluoride, synthetic quartz and mullite. Mass fractions of the crystalline phases in the mixture can be interpreted from the calibration lines by measuring the intensity ratio of the phase(s) to the internal standard. Figure 2.33 shows the diffraction peaks of interest for quantitative analysis lying between 15° and 40° of the diffraction angle 2θ. The figure shows the reflections of the (111) plane of calcium fluoride, (200) plane of mullite, and the (100) plane of quartz. In order to calculate the mass fractions of quartz and mullite in the mixture, the height of the chosen diffraction peak and its width at half-height were measured from the diffraction spectrum. The product of these two measures were then compared with that of the internal standard, and the resultant intensity ratio was used to find the exact mass fraction of the phase(s) measured in the glass that was subjected to x-ray diffraction.

2.4.2.3 Grinding Wheel Performance

A series of grinding wheel experiments were conducted in order to show the difference between bonding systems with different levels of quartz content contained in their bonding bridges. The experiments were conducted using high-speed steels and a high chromium content hypereutectoid steel (AISI 52100) in order to compare with field trials conducted using commercially available vitrified aluminum oxide grinding wheels. A series of controlled experiments were designed to compare grinding wheels under increasing rates of metal removal. The experiments were terminated when a condition of severe burn, chatter, or wheel breakdown was observed. The initial experimental wheels used were: angular white alumina with a low temperature bonding system (wheel specification A); a sol-gel alumina abrasive wheel with angular white abrasive mixed in a one-to-one proportion bonded with a low temperature fired bonding system (wheel specification B); and a monocrystalline alumina wheel with a low temperature fired bonding system (wheel specification C). All wheels were manufactured with a vitreous bond, 60 mesh size abrasive grain (approx. 220 μm in diameter) material, J-hardness grade, and a fairly open structure. Experiments were performed on a Jones and Shipman series 10 cylindrical grinding machine using a 450 mm diameter grinding wheel rotating at 33 m s^{-1} surface speed. The wheel was dressed using a diamond blade tool using a depth of

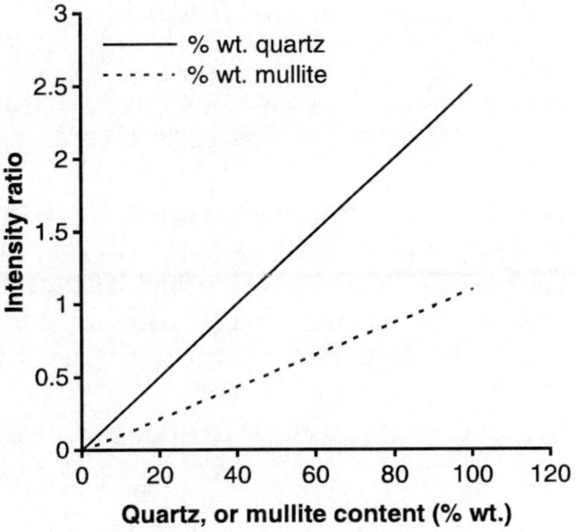

Fig. 2.32 Calibration curve for quantitative analysis of x-ray determined quartz and mullite using the CaF_2 (111) plane generated by the internal standard

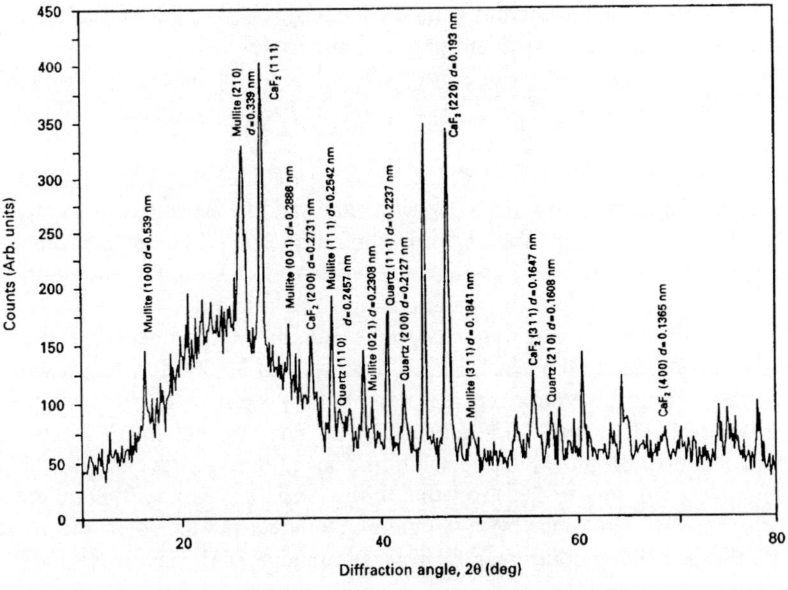

Fig. 2.33 X-ray diffraction spectrum of a vitrified bonding system showing the interplanar distances of crystallographic planes of mullite, quartz and calcium fluoride. Scan rate was 2° per minute

cut of 30 μm at a feed rate of 0.15 mm rev^{-1}, and a final dressing depth of cut of 15 μm prior to grinding workpieces. The amount of material removed was 250 μm per grinding stroke. The coolant flow pressure was 0.5 bar at a flow rate of 15 L min^{-1} using a 2% concentrated solution of oil in water.

2.4.3 Experimental Results

2.4.3.1 Silicon Carbide Bonding Systems: Verification and Comparison of Dissolution Models for Quartz

In addition to comparing the experimental results to the dissolution model, results published in the literature were also used to test the accuracy of the model. The composition of the experimental mixtures was matched to those specified by Lundin [12]. Lundin's experimental mixtures were composed of 25 wt.% quartz (13.2 μm particle size), 50 wt.% clay (kaolin), and 25 wt.% flux (potassium feldspar – 25 μm average particle size).

The constants A and B for the sintering bonding system were calculated,

$$A = 5.62x10^8 \tag{2.3}$$

$$B = 33374 \tag{2.4}$$

From which the experimental activation energy, Q, is 132.65 kcal mol^{-1}. The residual quartz content for the sintering bonding system is,

$$m_{T,t} = 26.25.\exp\left[-5.62x10^8.t^{1/2}.e^{\frac{-33374}{T}}\right] \tag{2.5}$$

The data comparing Lundin's experimental results, the author's experimental results, and the dissolution model is shown in Table 2.5. When the data is plotted as the logarithm of (-ln[m/M]/t$^{1/2}$) versus the reciprocal of absolute temperature, 1/T, then all data fits a straight-line relationship. The gradient was calculated to be 33,374, the constant B, using two data points. Lundin's experimental gradient gave a value of 32,962 using the least squares method, and 34,000 for the present work. The corresponding activation energies for both systems are 131 kcal mol^{-1} for Lundin's work [12], and 135 kcal mol^{-1} for the present work, respectively. Figures 2.34 and 2.35 show the effects of time on residual quartz content at different temperatures according to (2.5) together with comparative experimental data.

A comparison was made with dissolution models published in the literature. One of the earliest models was derived by Jander [35]. The equation can be expressed,

$$\left(1-\sqrt[3]{1-Z}\right)^2 = \left\{\frac{C_1.D}{r^2}\right\}.t \tag{2.6}$$

Where Z is the volume of quartz that has been dissolved, r is the original particle radius, and D is the diffusion coefficient for the diffusing species. This equation can be transformed into mass fractions using Archimedes' law, thus,

Table 2.5 Residual quartz content of a sintering bonding system at various vitrification temperatures

Temp. (°C)	Time (h)	Lundin's experimental result (wt.%)	Experimental result (wt.%)	Jackson & Mills' [32] result (wt.%)
1,200 (1,473 K)	1	24.1	24.2	24.2
1,200	1	24.7	24.3	24.2
1,200	1	26.1	24.8	24.2
1,200	2	23.7	23.8	23.4
1,200	2	23.6	23.9	23.4
1,200[a]	2	23.4	23.4	23.4
1,200	4	21.3	22.2	22.3
1,200	8	20.3	20.9	20.8
1,200	18	19.0	18.5	18.6
1,200	18	18.9	18.6	18.6
1,200	48	15.2	15.1	14.9
1,250 (1,523 K)	1	22.7	22	22.1
1,250[a]	2	20.6	20.6	20.6
1,250	4	18	18.5	18.6
1,250	8	15.5	16	16.2
1,250	18	12.6	12.5	12.6
1,250	48	8.3	7.8	8.0
1,300 (1,573 K)	0.5	22.6	20.4	20.6
1,300	0.5	21	20.9	20.6
1,300	1	20	18.3	18.6
1,300	2	16.1	15.9	16.2
1,300	4	13.4	12.8	13.2
1,300	8	10	9.7	9.9
1,300	18	5.9	5.8	6.1
1,300	50	1.6	1.8	2.3
1,300	120	0.3	0.2	0.6

Lundin's [12] experimental data is compared with the author's experimental data and the model [32]
[a]Values used for deriving the constants used in the theoretical model

$$\left(1 - \sqrt[3]{\frac{m}{M}}\right)^2 = C_2.t \tag{2.7}$$

Where, C is a constant dependent on soaking temperature and initial particle size of quartz. Krause and Keetman [36] expressed the dissolution of quartz as a function of isothermal firing time, viz,

$$M - m = C_3.\ln t \tag{2.8}$$

Where M is the initial quartz content, m is the residual quartz content after time, t. The unit of time here is seconds such that after 1 s of firing the residual quartz content is equal to the initial quartz content. Monshi's dissolution model [37] can be transformed into the following equation assuming isothermal firing conditions,

Fig. 2.34 Effect of time on residual quartz content of a sintering bonding system according to Jackson and Mills model [32], and compared with Lundin's experimental data [12]

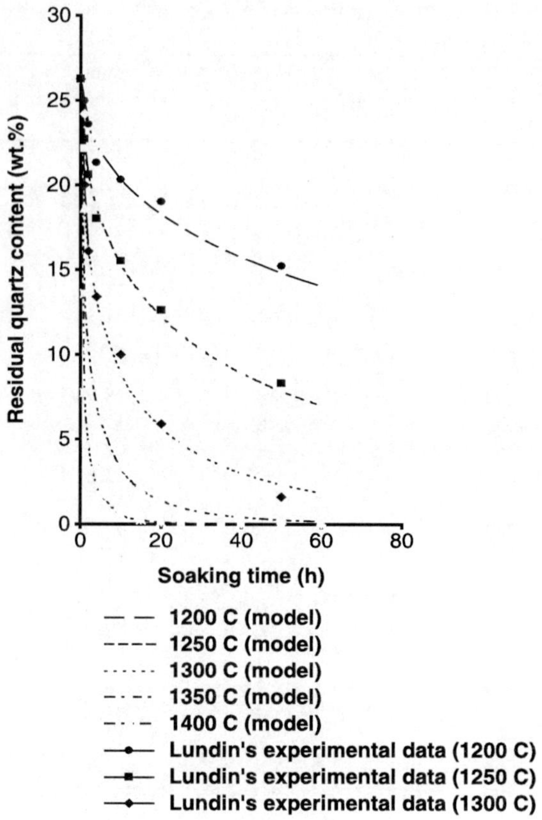

$$\ln\left\{\frac{m}{M}\right\} = -C_6\sqrt{t} \tag{2.9}$$

Jackson and Mills' model [29] for isothermal firing conditions is transformed into,

$$\ln\left\{\frac{m}{\gamma.M}\right\} = -C_7\sqrt{t} \tag{2.10}$$

Where γ is the ratio of densities of $\beta-$ and $\alpha-$ quartz. Constants for all the equations presented here are calculated using quartz mass fraction data after 18 h firing. The constants are dimensioned in seconds. The equations shown were compared with experimental data generated by Lundin [12] for a clay-based material containing 40 wt.% kaolin, 40 wt.% quartz, and 20 wt.% feldspar. According to the transformed equations, the mass fraction of quartz can be calculated as follows,

Jander's model [35]

$$m = 41.9.\left(1 - \left\{1.55x10^{-6}.t\right\}\right)^{3/2} \tag{2.11}$$

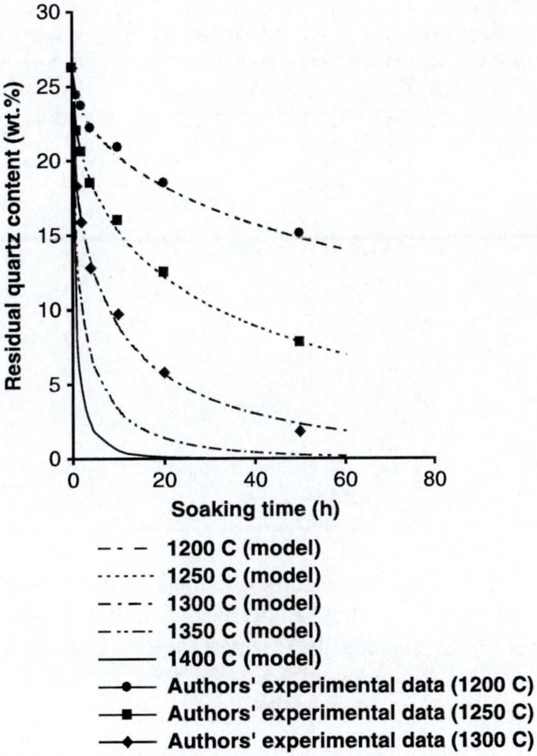

Fig. 2.35 Effect of time on residual quartz content of a sintering bonding system according to Jackson and Mills' model [32] and compared with the authors' experimental data

Krause and Keetman's model [36]

$$m = 41.9 - (2.58.\ln t) \tag{2.12}$$

Monshi's model [37]

$$m = 41.9.e^{-4.5x10^{-3}\sqrt{t}} \tag{2.13}$$

Jackson and Mills' model [32]

$$m = 41.73.e^{-4.5x10^{-3}\sqrt{t}} \tag{2.14}$$

The transformed equations are then tested using data provided by Lundin [12]. Referring to Table 2.6, it can be shown that the mass fraction of quartz obtained using the equations derived by Jander [35] and Krause and Keetman [36] did not agree with Lundin's experimental results [12]. The results obtained using Monshi's model [37] are in much better agreement compared to Lundin's data.

Table 2.6 Residual quartz content for different soaking times at 1,300°C for a sintering bonding system composed of 40 wt.% kaolin, 40 wt.% quartz, and 20 wt.% feldspar (Lundin's mixture number M21 [12]) compared with other dissolution models

Time (h)	Lundin's experimental data [12]	Jander [35]	Krause & Keetman [36]	Monshi [37]	Jackson & Mills [32]
0	41.9	41.9	0.00	41.9	41.9
0.5	35.9	41.72	22.55	34.61	34.76
1	32.8	41.54	20.76	31.97	32.12
2	29.2	41.19	18.97	28.58	28.72
4	23.2	40.49	17.18	24.39	24.51
8	19.5	39.11	15.39	19.49	19.59
18	13.3	35.72	13.30	13.30	13.36
24	10.7	33.74	12.56	11.13	11.19
48	6.9	26.18	10.77	6.43	6.51
120	3.6	7.85	8.96	2.17	2.17
190	2.7	0.00	7.22	1.00	1.01
258	2.0	0.00	6.43	0.54	0.55

However, the results obtained using Jackson and Mills' model [32] is more accurate at predicting the mass fraction of quartz remaining owing to the differences in the density of quartz. After long periods of heat treatment, the model predicts lower magnitudes of mass fractions of quartz when compared to Lundin's experimental results [12].

2.4.3.2 Aluminum Oxide Bonding Systems: Verification and Comparison of Dissolution Models for Quartz

The constants, A and B, for the fusible bonding system were determined using time and temperature couples at 2 and 10 h and were calculated to be, -5.2×10^8 and $-33{,}205$, respectively. The dissolution equation then becomes,

$$m_{T,t} = 10.06 \exp\left[-5.2x10^8 t^{1/2}.e^{\frac{-33205}{T}}\right] \tag{2.15}$$

Equation (2.15) is used to compare the experimentally determined mass fraction of quartz remaining after heat treatment with the predicted values. The calculated mass fraction of quartz remaining after a period of heat treatment is calculated using the equation derived by Jackson and Mills [32]. The results of the dissolution model compare well with the experimental data over short periods of time. However, over longer periods of heat treatment the model tends to become less accurate (Fig. 2.36). A comparison was made with published dissolution models. The equations shown were compared with experimental data at 1,050°C. According to the transformed equations, the mass fraction of quartz can be calculated as follows,

Fig. 2.36 Effect of time on residual quartz content of a fusible bonding system according to Jackson and Mills' model [32] and compared with the author's experimental data

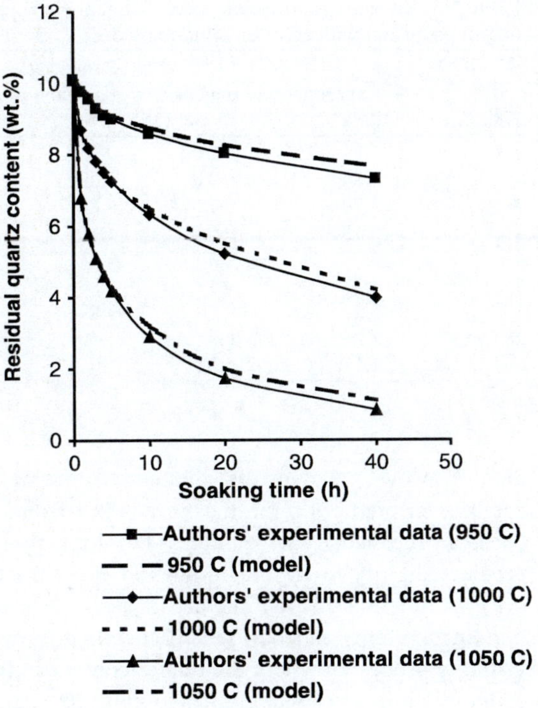

Jander's model [35]

$$m = 10.1.\left(1 - \left\{3.44x10^{-6}.t\right\}\right)^{3/2} \tag{2.16}$$

Krause and Keetman's model [36]

$$m = 10.1 - (0.59.\ln t) \tag{2.17}$$

Monshi's model [37]

$$m = 10.1.e^{-6.4x10^{-3}\sqrt{t}} \tag{2.18}$$

Jackson and Mills' model [32]

$$m = 10.06.e^{-6.37x10^{-3}\sqrt{t}} \tag{2.19}$$

With reference to Table 2.7, it can be shown that the mass fraction of quartz obtained using the equations derived by Jander [35] and Krause and Keetman [36], did not agree with the experimental results at 1,050°C. The results obtained using Monshi's model [37] are in much better agreement compared to the experimental

Table 2.7 Residual quartz content for different soaking times at 1,050°C for a fusible bonding system compared with other dissolution models

Time (h)	Experimental data	Jander model [35]	Krause & Keetman model [36]	Monshi's model [37]	Jackson & Mills' model [32]
0	10.1	10.1	0	10.1	10.1
1	6.84	9.91	5.23	6.88	6.86
2	5.79	9.72	4.82	5.87	5.86
3	5.13	9.54	4.58	5.21	5.19
4	4.7	9.36	4.41	4.7	4.68
5	4.28	9.18	4.28	4.28	4.28
10	3.2	8.28	3.87	2.99	3
20	2	6.6	3.46	1.81	1.82
40	1.1	3.62	3.04	0.89	0.89

data. However, the results obtained from Jackson and Mills' model are more accurate at predicting the mass fraction of quartz remaining owing to the differences in the density of quartz. After long periods of heat treatment, the model predicts slightly lower magnitudes of mass fractions of quartz when compared to the experimental results. The use of x-rays to predict the level of quartz in vitrified bonding systems can be used to specifically design grinding wheels for specific grinding processes where the quartz content in the bonding system will reduce the economic impact of using vitrified alumina grinding wheels.

2.4.3.3 Grinding Wheel Experiments

The dissolution of quartz during heat treatment has a significant effect on the wear of vitrified grinding wheels. Figure 2.37 shows the effect of using a high and a low quartz content bonding system on the wear of vitrified aluminum oxide grinding wheels grinding a large number of tool steel materials [38]. The classification of tool steels is in the form of an abrasive hardness number, which is a weighted average of the number of carbides contained within the tool material. As shown in Fig. 2.37, the grinding ratio, or G-ratio, is a measure of the efficiency of the grinding wheel. It is the quotient of the volume of workpiece material removed and the volume of the wheel material removed. The figure demonstrates the effectiveness of reducing the quartz content of the bonding system. X-ray diffraction techniques have been used to characterize the bonding system and is an effective method in the selection of raw materials used for high efficiency grinding wheels.

Further experimental results using hypereutectoid steels were compared with field experiments using camshaft and crankshaft grinding operations as the basis for comparison. Forty workpiece samples were ground and the results of the initial experiments are shown in Figs. 2.38–2.40. None of the wheels used produced chatter vibrations or burned the workpieces, which was represented by a change in the level of grinding power. The grinding wheels gave a similar performance

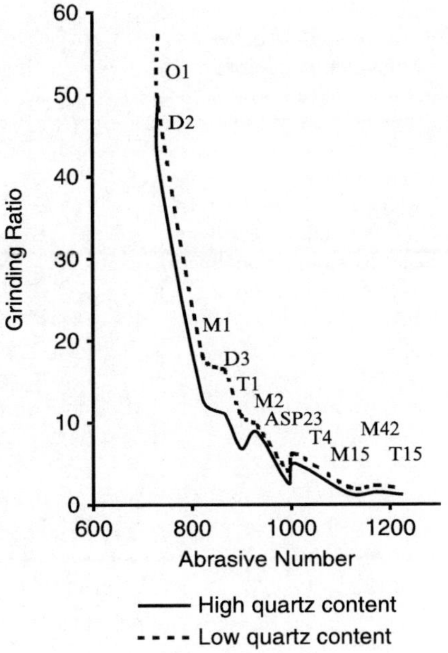

Fig. 2.37 Effect of the abrasive number on the grinding ratio for a high-quartz content and a low quartz-content bonding system grinding tool steel materials in the cylindrical surface grinding mode. Tool steel material is marked on the trend lines, and is a function of the carbide content in their microstructure expressed as an abrasive number

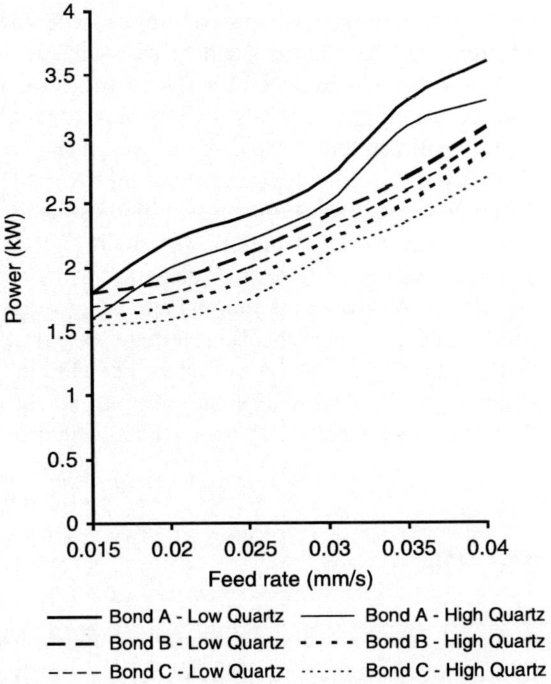

Fig. 2.38 Grinding power as function of feed rate for vitrified aluminum oxide grinding wheels with different types of abrasive and bond. Grinding conditions: wheel surface speed = 33 m s^{-1}; depth of cut = 0.25 mm; dress conditions = diamond blade traversed at 0.15 mm rev^{-1} at a depth of cut of 0.03 mm

Fig. 2.39 Surface roughness as function of feed rate for vitrified aluminum oxide grinding wheels with different types of abrasive and bond. Grinding conditions: wheel surface speed = 33 m s^{-1}; depth of cut = 0.25 mm; dress conditions = diamond blade traversed at 0.15 mm rev^{-1} at a depth of cut of 0.03 mm

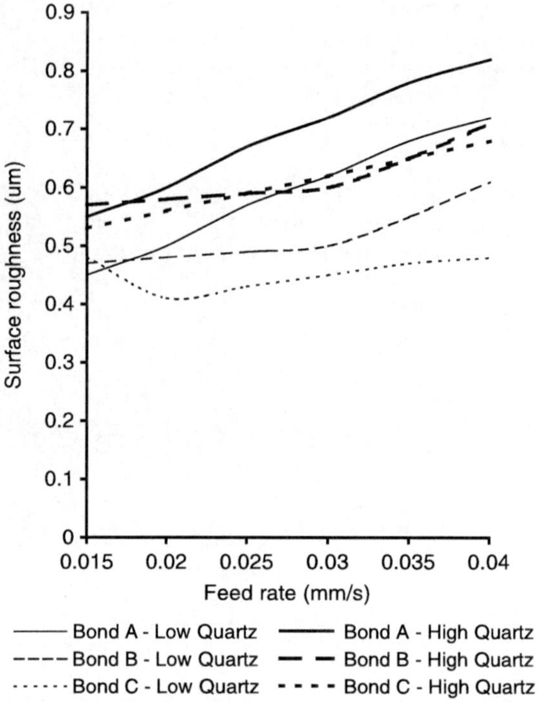

level, surface roughness, and grinding ratio. At low grinding rates, it was possible to differentiate between the grinding performance of each wheel.

The sol-gel wheel gave the best performance in terms of surface roughness and the highest grinding ratio whilst the angular white alumina wheel provided the worst results that resulted in rapid wheel wear. The use of monocrystalline abrasives at moderate metal removal rates ground approximately 30 workpiece materials before it started to breakdown. From these initial results, it can be shown that sol-gel abrasive wheels operate well at high power levels and can grind more efficiently at higher metal removal rates compared to angular white and monocrystalline abrasive grinding wheels. At low metal removal rates, there appears to be no significant difference between the grinding wheels used in the experiments. These experiments also show that bonding systems that are rich or depleted in quartz particles have a significant effect on the breakdown of the grinding wheel during grinding experiments – a significant factor in the subsequent change in grinding ratio.

2.5 Discussion

A model describing the dissolution of quartz has been used to predict the mass fraction of quartz remaining after a period of heat treatment in vitrified grinding

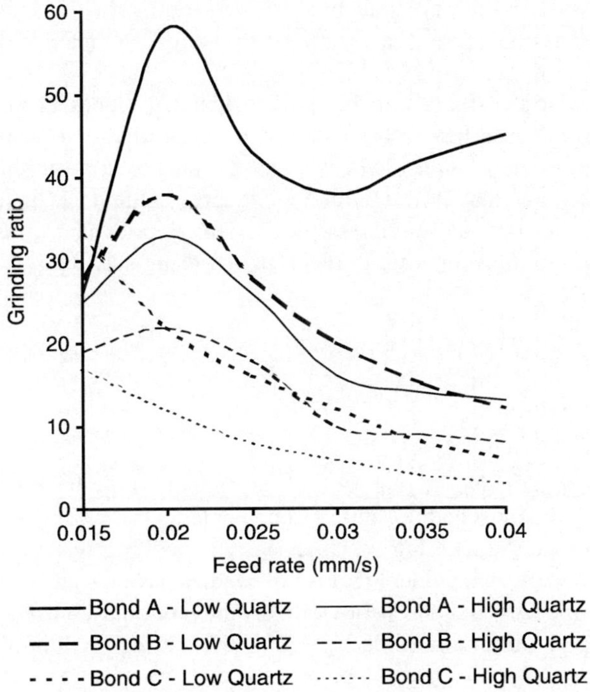

Fig. 2.40 Grinding ratio as function of feed rate for vitrified aluminum oxide grinding wheels with different types of abrasive and bond. Grinding conditions: wheel surface speed = 33 m s^{-1}; depth of cut = 0.25 mm; dress conditions = diamond blade traversed at 0.15 mm rev^{-1} at a depth of cut of 0.03 mm

wheel bonding systems. The model has been compared with experimentally determined mass fractions of quartz in an industrial vitrified bonding system. However, the assumptions made when formulating the model may invalidate its wider application. In diffusion controlled processes, the total flux per unit time is proportional to the total surface area available and the concentration gradient at the interface between quartz particle and viscous melt. The model assumes that a linear concentration gradient exists across the spherical shell of reaction products that invalidates the real situation where quartz particles are anything but perfect spheres.

The model does obey the parabolic law for diffusion around a sphere that can only apply when the sphere has a very low solubility in the solvent. However, this is not the case for quartz in most silicate systems. Also, the bonding between quartz particle and viscous glass melt is moving and not stationary. This implies that the concentration gradient is transient and not constant. The experimental data used to justify the accuracy of the dissolution model do not meet the conditions implicit in the model. The reactions between the three basic components of the bonding system, i.e. clay, feldspar, and quartz, cannot solely be described as diffusion-controlled dissolution of quartz in a liquid phase of constant composition and properties. It seems unlikely that the model can predict the mass fraction of quartz in any silicate system with great accuracy. However, over short soaking periods at the vitrification temperature, the results of the model compare well with experimental data.

X-ray diffraction methods can be used to identify planes in vitrified bonding systems during and after heat treatment, which can be used to determine the correct bond formulations for grinding wheels used in specific applications. The experimental results show that bonds with high quartz contents lead to much lower grinding ratios and surface roughness levels. This implies that x-ray methods can be used to increase the economic productivity of using such tools.

2.6 Conclusions

The behaviour of abrasive cutting tools is dependent on the type of abrasive grain used, its heat treatment schedule, and its bond content. The vitrification behaviour is dependent on certain processing variables and the bonding systems used for precision grinding wheels. It is evident that rutile needles have a significant effect on the mechanical properties and the behaviour of conventional abrasive cutting tools. If titanium-doped aluminum oxide is used as the cutting medium then heat treatment cycles should be designed that prevent the growth of the needles into the bonding system. The development of strength during heat treatment depends upon the nature of the compounds formed at the interfacial layer between abrasive grain and bonding material. Further studies are required to understand the nature of this bonding with new abrasive grains and glass ceramic bond formulations.

Vitrified cBN grinding wheels tend to fracture at the bond bridge rather than at the interface between cBN and its glass bonding bridge. This implies that the relatively thin layer of boric oxide is stronger than the bonding bridge material. However, the choice of bonding composition may significantly influence the thickness of the interfacial layer that may become detrimental to the performance of the grinding wheel. This is particularly important if the failure mode during grinding is interfacial fracture between bond and abrasive grain.

The dissolution model derived by Jackson and Mills has been compared with experimental data using sintering and fusible vitrified bonding systems that are used extensively with high performance grinding wheels. The results predicted by the model compare well with the experimental results presented in this paper. However, over longer periods of isothermal vitrification, the model becomes less accurate due to the assumptions made in the dissolution model. The model may be of use when predicting the mass fraction of quartz using high temperature firing cycles that are characterised by short soaking periods. The models used are useful where bond bridge fracture wear is the dominant mode of grinding wheel wear.

Acknowledgements The author thanks Professor Givi Bockuchava for permission to use micrographs of abrasive materials from his extensive works on the wear of grinding wheels and grinding wheel structure. The author also thanks the late Professor David Tabor, FRS, for advice and direction on the tribology of abrasive materials and the structure of diamond and cBN, whilst the author was a research fellow at the Cavendish Laboratory, University of Cambridge.

References

1. M.J. Jackson, J. Manuf. Proc., (2001), **3**, 17–28.
2. E.J. Krabacher, Trans. A.S.M.E. J. Eng. Ind., (1959), **81**, 187–200.
3. G.J. Geopfert and J.L. Williams, Mech. Eng., (1959), **81**, 69–73.
4. R.S. Hahn in, 'Proceedings of the 3rd International Machine Tool Design and Research Conference', April 1962, Manchester, UK, (Pergamon Press, UK, 1962), 129–154.
5. L.P. Tarasov in, 'International Research in Production Engineering – American Society of Mechanical Engineers', (A.S.M.E., U.S.A., 1963), Paper No. 21, p. 196.
6. S.K. Bhattacharyya, H. Grisbrook, and H. Moran, Microtechnic, (1965), **22**, 114–116.
7. S. Malkin and N.H. Cook, Trans. A.S.M.E. J. Eng. Ind., (1971), **93**, 1120–1128.
8. M.J. Jackson, Trans. North Am. Manuf. Res. Inst. Soc. Manuf. Engrs., (2002), **30**, 287–294.
9. A. Decneut, R. Snoeys, and J. Peters, "*Sonic testing of grinding wheels*", Report MC 36, November 1970, Centre de Recherces Scientifiques et Techniques de L'industrie des Fabrications Metalliques, University of Louvain, Belgium, 1970.
10. E.R. Winkler, J.F. Sarver, and I.B. Cutler, J. Am. Ceram. Soc., (1966), **49**, 634–637.
11. J.C. Walmsley and A.R. Lang, in 'Advances in Ultrahard Materials Application and Technology', Edited by Chris Barrett, (De Beers Industrial Diamond Division, UK, 1988), 61–75.
12. S.T. Lundin, *Studies on Triaxial Whiteware Bodies*, (Almqvist and Wiksell), Stockholm, Sweden, 1959.
13. S. Malkin and N.H. Cook, Trans. A.S.M.E. J. Eng. Ind., (1971), **93**, 1120–1128.
14. H. Yoshikawa, *Fracture wear of grinding wheels,* International Research in Production Engineering – American Society of Manufacturing Engineers, (1963), Paper No. 23, p. 209.
15. L.P. Tarasov, *Grinding wheel wear grinding tool steels,* International Research in Production Engineering – American Society of Manufacturing Engineers, (1963), Paper No. 21, p. 196.
16. H. Yoshikawa and T. Sata, Trans. A.S.M.E. J. Eng. Ind., (1963), **85**, 39–43.
17. H. Tsuwa, *On the behaviour of abrasive grains in the grinding process,* Technical Report of Osako University, Japan, (1960), **10**, 733–743.
18. H. Tsuwa, *On the behaviour of abrasive grains in the grinding process – part 2,* Technical Report of Osako University, Japan, (1961), **11**, 287–298.
19. H. Tsuwa, *On the behaviour of abrasive grains in the grinding process – part 3,* Technical Report of Osako University, Japan, (1961), **11**, 299–309.
20. Y. Tanaka and N. Ikawa, *Behaviour of abrasive grains on the diamond wheel,* Technical Report of Osako University, Japan, (1962), **12**, 345–354.
21. G.J. Geopfert and J.L. Williams, Mech. Eng., (1959), **81**, 69–73.
22. H. Tsuwa, Trans. A.S.M.E. J. Eng. Ind., (1964), **86**, 371–382.
23. G.K. Lal and M.C. Shaw, *Wear of single abrasive grains in fine grinding,* Proceedings of the International Grinding Conference, Pittsburgh, USA, (1972), 107–126.
24. N.S. Eiss, Trans. A.S.M.E. J. Eng. Ind., (1967), **89**, 463–470.
25. S.K. Bhattacharyya, H. Grisbrook, and H. Moran, Microtechnic, (1965), **22**, 114–116.
26. W. Mohun, Trans. A.S.M.E. J. Eng. Ind., (1962), **84**, 431–482.
27. K. Saito and T. Kagiwada, Bull. Jap. Soc. Prec. Eng., (1974), **8**, 125–126.

28. W. Storch, H. Ruf, and H. Scholze, Berichte Deut. Keram. Ges., (1984), **61**, 325.
29. E. Binns, Science of Ceramics, (1962), **1**, 315.
30. W.F. Ford, and J. White, Trans. J. Brit. Ceram. Soc., (1951), **50**, 461.
31. G. Kirchoff, W. Pompe, and H.A. Bahr, J. Mat. Sci., (1982), **17**, 2809.
32. M.J. Jackson, and B. Mills, J. Mat. Sci., (1997), **32**, 5295–5304.
33. I.E. Alexander, and H.P. Klug, Anal. Chem., (1948), **20**, 886.
34. S.K. Khandelwal, and R.L. Cook, Am. Ceram. Soc. Bull., (1970), **49**, 522–526.
35. W. Jander, Z. Anorg. U. Allgem. Chem., (1927), **163**, 1–30.
36. P. Krause, and E. Keetman, Sprechsaal, (1936), **69**, 45–47.
37. A. Monshi, *Investigation into the strength of whiteware bodies,* Ph. D. Thesis, University of Sheffield, UK, 1990.
38. M.J. Jackson, *A study of vitreous-bonded abrasive materials,* Ph. D. Thesis, Liverpool John Moores University, UK, December 1995.

Chapter 3
Grinding Wheel Safety and Design

Mark J. Jackson

Abstract Grinding wheels are designed to withstand a variety of loads during the grinding of complex shapes. This chapter highlights the fundamental equations that are applied to calculate the highest stresses that may be encountered should the wheel rotate freely on its spindle. These are used to calculate the most conservative safety factors that are required in order to protect the grinding machine operator from wheel burst. The chapter focuses on new grinding wheel designs that are used in processes such as creep freed and VIPER grinding in addition to plain, parallel sided grinding wheels. Implications for safe practice during grinding are discussed.

Keywords Grinding wheels · Safety · Mechanical design · Analysis

3.1 Introduction

To achieve a high performance from a grinding wheel it must stay sharp, and must have the ability to absorb a high volume of metal chips. Therefore, the grinding wheel must be porous and must be able to withstand high grinding loads that are placed on the abrasive grains and on the bonding bridges that hold the grains in position. The nature of the properties of the grinding wheel at the interface between the bonding bridge and the abrasive grains is very important when one considers how forces are transmitted into the bonding bridges through the interfacial layer. Vitrified bonds are typically used for high performance grinding processes and in comparison with other types of bonds, vitrified bonds permit easy dressing while at

M.J. Jackson (✉)
MET, College of Technology, Purdue University, 401 North Grant Street,
West Lafayette, IN 47907, USA
e-mail: jacksomj@purdue.edu

M.J. Jackson and J.P. Davim (eds.), *Machining with Abrasives*,
DOI 10.1007/978-1-4419-7302-3_3,

the same time possess high levels of resistance to wear [1–3]. In recent years, the development of high speed grinding wheels has been focused on high-speed creep feed grinding processes, or high efficiency deep grinding processes [4–8]. These studies have led to the development of novel designs of grinding wheels that overcome the severe limitations that are presented when using plain grinding wheels. In addition to vitrified segmented wheel designs, single layer electroplated wheels are also gaining popularity especially when applied to high-speed form grinding applications. Therefore, special consideration must be spent on the design of abrasive grinding wheels.

3.2 Rotational Stresses

Stresses that occur in rotating grinding wheels arise from six principal sources, namely: centripetal forces as a result of rotation at a steady operating speed; clamping loads at the bore of the wheel; grinding forces at the point of wheel-workpiece contact; acceleration to normal operating speed, and subsequent retardation; residual out-of-balance forces in the wheel itself; and thermal stresses from heat generated when grinding. The grinding wheel must grind without significant heat generation or the effects of vibration during precision grinding. This avoids thermal and mechanical damage to the workpiece so that wheels must be selected in order to minimize grinding forces and heat generation. The stresses of real concern during high speed grinding results from centripetal loading at the maximum operating speed. The application of Lamé's equations to the solution of rotational stresses in grinding wheels has been investigated by Munnich [9]. Munnich's results confirmed the validity of using the equations of elasticity to solve the magnitude of rotational stress in grinding wheels, especially for porous vitrified wheels.

Vitrified grinding wheels are brittle bodies and can be treated as perfectly elastic until the point of fracture is reached. Stresses occurring during rotation are the same for those in any rotating disc. Assuming that the rotational stresses do not vary over the thickness of the wheel, the governing equation for radial displacement is,

$$r^2.\frac{d^2u}{dr^2}+r\frac{du}{dr}-u+\frac{r}{h}.\frac{dh}{dr}\left(r.\frac{du}{dr}+\nu u\right)=-\frac{1-\nu^2}{E}.\rho\omega^2r^3 \tag{3.1}$$

For parallel and trapezoidal shaped grinding wheels, the thickness at any radius is,

$$h=A+B.r \tag{3.2}$$

The governing (3.1) simply becomes,

$$\frac{d^2u}{dr^2}+\frac{A+2Br}{Ar+Br^2}.\frac{du}{dr}-\frac{A+(1-\nu)Br}{Ar^2+Br^3}.u=-\frac{1-\nu^2}{E}.\rho\omega^2r \tag{3.3}$$

This equation can be solved using finite difference approximations to give radial displacement and derivatives at any radius. The rotational radial and circumferential stresses are,

$$\sigma_{rr} = \frac{E}{1-\nu^2} \cdot \left(\frac{du}{dr} + \nu \frac{u}{r} \right) \tag{3.4}$$

And,

$$\sigma_{\theta\theta} = \frac{E}{1-\nu^2} \cdot \left(\frac{u}{r} + \nu \frac{du}{dr} \right) \tag{3.5}$$

These solutions are applied to plane stress conditions. For grinding wheels whose peripheral diameters are less than ten times that of their thickness, (3.4) and (3.5) can be modified by replacing ν with $\frac{\nu}{1-\nu}$ to account for axial stress. These conditions are known as plane strain conditions. It is usual in the design of grinding wheels to assume the boundary condition of free radial displacement since this condition gives the highest level of stress that is circumferential and acts at the bore. This stress is tensile in nature and is used to find the maximum operating speed of the grinding wheel. Modification of the governing equation for the relevant boundary conditions for a parallel-sided disc produces,

$$\frac{d^2u}{dr^2} + \left(\frac{1}{r} + \frac{1}{h} \cdot \frac{dh}{dr} \right) \frac{du}{dr} + \left(\frac{\nu}{hr} \cdot \frac{dh}{dr} - \frac{1}{r^2} \right) u + \frac{(1-\nu^2)}{E} \rho\omega^2 r = 0 \tag{3.6}$$

For a parallel-sided disc, h is constant, therefore, dh/dr is equal to zero. Thus, the governing equation becomes,

$$\frac{d^2u}{dr^2} + \frac{1}{r} \cdot \frac{du}{dr} - \frac{u}{r^2} + \frac{(1-\nu^2)}{E} \rho\omega^2 r = 0 \tag{3.7}$$

The displacement is calculated by solving the second order differential in (3.7), thus,

$$u = \frac{r}{E} \cdot \left[(1-\nu)C_1 - (1+\nu)\frac{C_2}{r^2} - \frac{(1-\nu^2)}{E} \rho\omega^2 r^2 \right] \tag{3.8}$$

Where C_1 and C_2 are constants found from applying the boundary conditions. The resulting radial and circumferential stresses, for plane stress conditions, are,

$$\sigma_{rr} = C_1 + \frac{C_2}{r^2} - \frac{(3+\nu)}{8} \rho\omega^2 r^2 \tag{3.9}$$

$$\sigma_{\theta\theta} = C_1 - \frac{C_2}{r^2} - \frac{(1+3\nu)}{8} \rho\omega^2 r^2 \tag{3.10}$$

The solutions to (3.9) and (3.10) for free radial displacement at the bore for plane stress conditions are,

$$\sigma_{rr} = \frac{(3+v)}{8}\rho\omega^2\left(r_1^2 + r_2^2 - \frac{r_2^2.r_1^2}{r^2} - r^2\right) \tag{3.11}$$

$$\sigma_{\theta\theta} = \frac{(3+v)}{8}\rho\omega^2\left(r_2^2 + r_1^2 - \frac{r_1^2.r_2^2}{r^2} - \left(\frac{1+3v}{3+v}\right)r^2\right) \tag{3.12}$$

For plane strain conditions, the resulting radial and circumferential stresses are,

$$\sigma_{rr} = C_1 + \frac{C_2}{r^2} - \frac{(3-2v)}{8(1-v)}\rho\omega^2 r^2 \tag{3.13}$$

$$\sigma_{\theta\theta} = C_1 - \frac{C_2}{r^2} - \frac{(1+2v)}{8.(1-v)}\rho\omega^2 r^2 \tag{3.14}$$

The solutions to (3.13) and (3.14) for free radial displacement at the bore for plane strain conditions are,

$$\sigma_{rr} = \frac{(3-2v)}{8.(1-v)}\rho\omega^2\left(r_1^2 + r_2^2 - \frac{r_2^2.r_1^2}{r^2} - r^2\right) \tag{3.15}$$

$$\sigma_{\theta\theta} = \frac{(3-2v)}{8.(1-v)}\rho\omega^2\left(r_2^2 + r_1^2 - \frac{r_1^2.r_2^2}{r^2} - \left(\frac{1+2v}{3-2v}\right)r^2\right) \tag{3.16}$$

If the radial displacement of the grinding wheel is restrained from movement, the circumferential stress is reduced while the radial stress increases until, with full restraint, the radial stress exceeds the circumferential stress.

3.3 Factor of Safety

The calculation of the factor of safety for the grinding wheels discussed in this paper is dependent upon the type of material that the wheel is composed of. In the case of reinforced, or composite, wheels, the safety factor is calculated using a number of different failure criteria. For brittle materials, the modified Mohr theory is used as it produces the most conservative estimate of safety factor. This involves considering the maximum level of stress that exists in an uneven material using a number of Dowling factors to provide the most conservative factor of safety. For ductile materials, the distortion-energy theory or von Mises–Hencky theory is used to observe the most conservative value of safety factor in the material that takes account of an effective stress as a function of the distortion energy experienced by the stressed volume of material subjected to centripetal loading when dynamically loaded at very high spindle speeds. The grinding wheel industry uses a factor of safety of three unless the wheel is completely guarded, then a factor of safety of two is used.

3.4 Segmented Grinding Wheels

The magnitude of the circumferential stress due to centripetal loading is a direct result of radial expansion that leads to "circumferential stretching" of the abrasive layer. The magnitude of circumferential stretching can be reduced by replacing the continuous layer of abrasive, which is attached to the outer surface of the center section, with a set of discontinuous segments.

Although the linear theory of elasticity can be used to predict the level of stress due to centripetal loading in a continuous grinding wheel, a numerical technique such as the finite element method is required to predict stress levels in a segmented grinding wheel. In order to allow the penetration of lubricants to the grinding zone in creep feed grinding, conventional wheels are slotted and give rise to localized stress concentrations at the root of the slot that is far away from the bore. An example of this is shown in Fig. 3.1. Here, the slots are molded into the face of the grinding wheel and are usually present in conventional aluminum oxide grinding wheels.

The design of these particular wheels are subjected to special treatment and are not discussed here as they are limited to maximum operating speeds of 35 m/s at their surface. However, abrasive segments bonded to metal discs are considered here as they operate at much higher speeds. These are arranged as a series of segments adhesively bonded to the solid disc that is usually metallic. In order to examine the influence of the number and the size of the abrasive segment on the

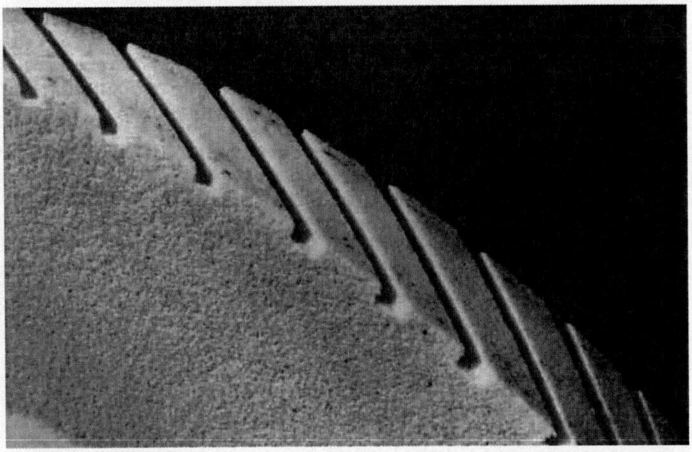

Fig. 3.1 Peripheral slotting on the surface of a conventional grinding wheel (courtesy of S. A. Jones, Unicorn Abrasives)

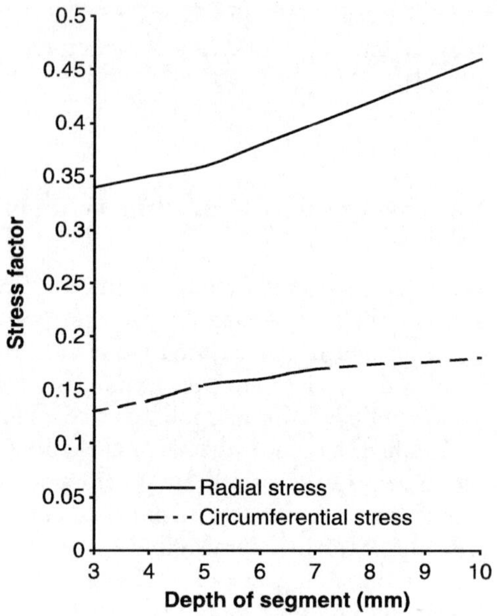

Fig. 3.2 Effect of depth of the segments on the stress levels in the abrasive part of the grinding wheel. Used with permission from Inderscience Publishers (2010)

levels of stress, a finite element model of a segmented grinding wheel was created. The abrasive material is a vitreous-bonded cubic boron nitride that has a density of 2,270 kg/m^3, a modulus of elasticity of 8.6 GN/m^2, and a Poisson's ratio of 0.2. The material used for the reinforcing center section was En 24T steel.

The segments were bonded to the steel center section using an adhesive with a density of 1,700 kg/m^3, modulus of elasticity of 1.5 GN/m^2, and a Poisson's ratio of 0.4. The edges of the model shown were constrained from movement in the "Y" and "Z" directions in order to simulate connectivity with the remaining parts of the wheel. All other nodes were allowed to move freely, and the size and number of segments were varied according to the nature of the grinding operation. Figure 3.2 shows the influence of the depth of the abrasive segments on the circumferential and radial stresses in the abrasive part of the grinding wheel. The number of segments was constant at this stage of the analysis.

The number of segments used in industrial practice for this particular application was 60. Figure 3.3 shows the influence of the number of segments on the circumferential and radial stresses in the abrasive part of the wheel. The depth of the abrasive segment has a significant effect on the levels of stress in a segmented grinding wheel [5, 7]. For this reason, the depth of the abrasive segment was 4 mm. The abrasive segment consists of a 3-mm thickness usable layer and a 1-mm backing layer that has a coefficient of thermal expansion that is matched to that of the abrasive layer.

For the grinding wheel considered here, it is shown that the smaller the depth of the abrasive segment and the higher the number of segments leads to a reduction in the level of stress in the abrasive part of the grinding wheel. The greatest effect on

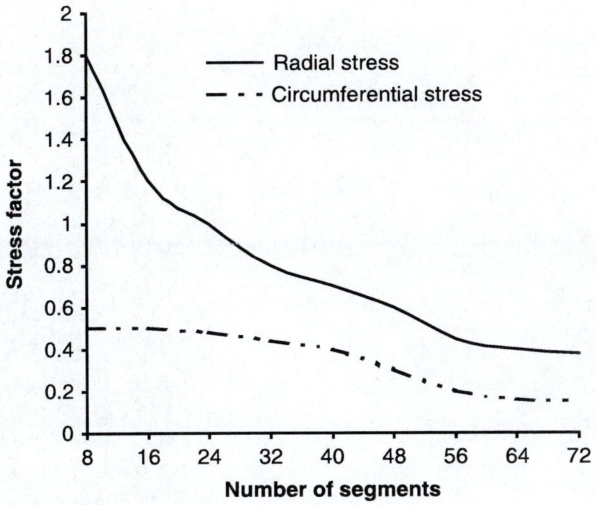

Fig. 3.3 Effect of the number of segments on the stress levels in the abrasive layer of the grinding wheel. Used with permission from Inderscience Publishers (2010)

the magnitude of stresses in segmented grinding wheels is clearly the number of segments. It can be shown that the maximum stress in the abrasive segments is in the radial direction, whereas the maximum stress in the continuous layer of abrasive is in the circumferential direction.

3.5 High Speed Segmented Grinding Wheels

3.5.1 Grinding Wheels Capable of Being Dressed

Very high grinding wheel speeds can be achieved (up to 500 m/s) if more exotic designs are used that effectively concentrate mass at, or near to, the bore of the wheel [5]. Figure 3.4 shows a wheel design that was developed to reduce the mass of the reinforced center section. This design allows a wider segment to be used that allows direct plunge-grinding operations to be undertaken at higher metal removal rates. Hyperbolic- and trapezoidal-shaped grinding wheels use narrow abrasive segments, which means that plunge grinding operations are accompanied by traverse movements in order to remove the same stock of material compared to the ultra high speed grinding wheel shown in Fig. 3.4. The finite element model represents one half of the wheel that has symmetry about the "Y" axis. The corners of the wheel are restrained from movement in the "Y" direction in order to simulate connectivity. The properties of the grinding wheel are the same as those stated earlier, in order to compare with the segmented grinding wheel.

Figure 3.5 shows the results of the finite element analysis carried out on this particular geometry. The results are compared with the results of the segmented

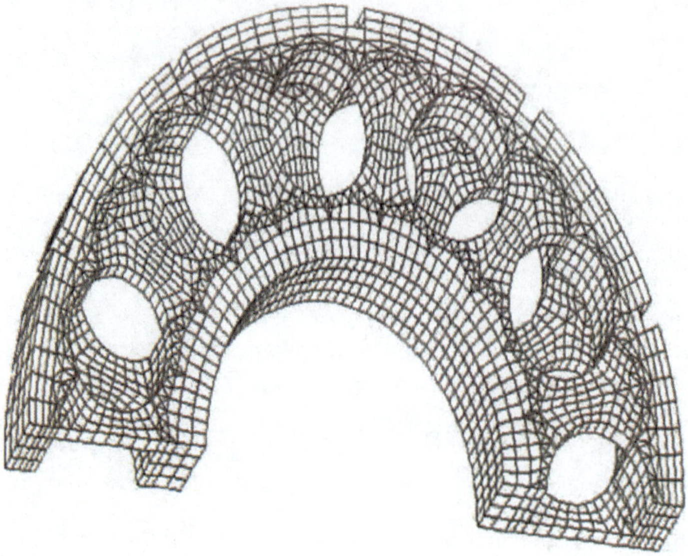

Fig. 3.4 Finite element model of the ultra high-speed grinding wheel showing abrasive segments bonded to the "porous" steel reinforced center. Used with permission from Inderscience Publishers (2010)

grinding wheel. Figure 3.5 shows the effect of the number of segments on the circumferential and radial stress in the abrasive part of the grinding wheel shown in Fig. 3.4.

The depth of the abrasive segments was maintained at 4 mm. It is shown that the ultra high-speed grinding wheel with the "porous" reinforced steel center exhibits a lower stress factor due to centripetal loading. For the same level of safety experienced by segmented grinding wheels, higher wheel speeds can be achieved using the novel design shown in Fig. 3.4. Once again, the maximum stress observed is in the radial direction and is tensile in nature.

3.5.2 *Electroplated Grinding Wheels*

Electroplated wheels can be rotated at much higher speeds than vitrified wheels owing to the nature of the grinding layer that is on the wheel body itself. The stochastic nature of strength in vitrified segments means that speeds are limited due to the brittleness of the abrasive layer. However, in practice speeds are lower so that tolerances are achieved on engineering components. Higher speeds lead to distortion of the body that impairs tolerances on components especially run-out tolerances on camshafts and crankshafts. Figure 3.6 shows a typical cross-section of a

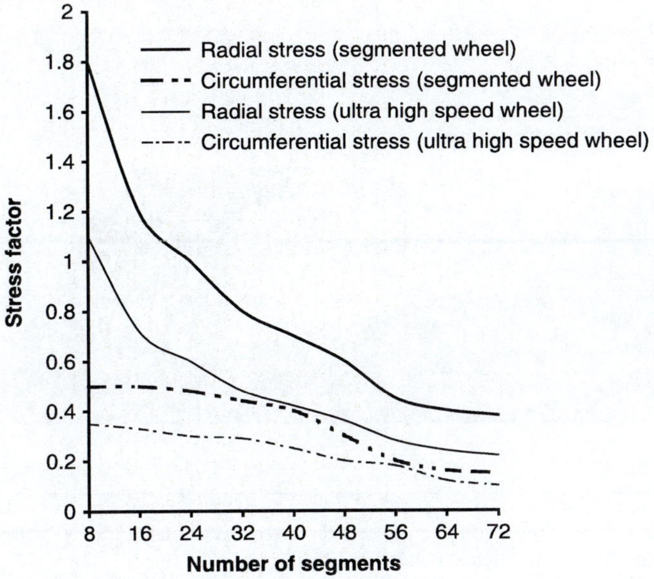

Fig. 3.5 Influence of number of segments on stress levels in the abrasive layer for segmented and ultra high speed grinding wheels. Used with permission from Inderscience Publishers (2010)

single-layer electroplated grinding wheel that supplies coolant directly through the wheel and into the contact zone. The coolant channel is drilled with a series of through holes and delivers coolant between the spaces of the abrasive grain.

Figure 3.7 shows the node numbers of the finite element model of the grinding wheel. These numbers are used to identify points of stress when the wheel is rotated at various speeds using different materials.

Figure 3.8 shows the magnitude of maximum stress (circumferential, or hoop stress) assuming that the wheel is rotating freely at the bore, i.e. the maximum stress condition. The speed of rotation is 45 m/s. At this speed, there is no appreciable difference in the stress factor as the material of the wheel changes (Fig. 3.9). Figure 3.9 also shows that the maximum stress occurs at nodes 2 and 48, which are the positions of the bore and the point at which the coolant is channeled to the contact zone. At 100 m/s, the stresses have increased (Fig. 3.10) and the stress factors at node 2 and 48 are still highest (Fig. 3.11).

However, it is shown that the use of carbon fiber wheel material has slightly decreased the stress factor experienced by the rotating wheel. A further increase in speed to 200 m/s has raised the stresses significantly (Fig. 3.12) and the maximum stress still occurs at node positions 2 and 48. The use of carbon fiber has again reduced the stress factor significantly (Fig. 3.13) at the highest speed of 200 m/s.

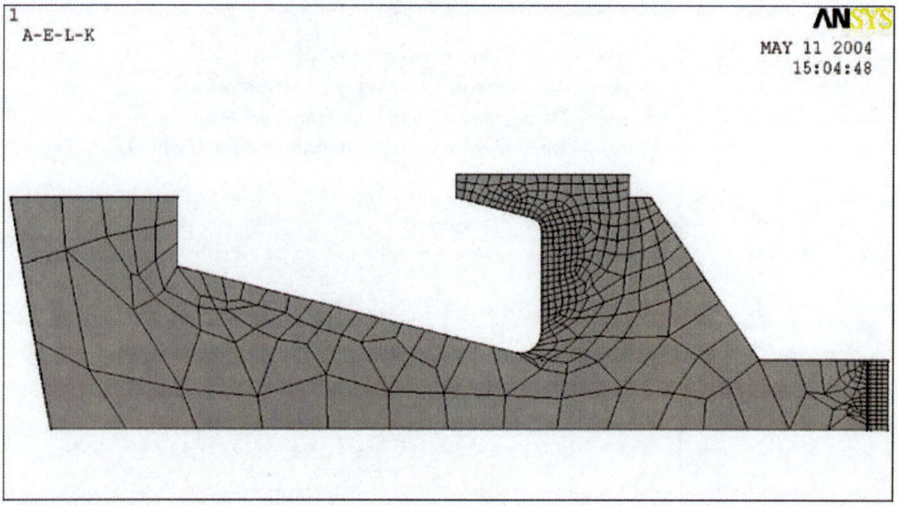

Fig. 3.6 Finite element mesh of the cross-section of the electroplated cBN grinding wheel. Used with permission from Inderscience Publishers (2010)

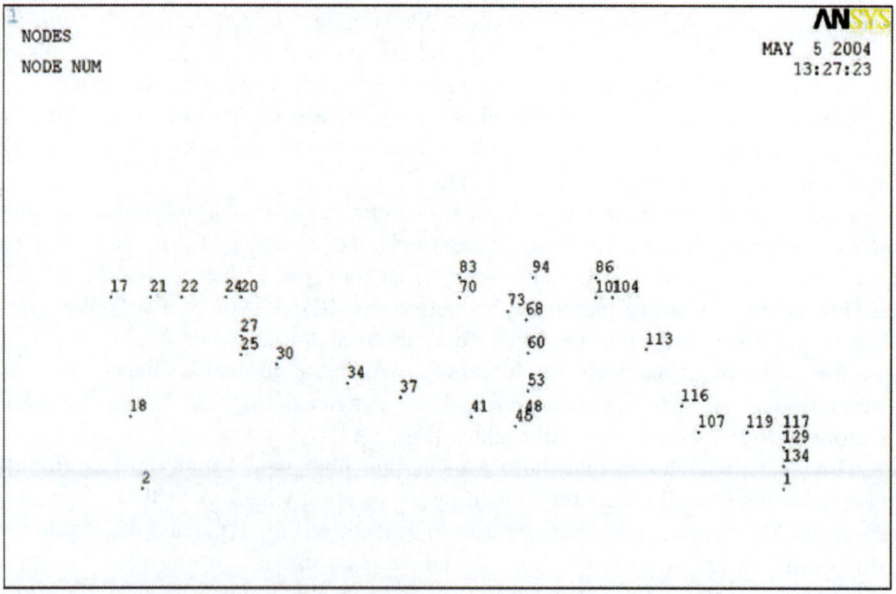

Fig. 3.7 Node numbers associated with the finite element model shown in Fig. 3.5. Used with permission from Inderscience Publishers (2010)

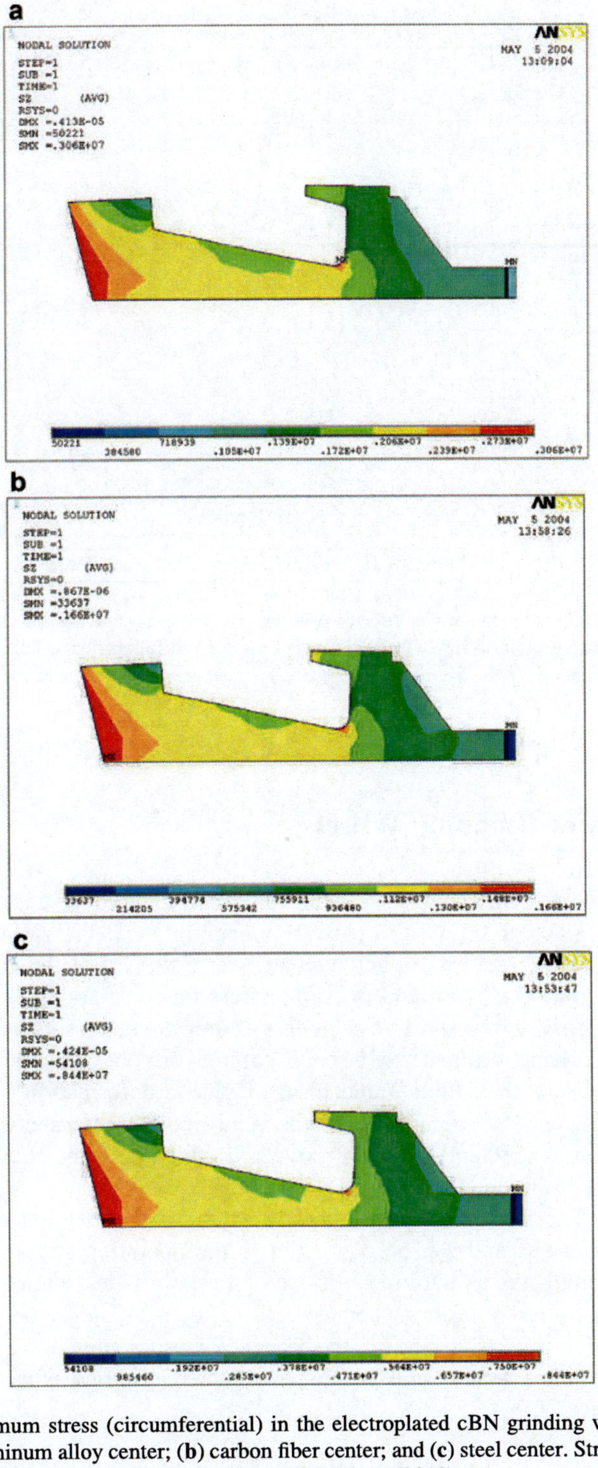

Fig. 3.8 Maximum stress (circumferential) in the electroplated cBN grinding wheel rotating at 45 m/s: (**a**) aluminum alloy center; (**b**) carbon fiber center; and (**c**) steel center. Stress scale is in N/mm^2. Used with permission from Inderscience Publishers (2010)

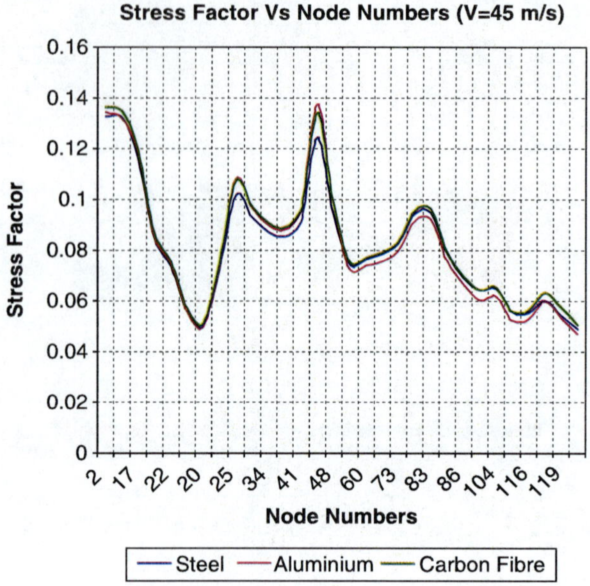

Fig. 3.9 Stress factor at various nodal positions on the cross-section of the electroplated cBN grinding wheel rotating at 45 m/s. Used with permission from Inderscience Publishers (2010)

3.6 Safety of Grinding Wheels

The results of the F. E. analysis provide the design engineer with the ability to calculate the factor of safety at a particular operating speed. When using certain yield criteria, safety factors can be calculated for each part of the grinding wheel. It is usual to apply the maximum principal stress theory, sometimes referred to as the Rankine theory, to the abrasive segments as they are classed as brittle materials. The maximum shear strain energy per unit volume, or von Mises, theory is applied to ductile materials that allow some form of plastic deformation before fracture occurs. This theory can be applied to the reinforcing center section of the segmented grinding wheel, and to the adhesive layer that bonds the abrasive segments to the center section.

These two theories allow the designer to calculate the yielding strength of the composite materials, and can be compared to the operating stress at a particular operating speed to give a factor of safety. In order to provide a comparison between materials used for the reinforcing center section, Fig. 3.14 shows the effect of using different reinforcing materials on the safety factor exhibited by the standard, segmented grinding wheels and porous, segmented grinding wheels at different surface speeds.

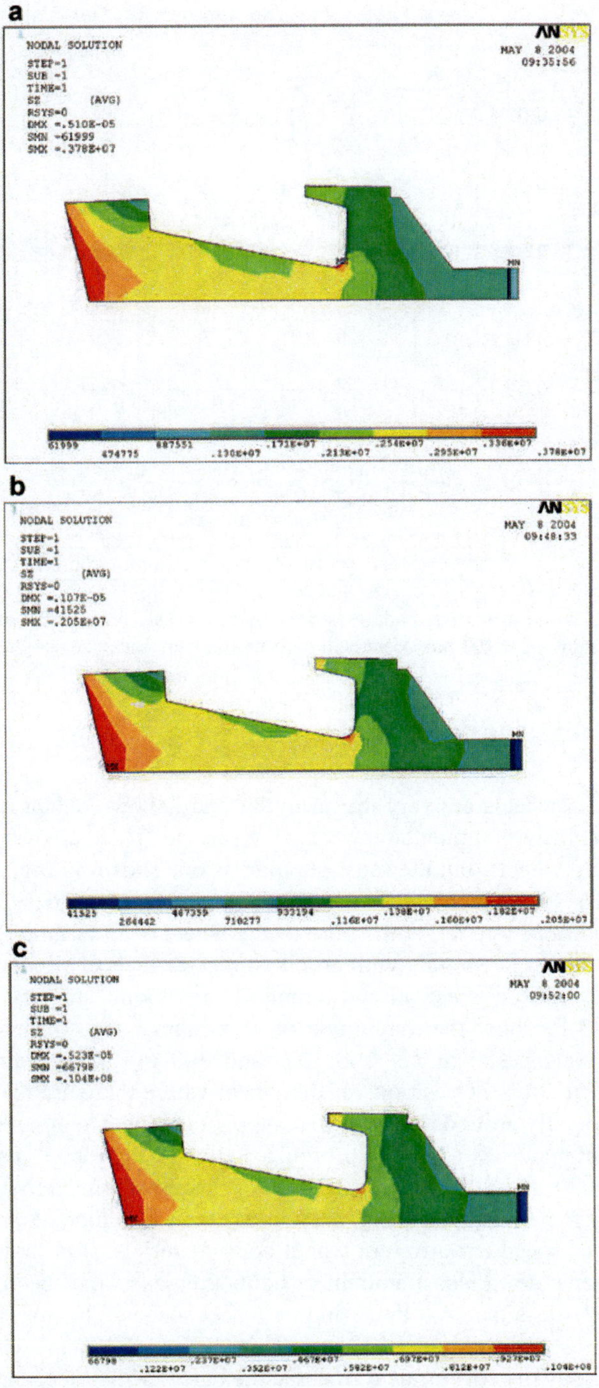

Fig. 3.10 Maximum stress (circumferential) in the electroplated cBN grinding wheel rotating at 100 m/s: (**a**) aluminum alloy center; (**b**) carbon fiber center; and (**c**) steel center. Stress scale is in N/mm^2. Used with permission from Inderscience Publishers (2010)

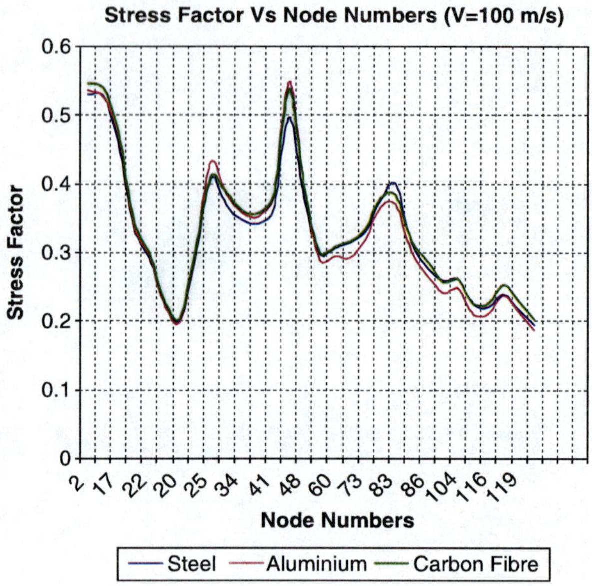

Fig. 3.11 Stress factor at various nodal positions on the cross-section of the electroplated cBN grinding wheel rotating at 100 m/s. Used with permission from Inderscience Publishers (2010)

3.7 Slotted Grinding Wheels

Slotted grinding wheels are very useful in that they allow coolant to penetrate the gap between abrasive segment and workpiece surface. There are various designs of slotted grinding wheels, but the most popular type is shown in Fig. 3.15.

Figure 3.16 shows a three-dimensional computer aided design model with boundary conditions applied at the bore of the wheel. It is assumed that the wheel is freely rotating on its spindle with no clamping, i.e., wheel slippage is occurring that generates the highest circumferential, or hoop, stresses at the bore. Figures 3.17–3.19 show the magnitude of maximum hoop stress at the bore for wheel peripheral speeds of 45, 100, 200 and 300 m/s. There appears to be a concentration of stress at positions of the wheel where there are discontinuities in geometry especially around the holes that supply coolant. The figures represent the use of different materials for the reinforcing body, i.e., aluminum, carbon fiber and steel. Figure 3.20 shows the magnitude of hoop stress as a function of speed for the different body materials, whilst Fig. 3.21 shows the magnitude of stress factor also as a function of speed for different wheel body materials. It is shown that carbon fiber bodies generate the least amount of hoop stresses as the speed increases, but has the highest stress factor at low grinding wheel speeds. This implies that carbon fiber bodies must used at peripheral speeds in excess of 100 m/s (or significantly re-design the body to concentrate its mass at the bore), whilst steel and aluminum be used at much lower peripheral speeds.

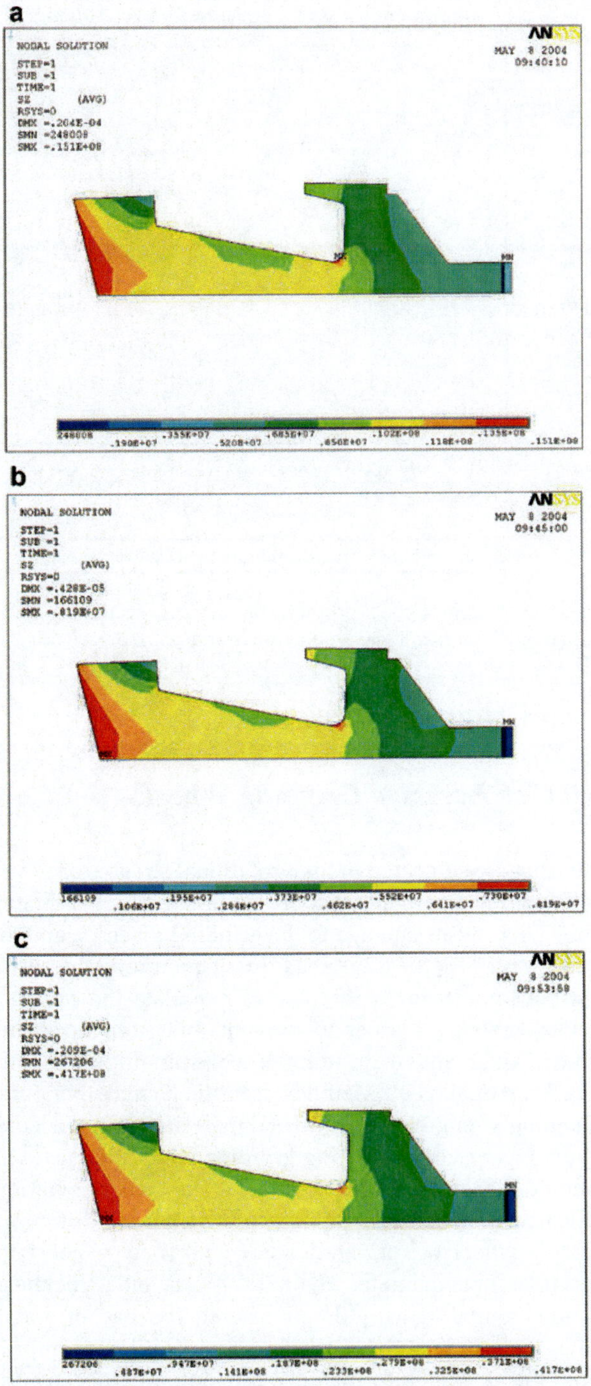

Fig. 3.12 Maximum stress (circumferential) in the electroplated cBN grinding wheel rotating at 200 m/s: (**a**) aluminum alloy center; (**b**) carbon fiber center; and (**c**) steel center. Stress scale is in N/mm. Used with permission from Inderscience Publishers (2010)

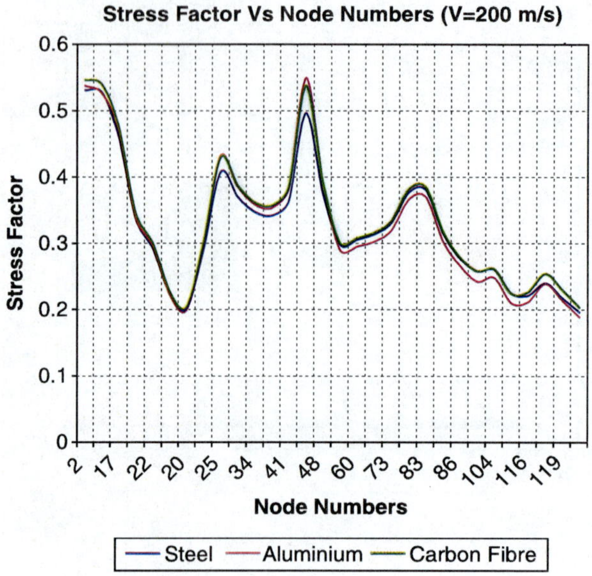

Fig. 3.13 Stress factor at various nodal positions on the cross-section of the electroplated cBN grinding wheel rotating at 200 m/s. Used with permission from Inderscience Publishers (2010)

3.8 Recessed Grinding Wheels

3.8.1 Small Cup Recessed Grinding Wheels

Grinding wheels that are subjected to high rotational stresses can eventually "burst" by breaking into smaller fragments that are dominated by bending moments at the onset of fracture. This causes damage to the grinding machine and to the operator if guards inadequately prevent the fragments from penetrating the case of the machine tool. It is of prime importance to be able to calculate the bursting speed of the grinding wheel accurately in order to prevent this from occurring. The energy contained within a fragment of grinding wheel spinning at a certain speed was calculated by Jackson et al. [10]. Methods used to calculate bursting speed that are based on the bending strength of the vitrified material are not accurate and deviate from the true bursting speeds of rotating grinding wheels. Deviations of up to 20% have been observed by Munnich [11]. Table 3.1 shows the deviations for a variety of grinding wheels with constant bonding formulation, but different abrasive mesh sizes. The grinding wheel was a parallel-sided vitrified wheel with an outer diameter, D, of 610 mm, inner diameter, H, of 304.8 mm, and a thickness, T, of 20 mm.

Thies [12] and Pompe et al. [13] considered the use of fracture mechanics coupled with the stochastic nature of vitrified compositions to improve the calculation of bursting speed, but unfortunately did not provide any improvement in accuracy. Mewes et al. [14] developed correction functions using linear regression

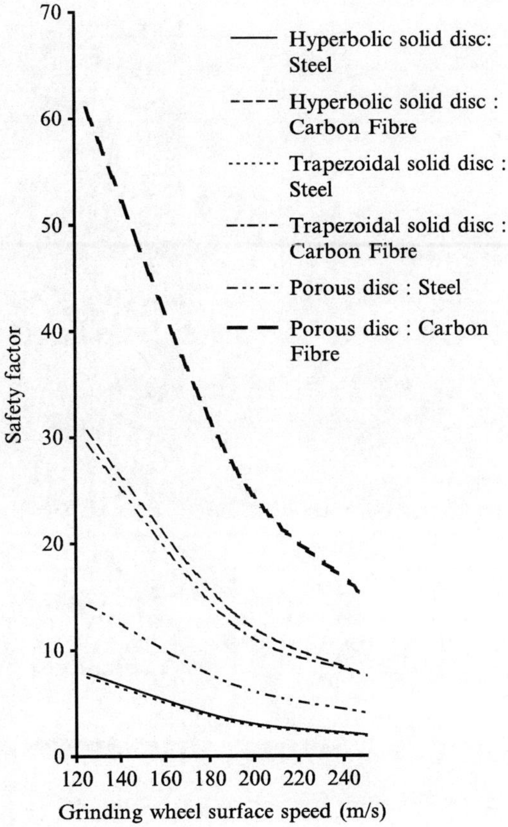

Fig. 3.14 Effect of material and geometric shape of the reinforcing center-section on the safety factor of vitrified, segmented grinding wheels. Used with permission from Inderscience Publishers (2010)

techniques, but they are not applicable because of non-uniform deviation of the calculated bursting speed compared to observed bursting speeds. In a recent paper by Behrens and Kammler [15], a procedure is described where bursting speed is calculated using linear elastic fracture mechanics coupled with the determination of rotational stresses using finite element calculations. In their approach, the critical pore size that triggers failure is calculated for a 36-mesh and an 80-mesh vitrified grinding wheel. The stress intensity at the flaw tip is calculated using the mode I stress intensity factor, and is used to calculate the fracture toughness of the abrasive body. The experimental fracture toughness is measured using a single edge v-notch beam loaded in the four-point bending mode to failure. The results are then used in concert with a set of modifying functions that are incorporated into Munnich's equation for calculating bursting speed. The modified equation is:

$$v_{bursting} = \sqrt{\frac{4\sigma_{fracture}}{3 + \nu + (1 - \nu).\frac{H^2}{D^2}.\rho.f_\sigma}} \quad (3.17)$$

Fig. 3.15 Slotted grinding wheels showing mechanism of lubricant penetration (courtesy of Dr. M. W. Bailey, De Beers Diamond Company)

where $v_{bursting}$ is the bursting speed, $\sigma_{fracture}$ is the fracture strength of the grinding wheel, ν is Poisson's ratio, H is diameter of the bore of the grinding wheel, D is the outer diameter of the grinding wheel, ρ is the density of the vitrified material, and f_σ is the stress intensity function at the maximum tangential stress at the outer edge of the bore as a function of the normalized diameter and the depth of the recess. The function f_σ is calculated using the following equation:

$$f_\sigma = -3.00102x^3 + 4.49064x^2 - 2.58489x + 0.640967y^2 - 2.05676y + 4.98248x^3y^3 - 8.56909x^2y^2 + 6.31512xy + 1.59627 \quad (3.18)$$

where x = P/D, y = F/T, P is the diameter of the recess, T is the thickness of the grinding wheel, F is the depth of the recess, x is the normalized recess diameter, and y is the normalized recess depth. Equations (3.17) and (3.19) are valid for the range of F/T between 0.13 and 0.88, and for the range of P/D between 0.36 and 0.96.

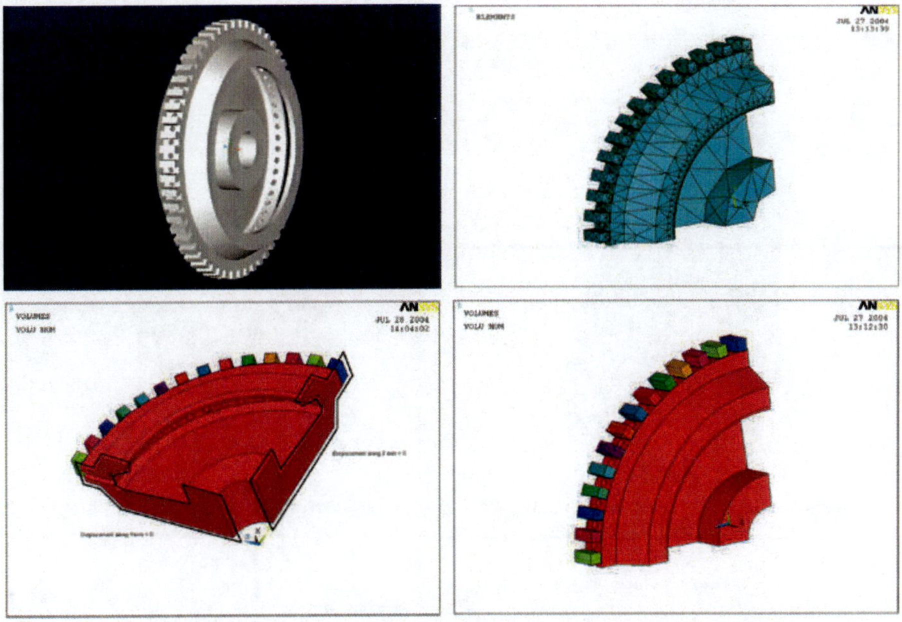

Fig. 3.16 Finite element model of a slotted grinding wheel showing boundary conditions and elements in various orientations. The images show the inclusion of coolant holes in the body of the wheel that are used to direct coolant into the slots that separate the abrasive segments. Used with permission from Inderscience Publishers (2010)

Equations (3.17) and (3.18) are used in the present work for calculating bursting speeds for a variety of different recessed grinding wheels and are compared with experimental bursting speed data.

3.8.1.1 Computational Analysis

Dynamic loads on a rotating grinding wheel is defined in the rotating co-ordinate system and while the stiffness and damping terms are the same as those measured in the stationary system, the terms due to inertial resistance are dependent upon the rotation of the grinding wheel. Owing to the porous nature of the grinding wheel body, the dynamic imbalance is magnified in certain parts of the grinding wheel when rotating at certain speeds. The inertial dependent terms need to be determined for a particular wheel structure and are then added to the total impedance of the structure before static and dynamic analyses in the rotating wheel can be performed.

The approach in this study assumes that the co-ordinate system used will rotate at a constant rate about a fixed axis. Displacements associated with the structure and forces applied to the structure are measured in the rotating system. The system is accelerating relative to a stationary inertial system and as such, the mass dependent impedance cannot be directly calculated in the rotating system. Therefore, the impedance is calculated for the stationary system then transformed to the rotating

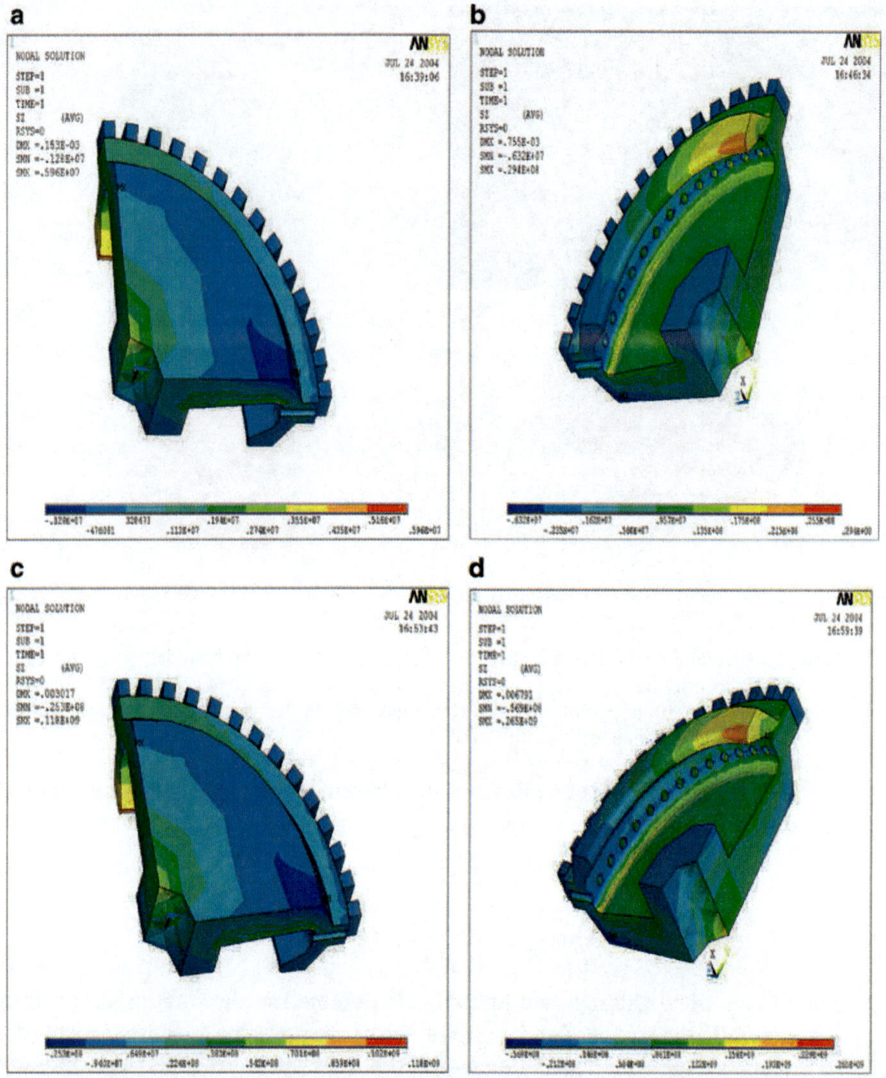

Fig. 3.17 Finite element hoop stress plots showing an aluminum body in various configurations at maximum peripheral operating speed of (**a**) 45, (**b**) 100, (**c**) 200, and (**d**) 300 m/s. Used with permission from Inderscience Publishers (2010)

system. The development of general transformations is required between stationary and rotating systems. This is accomplished by describing the general vector transformations between stationary and rotating systems, defining the inertial terms in the rotating co-ordinate system, applying these results to develop impedance in the rotating system, and finally, the gyroscopic terms are added to the structural matrices for analyzing the rotating system.

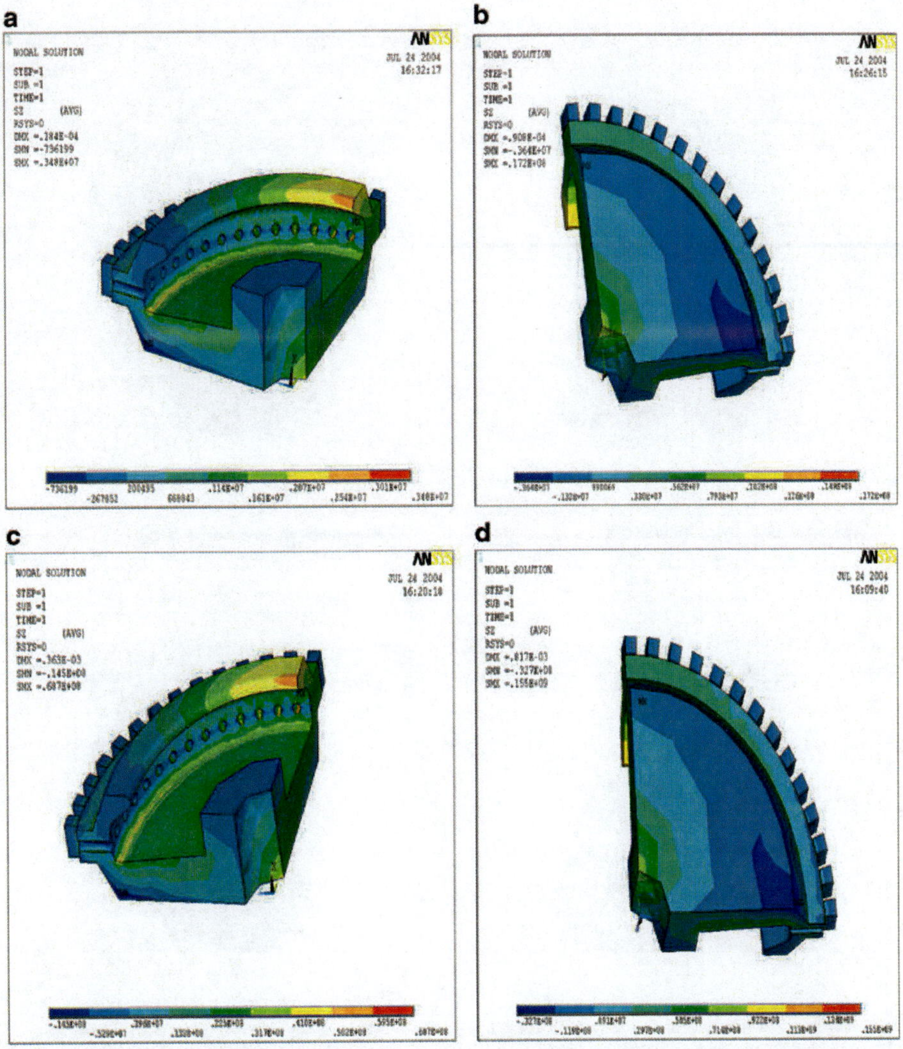

Fig. 3.18 Finite element hoop stress plots showing a carbon fiber body in various configurations at maximum peripheral operating speed of (**a**) 45, (**b**) 100, (**c**) 200, and (**d**) 300 m/s. Used with permission from Inderscience Publishers (2010)

The general transformation of a time-dependent vector from a stationary to a rotating co-ordinate is given by the following:

$$\{v(t)_r\} = [A(t)]\{v(t)_s\} \tag{3.19}$$

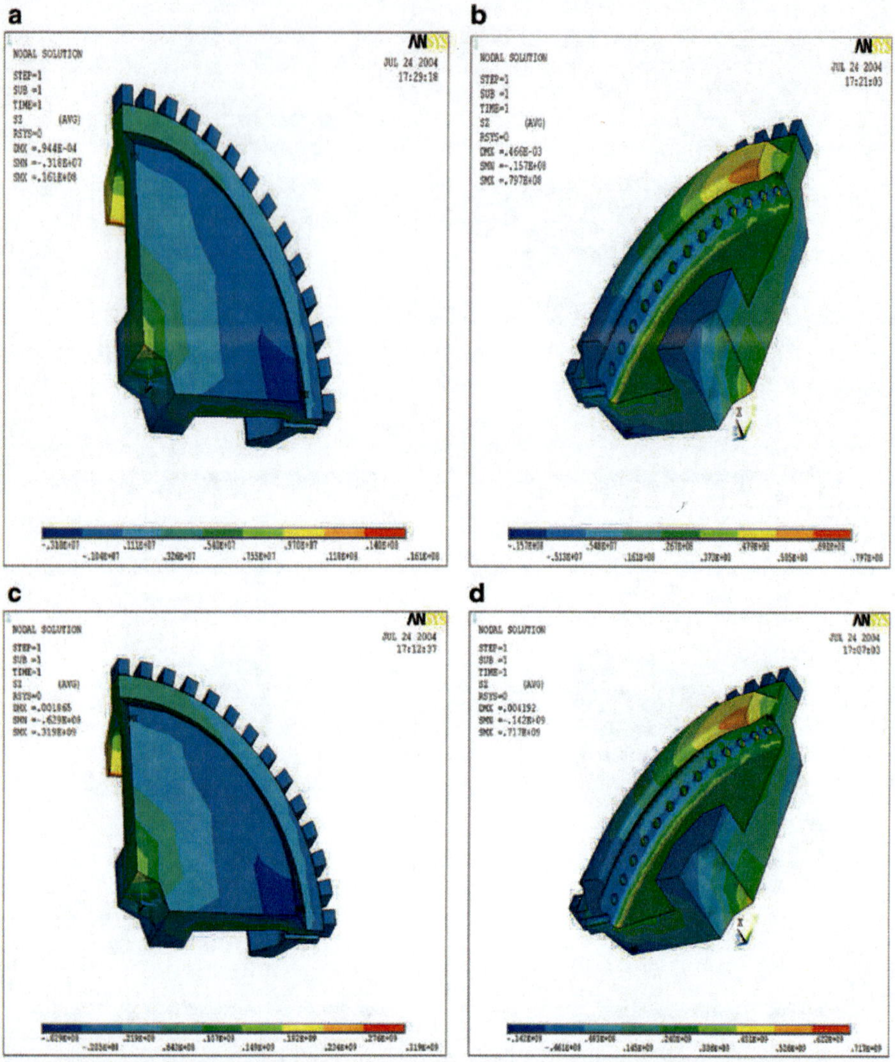

Fig. 3.19 Finite element hoop stress plots showing a steel body in various configurations at maximum peripheral operating speed of (**a**) 45, (**b**) 100, (**c**) 200, and (**d**) 300 m/s. Used with permission from Inderscience Publishers (2010)

where $\{v(t)_r\}$ is the rotating co-ordinate system of the time-dependent vector, $[A(t)]$ is the time-dependent transformation matrix from the rotating to the stationary system, and $\{v(t)_s\}$ is the stationary co-ordinate system of the time-dependent vector. The transformation is valid for any vector, both real and complex. Transformations using complex identities for small angular deviations allows the transformation vector to be described in the x-y-z co-ordinates:

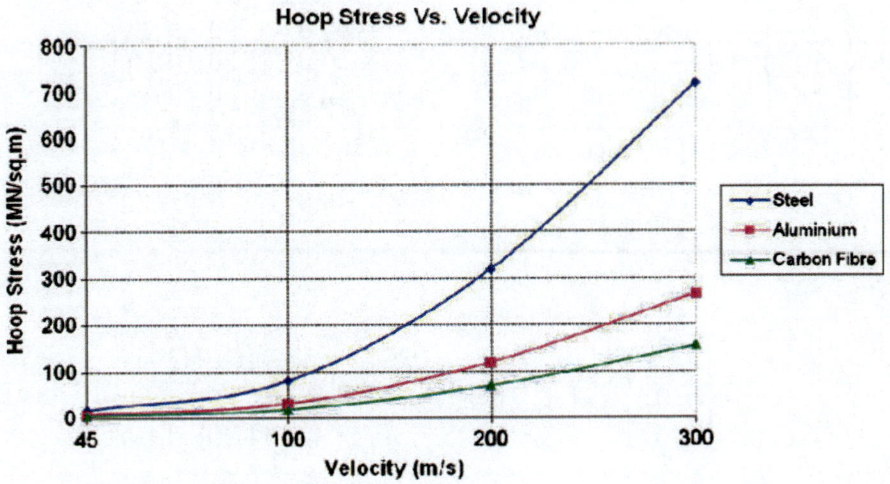

Fig. 3.20 Calculated hoop stress versus peripheral velocity for the various types of slotted grinding wheels. Used with permission from Inderscience Publishers (2010)

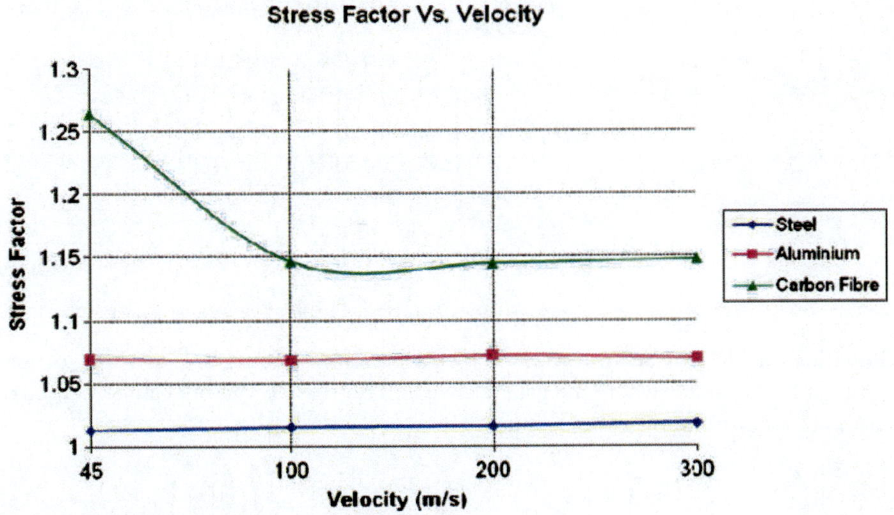

Fig. 3.21 Stress factor versus peripheral velocity for various types of slotted grinding wheels. Used with permission from Inderscience Publishers (2010)

Table 3.1 Calculated and measured bursting speeds of a parallel-sided vitrified grinding wheel

Abrasive mesh size	Calculated bursting speed (m/s) (Munnich [11])	Measured bursting speed (m/s)
36	140.7	119.6
54	156.5	133
70	166.9	142.8
100	183.8	153.9
150	196.9	164.6

$$\begin{Bmatrix} x(t)_r \\ y(t)_r \\ z(t)_r \end{Bmatrix} = \left(\frac{e^{i\theta}}{2} \begin{bmatrix} 1 & -i & 0 \\ i & 1 & 0 \\ 0 & 0 & 0 \end{bmatrix} + \frac{e^{-i\theta}}{2} \begin{bmatrix} 1 & i & 0 \\ -i & 1 & 0 \\ 0 & 0 & 0 \end{bmatrix} + \begin{bmatrix} 0 & 0 & 0 \\ 0 & 0 & 0 \\ 0 & 0 & 1 \end{bmatrix} \right) \times \begin{Bmatrix} x(t)_s \\ y(t)_s \\ z(t)_s \end{Bmatrix} \tag{3.20}$$

Or,

$$\begin{Bmatrix} x(t)_r \\ y(t)_r \\ z(t)_r \end{Bmatrix} = \left(\frac{e^{i\theta}}{2} [T_1] + \frac{e^{-i\theta}}{2} [T_1]^* + [T_o] \right) \begin{Bmatrix} x(t)_s \\ y(t)_s \\ z(t)_s \end{Bmatrix} \tag{3.21}$$

And the conjugate matrix is:

$$[T_1] = \frac{1}{2} \begin{bmatrix} 1 & -i & 0 \\ i & 1 & 0 \\ 0 & 0 & 0 \end{bmatrix} \tag{3.22}$$

$$[T_1]^* = \frac{1}{2} \begin{bmatrix} 1 & i & 0 \\ -i & 1 & 0 \\ 0 & 0 & 0 \end{bmatrix} \tag{3.23}$$

$$[T_o] = \begin{bmatrix} 0 & 0 & 0 \\ 0 & 0 & 0 \\ 0 & 0 & 1 \end{bmatrix} \tag{3.24}$$

Equation (3.21) can be re-formulated as a transformation from rotating to non-rotating co-ordinates where the progressive and regressive vectors can be written in terms of a non-rotating system, thus,

$$\begin{Bmatrix} x(t)_s \\ y(t)_s \\ z(t)_s \end{Bmatrix} = \left(\frac{e^{i\theta}}{2} [T_1]^* + \frac{e^{-i\theta}}{2} [T_1] + [T_o] \right) \begin{Bmatrix} x(t)_r \\ y(t)_r \\ z(t)_r \end{Bmatrix} \tag{3.25}$$

Inertial forces are calculated using Newton's laws of motion, whilst particles contained within the body are defined as position vectors. The inertial force on a particle whose position is measured relative to a rotating co-ordinate system is,

$$\{F(t)_{inertial}\}_r = -\left([M]\left(\frac{d^2\{\rho(t)_r\}}{dt^2}\right) - \Omega[B^c]\left(\frac{d\{\rho(t)_r\}}{dt}\right) - \Omega^2[K^c]\{\rho(t)_r\}\right) \tag{3.26}$$

where,

$$[M] = \begin{bmatrix} m & 0 & 0 \\ 0 & m & 0 \\ 0 & 0 & m \end{bmatrix} \tag{3.27}$$

$$[B^c] = \begin{bmatrix} 0 & 2m & 0 \\ -2m & 0 & 0 \\ 0 & 0 & 0 \end{bmatrix} \tag{3.28}$$

$$[K^c] = \begin{bmatrix} m & 0 & 0 \\ 0 & m & 0 \\ 0 & 0 & 0 \end{bmatrix} \tag{3.29}$$

The structural impedance of the rotating structure requires inertial force terms that are functions of the time varying position of mass of the particle. Therefore, the position vector can be written in terms of an initial component and a time-varying component. The equation of motion for a particle in a rotating elastic disk with no externally applied forces is given by,

$$\Omega^2[K^c]\{\rho_r\} = \left([M]\left(\frac{d^2\{u(t)_r\}}{dt^2}\right) + ([B] - \Omega[B^c])\left(\frac{d\{u(t)_r\}}{dt}\right) + ([K] - \Omega^2[K^c])\{u(t)_r\}\right) \tag{3.30}$$

Equation (3.30) shows that the centripetal loading is always present in the rotating structure. For accurate analysis of the rotating structure, a differential stiffness term is added,

$$[M]\left(\frac{d^2\{\Delta\rho(t)_r\}}{dt^2}\right) + ([B] - \Omega[B^c])\left(\frac{d\{\Delta\rho(t)_r\}}{dt}\right) + ([K] + \Omega^2[K^d] - \Omega^2[K^c]\{\Delta\rho(t)_r\}) = 0 \tag{3.31}$$

Equation (3.31) is used to determine the motion of a rotating structure about the deformation due to centripetal loading. For the case of a rotating recessed grinding wheel, finite element computations are modeled as rigid bodies with added gyroscopic effects at critical speeds of rotation. At selected speeds, dynamic out-of-balance forces are predicted using rotation-dependent terms. The finite element program used in the present work is MSC. Nastran™ and the procedure to add rotation dependent terms is done by calculating the terms then adding them to the structural damping and stiffness matrices, and by starting the analysis with the modified matrices.

3.8.1.2 Experimental Methods

The experimental part of the current work is focused on demonstrating the magnitude of stresses in recessed vitrified grinding wheels, calculating bursting speeds of the said wheels, and comparing those speeds with experimentally determined bursting speeds. The experimental section also focuses on measuring selected mechanical properties in order the calculate factors of safety for a variety of small cup grinding wheels.

Computational Stress Analysis

A solid model of a parallel-sided grinding wheel was created using the Solid Works™ software package. The dimensions of the wheels and associated solid models are shown in Fig. 3.22. The grinding wheel possessed an outer diameter of 508 mm and an inner diameter of 304.8 mm. The grinding wheel was composed of a mixture of chromium-doped alumina abrasive grain mixed with a glass-clay bonding agent. The grade of the wheel was RA36J8V, indicating a 36-mesh grain size, fairly open eight structure, J-grade vitrified structure. The model was imported into the MSC. Patran™ software program in the form of a .xmt file, or Parisolid Model. The model was then meshed and the global edge length to optimize the mesh density of the solid was determined by means of convergence. The convergence model is shown in Fig. 3.23. Material properties (Table 3.2) were assigned to the solid model of the wheel and constraints and boundary conditions were applied to the model. The center of the wheel was constrained in the translational x-, y-, and z-directions as well as in the rotational x- and z-directions. The y-rotational direction was left unconstrained because it is the axis in which the wheel rotates. An inertial velocity was then applied to the entire solid in order to simulate the rotation of the grinding wheel about the y-axis. Visual results, or fringe plots, showing where the maximum principal stress occurred and the quantity of that stress, was the final output of the analysis. The results are shown in Sect. 3.8.1.3 of the chapter.

Figure 3.24 shows the dimensions and the solid model of a recessed, small cup vitrified grinding wheel. The grinding wheel possessed an outer diameter of 250 mm and an inner diameter of 76 mm. The grinding wheel was composed of a mixture of chromium-doped alumina abrasive grain mixed with a glass-clay bonding agent. The grade of the wheel was RA36J8V, indicating a 36-mesh grain size, fairly open eight structure, J-grade vitrified structure. The recessed model was also imported into the MSC. Patran™ software program in the form of a .xmt file, or Parisolid Model. Again, material properties (Table 3.1) were assigned to the solid model of the wheel and constraints and boundary conditions were applied to the

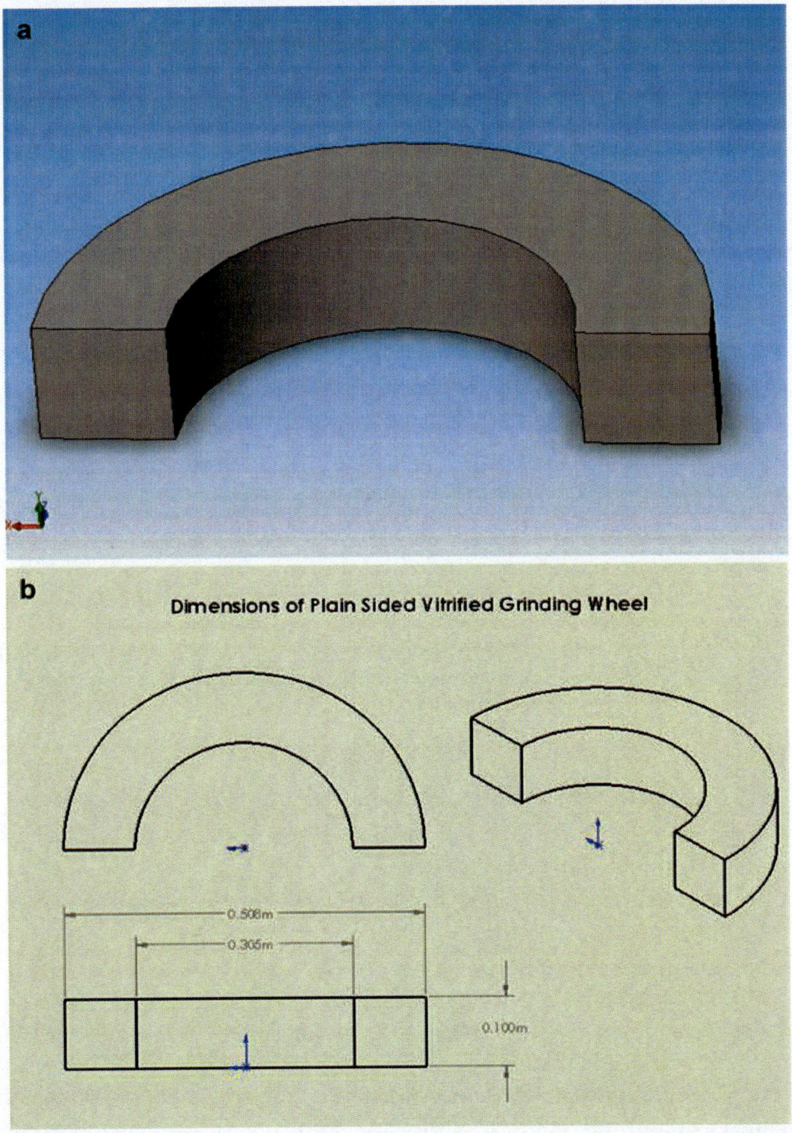

Fig. 3.22 (**a**) Solid Works™ model of the parallel-sided grinding wheel, (**b**) and dimensions associated with the parallel-sided grinding wheel. Used with permission from Inderscience Publishers (2010)

model. The center of the wheel was constrained in the translational x-, y-, and z-directions as well as in the rotational x- and z-directions.

The y-rotational direction was left unconstrained because it is the axis in which the wheel rotates. An inertial velocity was then applied to the entire solid in order to simulate the rotation of the grinding wheel about the y-axis. Visual results, or fringe

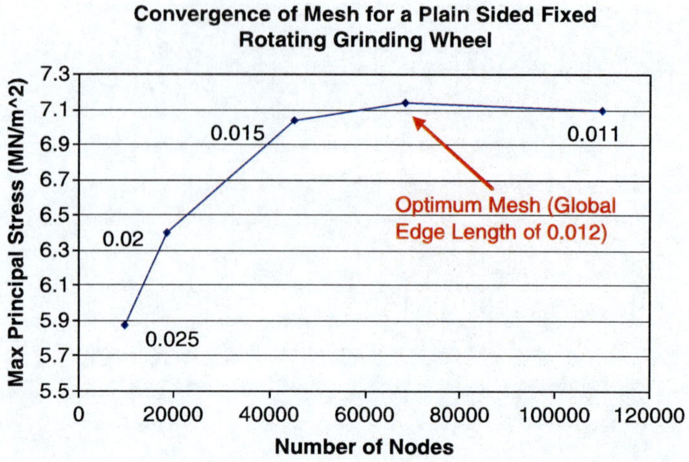

Fig. 3.23 Convergence of the solid model mesh showing an optimized global edge length of 0.012 mm. Used with permission from Inderscience Publishers (2010)

Table 3.2 Properties of the vitrified grinding wheel structure

Property	Value
Young's modulus of elasticity (N/m)	57.2×10^9
Poisson's ratio	0.2
Density (kg/m^3)	2,257

plots, showing where the maximum principal stress occurred and the quantity of that stress, was the final output of the analysis.

Determination of Bursting Speed

In order to measure the bursting speed of a grinding wheel, it is necessary to rotate the wheel at its maximum speed when failure occurs. In order to achieve this in a safe way, the grinding wheel is located in rotating frame that it buried in the ground surrounded by reinforced concrete baffles that absorbs the energy of impacting parts of the broken grinding wheel. The containment structure was lined with an energy absorbing foam, of which the top could be opened for access. The enclosure contained a roll grinding wheel head with the workpiece drive and in-feed mechanism removed. The wheel head was modified and fitted with a spindle cartridge designed for speeds up to 3,000 rpm. Once the grinding wheels were mounted and bolted on the spindle housing, the speed of the wheel was increased until the wheel fractured about the spindle, usually into three to four parts. The grinding wheel was inspected for failure and the failure speed was recorded. The grinding wheel was photographed and points of failure recorded for posterity.

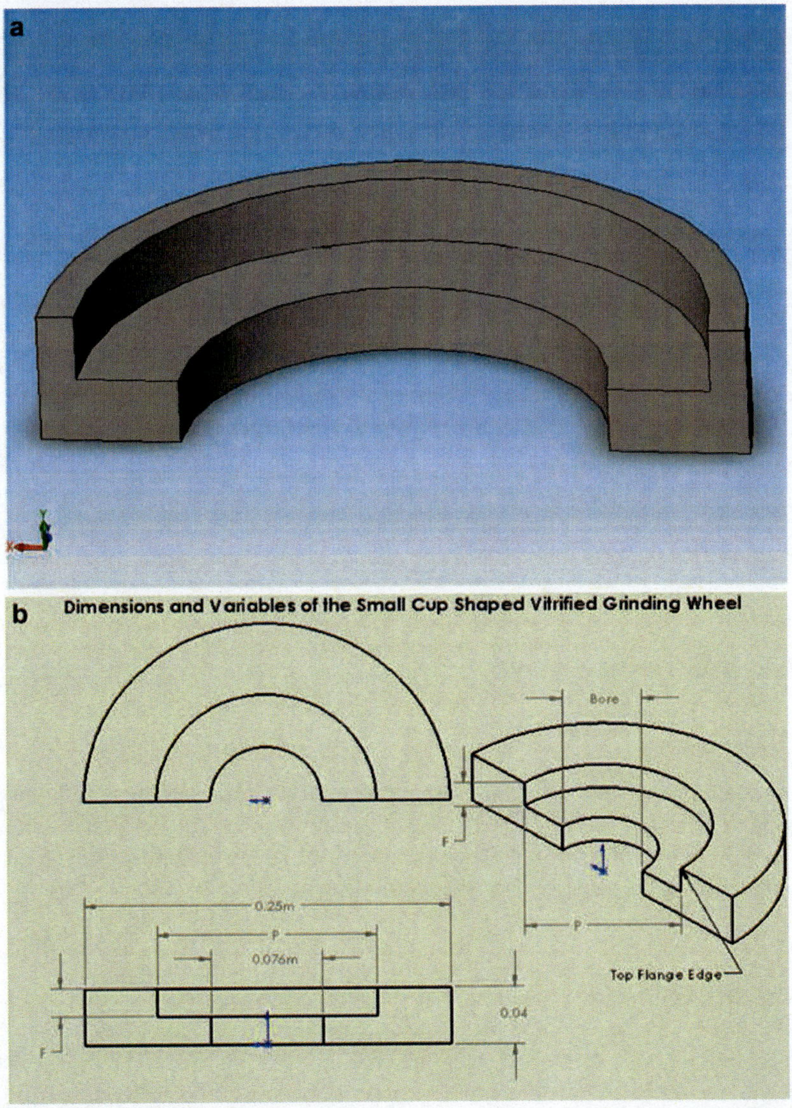

Fig. 3.24 (**a**) Solid Works™ model of the recessed, small cup grinding wheel, (**b**) and dimensions associated with the recessed, small cup grinding wheel. Used with permission from Inderscience Publishers (2010)

Determination of Mechanical Properties of Grinding Wheels

Two types of grinding wheels were used to determine the mechanical properties of the abrasive structures. The experimental determination of density and Poisson's ratio was demonstrated in a previous paper by the author [10], and involves using a pycnometer and a universal tensile testing machine. The grinding wheel structures were manufactured in the form of tensile test bars and parallel-sided and recessed grinding wheels. The grains of abrasive were mixed with a clay-glass bonding formulation, molded, and pressed to the correct density to provide the correct amount of porosity and structure. The samples were fired in an electric furnace to a temperature of approximately 1,080°C. The samples were cooled slowly until their appropriate hardness is reached. Hardness was measured by measuring the torque resistance of an indenting chisel that was rotated at a prescribed force. Once cooled, samples of grinding wheel were polished and mounted in resin moulds. Figure 3.25 shows the polished microstructure of a vitrified grinding wheel of known specification, RA36J8V, whilst Fig. 3.26 shows a microstructure with a known specification of RA80J8V. From the micrographs the critical pore size is found by measuring the distance between abrasive grains. The critical pore size allows one to calculate the fracture toughness of the abrasive structures. The measured fracture toughness is determined by stressing an abrasive structure to failure using notched test bars in the four-point bending mode I using a universal testing machine.

3.8.1.3 Experimental Results

Computational Stress Analysis

The computational stress analysis was performed on both parallel-sided and small cup grinding wheel structures and the results are shown in this section of the paper. Figure 3.27 shows the computed circumferential stress as a function of distance from the bore to the edge of the wheel spinning at 35 m/s.

Parallel-Sided Grinding Wheels

The maximum stress at the bore is approximately 3.75 MN/m^2. Figure 3.28 shows the finite element mesh and boundary conditions for the parallel-sided grinding wheel spinning at 45 m/s. The associated fringe plot showing the magnitude of circumferential stress is shown in Fig. 3.29. The indicated maximum stress is approximately 7.14 MN/m^2.

Figures 3.30–3.33 show the magnitudes of stress throughout a parallel-sided grinding wheel for a wheel rotating at 45, 60, 80, 100, and 120 m/s. The maximum circumferential stress occurs at the bore and the stresses are approximately, 6, 13, 19, 30, and 43 MN/m^2, respectively.

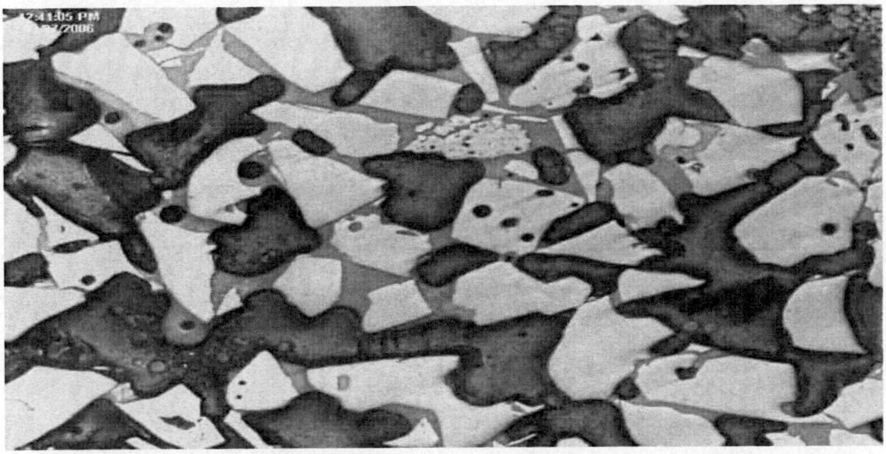

Fig. 3.25 Polished microstructure of the RA80J8V grinding wheel. *Light areas* indicate the abrasive grain, *grey areas* show the vitrified bonding system, and the *darker areas* show porosity. *Bubbles* within the pores are features associated with the resin that has infiltrated the *open pores*. *Closed pores* are shown with the abrasive grains and the bonding bridges. Average distance between grains is approximately 1.8 mm. Used with permission from Inderscience Publishers (2010)

Fig. 3.26 Polished microstructure of the RA36J8V grinding wheel. *Light areas* indicate the abrasive grain, *grey areas* show the vitrified bonding system, and the *darker areas* show porosity. *Bubbles* within the pores are features associated with the resin that has infiltrated the *open pores*. *Closed pores* are shown with the abrasive grains and the bonding bridges. Average distance between grains is approximately 0.7 mm. Used with permission from Inderscience Publishers (2010)

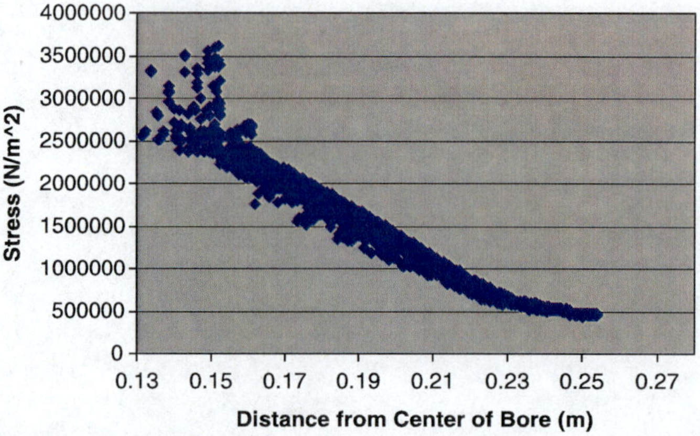

Fig. 3.27 Calculated circumferential stress distribution as a function of distance from the center of the bore of the grinding wheel spinning at 35 m/s. Used with permission from Inderscience Publishers (2010)

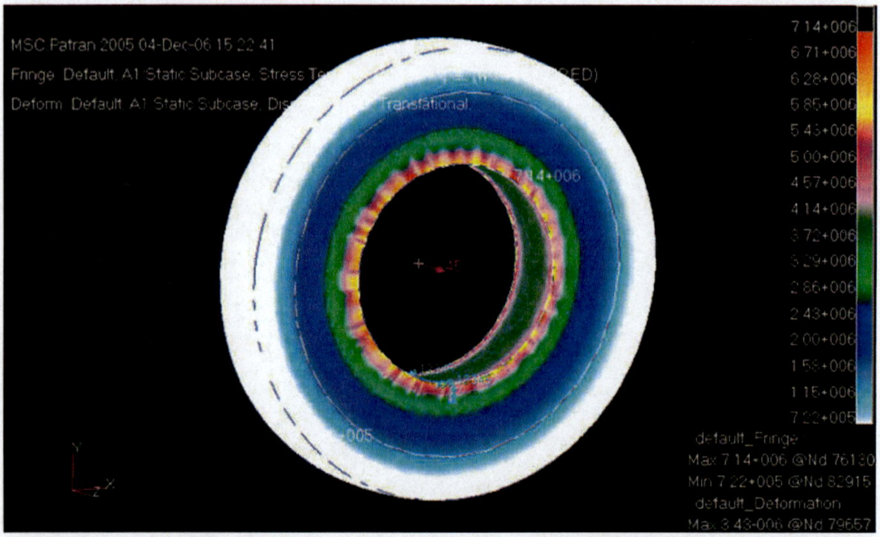

Fig. 3.28 Fringe plot of a plain sided grinding wheel spinning at 45 m/s. Used with permission from Inderscience Publishers (2010)

Small Cup Recessed Grinding Wheels

In contrast to parallel-sided grinding wheels, MSC Patran was used to calculate the maximum principal stress for a variety of small cup recessed grinding wheels. Figure 3.34 shows a fringe plot for a wheel spinning at 45 m/s ($P = 100$ and $F = 10$). The maximum stress occurs at the bore and is circumferential in nature.

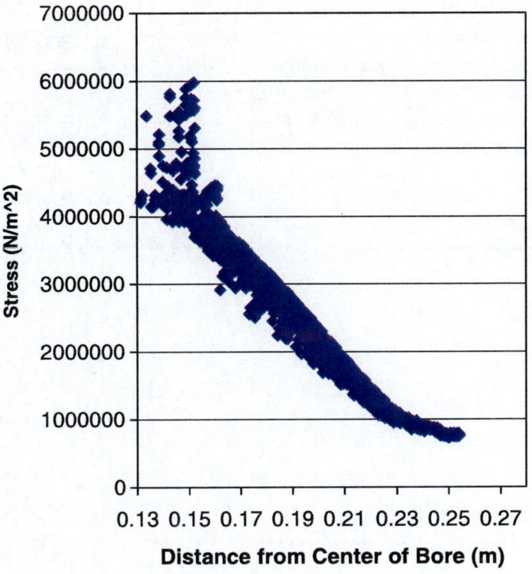

Fig. 3.29 Calculated circumferential stress distribution as a function of distance from the center of the bore of the grinding wheel spinning at 45 m/s. Used with permission from Inderscience Publishers (2010)

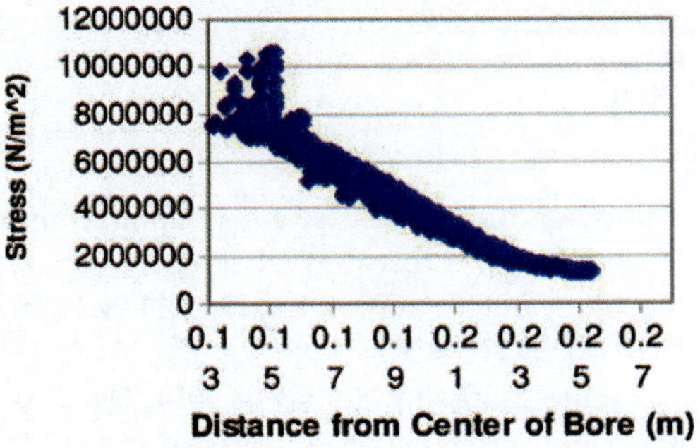

Fig. 3.30 Calculated circumferential stress distribution as a function of distance from the center of the bore of the grinding wheel spinning at 60 m/s. Used with permission from Inderscience Publishers (2010)

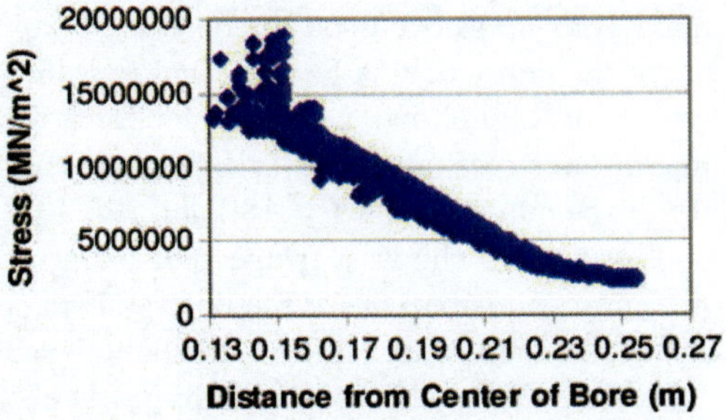

Fig. 3.31 Calculated circumferential stress distribution as a function of distance from the center of the bore of the grinding wheel spinning at 80 m/s. Used with permission from Inderscience Publishers (2010)

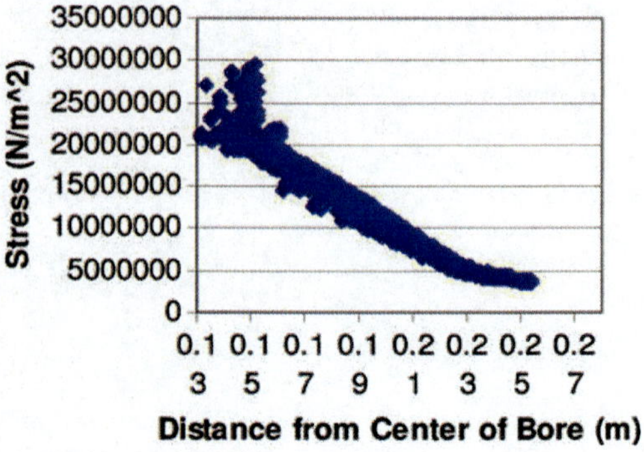

Fig. 3.32 Calculated circumferential stress distribution as a function of distance from the center of the bore of the grinding wheel spinning at 100 m/s. Used with permission from Inderscience Publishers (2010)

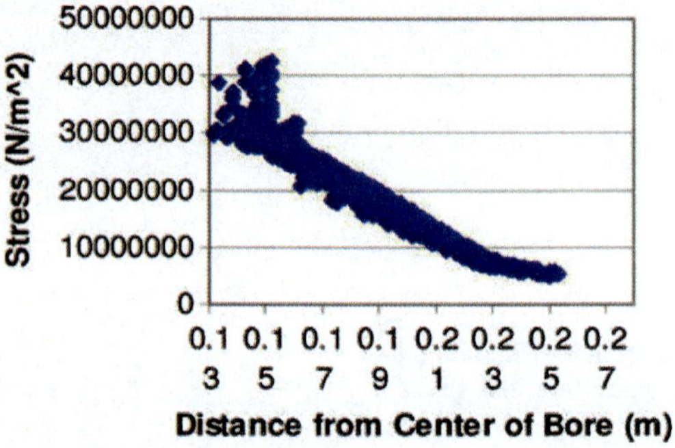

Fig. 3.33 Calculated circumferential stress distribution as a function of distance from the center of the bore of the grinding wheel spinning at 120 m/s. Used with permission from Inderscience Publishers (2010)

The magnitude of the stress is approximately 3.5 MN/m^2. Figure 3.35 shows a fringe plot for a recessed wheel spinning at 45 m/s. The maximum stress occurs at the bore and is approximately 4.4 MN/m^2. The geometry of the cup wheel is P = 100 and F = 20.

Figure 3.36 shows a fringe plot for a recessed wheel spinning at 45 m/s. The maximum stress occurs at the bore and is approximately 3.8 MN/m^2. The geometry of the cup wheel is P = 150 and F = 10.

Figure 3.37 shows a fringe plot for a recessed wheel spinning at 45 m/s. The maximum stress occurs at the bore and is approximately 3.69 MN/m^2. The geometry of the cup wheel is P = 150 and F = 20.

Figure 3.38 shows a fringe plot for a recessed wheel spinning at 45 m/s. The maximum stress occurs at the bore and is approximately 3.19 MN/m^2. The geometry of the cup wheel is P = 150 and F = 30.

Figure 3.39 shows a fringe plot for a recessed wheel spinning at 45 m/s. The maximum stress occurs at the bore and is approximately 3.54 MN/m^2. The geometry of the cup wheel is P = 180 and F = 20.

Figure 3.40 shows a fringe plot for a recessed wheel spinning at 45 m/s. The maximum stress occurs at the bore and is approximately 3.39 MN/m^2. The geometry of the cup wheel is P = 200 and F = 30. Once the rotational stresses were calculated, the small cup grinding wheels were rotated to failure. Figure 3.41

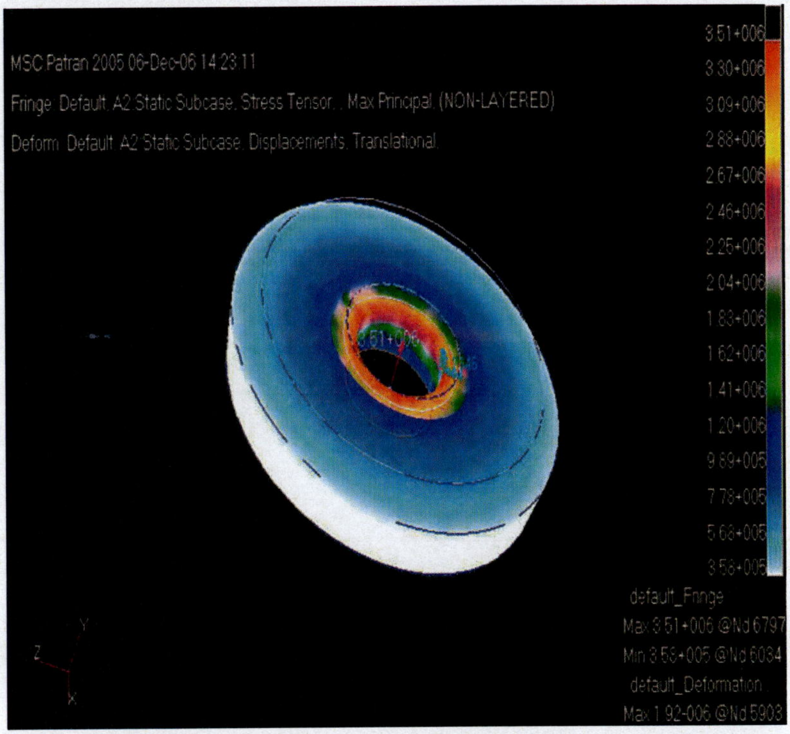

Fig. 3.34 Fringe plot of small cup shaped grinding wheel spinning at 45 m/s (P = 100, F = 10). Used with permission from Inderscience Publishers (2010)

shows a fractured small cup recessed grinding wheel. The speed at which failure occurred was recorded and compared to the analytical calculation. Factors of safety for each wheel were also recorded and the complete set of data is shown in Table 3.3. Micrographs of fractured surfaces are shown in Figs. 3.42–3.44 at various magnifications. The images show that failure occurred by grain fracture and by fracture at the bond bridge interface.

3.8.2 Large Cup Recessed Grinding Wheels

3.8.2.1 Computational Stress Analysis

The computational stress analysis was performed using a large cup grinding wheel structure and the results are shown in this section of the paper. MSC Patran was used to calculate the maximum principal stress for a variety of large cup recessed grinding wheels. Figure 3.45 shows a fringe plot for a wheel spinning at 45 m/s (P = 200 and F = 25). The maximum stress occurs at the bore and is

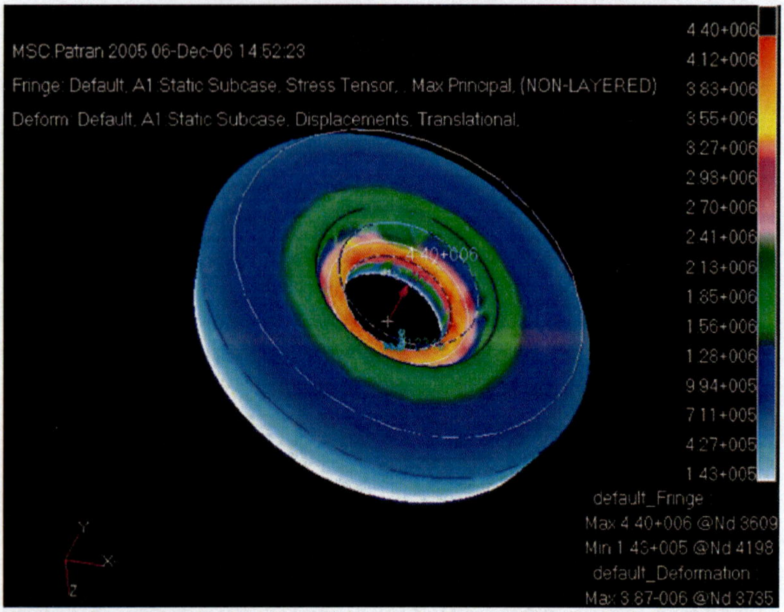

Fig. 3.35 Fringe plot of small cup shaped grinding wheel spinning at 45 m/s (P = 100, F = 20). Used with permission from Inderscience Publishers (2010)

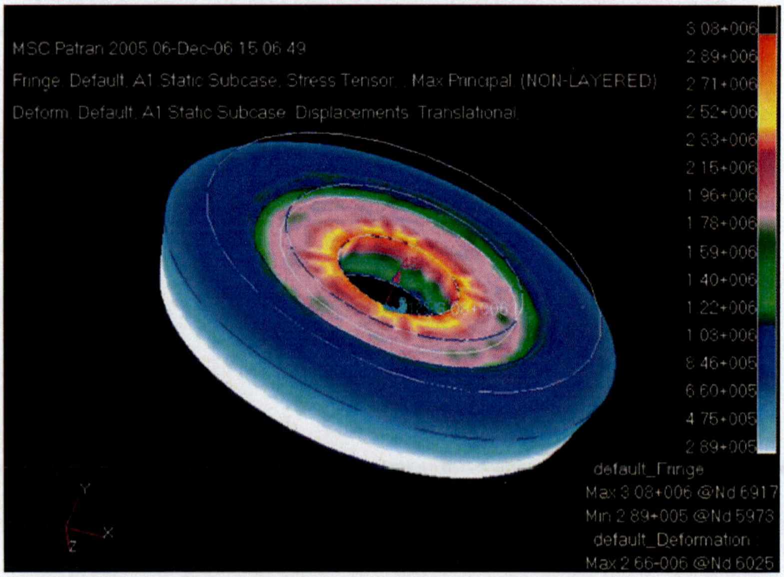

Fig. 3.36 Fringe plot of small cup shaped grinding wheel spinning at 45 m/s (P = 150, F = 10). Used with permission from Inderscience Publishers (2010)

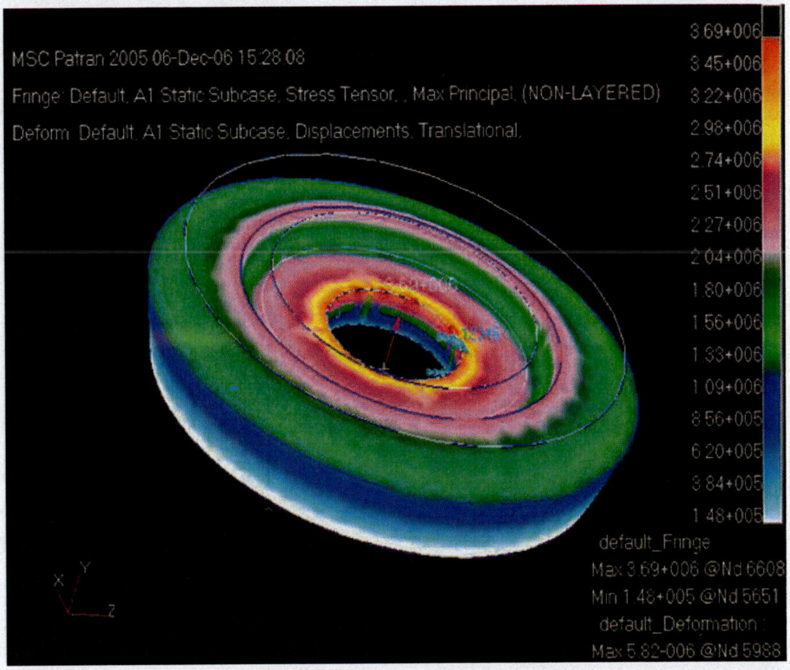

Fig. 3.37 Fringe plot of small cup shaped grinding wheel spinning at 45 m/s (P = 150, F = 20). Used with permission from Inderscience Publishers (2010)

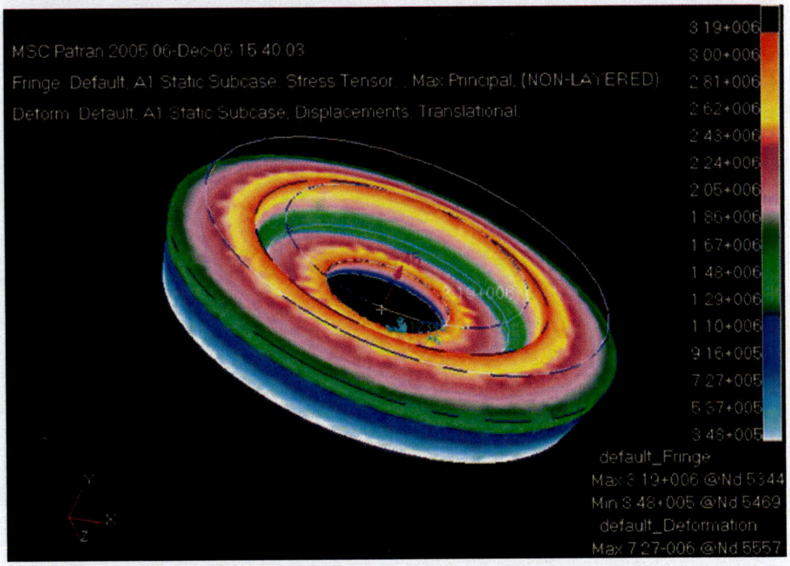

Fig. 3.38 Fringe plot of small cup shaped grinding wheel spinning at 45 m/s (P = 150, F = 30). Used with permission from Inderscience Publishers (2010)

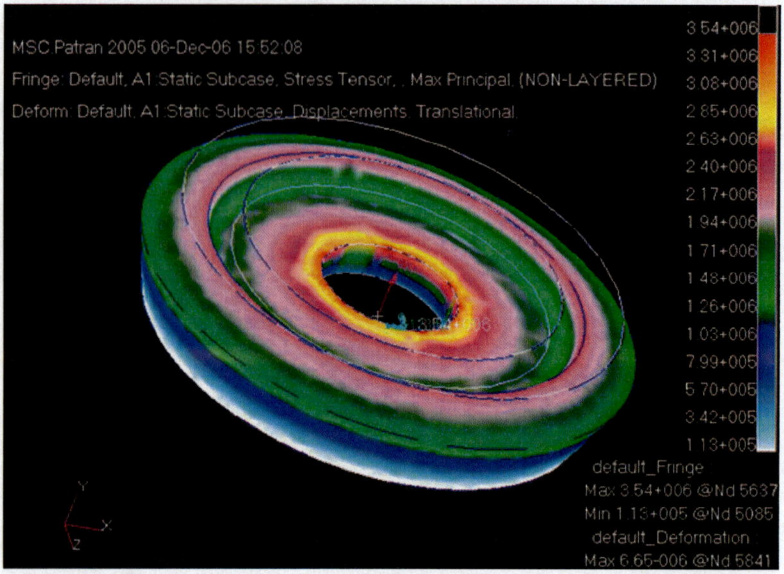

Fig. 3.39 Fringe plot of small cup shaped grinding wheel spinning at 45 m/s (P = 180, F = 20). Used with permission from Inderscience Publishers (2010)

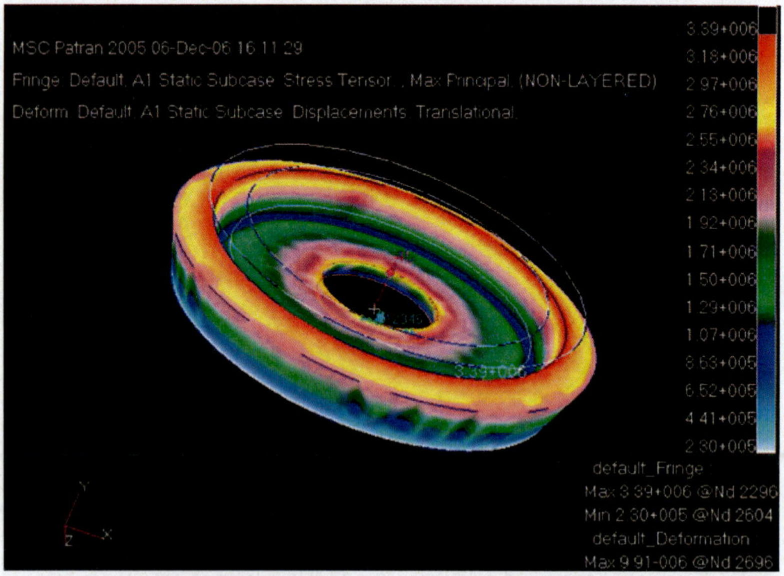

Fig. 3.40 Fringe plot of small cup shaped grinding wheel spinning at 45 m/s (P = 200, F = 30). Used with permission from Inderscience Publishers (2010)

Fig. 3.41 Exploded view of cup shaped vitrified grinding wheel After Recovery From the Burst Chamber. Used with permission from Inderscience Publishers (2010)

circumferential in nature. The magnitude of the stress is approximately 11.4p;MN/m^2. Figure 3.46 shows a fringe plot for a recessed wheel spinning at 45 m/s. The maximum stress occurs at the bore and is approximately 19.2 MN/m^2. The geometry of the cup wheel is P = 200 and F = 50.

Figure 3.47 shows a fringe plot for a recessed wheel spinning at 45 m/s. The maximum stress occurs at the bore and is approximately 11 MN/m^2. The geometry of the cup wheel is P = 250 and F = 25. Used with permission from Inderscience Publishers (2010).

Figure 3.48 shows a fringe plot for a recessed wheel spinning at 45 m/s. The maximum stress occurs at the bore and is approximately 15.6 MN/m^2. The geometry of the cup wheel is P = 250 and F = 50.

Figure 3.49 shows a fringe plot for a recessed wheel spinning at 45 m/s. The maximum stress occurs at the bore and is approximately 15.1 MN/m^2. The geometry of the cup wheel is P = 250 and F = 75.

Figure 3.50 shows a fringe plot for a recessed wheel spinning at 45 m/s. The maximum stress occurs at the bore and is approximately 13.8 MN/m^2. The geometry of the cup wheel is P = 275 and F = 50. Used with permission from Inderscience Publishers (2010).

Figure 3.51 shows a fringe plot for a recessed wheel spinning at 45 m/s. The maximum stress occurs at the bore and is approximately 13.9 MN/m^2. The geometry of the cup wheel is P = 325 and F = 75. Once the rotational stresses were calculated, the small cup grinding wheels were rotated to failure. The speed at which failure occurred was recorded and compared to the analytical calculation.

Table 3.3 Mechanical property data concerning the operating conditions for various grinding wheels with varying geometry and rotational speeds

Cup geometry (mm)	Maximum circumferential stress (MN/m^2)	Factor of safety	x	y	f_σ	Calculated bursting speed (m/s)	Experimental bursting speed (m/s)
Rotational speed = 45 m/s							
P = 100, F = 10	3.51	16	0.4	0.25	1.16	480	456
P = 100, F = 20	4.4	12.7	0.4	0.5	1.18	476	461
P = 150, F = 10	3.08	18.2	0.6	0.25	1.3	452	460
P = 150, F = 20	3.69	15.2	0.6	0.5	1.4	437	450
P = 150, F = 30	3.19	17.5	0.6	0.75	1.39	439	452
P = 180, F = 20	3.54	15.8	0.72	0.5	1.47	427	432
P = 200, F = 30	3.39	16.5	0.8	0.75	1.46	428	430
Rotational speed = 60 m/s							
P = 100, F = 10	6.25	9	0.4	0.25	1.16	480	456
P = 100, F = 20	7.83	7.1	0.4	0.5	1.18	476	461
P = 150, F = 10	5.47	10.24	0.6	0.25	1.3	452	460
P = 150, F = 20	6.56	8.54	0.6	0.5	1.4	437	450
P = 150, F = 30	5.66	9.9	0.6	0.75	1.39	439	452
P = 180, F = 20	6.29	8.9	0.72	0.5	1.47	427	432
P = 200, F = 30	6.03	9.3	0.8	0.75	1.46	428	430
Rotational speed = 90 m/s							
P = 100, F = 10	14.1	4	0.4	0.25	1.16	480	456
P = 100, F = 20	17.6	3.2	0.4	0.5	1.18	476	461
P = 150, F = 10	12.3	4.5	0.6	0.25	1.3	452	460
P = 150, F = 20	14.8	3.8	0.6	0.5	1.4	437	450
P = 150, F = 30	12.7	4.4	0.6	0.75	1.39	439	452
P = 180, F = 20	14.2	4	0.72	0.5	1.47	427	432
P = 200, F = 30	13.6	4.1	0.8	0.75	1.46	428	430
Rotational speed = 125 m/s							
P = 100, F = 10	27.1	2	0.4	0.25	1.16	480	456

(continued)

Table 3.3 (continued)

Cup geometry (mm)	Maximum circumferential stress (MN/m^2)	Factor of safety	x	y	f_σ	Calculated bursting speed (m/s)	Experimental bursting speed (m/s)
P = 100, F = 20	34	1.6	0.4	0.5	1.18	476	461
P = 150, F = 10	23.7	2.4	0.6	0.25	1.3	452	460
P = 150, F = 20	28.5	2	0.6	0.5	1.4	437	450
P = 150, F = 30	24.6	2.3	0.6	0.75	1.39	439	452
P = 180, F = 20	27.3	2	0.72	0.5	1.47	427	432
P = 200, F = 30	26.2	2.15	0.8	0.75	1.46	428	430

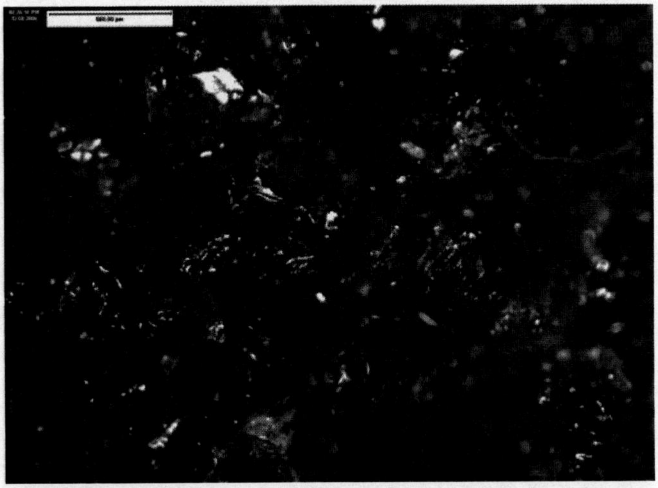

Fig. 3.42 Micrograph of vitrified grinding wheel magnified at ×5 magnification. Used with permission from Inderscience Publishers (2010)

Factors of safety for each wheel were also recorded and the complete set of data is shown in Table 3.4.

For recessed grinding wheels, principal stresses were computed using a finite element program were rotation dependent terms were modified by adding structural damping and stiffness matrices. It was discovered that the maximum principal stress occurs at the bore of the small cup-shaped recessed grinding wheels and

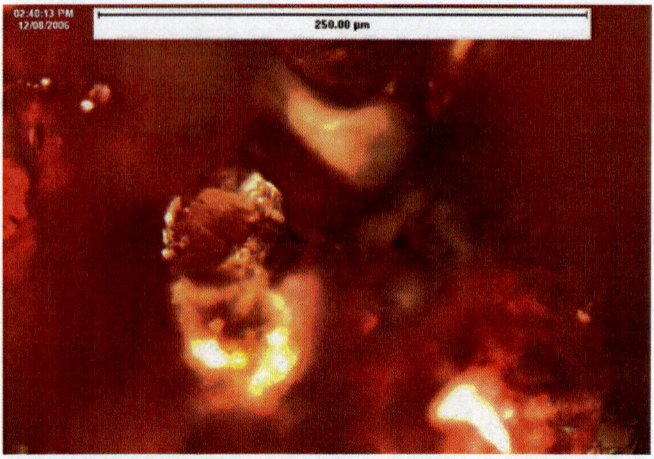

Fig. 3.43 Micrograph of cup shaped vitrified grinding wheel magnified at ×10 magnification. Failure has occurred by failure at the bond bridge interface. The area that is focus shows the end of a bond bridge that was previously connected to an abrasive grain. Used with permission from Inderscience Publishers (2010)

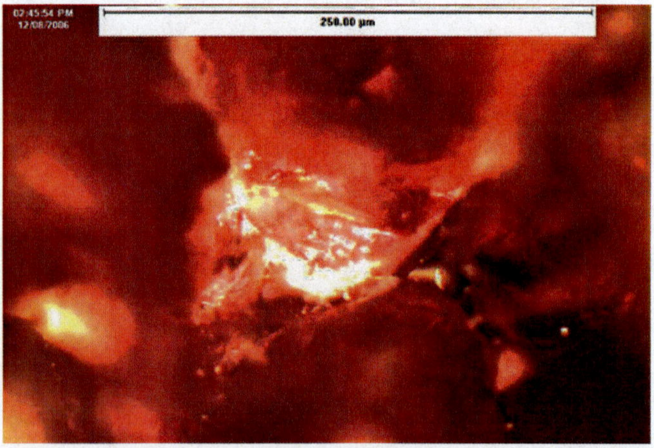

Fig. 3.44 Micrograph of cup shaped vitrified grinding wheel magnified at ×10 magnification. The micrograph shows a partially fractured abrasive grain. Used with permission from Inderscience Publishers (2010)

increases with rotational speed. Tables 3.3 and 3.4 show the magnitude of those stresses for various geometry and peripheral operating speed. Convergence of the finite element results occurred with a mesh that possesses a global edge length of 0.012 mm with approximately 68,000 nodes for the recessed cup shape. The range

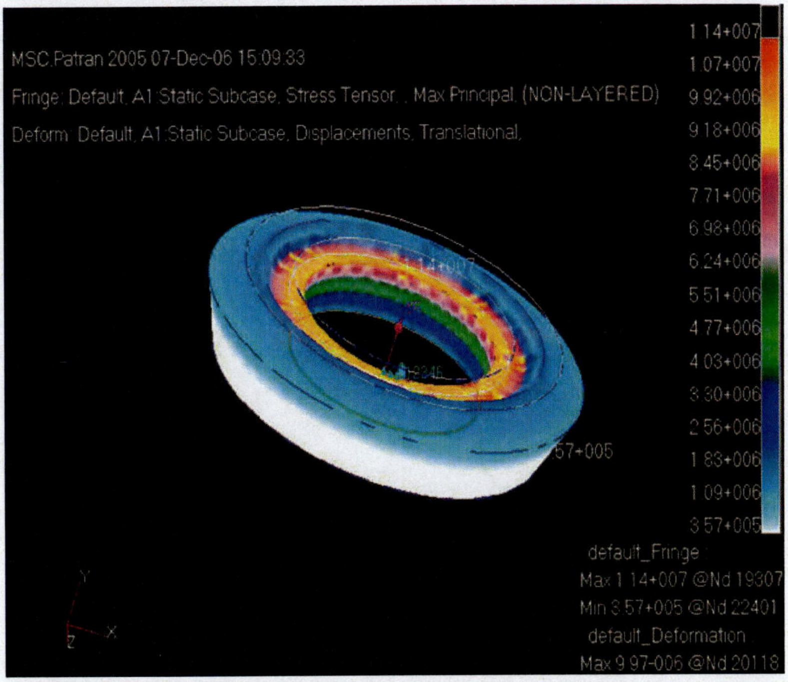

Fig. 3.45 Fringe plot of large cup shaped grinding wheel spinning at 45 m/s (P = 200, F = 25). Used with permission from Inderscience Publishers (2010)

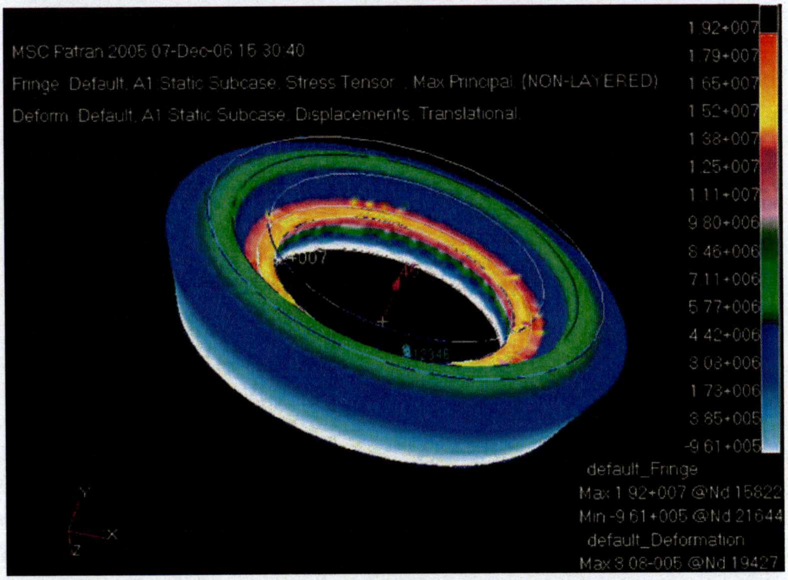

Fig. 3.46 Fringe plot of large cup shaped grinding wheel spinning at 45 m/s (P = 200, F = 50). Used with permission from Inderscience Publishers (2010)

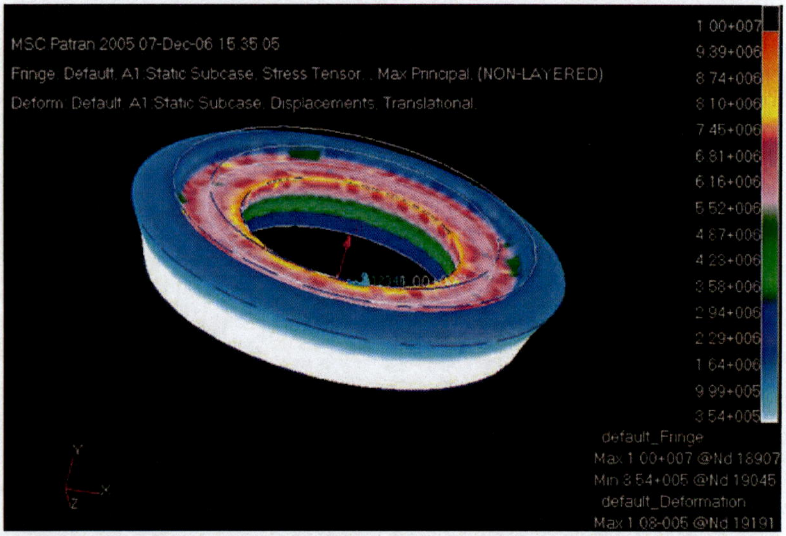

Fig. 3.47 Fringe plot of large cup shaped grinding wheel spinning at 45 m/s (P = 250, F = 25). Used with permission from Inderscience Publishers (2010)

of stresses increased from 3.51 MN/m^2 to a maximum of 34 MN/m^2 as a function of rotational speed. The measured tensile strength of the abrasive grain material used was approximately 56 MN/m^2. This yielded safety factors in the range between 1.6 and 18. The calculated bursting speeds are also shown in Table 3.3 and 3.4 and are compared with experimentally determined bursting speed. The calculated bursting speeds are within ±5% of the measured bursting speeds and show remarkable accuracy when using the method developed by Behrens and Kammler [15]. Their method takes account of stress concentrations that occur at the recesses of cup-shaped grinding wheels. Polished micrographs of the structure of cup wheels also allowed the author to measure the critical pore size in order to measure the fracture toughness of the cup wheels. The measured fracture toughness for the 80-grain size cup wheel was 2.14 MPa m, and 1.8 MPa m for a 36-grain size cup wheel. The standard deviation was 10 and 12%, respectively.

When considering conventional grinding wheels, a continuous layer of abrasive adhered to the periphery of the hub material generates a maximum stress due to centripetal loading that acts in the circumferential direction and is tensile in nature. The circumferential stress is highest in all components of the grinding wheel, i.e. abrasive layer, adhesive bonding layer, and the hub material. This magnitude of stress is used to ascribe the maximum operating speed to a particular grinding wheel. A clamping arrangement is used that prevents radial displacement at the bore so that grinding wheels operate within their recommended safe operating limits. The results obtained from finite element analyses of segmented grinding wheels suggest that the maximum rotational stress in the grinding wheel is in the radial direction and is tensile in nature. The change in direction is caused by the

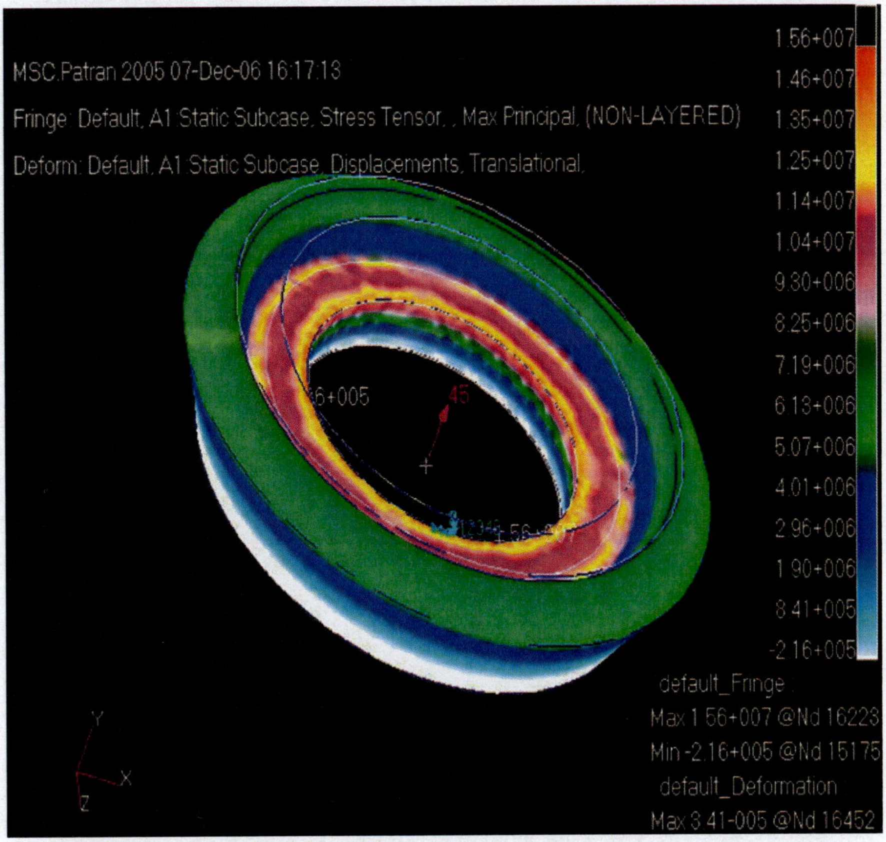

Fig. 3.48 Fringe plot of large cup shaped grinding wheel spinning at 45 m/s (P = 250, F = 50). Used with permission from Inderscience Publishers (2010)

radial movement of the abrasive segments without significant movement in the circumferential direction, thus reducing the maximum circumferential stress in the abrasive segment and in the adhesive bonding layer. This means that segmented vitrified and electroplated single-layer grinding wheels can be used at much higher operating speeds than conventional and reinforced grinding wheels.

3.9 Conclusions

The levels of stress in a cylindrical grinding wheel due to centripetal loading can be reduced by using a reinforced center made from a material that is rigid, lighter, and is stronger than the abrasive material. Replacing the continuous layer of abrasive with a layer of discontinuous segments can reduce the levels of stress due to centripetal loading further. The depth and number of segments has a significant influence on the

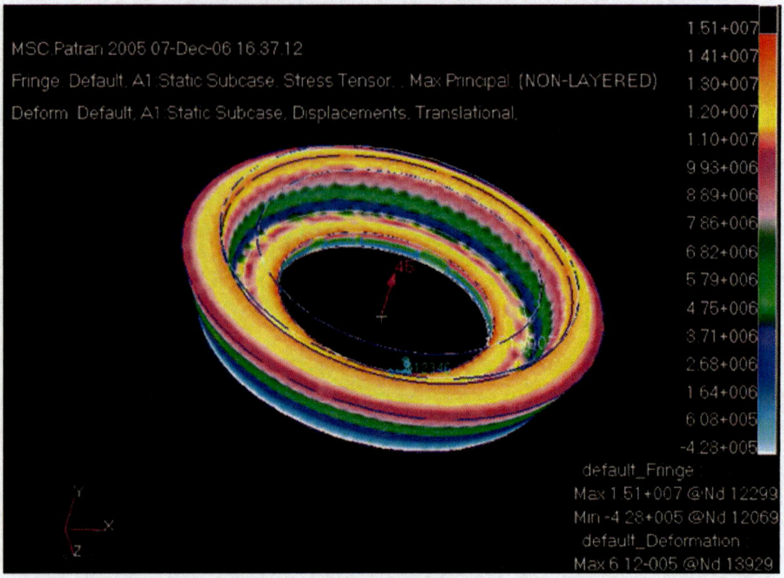

Fig. 3.49 Fringe plot of large cup shaped grinding wheel spinning at 45 m/s (P = 250, F = 75). Used with permission from Inderscience Publishers (2010)

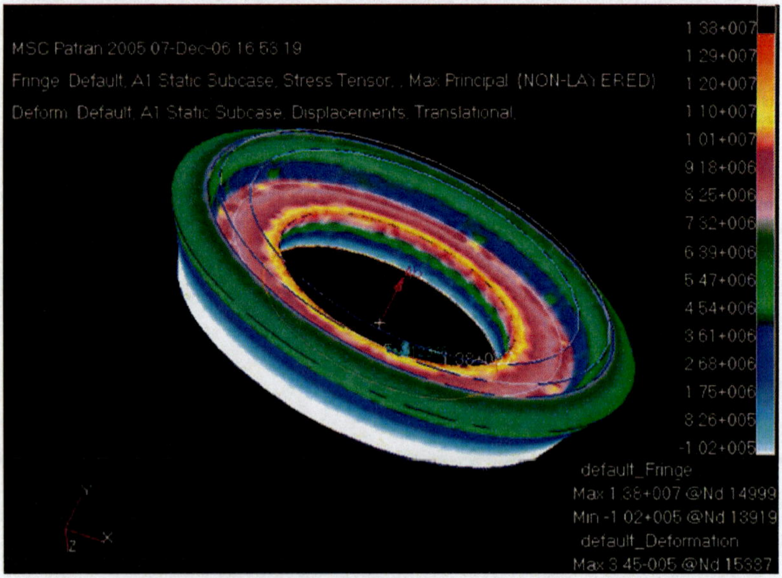

Fig. 3.50 Fringe plot of large cup shaped grinding wheel spinning at 45 m/s (P = 275, F = 50). Used with permission from Inderscience Publishers (2010)

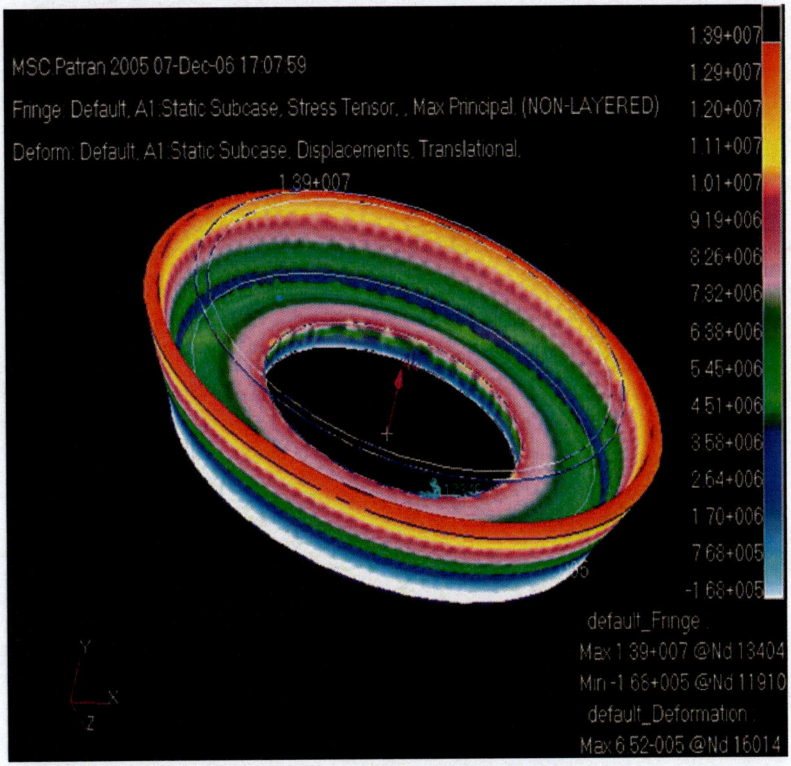

Fig. 3.51 Fringe plot of large cup shaped grinding wheel spinning at 45 m/s (P = 325, F = 75). Used with permission from Inderscience Publishers (2010)

levels of stress in a rotating grinding wheel. As a general rule, increasing the number of segments and reducing the depth of the abrasive segments can reduce stress levels. The maximum stress in an abrasive segment due to centripetal loading is in the radial direction. Stress factors in high-speed segmented grinding wheels can be reduced further by reducing the mass of the wheel by redesigning the center section of the wheel using the new "porous" design. Stresses can be reduced even further when using single-layer abrasive grinding wheels. It appears that carbon fiber is suited to high speed grinding operations (in excess of 100 m/s), whilst steel and aluminum bodies can be used at much lower peripheral speeds. For small recessed grinding wheels, the following conclusions are applicable:

1. For both parallel-sided wheels and recessed cup wheels, the maximum principal stress acted circumferentially at the bore of the grinding wheel;
2. For the recessed cup-shaped grinding wheels, the factor of safety varied from 1.6 to 18 depending on their rotational speed;

Table 3.4 Mechanical property data concerning the operating conditions for various grinding wheels with varying geometry and rotational speeds

Cup geometry (mm)	Maximum circumferential stress (MN/m^2)	Factor of safety	x	y	f_σ	Calculated bursting speed (m/s)	Experimental bursting speed (m/s)
Rotational speed = 45 m/s							
P = 200, F = 25	9.92	5.65	0.39	0.25	1.16	243	233
P = 200, F = 50	10.2	5.5	0.39	0.5	1.17	242	238
P = 250, F = 25	8.75	6.4	0.49	0.25	1.24	235	241
P = 250, F = 50	13.5	4.1	0.49	0.5	1.29	231	225
P = 250, F = 75	14.1	4	0.49	0.75	1.29	231	225
P = 275, F = 50	12	4.7	0.54	0.5	1.35	226	233
P = 325, F = 75	13.9	4	0.64	0.75	1.42	220	212
Rotational speed = 60 m/s							
P = 200, F = 25	17.6	3.18	0.39	0.25	1.16	243	233
P = 200, F = 50	17.5	3.2	0.39	0.5	1.18	242	238
P = 250, F = 25	17.8	3.15	0.49	0.25	1.3	235	241
P = 250, F = 50	24.1	2.3	0.49	0.5	1.4	231	225
P = 250, F = 75	23.2	2.4	0.49	0.75	1.39	231	225
P = 275, F = 50	21.3	2.6	0.54	0.5	1.47	226	233
P = 325, F = 75	24.7	2.3	0.64	0.75	1.46	220	212
Rotational speed = 90 m/s							
P = 200, F = 25	39.7	1.41	0.39	0.25	1.16	243	233
P = 200, F = 50	40.1	1.4	0.39	0.5	1.18	242	238
P = 250, F = 25	35	1.6	0.49	0.25	1.3	235	241
P = 250, F = 50	54.1	1.04	0.49	0.5	1.4	231	225
P = 250, F = 75	52.1	1.07	0.49	0.75	1.39	231	225
P = 275, F = 50	47.8	1.17	0.54	0.5	1.47	226	233
P = 325, F = 75	55.5	1.01	0.64	0.75	1.46	220	212

(continued)

Table 3.4 (continued)

Cup geometry (mm)	Maximum circumferential stress (MN/m^2)	Factor of safety	x	y	f_σ	Calculated bursting speed (m/s)	Experimental bursting speed (m/s)
Rotational speed = 125 m/s							
P = 200, F = 25	27.1	2	0.4	0.25	1.16	243	233
P = 200, F = 50	34	1.6	0.4	0.5	1.18	242	238
P = 250, F = 25	23.7	2.4	0.6	0.25	1.3	235	241
P = 250, F = 50	28.5	2	0.6	0.5	1.4	231	225
P = 250, F = 75	24.6	2.3	0.6	0.75	1.39	231	225
P = 275, F = 50	27.3	2	0.72	0.5	1.47	226	233
P = 325, F = 75	26.2	2.15	0.8	0.75	1.46	220	212

3. The measured fracture toughness for the 80-grain size cup wheel was 2.14 M Pa m, and 1.8 M Pa m for a 36-grain size cup wheel. The standard deviation was 10 and 12%, respectively.

For large recessed cup wheels, the maximum principal stress acted circumferentially at the bore of the grinding wheel and ranged from 8.75 to 55.5 MN/m^2 depending upon the operating speed. The following conclusions can be applied to large recessed cup wheels:

1. For large recessed cup-shaped grinding wheels, the factor of safety varied from 1.01 to 6.4 depending on their rotational speed;
2. The measured fracture toughness for the 80-grain size cup wheel was 2.14 M Pa m, and 1.8 M Pa m for a 36-grain size cup wheel. The standard deviation was 10 and 12%, respectively. And,
3. The method developed by Behrens and Kammler [15] proved to yield accurate values of bursting speed when compared to experimentally determined bursting speeds.

Acknowledgements The author acknowledges support from Vitrified Technologies, Inc., and thanks Ameya Deshpande for contributions made to the finite element analysis of the electroplated grinding wheel design. The author wishes to thank Dr. Malcolm Bailey, Alan Jones, Roland Wakefield and Peter Derbyshire for providing grinding wheel material and data. The author wishes to thank Inderscience publishers for allowing the author to reproduce the author's material from papers published in the *International Journal of Computational Materials Science and Surface Engineering* and the *International Journal of Manufacturing Technology and Management.*

References

1. M. J. Jackson, C. J. Davis, M. P. Hitchiner, and B. Mills, *High-Speed Grinding with c.B.N. Grinding Wheels – Applications and Future Developments*, Journal of Materials Processing Technology (2001), **110**, 78–88.
2. M. J. Jackson and N. Barlow, *Computer Aided Design of High-Performance Grinding Tools*, Proceedings of the Institution of Mechanical Engineers (London), Part B – Journal of Engineering Manufacture (2001), **215**, 583–588.
3. M. J. Jackson, *Fracture Dominated Wear of Sharp Abrasive Grains and Grinding Wheels*, Proceedings of the Institution of Mechanical Engineers (London), Part J – Journal of Engineering Tribology (2004), **218**, 225–235.
4. N. Barlow, M. J. Jackson, B. Mills, and W. B. Rowe, *Optimum Clamping of cBN and Conventional Vitreous-Bonded Cylindrical Grinding Wheels*, International Journal of Machine Tools and Manufacture (1995), **35**, 119–132.
5. M. J. Jackson, N. Barlow, B. Mills, and W. B. Rowe, *Mechanical Design Safety of Vitreous-Bonded Cylindrical Grinding Wheels*, British Ceramic Transactions (1995), **94**, 221–229.
6. W. Koenig and F. Ferlemann, "CBN grinding at 500 m/s", *In 'Ultrahard Materials in Industry – Grinding Metals'*, De Beers Industrial Diamond Company, UK (1991), pp. 58–65.
7. M. J. Jackson and N. Barlow, *Computer Aided Design of High Performance Grinding Wheels*, Computer Aided Production Engineering – Transactions of the Institution of Mechanical Engineers, London, UK (2000), **16**, 243–251.
8. T. Suto, T. Waida, T. H. Noguchi, and H. Inoue, "Wheel Designs for Grinding", *In 'Ultrahard Materials in Industry – Grinding Metals'*, De Beers Industrial Diamond Company, UK (1991), pp. 34–37.
9. H. Munnich, *High Speed Grinding Wheel Design*, Doctoral Thesis, Technische Hochschule, Hanover (1956).
10. M. J. Jackson, et al., *Mechanical Design Safety of Vitreous Bonded Cylindrical Grinding Wheels*, British Ceramic Transactions (1995), **94**(6), 221–229.
11. H. Munnich, *Beitrag zur Sicherheit von umlaufenden Schleifkoerpern von keramisch gebundenen Scleifscheiben*, PhD Thesis, University of Hannover, Hannover (1956).
12. J. Thies, *Bruchmechanische und zerstoerungsfreie Unterschungen en keramisch gebundenen Schliefschieben*, PhD Thesis, University of Hannover, Hannover (1986).
13. W. Pompe, et al., *Bruchstatische Aspekte der Festigkeit von Schliefscheiben*, Ceramic Forum International (1983), **60**(8), 296–300.
14. D. Mewes, et al., *Die Festigkeit von Schliefscheiben als Faktor fuer die Prozessicherheit*, Maschinenmarkt (2000), **40**, 38–41.
15. J. A. Behrens and M. Kammler, *Numerical and Experimental Analysis of the Breaking Behaviour of Vitrified Bonded Grinding Wheels*, Ceramic Forum International (2005), **82** (11), 47–52.

Chapter 4
Dressing of Grinding Wheels

Taghi Tawakoli and Abdolreza Rasifard

Abstract The appropriate dressing of the grinding wheel is one of the most important prerequisites for an efficient grinding process because, besides the generation of the required grinding wheel profile, it should also produce a suitable wheel topography. In this chapter, the terminology associated to the conditioning of the grinding wheels and the procedures of dressing will be first introduced. Diamond dressing tools will be then presented, to continue with a detailed description of the dressing with diamond dressers, including the ultrasonic assisted dressing as a new process. At the end of this chapter, the most important non traditional conditioning processes such as laser dressing and electrolytic in-process dressing (ELID) will be presented.

Keywords Dressing · Sharpening · Dressing tools · Diamond · Ultrasonic assistance · Laser dressing · ELID

4.1 Introduction

Grinding is generally used as a final machining step in the manufacture of components with high demands on manufacturing accuracy and surface quality. The grinding wheel performance has a great influence over the quality of the workpiece and process efficiency. The performance of the wheel is determined by a great variety of influencing factors. These include, among others, the preparation of the grinding wheel. The purpose of preparation is to produce the required

T. Tawakoli (✉)
Department of Mechanical and Process Engineering, Hochschule Furtwangen University, Villingen-Schwenningen, Germany
e-mail: tawakoli@hs-furtwangen.de

M.J. Jackson and J.P. Davim (eds.), *Machining with Abrasives*,
DOI 10.1007/978-1-4419-7302-3_4,

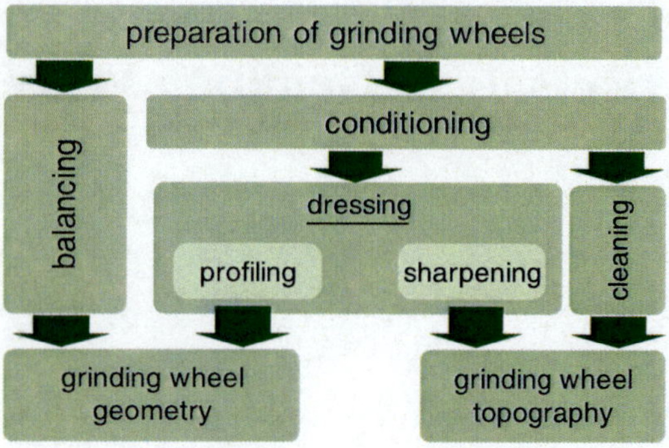

Fig. 4.1 Terms of the preparation of grinding wheels after Saljé [1]

geometrical and run-out accuracy as well as to generate a grinding wheel topography that is specific to the grinding process in question. During the grinding process, due to the wear of the grits and the bond, the grinding wheel is subject to macro-wear and micro-wear. Macro-wear alters the shape of the tool profile, which must always match the desired workpiece geometry. The micro-wear, however, leads to changes in the topography of the wheel, as a result of which the cutting surface of the wheel becomes either too smooth or too rough for the grinding task.

According to Saljé [1], preparing a grinding wheel includes two stages: balancing and conditioning (Fig. 4.1). Balancing serves to compensate for the form-induced and structure-induced imbalances of the rotating wheel, which cause dimensional and geometric errors on the workpiece [2]. The other operations involved in preparing wheels, which lead to changes in the macrotopography and/or the microtopography of the active surface of the wheel, come under the general term *conditioning*.

4.2 Grinding Wheel Conditioning

The purpose of conditioning is to produce, or restore, a wheel geometry and/or topography appropriate to the grinding task. Conditioning involves three different sub-processes: profiling, sharpening and cleaning, where profiling and sharpening come under the general term of *dressing* [1, 3].

With profiling, the required geometrical and run-out accuracy of the grinding wheel is produced. During the profiling, the bond and the grits are cut simultaneously. Therefore, the surface cutting of the grinding wheel possesses after profiling little or no grit protrusion, particularly after the profiling of resin bonded and metal bonded superabrasive wheels. In this condition the wheel doesn't posses a sufficient cutting ability. Hence the desired grit protrusions and the associated chip pockets have to be produced in the surface of the grinding wheel during a sharpening process after the profiling. During the sharpening, the grits of the top-most abrasive layer are exposed by cutting back the bond, whereby the chip pocket needed for the grinding process is generated [4–6].

During the grinding, some of the material being cut may be compressed and adhered into the space between grains. This phenomenon, called wheel loading, leads to a reduction of chip space, which can decrease the cutting ability of the wheel. With the cleaning process, the loaded chips and loosened grit and bond residues are removed from the wheel surface.

For conventional vitrified wheels with their porous structure the term *dressing* is often used for removal process carried out with a diamond tool, whereby the wheel is simultaneously profiled and sharpened. In scientific work and in the relevant specialist literature there are also other names for the terms profiling, sharpening and cleaning. In the English-speaking world, the profiling of a grinding wheel, for example, is also called *truing*. Another example is the use of the terms *conditioning* and *dressing* for the sharpening operation [7, 8].

The conditioning processes can be classified according to the mechanism of grit and/or bond removal during the process into two groups: the processes with the principle of mechanical material removal and the processes with the principle of non-mechanical material removal. In the first group, grits and/or the bond are removed by cutting or splitting the grits and/or the bond using a diamond or a diamond-free dressing tool. In the second group, the dressing process is carried out using non-mechanical forms of energy such as thermal and electrical energy. Laser dressing and electrolytic in-process dressing (ELID) are two examples of conditioning processes with the principle of non-mechanical material removal.

4.2.1 *Profiling*

In industry, the profiling of a wheel is carried out almost exclusively by a cutting process with a stationary or rotary dressing tool. The tools used for profiling are predominantly diamond dressers, in order to guarantee a high degree of geometric and dimensional accuracy on the wheel that is to be dressed. Depending on the kinematics of the dressing process, dressers can be divided into two groups: stationary dressers and rotary dressers.

4.2.1.1 Stationary Dressers

Stationary dressers have no movement in the circumferential direction of the grinding wheel. The wheel is dressed by axial movement of the dresser along the contour of the grinding wheel. Dressing with stationary dressers is therefore comparable to the turning process (Fig. 4.2). In industry, stationary diamond dressers are used in different forms and with different diamond abrasive coatings or tips (Fig. 4.3). Almost all of the stationary diamond dressers are only suitable for profiling or dressing rectilinear and simple wheel profiles. Furthermore, because Stationary diamond dressers contain only a small amount of diamond, when dressing superabrasive CBN and diamond wheels they are subject to a great deal of wear.

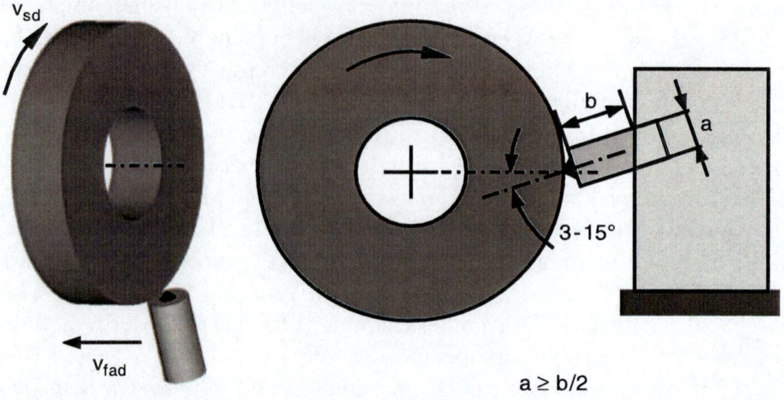

Fig. 4.2 Process kinematics in dressing with stationary dressers [9] used with permission copyright Winterthur Technology Group

dressing tool		grinding wheel profile	application
single-layered	single-point diamond 	rectilinear (cylindrical, taper) single-profile (convex and concave radii)	surface, cylindrical and centreless grinding single and small batch production
	profile diamond 	multi-profile (complex profile with steep sides and close radii)	surface, external cylindrical and centreless grinding single and small batch production
multilayered	multi-point diamond 	rectilinear (cylindrical, Taper)	surface and external cylindrical grinding single and small batch production
	dressing blade 	rectilinear (cylindrical, taper) single-profile (convex and concave radii)	surface, external cylindrical and centreless grinding single to mass production

Fig. 4.3 Examples of stationary dressers used with permission copyright Tyrolit Co

Therefore, to meet the demands on geometric and dimensional accuracy of the wheel profile, the dresser have to be changed frequently.

The generation of the wheel topography and also the wear on a stationary dresser are influenced to a great extent by the process parameters. The influence of these process parameters on the profiling and grinding processes will be discussed in detail later in Sect. 4.4.

4.2.1.2 Rotary Dressers

Compared with stationary dressers, rotary dressers have an additional movement of rotation. The relative speed in the contact zone is determined by the peripheral speed of both the dresser and the wheel. There are both diamond-free and diamond-containing rotary dressers. Nowadays diamond-free rotary dressers are only used in very few cases.

Rotary diamond dressers have a diamond layer on their periphery or front face. Therefore, compared with stationary diamond dressers, these dressers contain much more diamond, as a result of which they are subject to much less wear. Therefore, with rotary diamond dressers the desired geometric and dimensional accuracy of the wheel can be obtained over a much longer period of time. Furthermore, when profiling with rotary dressers, the topography of the grinding wheel, due to the kinematics of the process can be influenced in a broader spectrum. Common rotary diamond dressers are diamond profile rollers, diamond form rollers and diamond cup wheels (Fig. 4.4). In profiling with rotary diamond dressers, the process results are also strongly influenced by the process parameter. These parameters and their influence on the dressing and grinding processes are discussed in detail later in Sect. 4.5.

For the profiling of superabrasive grinding wheels with dressable types of bond, SiC wheels can also be used [2]. Currently, diamond grinding wheels are profiled predominantly with SiC wheels [10, 11]. When profiling with a SiC wheel the bond of the wheel is cut back by the abrasive action of the SiC grits so that the grits on the wheel that is to be dressed can fall out. The grit size of the SiC wheel is between 80 and 30 US Mesh; a larger grit size is recommended for dressing

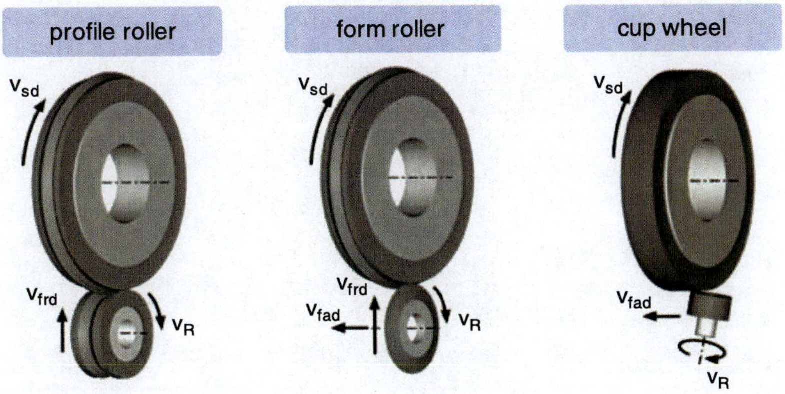

Fig. 4.4 Rotary dressers

diamond grinding wheels with a coarser grit [12]. After profiling with a SiC wheel the diamond grinding wheel has a smooth surface, so in many cases additional sharpening is needed [10, 11]. Due to the continuous-path controlled movement of the SiC wheel with an additional slewing motion, to a limited extent also non-rectilinear wheel profiles can be profiled. To generate a corresponding relative speed between the grinding wheel and dressing tool, the SiC wheel is either driven off course or braked [12].

4.2.2 Sharpening

After profiling, because the bond and the grits are cut simultaneously, the grits on the wheel surface exhibits little or no protrusion, particularly after the profiling of resin bonded and metal bonded superabrasive wheels. In this condition the wheel does not present a sufficient cutting ability. After profiling, the next step is sharpening. During this step, the grits of the top-most abrasive layer are exposed by cutting back the bond, whereby the chip pocket needed for the grinding process is produced (Fig. 4.5).

Sharpening can be carried out immediately after profiling or at the same time as the profiling process. Simultaneous sharpening while profiling reduces the wear on the dresser which is caused by the pressure and temperature during cutting through the grits and the bond [2].

A superabrasive grinding wheel can be sharpened with loose or bonded alumina or silicon carbide grits. Predominantly the sharpening process is carried out with a vitrified bonded alumina or silicon carbide block (called a sharpening block). The

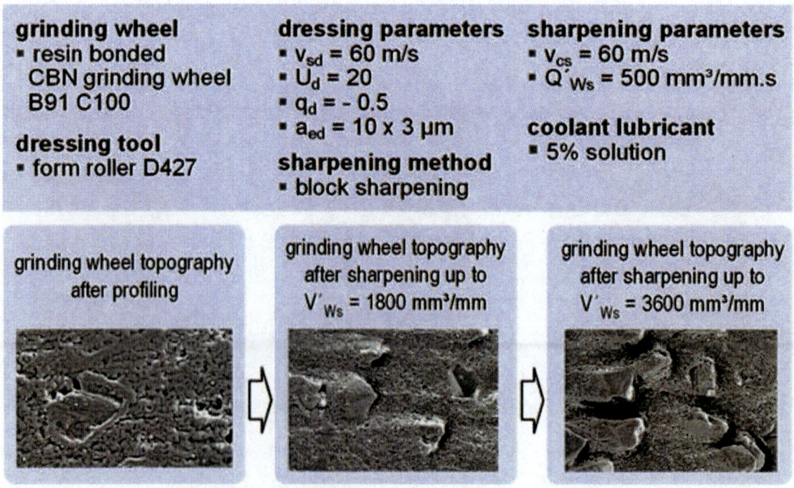

Fig. 4.5 Change in topography caused by sharpening

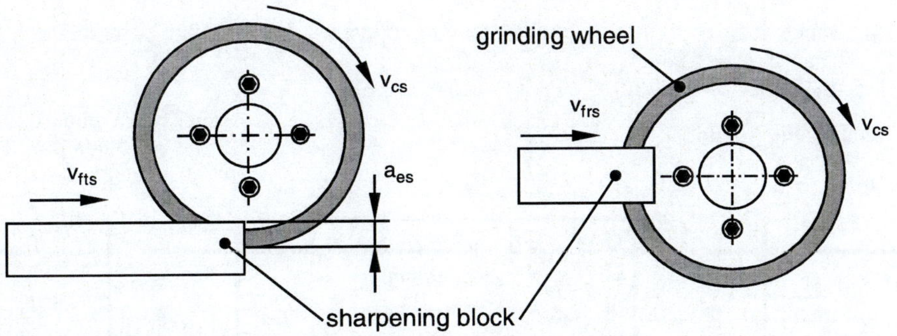

Fig. 4.6 Sharpening with a sharpening block

sharpening process can be carried out either by passing the wheel through the sharpening block or by plunging the wheel into the sharpening block (Fig. 4.6).

During the jet sharpening, loose abrasive grits are blasted with air and/or a carrier medium (usually coolant) at high pressure on to the abrasive layer of the grinding wheel. The sharpening with loose abrasives causes intense contamination inside the grinding machine. Hence this process is hardly ever used these days [13].

The topography of the wheel and consequently the grinding process are influenced by the sharpening conditions. Figure 4.7 shows the effect of specific sharpening material removal rate Q'_{Ws} on the grinding process when grinding with a resin bonded diamond wheel. Particularly high grinding forces occur with the unsharpened diamond wheel. Elements that are responsible for this are the very small chip pockets and the high cutting edge density because of the low depth of surface smoothness, which lead to increasing grinding force. Furthermore, the greater grit retention force due to the low grit protrusion leads to the diamond grits remaining in the bond longer and the grits being flattened as a result, which in turn causes the grinding forces to increase. The extended period in which the grit remains in the bond also leads to low radial wear. The workpiece roughness is also influenced strongly by specific sharpening material removal rate. The smaller chip thicknesses with greater cutting edge density, the greater covering of the grinding marks and also the grit flattening and the associated increase in the number of active cutting edges are responsible for the better quality of the surfaces with a smoother wheel topography, caused by a lower specific sharpening material removal rate [14].

Because of the porosity of vitrified bonded superarasive wheels, they can also be efficiently used in many cases without sharpening too. Figure 4.8 shows the effect of sharpening at different dressing conditions. The vitrified bonded CBN wheel used here was dressed with a diamond form roller. It can be seen that due to sharpening, the initial relief grinding *phase* is reduced. Relief grinding (also called the initial self-sharpening) occurs when grinding with vitrified bonded superabrasive wheels. During the initial relief grinding phase, the chip pockets between the grits in the vitrified bond become larger due to the removal of bond material by the abrasive action of the chips. The grinding process is unstable during

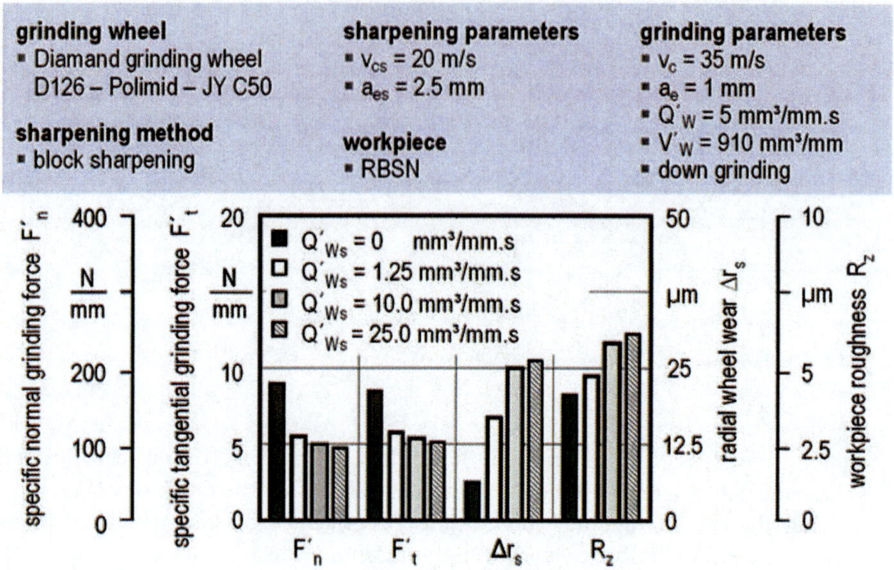

Fig. 4.7 Influence of sharpening conditions on the deep grinding of RBSN [14] used with permission copyright E. Uhlmann

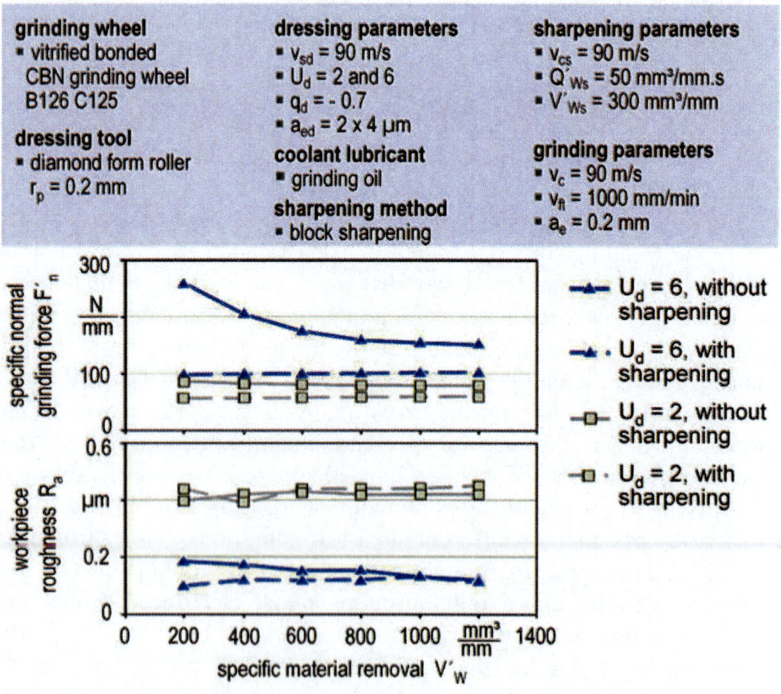

Fig. 4.8 Influence of sharpening at different dressing conditions

the relief grinding phase. On the other hand, however, for cost reasons it is desirable to avoid a separate sharpening process. A stable grinding process after dressing should therefore be achieved by appropriate selection of the dressing conditions. In the tests shown in Fig. 4.8, sharpening did not have any significant influence on workpiece roughness.

4.2.3 Cleaning

The cleaning of a grinding wheel serves mainly to remove the loading chips from the pores of a grinding wheel, but primarily without any intention to change the geometry or topography of the wheel. Wheel loading impairs the chip-forming process due to the reduction of the available chip spaces and the restriction of coolant lubricant in the grinding contact zone. Hence the wheel loading results in higher grinding forces and higher workpiece roughness as well as higher wheel wear [15, 16]. The wear on the diamond dressers is also higher during dressing of a loaded grinding wheel.

The tendency for the grinding wheel to get loaded depends on almost all the variables involved in the grinding process and in particular on the workpiece material. The loading occurs particularly when grinding ductile steels, aluminium alloys or titanium alloys [2]. During the grinding of tough materials such as nickel-chrome alloys, the grinding wheel becomes also very heavily loaded. The coolant that is used can also strongly affect loading. The use of water-miscible coolants results in a higher grinding wheel loading compared to when grinding oils are used instead.

Sharpening and dressing exhibit a cleaning effect, although in practice the wheels are cleaned rather with a high-pressure/high-speed jet of coolant during the grinding process itself. High-pressure/high-speed jetting with coolant is carried out with a special cleaning nozzle. The cleaning effect of jetting with coolant is with increasing jet pressure, velocity and flow rate higher. However, too high jet pressure must be avoided, because it can damage the abrasive layer [17, 18].

4.3 Diamond Dressing Tools

Diamond is a crystalline form of carbon, which is formed at high pressures and temperatures (Fig. 4.9). In a diamond crystal the carbon atoms are connected to each other by pure energy-rich covalent bonds in a narrow and extremely strong tetrahedral arrangement. As a result diamond is the hardest known natural material (Fig. 4.10).

The hardness of diamond depends on properties like purity, crystalline perfection and crystal direction [19]. Because of the different distances between the carbon atoms in the different crystal planes, the hardness of diamond varies in

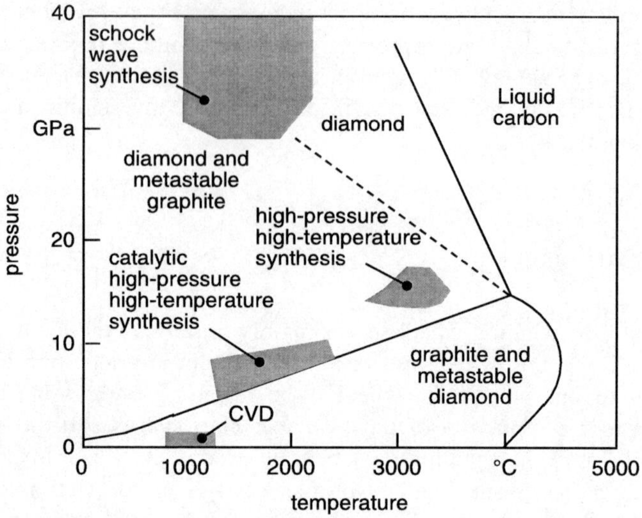

Fig. 4.9 Phase diagram of carbon [21] used with permission copyright American Chemical Society

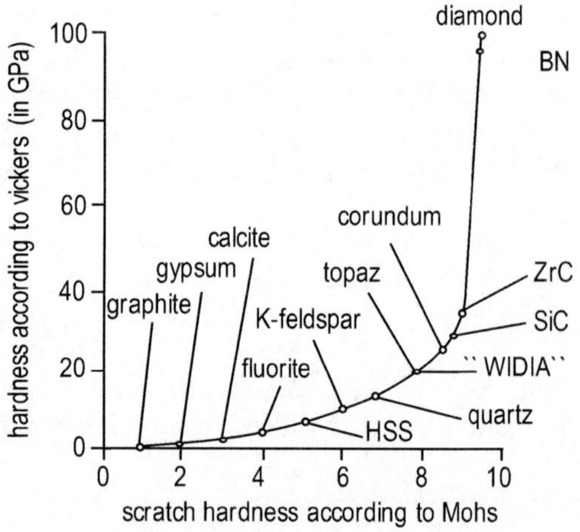

Fig. 4.10 Hardness values of different materials [22] used with permission copyright V. Buck

different crystal directions (anisotropy). The hardest is the <110> direction on the {111} plane [20].

As already mentioned, diamond is the hardest material known in nature. Therefore grinding wheels are predominantly dressed with diamond tools. As well as its great hardness, diamond also has good thermal conductivity. However, in dressing, the good thermal conductivity of diamond presents a danger to the diamond itself. As the materials that are to be dressed are generally not heat-sensitive themselves, a build-up of heat can develop in the diamond holder, which results in the diamond being destroyed. Hence temperatures above 900°C should be avoided, as otherwise, even with low pressures, graphitisation of the diamond occurs. To prevent graphitisation, there must be sufficient cooling during the dressing process [23]. The behaviour of a diamond dresser is strongly influenced by its diamond coating and its manufacturing method.

4.3.1 Diamond Coating

The concept of diamond coating encompasses the type of diamond used, the diamond grit size and concentration, as well as the pattern in which the diamonds are set [2, 23]. The types of diamond used by the manufacturing of diamond dressing tools can be divided into two main groups: single-crystal diamond and polycrystalline diamond (Fig. 4.11). Single-crystal diamond includes natural diamond and monocrystalline diamond (MCD). Depending on the manufacturing process, polycrystalline diamond (PCD) can be divided into the groups with and without binder phase [24].

Natural diamonds are used either in sorted form or as grits that are crushed mechanically into different grit sizes. In the case of natural diamonds, because of differences in shape, size and crystal purity as well as internal stresses and crystallographic irregularities, abrasion resistance or wear rate can vary relatively greatly [26]. Nevertheless, in these days diamond-dressing tools are made predominantly with single-crystal natural diamonds [7].

MCD is produced by synthesis from graphite in a complex process under high pressures and high temperatures using metal catalysts, and is characterised

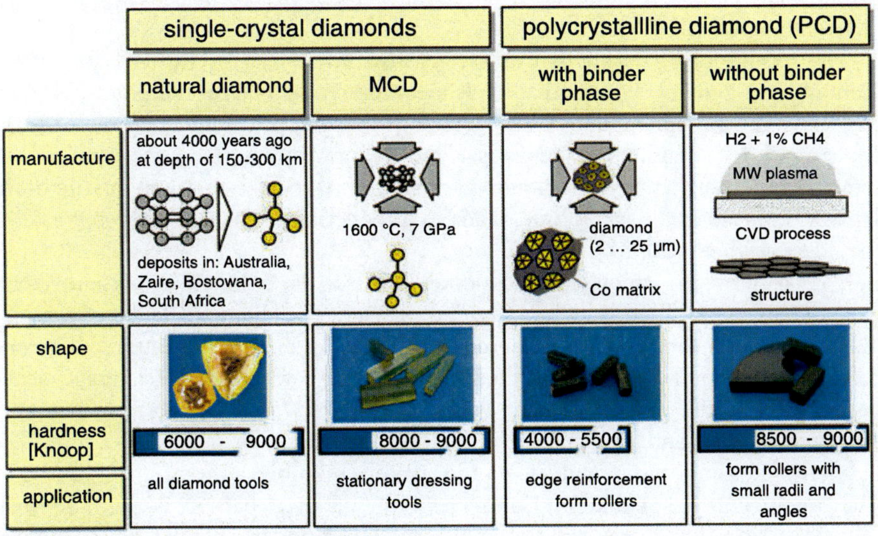

Fig. 4.11 Diamond for technical applications [25] used with permission copyright Dr. Kaiser

by extremely great hardness. Monocrystalline diamond logs are made by laser cutting monocrystalline diamond blanks. The homogenous crystal orientation of MCD leads to a constant abrasion resistance or wear rate along their length. In addition, as engineered products, synthetic diamonds have characteristics of reproducible quality. However, because of their anisotropy, the wear resistance of MCD logs, as with natural diamond crystals, is orientation-dependent. Therefore the MCD logs should be orientated in the dresser in their hardest direction [26]. Because of the high manufacturing costs, MCD is only used in tools where there is the risk of high-level wear on the tools, such as for example stationary dressers [24, 27].

Polycrystalline diamond with a binder phase consists of synthetic diamond particles which in a sintering process are bound with a binder matrix (cobalt or tungsten). The properties of the PCD are determined to a great extent by the grain size of the diamond particles. Typical grain sizes lie between 2 and 25 μm [27, 28]. This polycrystalline diamond is predominantly made in the form of log or disc using cobalt as the binder material [7]. Because of the random diamond orientation, PCD with binder phase forms as a hard, isotropic body, which does not have a preferred direction of cleavage and hardness. However, because of the binder phase element, PCD is the "softest" type of diamond. PCD is frequently used as edge protection in diamond in crushing rollers and in the frequently re-grindable PCD form rollers [28].

PCD without binder phase is produced by chemical vapour deposition (CVD) and is often referred to as CVD diamond. The great advantage of CVD diamond lies in its very high fracture strength because of its homogenous polycrystalline structure and the dense packing of its structure [28]. The properties of CVD diamonds are not dependent on their orientation, i.e. their wear resistance is the same in all directions [29]. Due to the possibility in the CVD process to deposit extremely thin layers, this method is suitable in particular for coating form rollers with small angles and radii [27–29].

The behaviour of a diamond dresser is also influenced by the grit size of the diamonds. In general, smaller diamonds are stronger than larger diamonds, because they have fewer and smaller crystal defects [8]. A further influencing factor is the concentration of diamond in the dresser. Increasing the diamond concentration of a rotary dresser with a constant diamond grit size leads to an increase in the life of the dresser, but also at the same time, leads to a reduction in the surface roughness of the dressed wheel [30, 31].

The pattern in which the diamonds are set is another important factor that influences the behaviour of diamond dresser. The diamond layer of a rotary diamond dresser can be either scattered or set (Fig. 4.12). With a scattered diamond coating the diamonds are distributed stochastically, whereby the average space between the grains depends on the diamond grit size [2]. Most form rollers with this type of diamond material have a radius in the range $R = 1.5$–5 mm. The diamond dressing tools with scattered pattern exhibit a good tool life [28]. The manufacturing of diamond dressers with scattered pattern is fast and cost-effective. However, form rollers with a scattered pattern cannot be re-ground. Scattered diamond coatings are made using natural diamonds [24].

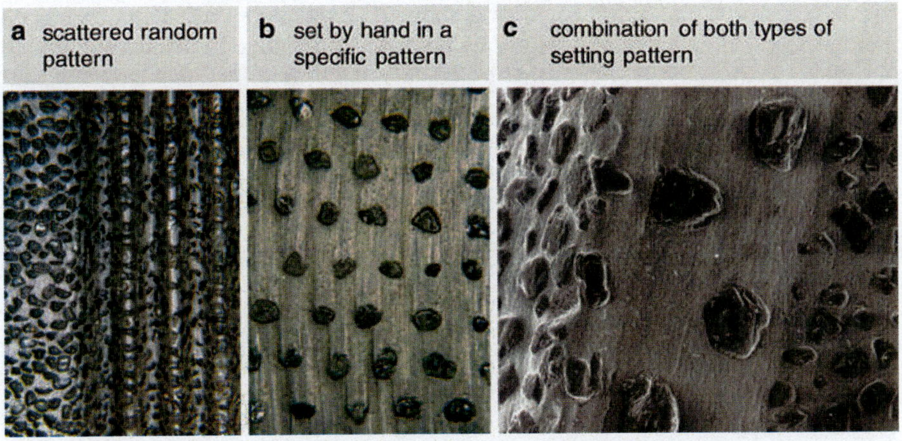

Fig. 4.12 Different types of diamond pattern [32] used with permission copyright Tyrolit

In set-diamond coatings, diamonds are set – commonly by hand – in a specific pattern. In this system, by appropriate choosing the position and the geometry of the diamonds, as well as their surface concentration on individual elements of the contour of the dresser, the topography of the dresser can be tailored to the profiling task [33, 34]. With this type of diamond coating it is possible to manufacture form rollers with a very small tip radius. Needle-shaped natural diamond and also monocrystalline or polycrystalline synthetic diamond logs can be used for set-diamond coatings. Form rollers with set-diamonds can be re-ground [24].

A combination of both types of patterns can also be useful for some profile rollers. For example, when grinding steep flanks, the small effective peak-to-valley roughness of the wheel causes thermal damage to the workpiece. A solution to this problem can be setting the diamonds by hand according to specific patterns on critical areas of the profile, as a result of which the effective peak-to-valley roughness of the wheel in these areas is increased [32]. A specific selection of the shape and the arrangement of diamond grits can be beneficial for stationary diamond tools as well. A defined contact width and a uniform wear characteristic, for example, can be achieved in blade dressers with needle diamonds through the array arrangement of the diamonds.

4.3.2 *Manufacture of Diamond Dressing Tools*

The Performance of a dressing tool can be influenced significantly by its manufacturing process. There are several manufacturing processes for both stationary and rotary diamond dressers. In stationary diamond dressers, the bonding of diamonds can be done mainly through an electroplated nickel layer or in a sintering process with a hard and wear-resistant binder matrix. Electroplaed bonds have little hardness and low heat conductivity, as well as a weak diamond binding; nevertheless, there is no thermal load on the diamond during manufacture. In sintered stationary diamond dressers, the sintered body is brazed to the steel base body. Brazing can, however, cause thermal damage to the diamond. Stationary diamond dressers with a brazed sintered body also have little bond hardness, but they can be produced readily. The whole dressing tool can also be made by sintering. These stationary dressers are characterised by their good bond hardness and high heat conductivity, as well as because of their strong diamond binding [25].

In the selection of stationary dressers, the specifications of the wheel that is to be dressed also have to be taken into account. For example, if a blade dresser with a soft bond is employed to dress sintered alumina or silicon carbide wheels, the bond matrix of the dresses is soon cut back. As a result, after dressing for a short time, the diamonds protrude from the tool and break out due to vibration or dressing forces. On the other hand, if a dresser with a highly hard bond is used to dress white-fused alumina wheels, the bond matrix does not get cut back enough. Here, the wheel is actually dressed also with the hard bond matrix, which results in a smooth grinding wheel [35].

The required manufacturing accuracy for rotary diamond dresser is very high (Fig. 4.13). The manufacturing accuracy of a rotary diamond dresser is influenced to a great extent by its manufacturing method. Diamond roller dressers are produced either in the positive method or the negative method (Fig. 4.14).

In the positive method (also called the direct method) both single-layer and also multi-layer diamond coatings can be produced (Fig. 4.15). In the manufacturing of

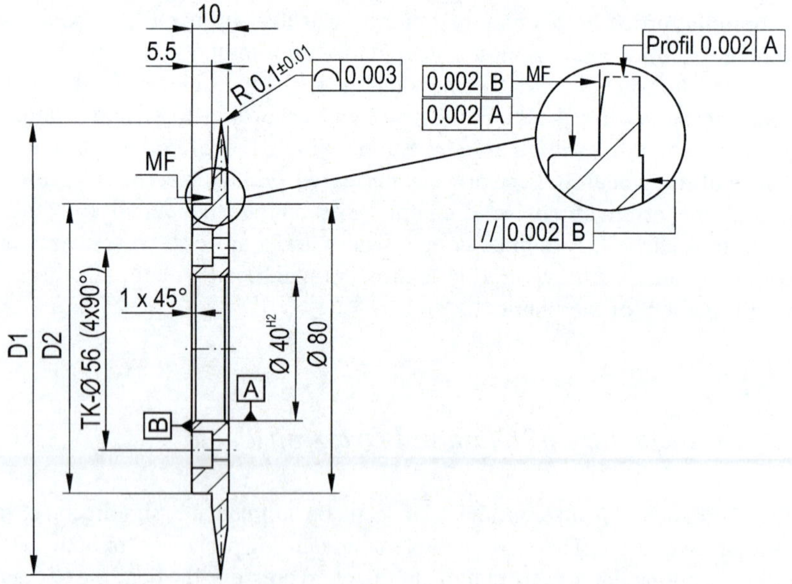

Fig. 4.13 Typical forms and tolerances of diamond form rollers used with permission copyright Dr. Kaiser

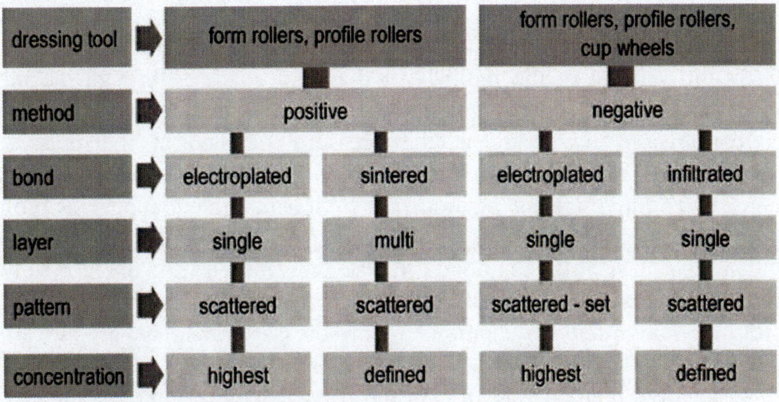

Fig. 4.14 Manufacturing methods for rotary diamond dressers

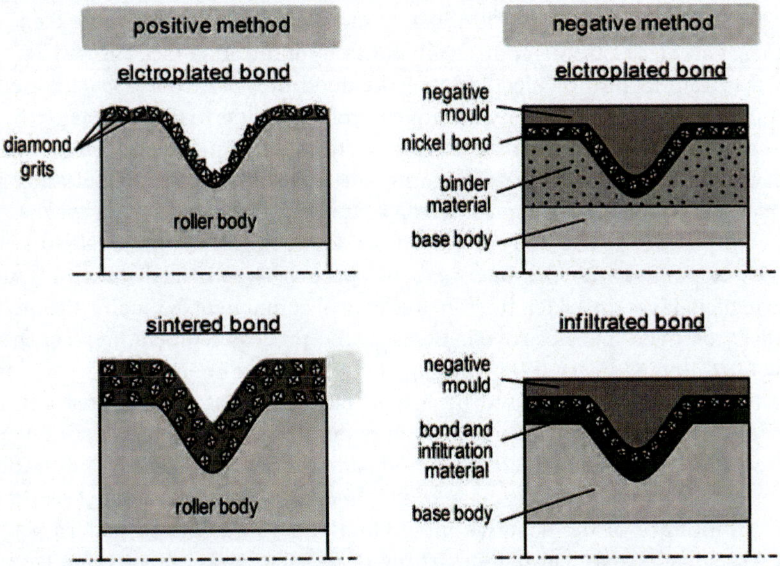

Fig. 4.15 Schematic illustration of the manufacturing methods for rotary diamond dressers [33]

single-layer diamond coatings using the positive method, the diamond grits are first scattered statistically directly on to the base body of the roller and joined then in an electroplating process with a single nickel layer to the base body [33]. The grit density is high and the geometric coating shape is determined by the scatter of the diamond grit size. These diamond rollers can be again re-coated after the end of their service life [24, 28]. However, because of the way in which they are made and the scatter of the diamonds in terms of size and shape, the surface of these tools is relatively rough and irregular. Hence these rollers should be ground before being used in the dressing process. This serves not only to achieve sufficient geometrical and dimensional accuracy, but also to generate a high contact area ratio on the roller right at the start of dressing, which avoids unstable dressing roller wear at the initial dressing phase [4].

The manufacturing of the multi-layer diamond coatings through the positive method is carried out by sintering, whereby a mixture of metal powder and diamond grits is compressed in a mould at high pressure and then, at high pressure and high temperature is connected to the tool body. With this method diamond layers between 2 and 5 mm thick can be produced. So it is possible to renew the profile on these diamond rollers by regrinding them. Also, the diamond concentration can be controlled regardless of the diamond grit size and geometry. As the relatively high sintering temperatures lead to warping and also non-uniform shrinkage of the layer, with this method often the diamond rotary dresser has to be ground after its manufacture [33].

In the negative method (also called the reverse method), the geometry of the roller is produced in a high-precision negative lost mould (made of graphite, aluminium or steel), which is destroyed afterwards. A single layer of diamonds is first applied to the surface of this negative mould and the diamonds are then joined together with an electroplated nickel bond or by infiltration (see Fig. 4.15).

In the manufacture of electroplated diamond dressing rollers by the negative method, the negative, extremely accurate mould, which is made of metal, is densely covered with a single layer of statistically distributed diamonds. The diamond grits are then joined together by electroplating an abrasion-resistant nickel alloy layer. The electroplating process can last for a number of days or even weeks. After the electroplating, a base body is placed inside the negative mould. Next, a binder material is poured into the intermediate space between the base body and the electroplated layer in order to join these two components. Finally the negative mould is removed and destroyed. Because the process temperatures can be kept below 60°C, there is very little warping and change in dimensions. These diamond roller dressers are characterised by a high percentage of contact area and better geometric and roundness accuracy, so they do not need to be ground after manufacture [24, 27, 33]. The negative method can also be carried out by infiltration. In this case the negative mould is made of graphite. The diamonds are first scattered on to the profile face of the negative mould and/or set in a defined pattern, and then fixed with an adhesive. During the setting of the diamonds, the position, geometry and type of the diamonds can be determined specifically. After laying the base body, which roughly fits the profile of the roller, the intermediate space between the base body and the diamond layer is filled with tungsten powder. After the compaction of the tungsten powder and the subsequent addition of the infiltration material (brazing solder), the mould is heated in an oven at atmospheric pressure, whereby, at high temperatures, the infiltration and the sintering processes take place. During the infiltration, the infiltration material penetrates into the capillaries of the compacted metal powder and joins the diamond layer with the roller body [33]. Due to the shrinkage of the binder matrix after sintering, the diamond layer becomes very solidly fixed to the tool body. Furthermore, because of its good thermal conductivity characteristics infiltration binding enables the diamonds to cool well, which results in longer tool life [28]. While the roller cools down, the bond shrinks in an uncontrolled way. In addition, the adhesive that is used to fix the diamond grits to the negative mould reduces the accuracy that can be achieved. Therefore, in order to achieve high tolerances, diamond rollers made by the negative method with infiltration have to be ground after manufacture.

4.4 Dressing with Stationary Diamond Dressing Tools

Dressing with stationary dressing tools is comparable to turning. When dressing with stationary dressers, the engagement conditions on the grinding wheel depend on the geometry of the dressing tool as well as on the dressing parameters (Fig. 4.16). The most important parameters are the depth of dressing cut a_{ed} and overlap ratio U_d. The dressing overlap ratio U_d is the quotient from the engagement width a_{pd} and the axial dressing feed f_{ad} and shows how often a point on the peripheral face of the wheel is engaged by the dressing tool. The permitted lower limit value for the overlap ratio is U_d=1.0, with which each peripheral point of the wheel is engaged once by the dressing tool. The dressing overlap ratio has a decisive effect on the generation of the topography of the wheel when dressing. A higher dressing overlap ratio results in a finer wheel topography. Therefore, as the dressing overlap ratio increases, the grinding forces augment and the workpiece roughness decreases (Fig. 4.17). Decreasing the U_d results in an increase in wheel wear. This can be due on the one hand to more intensive crack formation in the abrasive grits and the bond with a higher axial dressing feed, and on the other hand to the greater engagement forces on the individual grits when grinding with a coarser grinding wheel. As it can be seen from Fig. 4.17, an increase in U_d beyond a certain overlap ratio does not result in any significant change in the grinding process but in an unnecessary increase in dressing time. This maximum useful overlap ratio depends on the specification of the wheel that is being used [2]. The

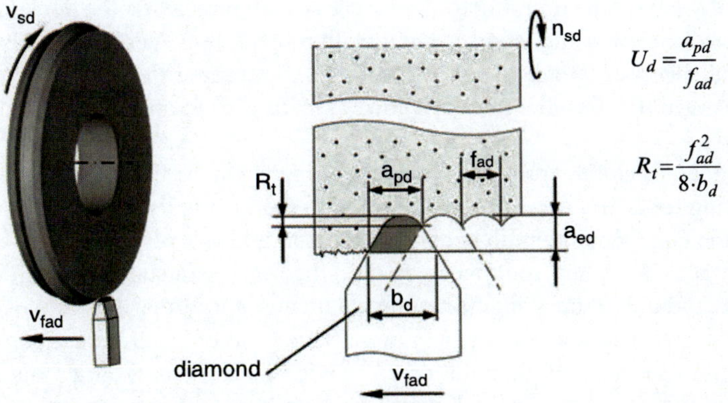

Fig. 4.16 Engagement conditions and parameters when dressing with stationary dressing tools

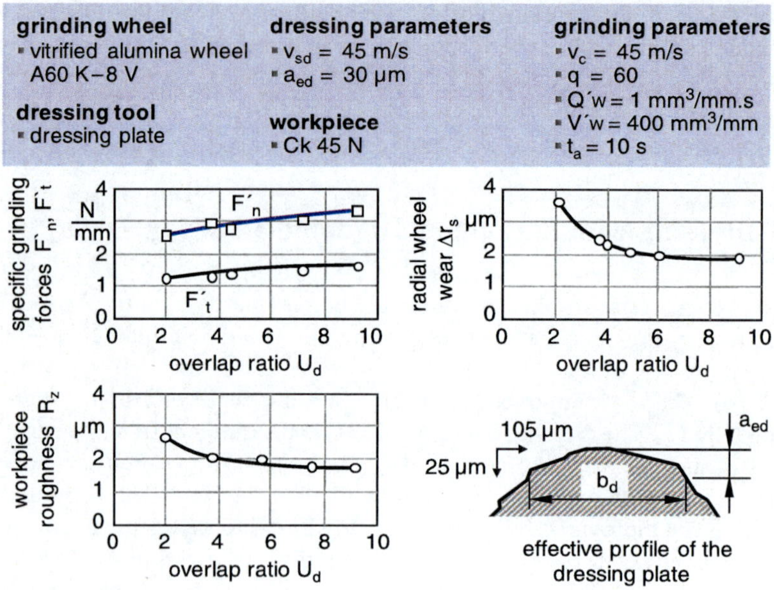

Fig. 4.17 Influence of dressing overlap ratio on the grinding process [2] used with permission copyright Springer-Verlag

grinding process is also influenced by depth of dressing cut. The influence of depth of dressing cut on the formation of the wheel topography is, however, less than that of the overlap ratio. A slight increase in the number of cutting edges with increasing depth of dressing cut when dressing with a blade dresser has been reported [33].

The dressing forces and the development of heat occurring in the dressing contact zone, and therefore dresser wear, are also influenced by the dressing parameters. As shown in Fig. 4.18, the dressing ratio G_d, which is calculated from the quotient of the volume of grinding wheel that has been dressed and the volume of wear on the dressing tool, decreases with increasing the depth of dressing cut a_{ed} and axial dressing feed f_{ad}. On the other hand, however, an increase in wheel peripheral speed v_{sd} at a constant U_d leads to an increase in the dresser wear ratio. An increase in the dressing overlap ratio U_d leads also to an increase in the wear of the dresser.

The generation of the grinding wheel topography can be influenced by the wear on the dressing tool. The wear of a diamond with the originally circular-arc-shaped tip leads to the engagement width becoming greater, and as a result to a reduction in the effective peak-to-valley roughness of the wheel at a constant axial dressing feed rate. The gradually increasing flattening wear causes irregularities in the results of a recurrently performed dressing operation. This type of wear behaviour cannot be avoided in the case of single point dressers with natural diamond. A solution to this problem is the use of synthetic diamond dresser logs with a constant cross-section, which despite wear have an engagement width that remains the same [26].

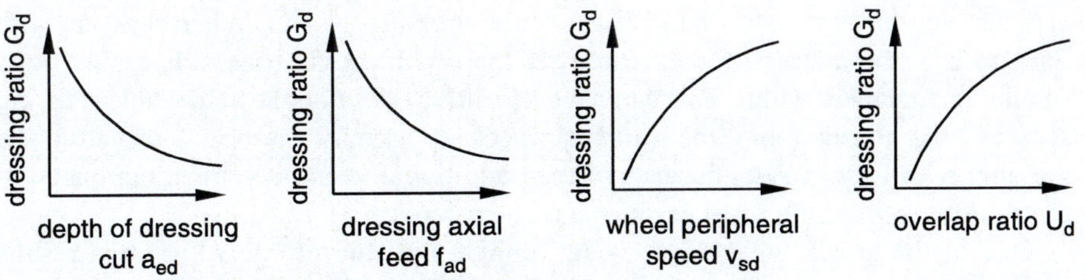

Fig. 4.18 Trend in the relationship between dressing parameters and dresser wear ratio [33] used with permission copyright Riegger Diamantwerkzeuge

4.5 Dressing with Rotary Diamond Dressing Tools

Rotary diamond dressing tools are extremely important in grinding, particularly when dressing of superabrasive CBN grinding wheels. Rotary diamond dressers contain much more diamond than stationary diamond dressers, which enables the wheel to be profiled with the required geometric and dimensional accuracies over a longer period. As Fig. 4.4 shows, depending on the way in which the profile is produced, there are essentially three types of rotary diamond dresser, namely:

- Diamond profile rollers
- Diamond form rollers
- Diamond cup wheels.

The choice of which dressing tool should be used depends, apart from what is technically possible with the machine being used, on the criteria of dressing time and tool costs as well as on the wheel profile that is to be dressed. While the area of application of the diamond cup wheel is limited to the dressing of rectilinear wheel profiles, complicated grinding wheel contours can also be dressed with diamond profile and form rollers. Dressing with profile rollers is a profile copying method, in which the profile roller has the negative form of the profile of the grinding wheel. The whole of the profile of the wheel is produced by plunging the rotating profile roller radially into the grinding wheel, which is also rotating. Diamond profile rollers are characterised by very short dressing times and also great profile accuracy. However, the use of a diamond profile roller is economically convenient only from a certain minimum batch size. For this reason diamond profile rollers are used mainly in mass production.

Dressing with form rollers is a profiling method in which the wheel profile is produced by the continuous-path controlled movement of a form roller with a small engagement width. Diamond form rollers are characterised by great flexibility. What is characteristic of a form roller is the tip radius of the roller, which determines, what shapes can be produced on the grinding wheel. The applications in which a diamond form roller can be used can be increased by an additional slewing movement [27, 33]. A problem with an additional slewing axis is, however, the fact that usually it involves greater costs and time in terms of programming and control, and also the fact that it reduces the stiffness of the dressing system. When dressing with form rollers, the dressing forces are lower than with profile rollers, because of the smaller engagement width. This can have a positive effect on the achievable dressing accuracy [4]. Furthermore, the grinding wheel topography can be influenced in the dressing with form rollers, because of their additional axial movement compared to profile rollers, over the overlap ratio in a wider range.

Due to the small engagement width, unlike with the profile roller, this results in longer dressing times. On the other hand, the cost of diamond form rollers, in contrast to diamond profile rollers are independent of the width of the grinding wheel. This can bring about cost advantages for form rollers even in mass production particularly in grinding with wide grinding wheels [4].

The dressing process is influenced by the specifications of the grinding wheel and the dresser, and also by the dressing parameters. The section below explains the effect of these influencing factors on the dressing and grinding results when dressing with rotary diamond dressing tools.

4.5.1 Diamond Profile Rollers

The kinematics of dressing with profile rollers are shown in Fig. 4.19. The most important parameters when dressing with profile rollers are:

- f_{rd}: radial dressing feed
- q_d: dressing speed ratio
- n_{rd}: number of roll-out revolutions.

The dressing speed ratio q_d is the ratio of the peripheral speed of the roller dresser v_R to the peripheral speed of the grinding wheel when dressing v_{sd}. The sign of the dressing speed ratio is defined by the speed directions of the active partners in the dressing contact zone. If the speed directions of the active partners in the contact zone are the same, the dressing speed ratio is called *down dressing* and is designated as positive. If the speed directions of the active partners in the contact zone are contrary to each other, the dressing speed ratio is called *up dressing* and is designated as negative.

In dressing with rotary diamond dressing tools the wheel topography is produced by engagements of the roller dresser diamonds with the surface of the wheel. When dressing with profile rollers the paths of the diamonds are determined by f_{rd} and q_d. Assuming a roller running in an ideal round movement, the path curve of a peripheral point of the rotating profile roller in the fixed-wheel coordinate system (*x*,*y*) can be described in accordance with (4.1) and (4.2). The angle φ_{RO} describes the starting angle of the point to be looked at in the polar coordinate system of the roller [5, 36].

$$x = ((r_{SO} + r_{RO}) - f_{rd} \cdot \frac{\phi_S}{2 \cdot \pi}) \cdot \cos\phi_S - r_R \cdot \cos(\phi_S \cdot (1 + q_d \cdot \frac{r_{SO}}{r_{RO}}) - \phi_{RO}) \quad (4.1)$$

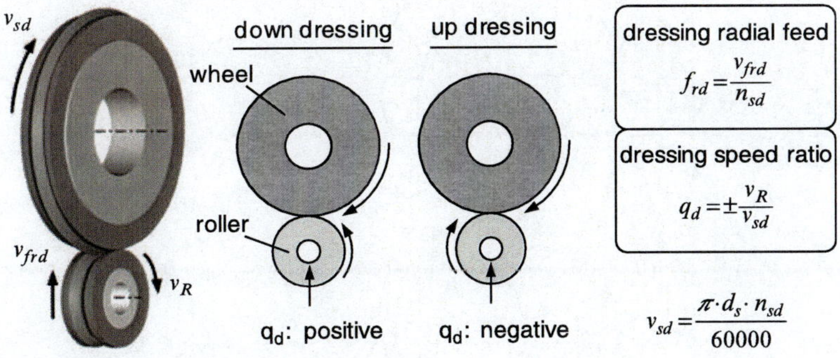

Fig. 4.19 Kinematics of dressing with profile rollers

$$y = ((r_{SO} + r_{RO}) - f_{rd} \cdot \frac{\phi_S}{2 \cdot \pi}) \cdot \sin \phi_S - r_R \cdot \sin(\phi_S \cdot (1 + q_d \cdot \frac{r_{SO}}{r_{RO}}) - \phi_{RO}) \quad (4.2)$$

Figure 4.20 shows some examples for path curves at different values of q_d and f_{rd}. As it can be seen from this figure, the dressing speed ratio q_d has a dominant influence not only on the relative speed between grinding wheel and dressing tool, but also on the shape of the engagement path curves, under which the diamond grits of the roller strike the surface of the grinding wheel. For values q_d that are common in practice ($0.1<|q_d|<1$), the engagement paths in up dressing are flatter than in down dressing.

The flatter diamond engagement paths and the resulting greater overlapping of the diamond grit engagements following each other lead in up dressing generally to a finer wheel topography, which results in a smaller initial effective peak-to-valley roughness of the grinding wheel (Fig. 4.21). A reduction in the radial dressing feed f_{rd} also results in a reduction in the starting effective peak-to-valley roughness [5, 30, 37].

As expected, the grinding process is influenced by the wheel topography generated during dressing operation (Fig. 4.22). After up dressing, because of its finer topography, the grinding wheel has a higher number of kinematic cutting edges. Consequently, the grinding forces are higher after up dressing and the workpiece roughness is lower. On the other hand, the topography of the wheel is itself influenced by the grinding conditions; therefore, after dressing an unstable initial grinding phase can occur. During this unstable initial grinding phase the effective peak-to-valley roughness of the grinding wheel seeks to achieve a quasi-stable value which particularly with conventional wheels depends only to a lesser extent on the initial effective peak-to-valley roughness (Fig. 4.23).

When down dressing, it is possible to vary the starting effective peak-to-valley roughness of the wheel in a greater range (see Fig. 4.21). However, the dressing forces in down dressing are higher than in up dressing (Fig. 4.24). Hence the wear of the diamond roller is in down dressing higher [33].

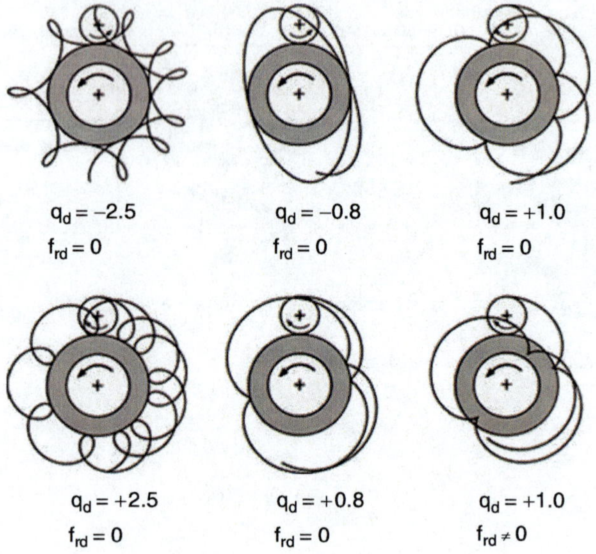

Fig. 4.20 Example of engagement paths of a peripheral point of a profile roller [36] used with permission copyright Vulkan-Verlag

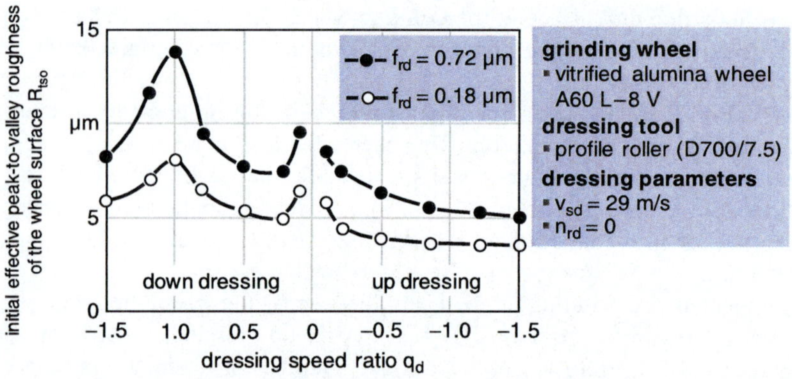

Fig. 4.21 Influence of the parameters q_d and f_{rd} on the initial effective peak-to-valley roughness of the grinding wheel [31] used with permission copyright R. Schmitt

When dressing with profile rollers, due to the radial feed, an Archimedes' spiral is produced on the periphery of the wheel. Therefore, if the profile roller is lifted after profiling immediately, a circular defect about the size of the feed amount remains on the grinding wheel. To avoid this error and to relieve the system, before lifting there should be a short roll-out, which corresponds to spark-out in grinding process. Also, by rolling-out, the true-running error

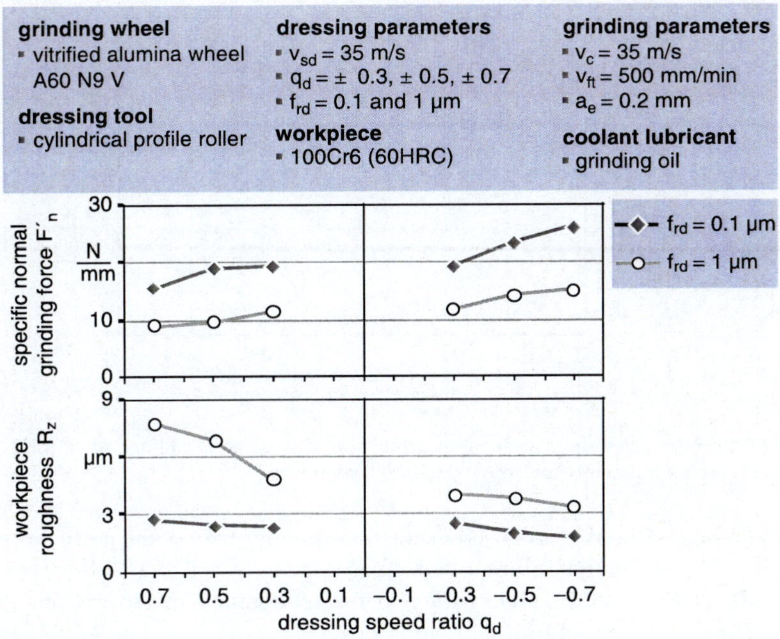

Fig. 4.22 Influence of dressing conditions on the grinding process

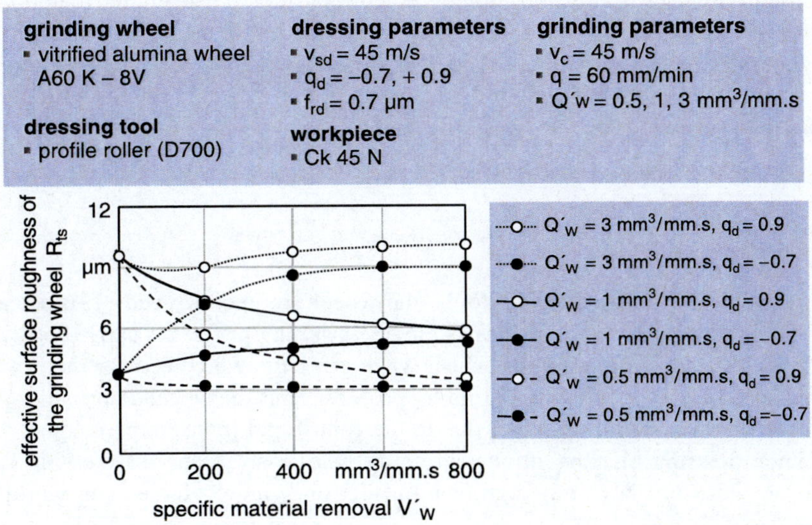

Fig. 4.23 Change over time of the effective peak-to-valley roughness of the grinding wheel as a function of dressing and grinding conditions [38]

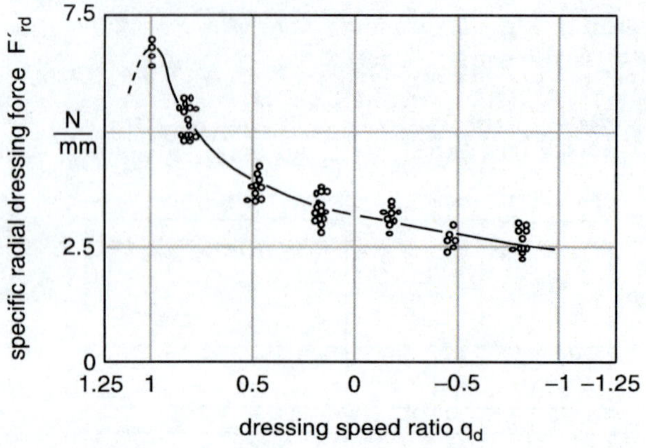

Fig. 4.24 Specific radial dressing force as a function of dressing conditions when using profile rollers [37] used with permission copyright H. Scheidermann

of the grinding wheel that is caused by the true-running error of the profile roller can be reduced. However, the rolling-out decreases the initial peak-to-valley roughness of the wheel (Fig. 4.25). The change in the topography of the grinding wheel depends on the number of roll-out revolutions n_{rd}. As shown in Fig. 4.25, the initial peak-to-valley roughness decreases strongly with the increasing the n_{rd}, especially in down dressing and asymptotically approaches a limit value. There is no benefit in there being more than 100 roll-out revolutions because after this point no significant change in the initial peak-to-valley roughness of the wheel is achieved [31].

The generation of the grinding wheel topography is also influenced by the diamond grit size and concentration of the profile roller. Reducing the diamond grit size or increasing the concentration leads to a reduction in the initial peak-to-valley roughness of the wheel. After Schmitt [31], the theoretical profile height of the wheel can be calculated as a function of the diamond grit size D and grit concentration K_D and also the dressing speed ratio q_d with (4.3).

$$H = \frac{D^2}{k \cdot K_D \cdot 8 \cdot r_R \cdot (q_d - 1)^2} \tag{4.3}$$

In dressing with rotary dressing tools, the wheel peripheral speed v_{sd} is a readily adjustable process parameter. If possible, the grinding process is carried out at the peripheral speed of the grinding wheel when grinding, which is, with the grinding wheel speed v_c of the following grinding process. Thus the wheel deformations that occur in dressing and in grinding due to the centrifugal forces remain equal.

When dressing of conventional grinding wheels at a constant dressing speed ratio, v_{sd} does not have any significant effect on grinding forces and workpiece roughness [31, 37].

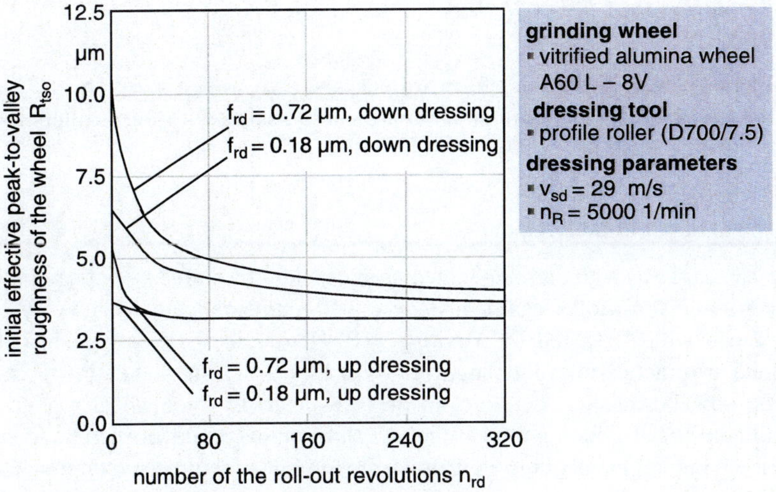

Fig. 4.25 Influence of roll-out revolutions on the starting effective peak-to-valley roughness of the wheel [31] used with permission copyright R. Schmitt

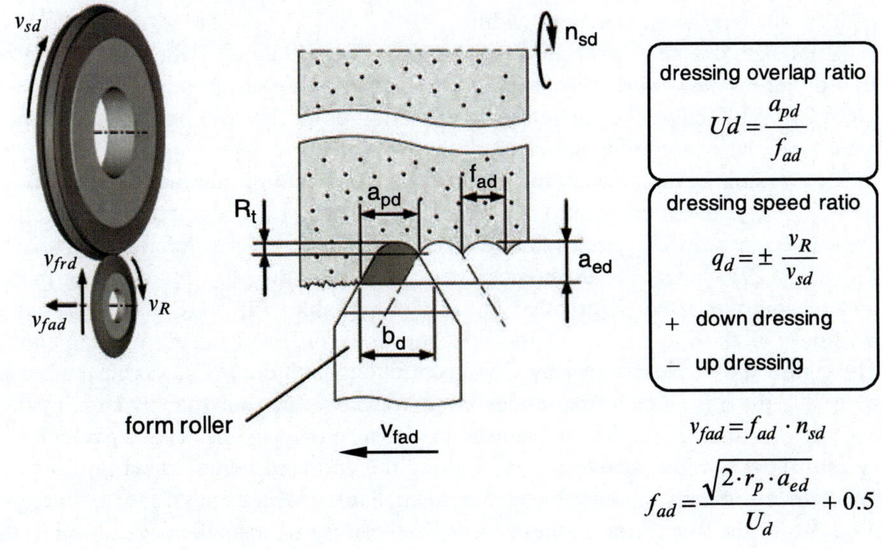

Fig. 4.26 Kinematics and parameters in dressing with form rollers

4.5.2 Diamond form Rollers

The pinciple of dressing with form rollers is shown in Fig 4.26. On the basis of the process kinematics, the main parameters in dressing with form rollers are:

- q_d: dressing speed ratio
- U_d: dressing overlap ratio
- a_{ed}: the depth of dressing cut.

In the same way as with dressing with a profile roller, the path curve of a point on the periphery of a form roller can be also described mathematically as an epicycloid in accordance with (4.1) and (4.2) with f_{rd}=0. Here also the engagement paths of diamond in practice-relevant range for q_d (0.1<$|q_d|$<1) are flatter in up dressing than in down dressing. Therefore up dressing leads to finer wheel topographies, resulting in higher grinding forces (Fig. 4.27). In practice, the vitrified bonded CBN wheels are predominantly dressed in down dressing, in order to avoid thermal damage to the workpiece.

The overlap ratio U_d is other important parameter in dressing with form rollers. As with stationary dressing tools, the dressing overlap ratio U_d is the value produced from the quotient of the engagement width of the form roller a_{pd} and the axial dressing feed of the roller per grinding wheel revolution f_{ad}. In practice and in research work, the overlap ratio is often calculated from the active width b_d – referred to here as $U_{d(bd)}$ – instead of a_{pd}. The $U_{d(bd)}$ is greater than the overlap ratio calculated with the engagement width a_{pd} [2, 39]. Increasing the f_{ad} increases the diffrence beetween the overlap ratios calculated from the active width and from the engagement width.

A higher dressing overlap ratio produces a smoother wheel topography and therefore less workpiece roughness, but also higher grinding forces (see Fig 4.27). This causes an increase in temperatures in the grinding contact zone, which may lead to thermal damages of the workpiece.

By dressing an inclined contour element of the wheel profile there are different ways to create the overlap ratio. Also the overlap ratio $U_{d(c)}$ along the contour, the axial overlap ratio $U_{d(a)}$ and radial overlap ratio $U_{d(r)}$ can be defined (Fig 4.28). As (Fig 4.29) shows, the differences between $U_{d(c)}$ and $U_{d(a)}$ for dressing parameters common used in practice are not remarkable. Fig 4.31 also shows the influence of the kinematics of the dressing units on the dressing overlap ratio. The figure shows clearly that by CNC continuous-path control at a constant path speed v_{fd}, there is a much smaller deviation in the overlap ratio than by copying the contour of a master template at a constant axial speed v_{fad} by means of a pantograph system [36]. A constant overlap ratio along the entire grinding wheel profile can result in a too fine topography at the sleep flanks, which may lead to thermal damages of the workpiece in these areas. In these cases, a smaller overlap ratio in the flank area is more advantageous [32].

When dressing an inclined contour element the process is also influenced by the direction of cut (Fig. 4.30). With a pulling dressing cut, the dressing engagement

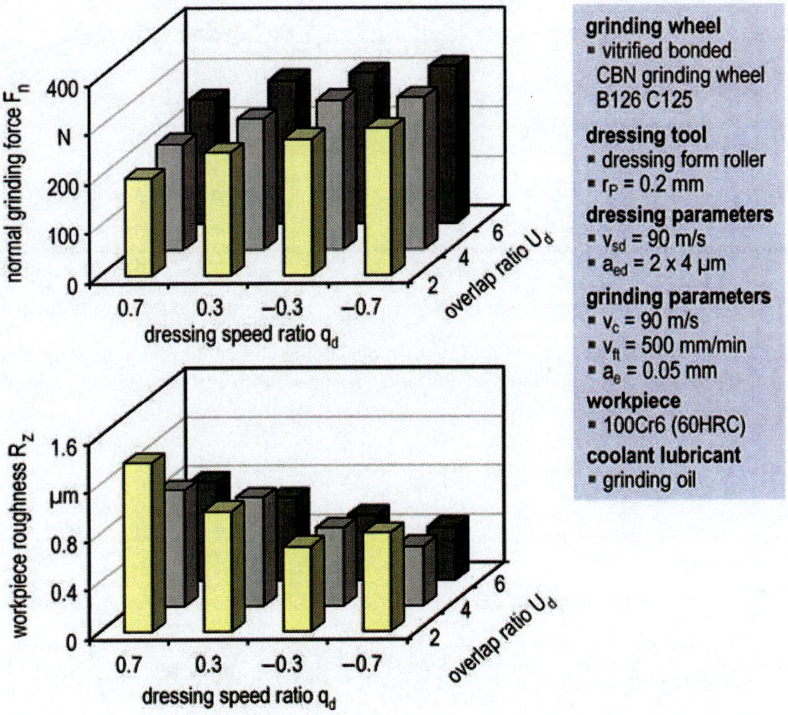

Fig. 4.27 Grinding forces and workpiece roughness as a function of dressing conditions when using form rollers

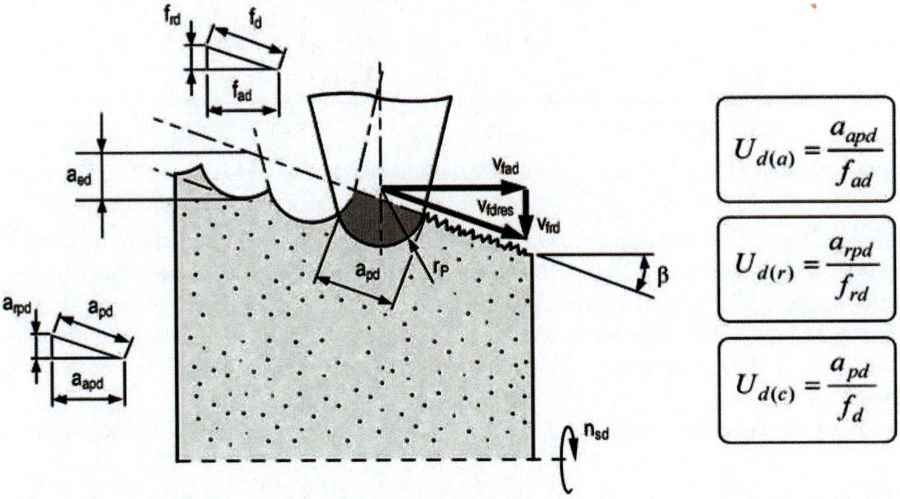

Fig. 4.28 Engagement conditions on an inclined contour element used with permission copyright [36] used with permission copyright Vulkan-Verlag

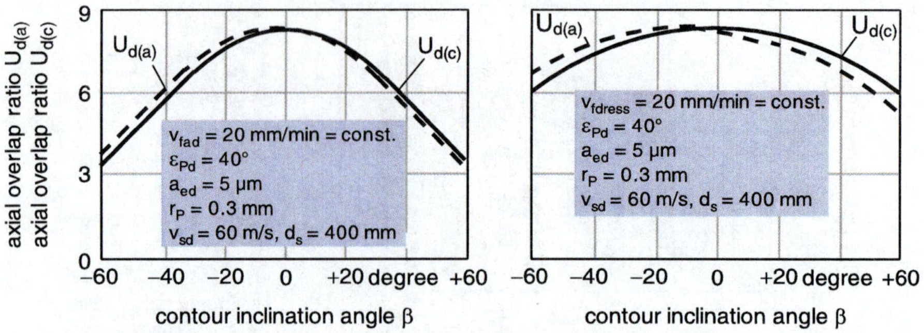

Fig. 4.29 Overlap ratio as a function of the contour inclination angle β at a constant axial feed speed (*left*) and at a constant path speed (*right*) [36] used with permission copyright Vulkan-Verlag

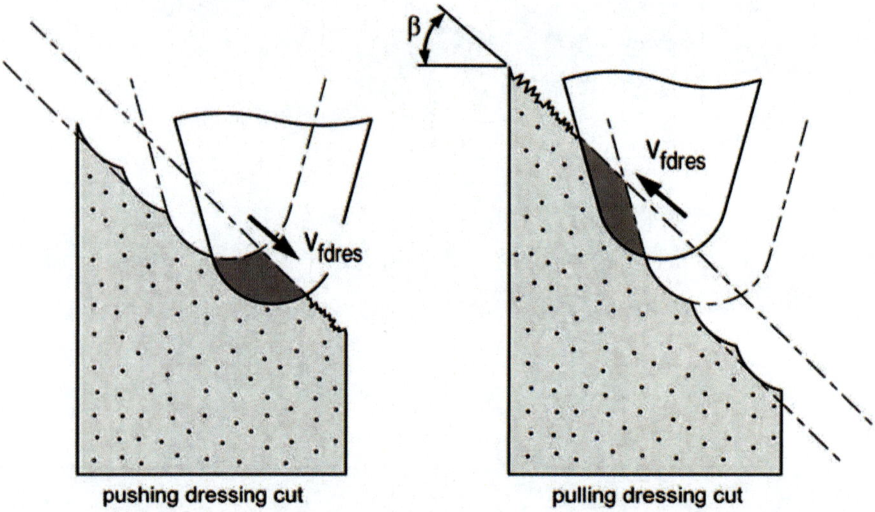

Fig. 4.30 Kinematics of engagement in the pressing and drawing dressing cut [5]

length is more in the flank area and with a pushuing dressing engagement length it is more in the front face area. Also, at a constant path speed, with the pulling dressing cut, because of the greater local chip thickness higher process forces occur, which on the one hand increase dressing roller wear and on the other hand increase the inducing of cracks in the grits and the bond of the grinding wheel and as a result increase wheel wear [4, 24].

The depth of dressing cut a_{ed} also affects the forming of the wheel topography and dressing roller wear. When the dressing of superabrasive CBN wheels, for technical and economic reasons, the depth of dressing cut is between 2 and 6 µm. Tests show that when dressing with a depth of dressing cut of less than 2 µm the CBN grits do not split completely, which results in the cutting ability of the grits

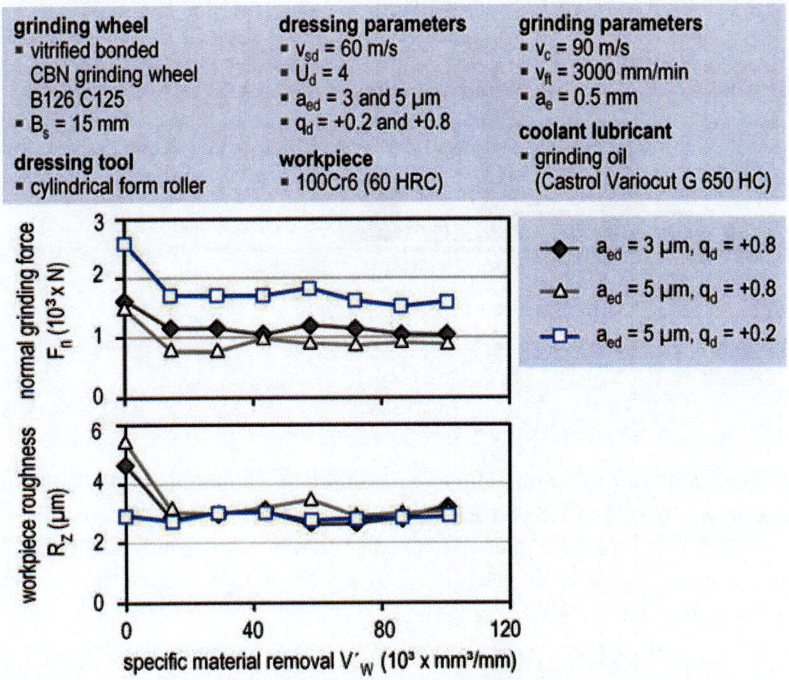

Fig. 4.31 Long-term wheel behaviour under different dressing conditions

being reduced [40]. While reducing the depth of dressing cut leads to an improvement in workpiece surface quality, because the wheel topography is levelled considerably, it does result in an increase in grinding forces and therefore in a reduction in the material removal rates that can be achieved [4]. Increasing depth of dressing cut leads to an increase in dressing roller wear [41].

The long-term behaviour of the wheel is also influenced by the dressing parameters. In the grinding tests illustrated in Fig. 4.31 workpiece roughness decreases at the beginning phase of the grinding operation. This is due to grit flatting or grit edge rounding. But on the other hand the grinding forces also fall at the start of the grinding process, which is due to an initial relief grinding phase. It can be established that after the unstable initial grinding phase the grinding process is only influenced to a lesser extent by the dressing parameters. The changes in grinding forces and workpiece roughness during the unstable initial grinding phase depend not only on the dressing parameters but also on the specifications of the grinding wheel and the dressing roller, so at the beginning of the grinding operation an increase in workpiece roughness or an increase in grinding forces can also occur [4, 42]. The dressing conditions can also have a significant influence on wheel wear, particularly when grinding with water-miscible cooling lubricants (Fig. 4.32).

The wear of a form roller is influenced by the dressing parameters. Generally, dressing roller wear is greater in down dressing than in up dressing. This may be due to the higher process forces and the resulting micro-splitting

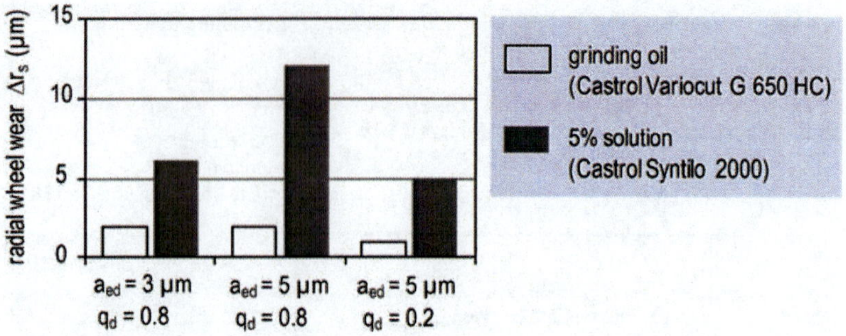

Fig. 4.32 Radial grinding wheel wear when grinding with different coolant lubricants and dressing parameters

of the diamond grits in down dressing [4, 23]. A reduction in U_d and also an increase in a_{ed} can also lead to increased dressing roller wear. Form roller wear can also be influenced by the size of the grits in the grinding wheel. When dressing superabrasive CBN wheels, with a coarser grit wheel there is greater wear on the form roller [4].

4.5.3 *Diamond Cup Wheels*

Diamond cup wheels can be used economically for the dressing of rectilinear small grinding wheels, which is for example nearly always the case when internal cylindrical grinding [43]. Diamond cup wheels are usually driven with an electric spindle motor, but for simple dressing tasks diamond cup with a rotary wing wheels can also be used (Fig. 4.33). Cup wheels with a rotary wing are driven by the grinding wheel through frictional contact or by the coolant or compressed air and braked by the wings at higher rotational speeds [44]. When dressing, the spindle motors are loaded axially and radially. To avoid high axial tilting moments on the spindle bearing, the external diameters of diamond cup wheels are kept small. As a result, only relatively low tool peripheral speeds can be achieved [33]. The dressing operation with diamond cup wheels is carried out by the axial feed of the cup wheel, whereby the cup wheel is tilted at an angle of 3–10° to the rotation axis of the grinding wheel [4]. The process parameters that can be set freely when dressing with diamond cup wheels are the dressing speed ratio q_d, the dressing overlap ratio U_d, the depth of dressing cut a_{ed} and the wheel peripheral speed v_{sd}. Like in dressing with diamond rollers, carrying out the dressing process in up dressing results in a finer wheel topograpy, which leads to higher grinding forces and a finer workpiece surface.

When dressing with diamond cup wheels the depth of dressing cut a_{ed} also plays a very important role. As Fig. 4.34 shows, with increasing a_{ed} the dressing normal forces increase more by superabrasive CBN wheels than by conventional alumina

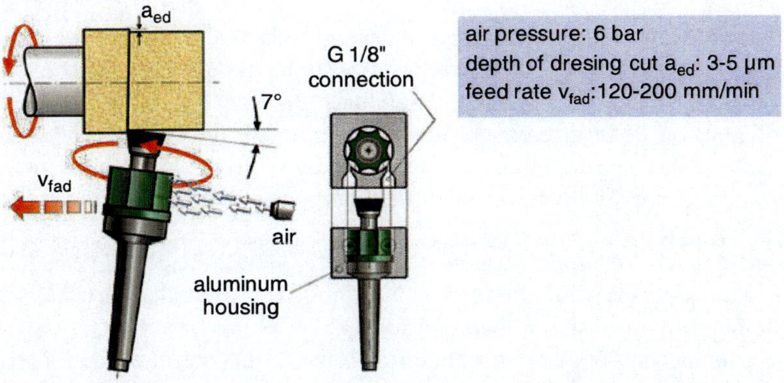

Fig. 4.33 Cup wheel with a rotary wing [9] used with permission copyright Winterthur Technology Group

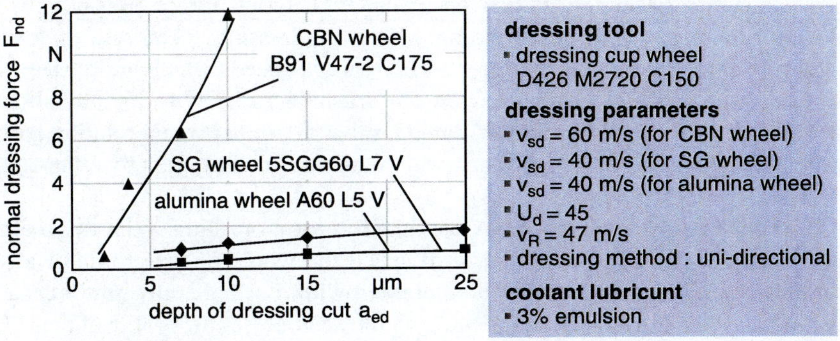

Fig. 4.34 Dressing normal forces with alumina, SG and vitrifiedbonded CBN grinding wheels [9] used with permission copyright M. Çinar

or SG grinding wheels. This is due to the greater compressive strength and greater hardness of the CBN grits, and also the greater strength of the vitrified bond used in CBN wheels. Higher dressing forces can lead to shape deformation particularly if the grinding systems are not sufficiently stiff, as for example by internal cylindrical grinding. Therefore CBN wheels are dressed with depth of dressing cuts that are as small as possible [43].

4.5.4 Crushing

Crushing is a special case of dressing with rollers, in which the dressing speed ratio q_d is equal to one. In this process, the dressing diamonds impact on the surface of the wheel at a penetration angle of 90°. In the absence of the relative speed between grinding wheel and dressing roll, the removal mechanism in crushing is based on the fracture of bond posts in the abrasive layer by applied normal forces. Therefore, crushing is only used for wheels with brittle bonds such as vitrified or brittle bronze bonds. When a diamond of the dressing roller engages the wheel surface, a normal force acts on the abrasive layer. When this normal force has exceeded a certain level, it causes the bond posts to break, thereby generating the wheel profile.

Depending on the process kinematics a distinction is made between full form crushing and point crushing. In full form crushing, a profile roller is used as dressing tool. In the point crushing, a diamond form roller is used as dressing tool, which produces the necessary wheel profile through the CNC control of the roller movement. Crushing can be carried out with both diamond dressing rollers and diamond-free rollers made of steel or tungsten carbide. Compared with point crushing, full form crushing is characterised by little being needed in terms of control and relatively short profiling times. However, in full form crushing, there is a relatively large contact cross-section, which results in high radial forces. Besides, in full crushing of a non-cylindrical grinding wheel the dressing speed ratio q_d can be equal to one only at one contact point. The relative speeds which are present over the rest of the profile lead to wear on the crushing roller [24, 36]. In full form crushing, the radial dressing feeds f_{rd} are less than 5 μm per wheel rotation and the peripheral speeds are less than 5 m/s, which are considerably less than the cutting speed in the grinding operation [24].

Point crushing is much more flexible than full form crushing. With this method, using a dressing spindle with control of speed the whole of the profile of a non-cylindrical grinding wheel can also be dressed with q_d=1, thereby minimising the wear of the roller and as a result maximising the accuracy of the profile of the wheel. In addition, compared with full crushing, because of the additional parameter of overlap ratio U_d, in the point crushing the wheel topography can be created over a wider range. Also, point crushing can be carried out at higher peripheral speeds than form full crushing. Point crushing with diamond form rollers has significant economic advantages particularly in the profiling of fine-grained diamond wheels for profile or tool grinding [24]. The effects of system variables and process parameters on the process rasults by point crushing of diamond grinding wheels are summarised in Fig 4.35.

4.5.5 Touch Dressing

Touch dressing is used for conditioning superabrasive CBN grinding wheels, whereby the wheel is dressed with a form roller with very small depth of dressing cuts (in the micrometer range). In touch dressing, the exposed peaks of the grits are only splintered in order to increase the number of dynamic cutting edges on the surface of the wheel and thereby reduce the roughness of the workpiece. On the other

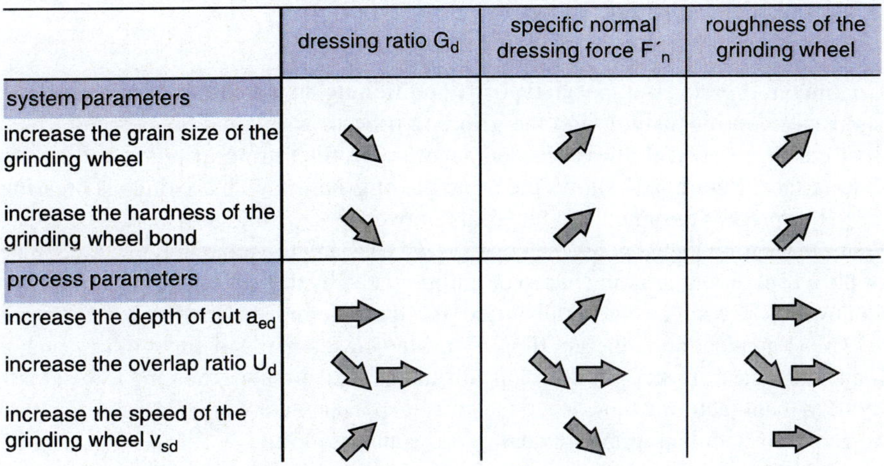

Fig. 4.35 Effect of system and tool setting parameters with the point crushing of diamond grinding wheels [25] used with permission copyright Dr. Kaiser Co

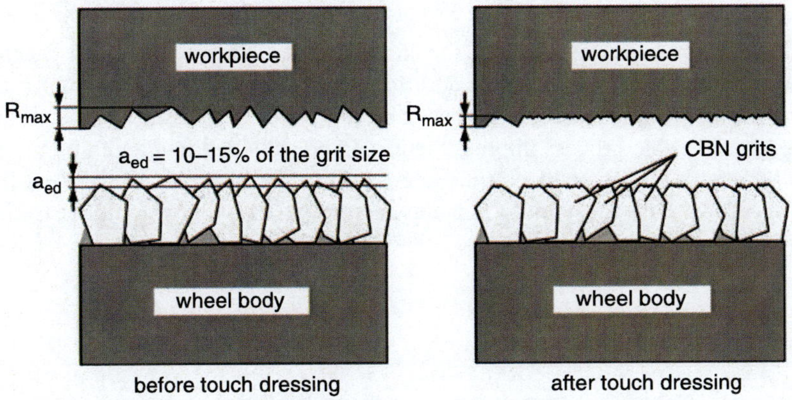

Fig. 4.36 Schematic of touch dressing an electroplated CBN grinding wheel

hand, however, increasing the number of dynamic cutting edges and reducing the grit protrusions and chip spaces by touch dressing leads to an increase in grinding forces, which must be taken into consideration in the design of the process. Touch dressing is mainly used for new electroplated CBN wheels, because in these wheels due to the different grit geometries and the scatter of the grit size the CBN grits protrude to a varying extent, as a result of which the workpiece roughness that is generated can be much higher than expected (Fig. 4.36).

When dressing multi-layer vitrified and metal bonded CBN wheels the attempt is also made to carry out the process with depth of dressing cuts that are as low as possible in order to avoid a serious reduction in the chip spaces between the abrasive grits.

4.5.6 Continous Dressing

Continuous dressing is a special type of conditioning, in which a conventional wheel is dressed continuously during the grinding process with a diamond profile roller. As a result, the wheel always has an accurate profile during grinding and a sharp cutting face. Figure 4.37 shows the principle of grinding with continuous dressing.

In this process the amount of material removed by dressing must always be greater than the wear on the wheel. When continuous dressing the reduction in the diameter of the wheel during grinding has to be compensated by the machine's control system. Continuous dressing is generally used for the grinding of hard-to-cut materials. A typical application is the grinding of turbine blades made of super alloys such as high nickel base alloys [45]. The main advantage of continuous dressing is the greatly reduced manufacturing time due to the greater specific material removal rates that can be achieved with high grinding quality and stability [46].

4.6 Ultrasonic Assisted Dressing

In ultrasonic assisted machining, the kinematics of the conventional machining process are superimposed with an additional vibration in the micrometer range at ultrasonic frequencies, namely higher than 16 kHz [47]. With the ultrasonic assistance, considerable process improvements can be achieved in machining processes with geometrically defined cutting edges as well with geometrically undefined cutting edges. The most important advantages of using ultrasonic assistance in traditional machining processes are [48–52]:

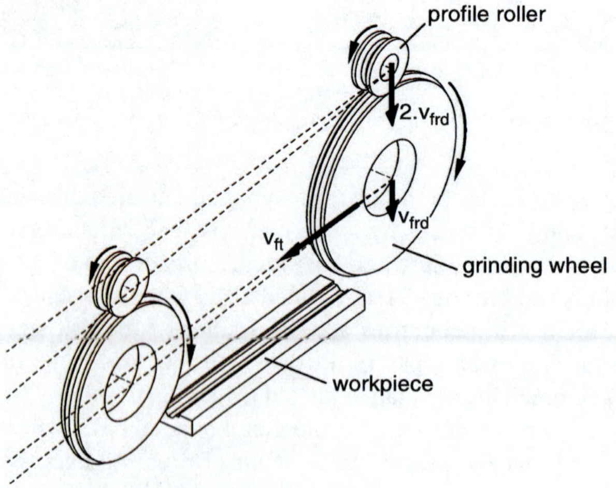

Fig. 4.37 Principle of continuous dressing

- Reduction in the machining forces involved;
- Improvement in the supply of coolant into the process zone;
- Reduction in friction between tool and chip;
- Reduction in tool wear;
- Reduction in microscopic and macroscopic material damage; and
- Increase in the dimensional accuracy of workpiece profiles.

Although ultrasonic assistance has been used successfully in many different machining processes for some time now, for technical and economic reasons as yet ultrasonic assisted dressing has not been used to any significant extent in industry. Nevertheless, the latest research reports indicate that with a suitable ultrasonic system and if the process parameters are set correctly, significant improvements can be achieved with ultrasonic assistance, particularly in the dressing of superabrasive CBN and diamond grinding wheels [11, 53–56].

4.6.1 Ultrasonic Vibration Systems

Elastomechanical longitudinal ultrasonic vibrations are produced by the conversion of electrical energy to mechanical vibration, for the most part in piezoceramic but also, in some cases, magnetostrictive altrasonic transducer. The high-frequency electric alternating voltage that is needed is produced by converting low-frequency mains voltage in a voltage generator. The longitudinal vibrations that are produced in the transducer are periodic, elastic deformations in the micrometer range at frequencies in the ultrasonic range. The ultrasonic amplitude produced in the ultrasonic transducer is often very small compared with the amplitudes needed in cutting process. The amplitude of the vibration can be increased considerably by a narrowing of the cross-section of a transition element fixed on the ultrasonic transducer (what is known as the booster) (Fig. 4.38).

Then, a sonotrode flanged to it transfers the vibration amplitudes through the tool into the process zone.

Decisive for the effectiveness of the ultrasonic vibration for the process is that all the elements, through which the vibration passes, exhibit suitable mode shape. In order to ensure that an increase in amplitude takes place that is largely free of losses, the system must vibrate at one of its natural frequencies (resonance). In terms of the design of the system this means that the lengths of the individual elements must correspond to the half wavelength of the vibrations or a full-number multiple of it, so that in the entire ultrasonic system a standing wave is generated [48].

4.6.2 Kinematics of Ultrasonic Assisted Dressing

In ultrasonic assisted dressing, ultrasonic vibrations are superimposed on the conventional acting movement of the process. It is possible to use ultrasonic with stationary as well as with rotary dressers (Fig. 4.39).

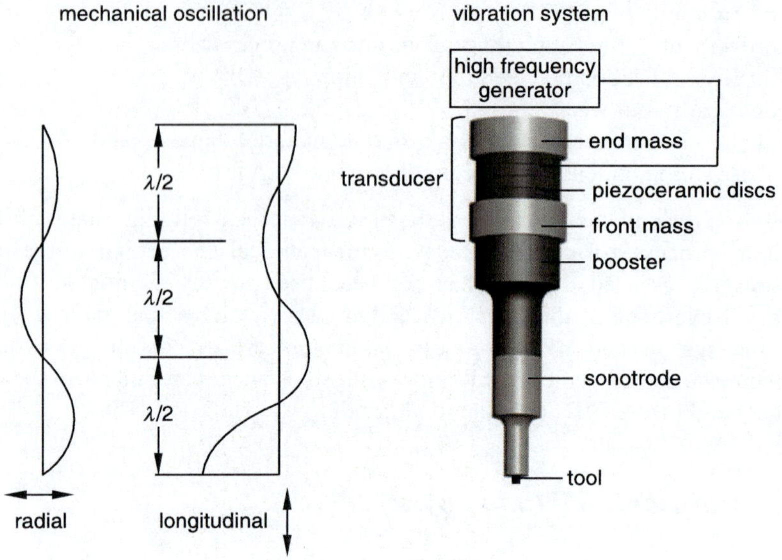

Fig. 4.38 Vibration chain in the forming of a fixed ultrasonic wave

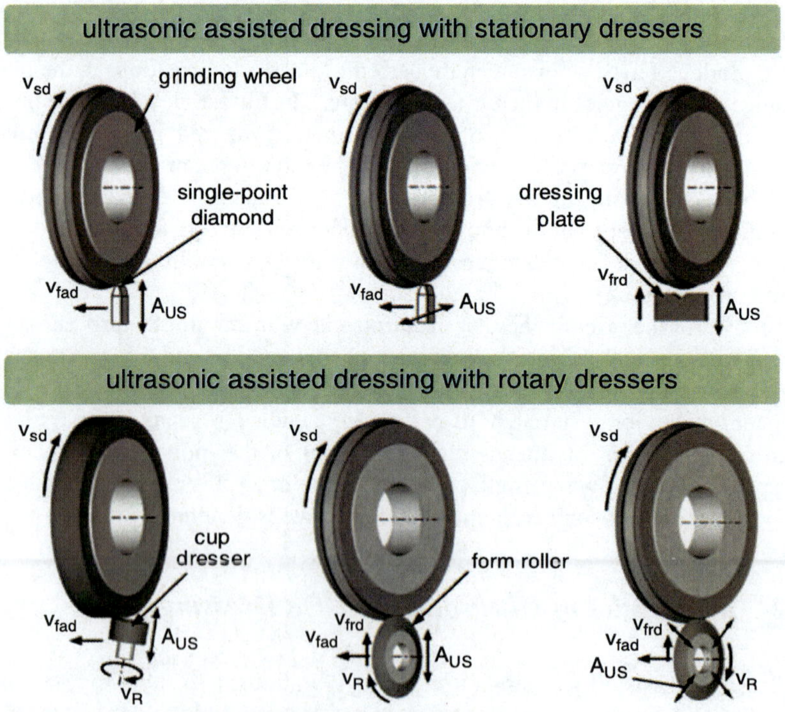

Fig. 4.39 Some ultrasonic assisted dressing methods

4.6.3 Ultrasonic Assisted Dressing with Stationary Dressers

The engagement conditions and the resulting splitting and fracture on the abrasive grits and the bond are different in conventional and ultrasonic assisted dressing. If the dressing tool vibrates in the radial direction of the grinding wheel y according to (4.4), the dresser strikes the wheel with a penetration angle according to (4.5), which causes the grits and brittle hard bond of the grinding wheel to split more intensively [57].

$$y(t) = A_{US} \cdot \sin(2 \cdot \pi \cdot f_{US} \cdot t) \tag{4.4}$$

$$\alpha_e = arctg\left(\frac{2.\pi.f_{US}.A_{US}.\cos(2.\pi.f_{US}.t_e)}{v_{sd}}\right) \tag{4.5}$$

Besides, the sinusoidal active path of the dressing tool in ultrasonic assisted dressing produces a rougher grinding wheel topography with a smaller number of kinematic cutting edges. The rougher wheel topography with sharper abrasive grits leads to lower grinding forces and greater roughness of the ground surface (Fig. 4.40). However, Jiao et al. have reported that in grinding with a resin bonds SiC wheel, after the unstable initial grinding phase the wheel dressed with ultrasonic assistance has better workpiece roughness. They also observed in their studies that the influence of ultrasonic assistance on workpiece roughness can also depend on the grinding depth of cut a_e, so that with depth of cuts of more than 0.03 mm, after ultrasonic assisted dressing in continuous mode, in which $a_{ed} \geq \mathbf{A}_{us}$, the resin bonded SiC wheel can produce the lowest surface roughness [58]. Ikuso et al. have

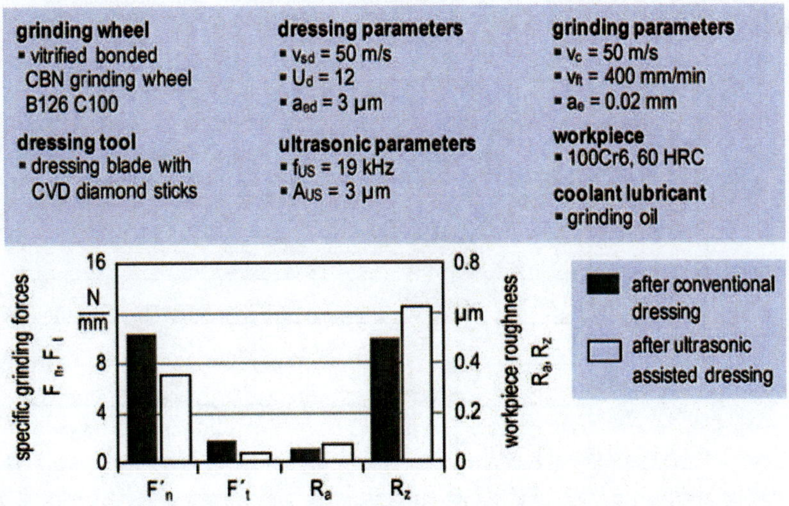

Fig. 4.40 Comparison of conventional and ultrasonic assisted dressing

also reported less workpiece roughness after ultrasonic assisted dressing when grinding with a vitrified bonded CBN wheel [53]. Therefore, the effect of ultrasonic assistance on the generation of the wheel topography during the dressing with stationary dressers and the resulting workpiece roughness clearly depends on the wheel specification and the grinding conditions. Ultrasonic assistance can also improve dressing accuracy (Fig. 4.41). This may be due to the non-sensitive vibration cutting mechanisms and the increase in the stiffness of the dressing unit due to the ultrasonic vibration [53]. According to Jiao et al., in discontinues dressing mode, in which $a_{ed} < A_{US}$, the process forces are on average smaller and more uniform, which results in a more even height of the micro cutting edges produced during dressing [58].

Diamond dresser wear can also be reduced by ultrasonic assistance (Fig. 4.42). The reduction in wear comes back to improved coolant supply and the reduction in

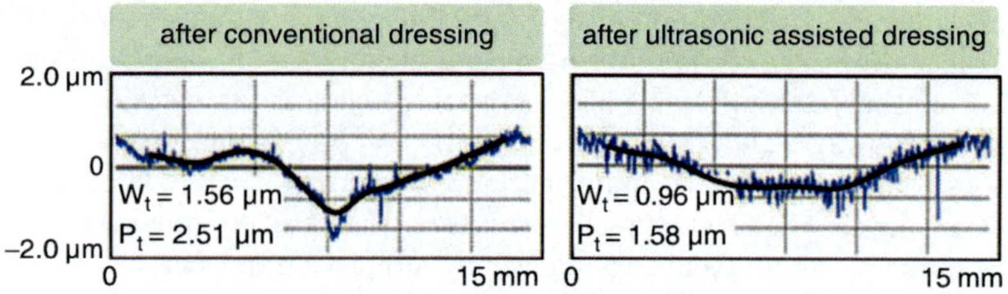

Fig. 4.41 Comparison of cross-sectional shapes of ground surface

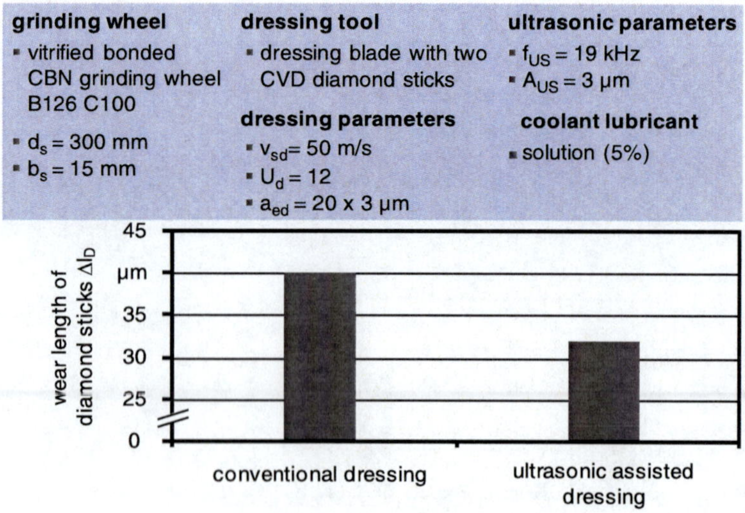

Fig. 4.42 Reduction in the wear lengths of the diamond logs of a CVD tile dresser due to ultrasonic assistance

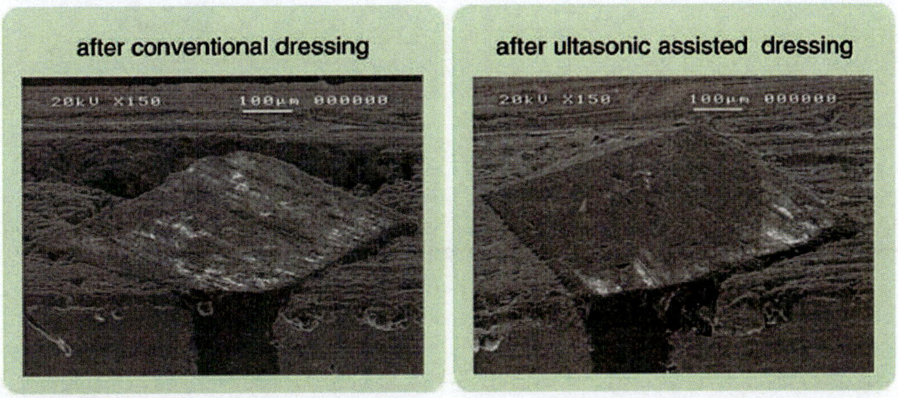

Fig. 4.43 SEMs of the diamond logs

friction coefficient with ultrasonic assistance, which reduces the heat generation in the dressing contact zone. As a result, the risk of graphitisation of the dressing diamond, which is directly related to the forming of wear on the dressing diamond, is reduced [23]. This explanation can be confirmed on the basis of SEMs of the CVD diamond logs of a blade dresser after conventional dressing and ultrasonic assisted dressing. As can be seen in Fig. 4.43, the surface of the diamond log after ultrasonic assisted dressing displays much more uniform wear with few abrasive traces and hardly any grit edge splitting compared with after conventional dressing. Furthermore, after ultrasonic assisted dressing the dressing tool body displays around the diamond log only slight damage. This comes back also to the less heat generation by applying ultrasonic vibration. Also, after conventional dressing there are a large number of white marks, which are clearly adhesive layers of the grit and binder elements. The abrasion and adhesion wear marks on the surface of the diamond after conventional dressing are due to the high development of temperature, caused by the continuous contact and the high friction between the grinding wheel and the stationary dresser.

4.6.4 Ultrasonic Assisted Dressing with Rotary Dressers

Superabrasive CBN and diamond grinding wheels are dressed almost exclusively with rotary diamond dressers, so ultrasonic assisted dressing with rotary dressers has attracted researchers' attention in the last decade as a new conditioning method. Liebe used the longitudinal ultrasonic vibrations in the profiling of vitrified and resin bonded grinding wheels with a diamond cup wheel, in which the angle between the axes of rotation of the grinding wheel and the cup wheel was 75°. With the ultrasonic assistance Liebe was able to achieve higher removal rates compared with conventional dressing (Fig. 4.44). During ultrasonic assisted

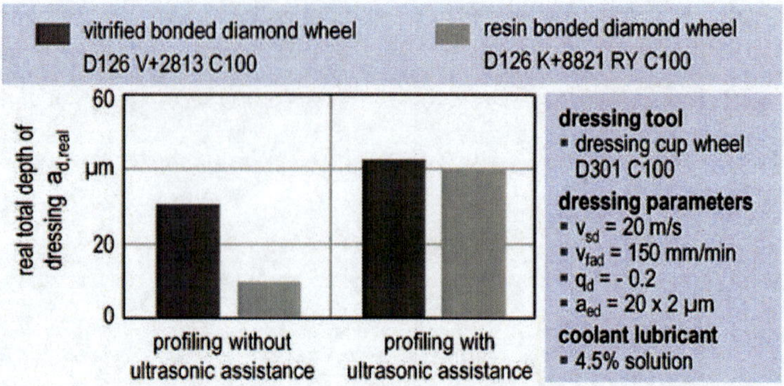

Fig. 4.44 Ultrasonic assisted dressing of diamond grinding wheels with a diamond cup wheel [11]

dressing of vitrified bonded wheel, however, cracks were induced in the bond by the vibration, which in subsequent grinding operetion led to a reduction in the grinding ratio. In subsequent grinding operation with resin bonded wheels, the increasing in grinding forces at the initial grinding phase was higher after ultrasonic assisted dressing, although the grinding forces were approximately at the same level over the course of the grinding operation. This is due to the sharp cutting edges becoming blunt relatively more quickly [11].

Sroka also conducted research concerning the dressing of resin bonded diamond grinding wheels with an ultrasonically vibrated diamond cup wheel. With the use of ultrasonic assistance he was able to increase the dressing ratio by a factor of 10 and to reduce dressing costs by 59% [55].

Nomura et al. investigated the effect of ultrasonic vibrations on the dressing of small CBN wheels in the internal cylindrical grinding of small bores. In their investigations, the longitudinal ultrasonic vibrations were exerted on the grinding tool. A diamond cup wheel was used for dressing of the grinding tool. As results, it was found that applying ultrasonic vibration decreased the dressing forces by more than 22%, and improved the run-out of the grinding wheel by 30% compared to the conventional dressing [54].

The use of ultrasonic assistance in dressing with diamond dressing rollers is of great interest in view of the widespread use of these dressers. But there are difficulties because of the dimensions and weight of the dressing rollers and spindles that are usual in industrial practice. An effective way of carrying out ultrasonic assisted dressing with a form roller is to generate the radial vibrations in the form roller. Here, the form roller is alternatingly expanded and compressed radially in the micrometer range at a ultrasonic frequency. This can be achieved by positioning the dressing roller on a nodal point of the sonotrode, whereby the sonotrode functions as the shaft of the dresser spindle. The change in cross-section of the sonotrode is greatest in the nodal points of the standing longitudinal vibration (see Fig. 4.38). This change in cross-section is amplified by the special shape of the dressing roller [57]. The radius of the dresser, which changes continuously due to the ultrasonic vibration, can be calculated with (4.6).

$$r_{R,US} = r_R + A_{US} \cdot \sin(2 \cdot \pi \cdot f_{US} \cdot t) \tag{4.6}$$

It is clear that the engagement path of the diamonds in the dressing roller are different in conventional and ultrasonic assisted dressing. This leads to the formation of different wheel topographies with conventional and ultrasonic assisted dressing, which in turn leads to different grinding results. As Fig. 4.45 shows, with ultrasonic assisted dressing the grinding forces are lower and the workpiece roughness is greater than with conventional dressing. This indicates that a rougher wheel topography is formed with ultrasonic assisted dressing.

With ultrasonic assistance, the lengths of the diamond engagements are shorter (Fig. 4.46). This leads to a greater theoretical effective peak-to-valley roughness on the grinding wheel. Besides, the penetration angles are greater with ultrasonic assisted dressing, resulting in greater normal loads and therefore more intensive splitting of the CBN grits when they impact with the diamond grits. As a result, the CBN wheels dressed with ultrasonic assistance have a sharper abrasive layer with a

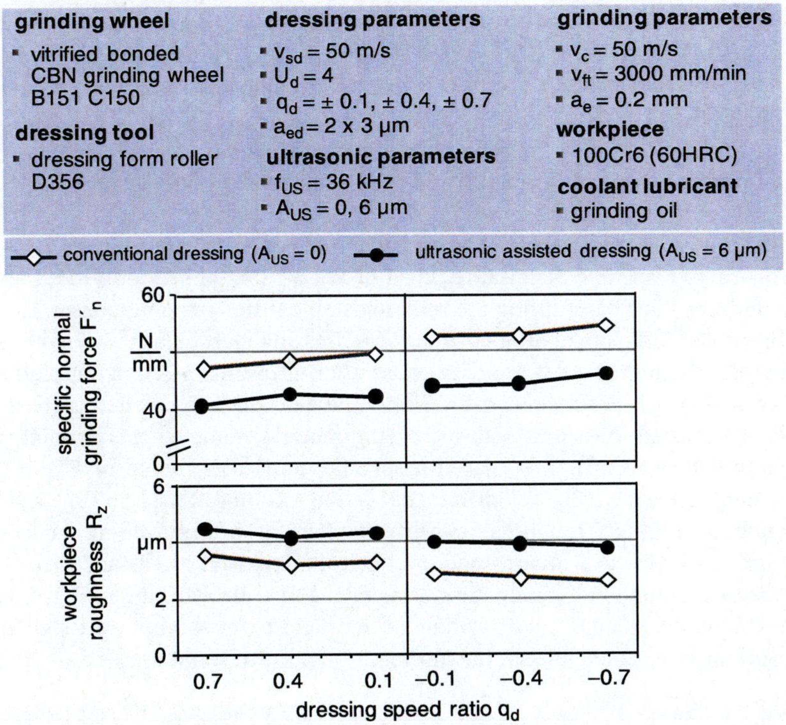

Fig. 4.45 Grinding forces and workpiece roughness after conventional and ultrasonic assisted dressing with a form roller

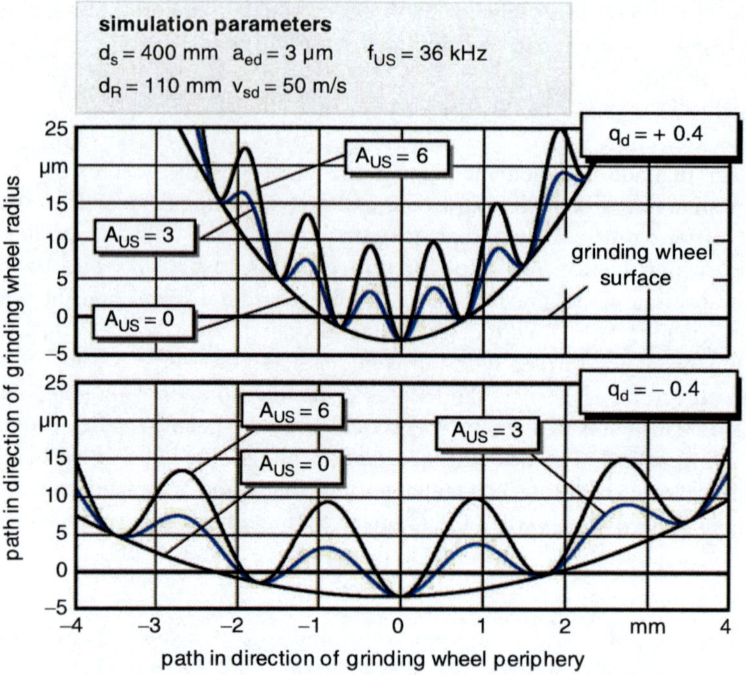

Fig. 4.46 Simulation of the engagement paths in ultrasonic assisted dressing with different dressing conditions

smaller number of kinematic cutting edges, which causes lower grinding forces and greater workpiece roughness.

Figure 4.47 shows wheel behaviour after conventional and ultrasonic assisted dressing up to a specific material removal of V'_W=12,000 mm^3/mm. In all tests the grinding forces decrease during the initial unstable relief grinding phase. The relief grinding phase lasts longer after conventional dressing, particularly when dressing in up dressing. A shorter relief grinding phase after ultrasonic assisted dressing comes back to higher grit protrusions caused by a greater splitting of bond material with ultrasonic assisted dressing. With increasing material removal the grinding forces gradually become similar, so that at a specific material removal of $V'_W = 12{,}000$ mm^3/mm there are hardly any differences in the grinding forces. In Fig. 4.47 too a reduction in workpiece roughess can be seen at the start of the grinding process in all the tests; this is due to flattening and/or edge rounding of the grits and the resulting increase in the number of kinematic cutting edges. It is also striking in Fig. 4.47 that, interestingly, the changes in the grinding forces and in the workpiece roughness are very similar after conventional down dressing and ultrasonic assisted up dressing, with q_d=+0.4 and q_d=−0.4 respectively.

Figure 4.48 shows the grinding ratio G after a specific material removal of V'_W=12,000 mm^3/mm. It is obvious that the wheel wear is greater after ultrasonic assisted dressing. This comes back to the greater load on the CBN layer

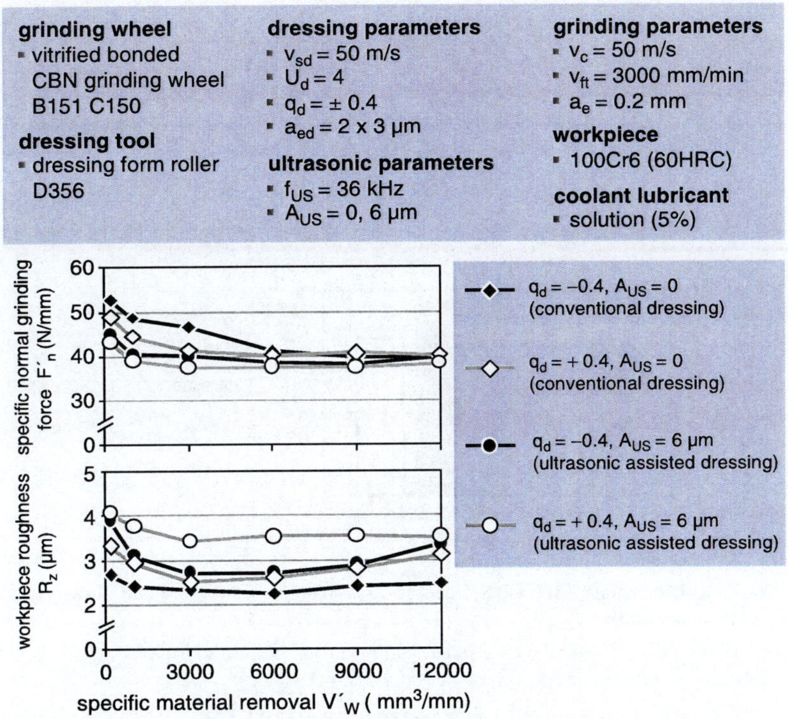

Fig. 4.47 Grinding wheel behaviour after conventional and ultrasonic assisted dressing

during the ultrasonic assisted dressing, as a result of which more intensive cracking occurs in the CBN grits and in the bond, which as weak points intensify the wheel wear. In addition, a greater effective peak-to-valley-roughness on the grinding wheel after ultrasonic assisted dressing leads to a reduction in the number of kinematic cutting edges and therefore to an increase in the individual chip thicknesses in the grinding process. Increase in the individual chip thicknesses leads to an increase in the cutting loads acted on the CBN grits in the grinding operation. As a result, in particular those CBN grits that are exposed and therfore are not as solidly fixed in the bond break out right at the start of the grinding process. Interestingly, the grinding ratios after conventional down dressing and ultrasonic assisted up dressing, with q_d=+0.4 and q_d=−0.4 respectively, are the same.

Figure 4.49 shows the radial form roller wear Δr_R in conventional and ultrasonic assisted dressing with $q_d = \pm 0.4$. With up dressing, ultrasonic assistance leads to a reduction in dresser wear, and with down dressing it leads to an increase. The different effects of ultrasonic assistance on the wear behaviour of the roller dresser in up dressing and down dressing are due to the different principal wear mechanisms that are dominant in the two dressing methods. While in down dressing the micro-splitting that results from the high mechanical loading is the principal wear mechanism in the development of wear on the dressing diamonds, in up

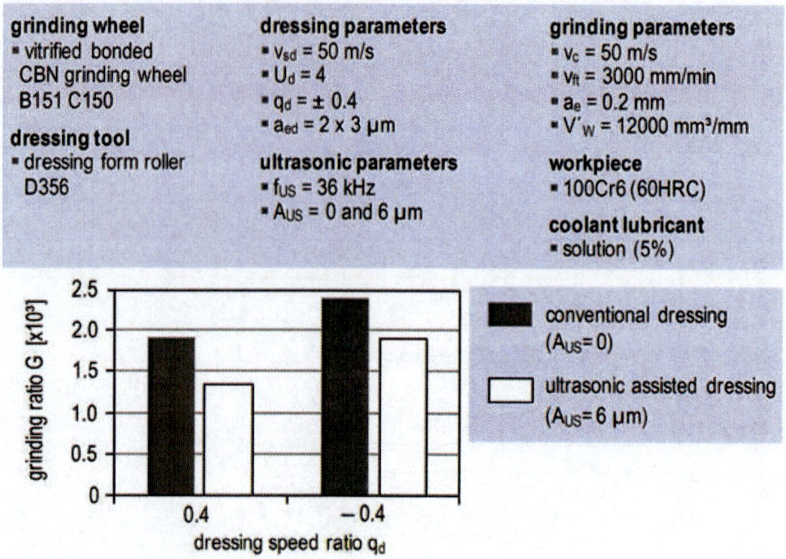

Fig. 4.48 Wear behaviour of the CBN wheel after conventional dressing and ultrasonic assisted dressing with a form roller

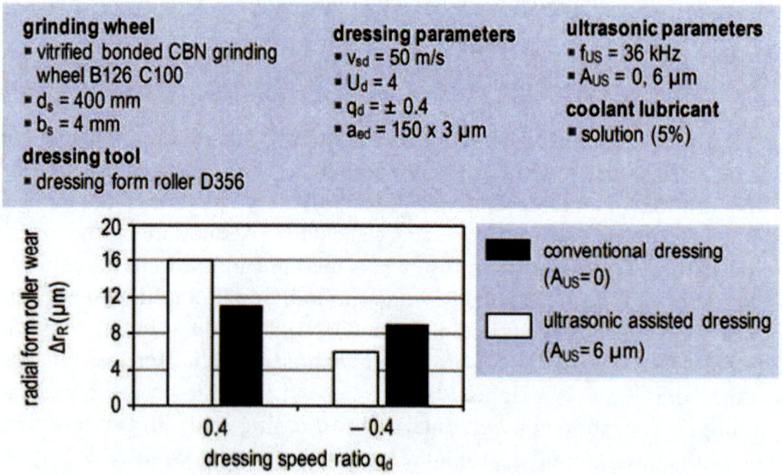

Fig. 4.49 Influence of ultrasonic assistance on the wear behaviour of the roller dresser

dressing the graphitisation of the dressing diamonds resulting from the abrasive friction is the principal mechanism of wear on the dressing diamonds [4, 23]. As already shown above, ultrasonic assistance leads to an increase in the kinematic penetration angle, as a result of which the mechanical load on the diamonds of the dressing tool increases. On the other hand, ultrasonic assistance improves the flow of coolant into the dressing contact zone and reduces the coefficient of friction between the active dressing partners, as a result of which the heat generation in the dressing contact zone, and therefore the intensity of graphitisation, decreases. Therefore, in down dressing ultrasonic assistance leads to an intensity in the principal mechanism of dressing diamond wear, and in up dressing it leads to a reduction in this mechanism the principal wear mechanism.

As mentioned above, in industry the conventional dressing of vitrified bonded CBN wheels is often carried out in down dressing mode, in order to avoid thermal damage to the workpiece cauesed by a too fine wheel surface. Ultrasonic assisted dressing tests with form rollers show, however, that with suitable ultrasonic assistance in up dressing (i.e. in the range $q_d<0$) a grinding wheel topography can be produced that is comparable to the topography produced after conventional down dressing. From roller wear with ultrasonic assistance in up dressing is, however, much less than with conventional dressing in down dressing.

4.7 Laser Dressing

With the further development of laser technology the areas of application for machining with laser beams have been increasing all the time in recent years. The laser has become established as a fast and versatile tool particularly for the machining of hard and brittle (difficult-to-machine) materials such as high-performance ceramics. As well as being extremely versatile, the laser has additional positive features such as its ability to introduce heat in a controlled condition, the small zone in which heat exerts an influence, machining without machining forces and the fact that it can be controlled easily in terms of space and time. The use of laser beams for conditioning grinding wheels enables the disadvantages of conventional conditioning methods arising from the tools and methods used to be avoided. These disadvantages include, for example, dresser wear when dressing with diamond dressers or the presence of process forces in profiling. The wear on an expensive diamond tool results not only in high tool costs but also in an unstable dressing process in terms of wheel geometry and topography. The process forces that arise in profiling with conventional conditioning methods can lead on the one hand to dimensional and profile errors on the geometry of the wheel, and on the other hand to the inducing of microcracks in the abrasive grits and the bond of the grinding wheel.

So far lasers have been used in various ways for conditioning grinding wheels. With a laser beam it is possible to clean the surface of grinding wheels non-aggressively. The laser beam can also be used to sharpen superabrasive grinding wheels, removing only the bond. If sufficient bond material is removed, the abrasive grits can also fall out. Influencing of both the abrasive grits and the bond at the same time with the laser beam and laser-assisted profiling are other methods that have been developed.

4.7.1 Principle of Laser Conditioning

Laser conditioning is a thermal machining method in which the surface of the grinding wheel is heated in controlled conditions by the laser beam. With this process the components of the abrasive layer and also the chips embedded in the pores of the abrasive layer can reach melting point. Laser conditioning is therefore influenced both by the specification of the wheel and also by the conditions of the laser irradiation. Laser conditioning is carried out best with a Nd:YAG laser because not only is it possible to achieve a high degree of focus with the beam of the Nd:YAG laser, but also it has the best absorption behaviour and therefore uses energy to optimum effect on the bond of the grinding wheel. Besides, by means of optical fibres the beam of the Nd:YAG laser can be directed safely and flexibly to the machining point, which is of great advantage particularly in industrial applications [13].

The laser irradiation of grinding wheels can be subdivided into three principal arrangements (Fig. 4.50). With a radial beam relative to the irradiated surface of the wheel, a Nd:YAG laser beam can produce a uniform irradiated area. This arrangement is therefore best suited for cleaning the wheel and for the controlled cutting back of bond material, namely for sharpening. For profiling with close tolerances on geometry, the radial irradiation would have to be carried out with a depth control and also with a cw-Q-switched laser beam (with pulses of a few nanoseconds) [36]. Also, unlike with conventional sharpening processes, with radial irradiation no bond bridges are produced behind the grits that have a supporting effect on the grit and therefore have a positive effect on wheel wear [59]. Another principal irradition arrangement is the beam being directed tangentially to the wheel. With this arrangement it is possible both to produce bond bridges and also to remove roundness errors. However, with this arrangement the beam power needed to

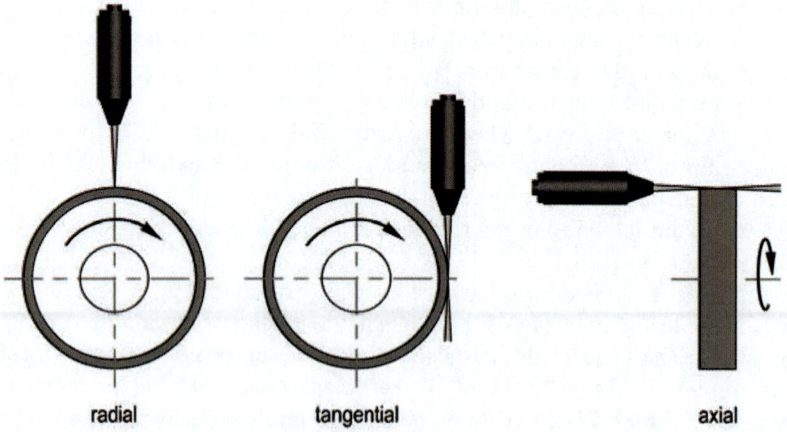

Fig. 4.50 Possible irradiation arrangements in laser conditioning [13] used with permission copyright J.-H. Timmer

achieve a process speed comparable to that achieved with radial irradiation is greater due to the larger irradiation area and because the focus of the beam is only partly used. The third arrangement that is possible with laser conditioning is in the axial direction, but because of the Gaussian beam optics it is not suitable for precision conditioning. With this arrangement, inevitably the profile of the wheel deteriorates [13].

In designing laser conditioning the influencing factors such as beam power, irradiation time and defocus distance should be set so that by heating the surface of the wheel, the desired melting and eventually the required removal of the elements of the abrasive layer and also of the chips embedded in the pores of the abrasive layer can be achieved. Therefore the relationship between influencing factors and the development of heat in laser conditioning has to be analyzed.

4.7.2 Thermal Consideration of Laser Conditioning

The heat induced in the surface of the wheel by laser irradiation leads to effects such as melting, evaporation and the forming of plasma, which enable the components of the abrasive layer and/or the chips embedded in it to be removed [60].

The development of heat in a surface irradiated with the laser depends on various influencing factors. These include in particular the power of the beam and the area that is to be machined by the laser. With laser conditioning the temperature distribution and eventually the depth to which material is melted can be described by (4.7) as the basic equation governing one-dimensional heat transfer [61].

$$\frac{\partial^2 T}{\partial z^2} = \frac{1}{\alpha}\frac{\partial T}{\partial t} \tag{4.7}$$

T is the temperature in a grinding wheel depth of z and t is the time after starting laser irradiation. α is the thermal diffusivity of the grinding wheel. If there is no loss of heat from the radiation and no melting, then (4.8) can be derived from (4.7) [60].

$$T(z,t) = \frac{2AP}{K} ierfc\left(\frac{z}{\sqrt{4\alpha t}}\right) \tag{4.8}$$

With pulsed laser irradiation, the laser beam intensity (P) is

$$P(t) = \begin{cases} P & 0<t<t_0 \\ 0 & t<0,\ t>t_0 \end{cases} \tag{4.9}$$

In (4.8) A is the material's absorption coefficient, and K is the thermal conductivity. As can be seen, at a constant P, increasing the irradiation time t_0 leads to an increase in the temperature in the lower layer of the grinding wheel. But this can affect the

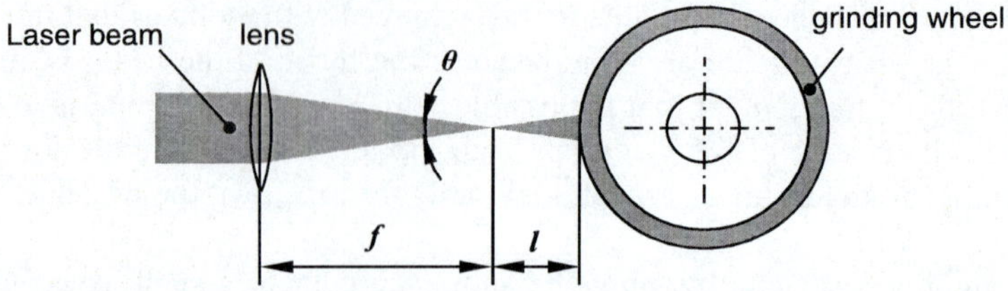

Fig. 4.51 Schematic of the laser conditioning arrangement

topography of the wheel. Therefore, in order to clean the wheel, a high laser power flux and short irradiation laser pulses should be set [61, 62]. The laser beam intensity depends on the irradiation arrangement, the average power P_m, the focus beam diameter d_b, the defocus distance z and the full divergent angle θ (Fig. 4.51). With radial irradiation, the laser power flux input can be calculated by (4.10) [60].

$$P(x, y, z) = \frac{8p_m}{\pi \cdot d(z)} \cdot \exp\left[-\frac{8(x^2 + y^2)}{(d(z))^2}\right] \tag{4.10}$$

$d(z)$ is the beam diameter at z from the focus plane along the propagating beam and is calculated according to (4.11). z is positive if the focus plane is outside the wheel. Conversely, it is negative if the focus plane is inside the wheel.

$$d(z) = d_b + 2z \cdot \tan(\theta/2) \tag{4.11}$$

As shown from (4.10) and (4.11), the laser power flux input into the wheel and therefore the development of heat in the surface of the wheel depends among other things on the defocus distance. The angle of incidence of the laser beam is another parameter that can influence laser conditioning. As the angle of incidence increases, the absorbed energy and consequently the process speed decrease [63].

4.7.3 Laser Conditioning of Conventional Grinding Wheels

With proper laser irradiation, conventional grinding wheels can be both cleaned and dressed. Fusion and the evaporation of chips are the main mechanisms in the laser cleaning of a wheel [61, 62]. The wheel can only be cleaned effectively and reliably by laser irradiation if all the process parameters are selected appropriately. The factors that lead to different influences on the components of the wheel and loosened chips by the laser irradiation have to be taken into consideration. These factors include, for example, melting point, optical reflectivity, thermal and optical conductivity, and specific heat capacities [61].

With laser dressing, a higher laser power flux input is generated compared with laser cleaning. This leads to melting (followed by re-solidification) and/or evaporation of the components of the grinding wheel, as a result of which the topography of the wheel changes in terms of its morphology and composition. Depending on the cooling speed, re-solidification produces new microstructures. As Fig. 4.52a shows, after laser dressing the alumina wheel has a compacted surface with reduced porosity [64]. In their studies, Jackson et al. also reported a reduction in porosity but without a full compaction of the surface of the wheel [61]. Figure 4.52b shows the microstructure of the grinding wheel surface before and after laser dressing. Although the grits after laser dressing are of very different sizes, from 5 to 100 μm, they are more regular and symmetrical in shape with well-defined edges and sides [61, 64]. The morphological features evolved during laser dressing reduce the surface roughness of the dressed wheel, which in turn affects the grinding performance [64]. Jackson et al. have found that there is little difference between grinding ratios for diamond-dressed and laser-dressed vitrified alumina grinding wheels, although the mechanisms of grinding wheel wear after dressing with a diamond tool and with the laser are different [61].

Fig. 4.52 Vitrified alumina grinding wheel surface before laser dressing (undressed) and after laser dressing with 750 W: **(a)** Optical micrographs of the grinding wheel surface, **(b)** Microstructure of the grinding wheel surface [64] used with permission copyright Elsevier

4.7.4 Laser Conditioning of Superabrasive Grinding Wheels

Laser conditioning is of great interest particularly for preparing superabrasive CBN and diamond grinding wheels. The laser can be used successfully for cleaning superabrasive wheels. For example, Chen et al. were able to clean a vitrified bonded CBN wheel successfully with laser irradiation. They found that the laser beam has to have greater laser energy for cleaning CBN wheels than alumina wheels due to the greater thermal conductivity of CBN [62].

Laser can also be used for dressing superabrasive grinding wheels with different types of bond. The influence on wheel topography depends not only on the irradiation parameters but also on the wheel specifications such as grit size and bond material. With the laser dressing of superabrasive grinding wheels the aim is to remove bond material in such a way that a new wheel topography is produced with the desired grit protrusion. The bond material should be removed without any thermal damage being caused to the abrasive grits, because this can increase the wear on the wheel. By irradiating the grits with individual laser pulses with different energy densities and then analysing the SEMs, the maximum permissible energy density for the laser irradiation can be determined. Any influence on or damage to the unclad grits can be seen in the form of discolouration and spalling. In the case of nickel-clad grits, if the energy density is too high, laser irradiation causes the nickel coating to melt (Fig. 4.53). Here, the maximum permissible energy density is defined as the energy density at which while most of the top side of the grit is exposed as far as the grit, the areas of grit to the side and below remain unaffected. With natural diamond as an example, Fig. 4.54 shows the boundary between damaged and undamaged grits calculated as a function of energy density and grit size. It can be seen that with increasing grit size, also, higher energy densities can be used for laser conditioning without damaging the grit [13]. The removal behaviour of the bond is another important influencing factor in the laser dressing of superabrasive grinding wheels. It could be established in

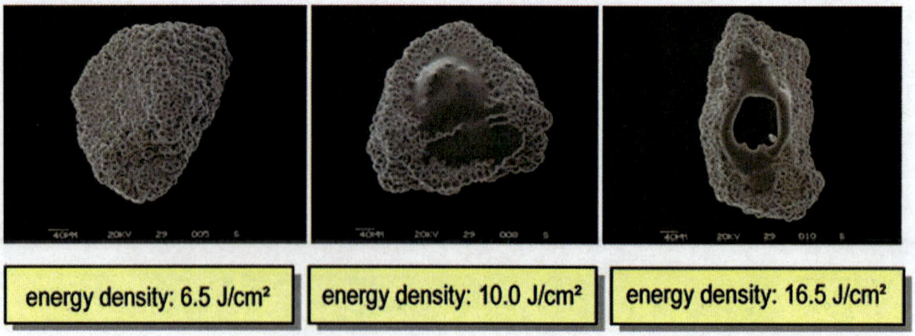

Fig. 4.53 Influence of laser beam energy density with nickel-clad diamond grits as an example [13] used with permission copyright institute of machine tools and production technology (IWF), technical university Braunschweig

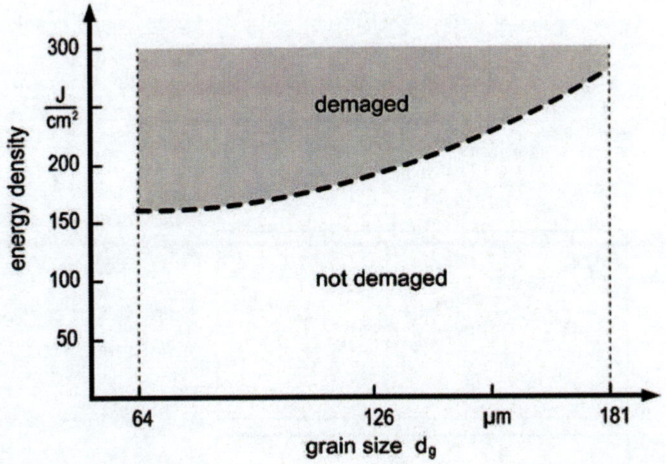

Fig. 4.54 Influence of grit size in the laser irradiation of natural diamond PDA 665 [13] used with permission copyright institute of machine tools and production technology (IWF), technical university Braunschweig

the laser irradiation of the flat sample of pure binder material that both with the vitrified and resin bond essentially there is the same removal behaviour and a linear relationship between energy density and the depth of material removal per laser pulse. This is why by setting the laser energy density adequately, a targeted removal of the bond and eventually a suitable grit protrusion (laser sharpening) or a desired grit removal (laser dressing) can be achieved. In general, resin bonds can be removed successfully with much lower energy densities than very strong vitrified bonds [59].

Similarly, an almost linear relationship between energy density and depth of removal per pulse can be established with the bronze bonded grinding wheel. However, greater laser energy density is needed for the removal of metal bonds, which in the case of diamond grinding wheels increases the risk of grit failure or graphitisation of the diamond particles. Hosokawa et al. have used an air jet directed on the laser irradiating spot in the laser dressing of a bronze bonded diamond wheel. The air jet blows away the molten binder before it solidifies again and thereby increases the efficiency of the process [66]. The colour of the bond also plays an important part for the quality of material removal. For example, by a change in colour, which does not influence the grinding process, a change in the absorption behaviour for the laser irradiation and therefore an improvement in the bond can be obtained [59].

A comparison between laser conditioning and conventional conditioning can be carried out on the basis of the grinding force ratio $\mu = F'_t/F'_n$. The higher μ is, the more favourable are the machining characteristics of the grinding wheel. Figure 4.55 shows the change in grinding force ratio in internal cylindrical grinding with a resin bonded CBN wheel that has been laser-conditioned and a wheel that has

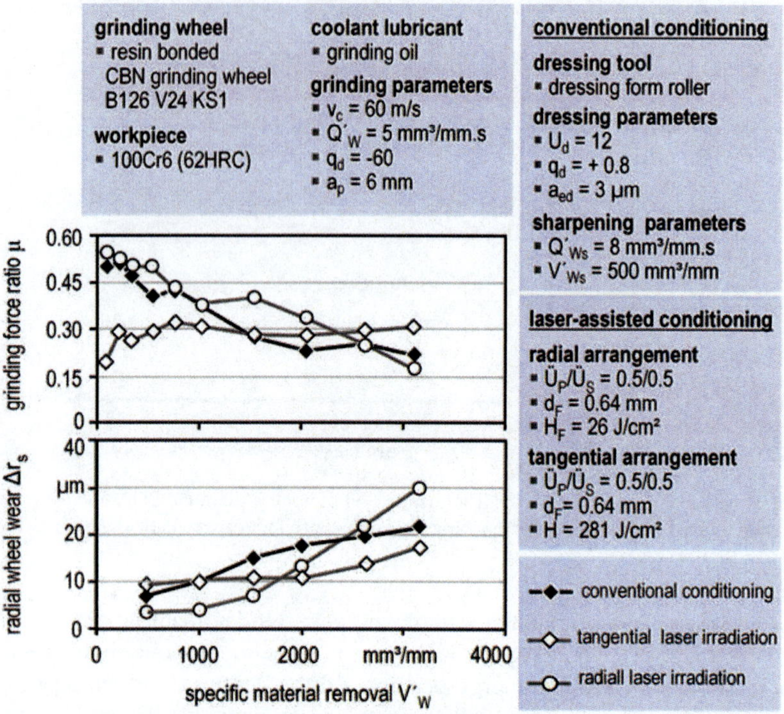

Fig. 4.55 Dependence of the grinding process on the conditioning method and material removal rate [66] used with permission copyright Vulkan-Verlag

been conditioned through the conventional method. With the wheel conditioned conventionally, the wheel was first dressed with a diamond cup wheel and then sharpened with an alumina sharpening block. At the start of the grinding tests, the wheels conditioned conventionally and by radial laser irradiation display higher grinding force ratios. However, with increasing special material removal the grinding force ratio for these two processes decreases drastically, so that at approx. 2,500 mm^3/mm the wheel conditioned with tangential laser irradiation has the highest grinding force ratio. This shows that while grinding after conventional and radial laser-irradiated conditioning, the topography of the wheel is changed considerably.

The efficiency of laser conditioning was also investigated on the basis of the wear behaviour of the wheel (Fig. 4.55). At the start of grinding, the radially irradiated wheel displays the least wear, but from a spesific material removal V'_w of approx. 1,500 mm^3/mm a clear and constant increase could be seen. At the end of the series of tests the lowest wheel wear was on the tangentially laser-sharpened CBN wheel [66].

The influence of the conditioning method on the change in wheel topography was also investigated on the basis of the effective peak-to-valley roughness of the wheel. As Fig. 4.56 shows, radial laser conditioning generates a high effective peak-to-valley roughness on the wheel, while on the other hand the roughness of a wheel produced by tangential laser conditioning lies in the range of that produced by conventional conditioning. The effective peak-to-valley roughness of all the wheels becomes smaller with increasing material removal. However, this change is greatest with the radially irradiated wheel, so that after a spesific material removal of 3,100 mm^3/mm the effective peak-to-valley roughness is almost the same on all three wheels.

Lasers can also be used to assist in conventional dressing with diamond dressers; here, the surface of the rotating grinding wheel is heated by a focused laser beam at a short distance in front of the dressing tool. This method, known as laser-assisted dressing, has been developed by Zhang and Shin for the conditioning of vitrified bonded CBN wheels with a single-point diamond dresser (Fig. 4.57). With the appropriate selection of laser parameters such as laser energy density and heating time, the brittle-hard vitrified bond of the wheel is softened and even melted, as a result of which the mechanism of material removal of the vitrified bond is changed from brittle fracture to ductile flow. This makes it

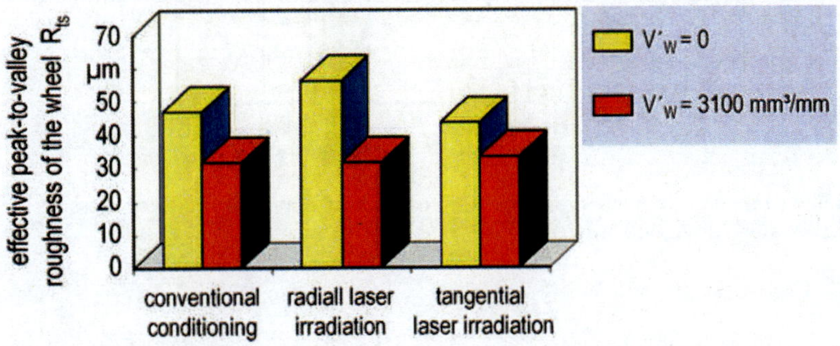

Fig. 4.56 Effective peak-to-valley roughness of a resin-bonded CBN grinding wheel as a function of the conditioning method [66] used with permission copyright Vulkan-Verlag

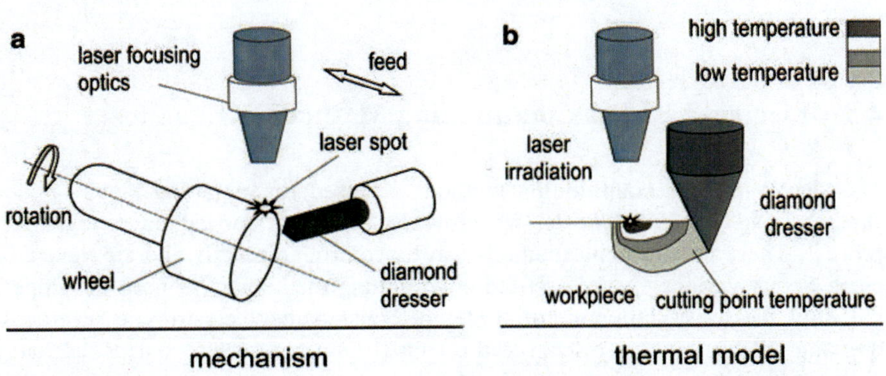

Fig. 4.57 Schematic of laser-assisted dressing [67] used with permission copyright Elsevier

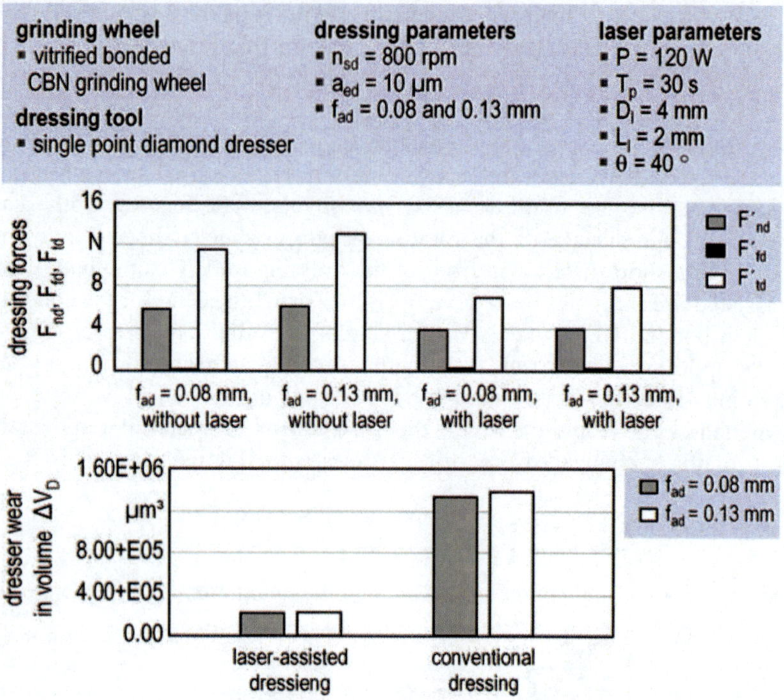

Fig. 4.58 Dressing forces and dresser wear in conventional and laser-assisted dressing [67] used with permission copyright Elsevier

easier for the bond material to be removed and as a result leads to a reduction in dressing forces and dresser wear (Fig. 4.58). The reduction in dresser wear brought about by laser assistance results in the production of wheel topographies that are more uniform with dressing operations carried out successively than those produced with conventional dressing [67].

4.8 Electro-Assisted Conditioning Methods

The electro-assisted conditioning methods are used for preparing metal bonded superabrasive grinding wheels, whereby the bond is removed by an electrical process. These conditioning methods can be controlled readily and are very efficient. So far various electro-assisted conditioning methods have been developed, but only the most important of these methods, namely electrolytic in-process dressing (ELID), electro-discharge dressing (EDD) and electrocontact discharge dressing (ECDD) are discussed in brief below.

4.8.1 *Electrolytic In-process Dressing*

Electrolytic in-process dressing (ELID) is an electro-assisted conditioning method that is in widespread use. With ELID the metal bond is removed continuously during the grinding process in accordance with the principle of anodic metal dissolution in electrolysis, whereby the blunt grits are continuously dissolved from the grinding wheel and the sharp grits below are exposed. ELID is used mainly for fine-grained metal bonded diamond grinding wheels [13]. It can also be used for fine-grained tough metal-bond CBN wheels in the grinding of ceramics. In this process, with ELID the protrusion of the CBN grits is controlled before and during grinding. Cast iron bonds and cast iron fibre-bonds are the most suitable bonds for ELID [7]. The ELID grinding setup and the mechanism of action of ELID are shown in Fig 4.59.

An ELID system essentially consists of a metal bonded grinding wheel, a counter-electrode and a DC pulse power supply, and also an electrolytic coolant with a high pH (Fig. 4.59a). The grinding wheel, which acts as the anode in the electrolysis cell, is connected via a brush contact on the wheel spindle to the positive pole of the power generator. The counter-electrode, which is adapted to the profile of the wheel, and acts as the cathode, is made of copper and is placed at a distance of 0.1–0.3 mm to the grinding wheel [7, 68]. The negative electrode must cover at least one sixth of the wheel's active surface and a width that is two millimetres wider than the wheel rim thickness [7]. The electrolytic coolant has low specific conductivity in the range from 1 to 3 mS/cm. As well as being supplied into the machining area itself, the coolant is injected directly into the gap between the wheel and the counter-electrode by additional nozzles [68].

The ELID grinding process consists of three stages: the set-up stage, the electrolytic pre-sharpening stage, and the ELID cycle stage [69]. The set-up

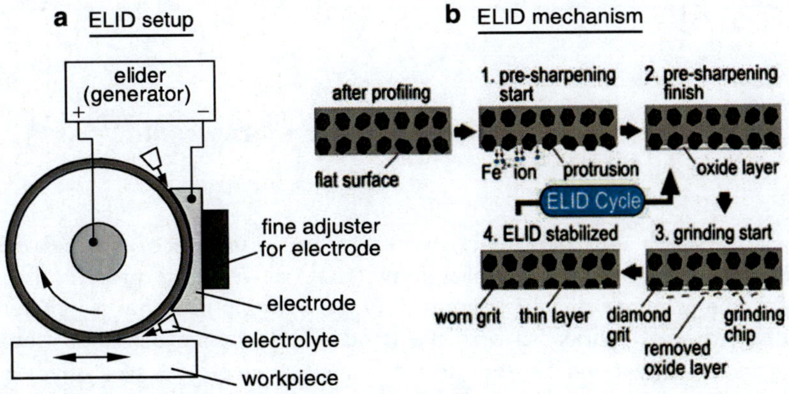

Fig. 4.59 Schematic of the ELID grinding setup and mechanism of action

stage takes place with a newly fitted grinding wheel, with balancing and precision profiling minimizing the run-out of the wheel. This stage should achieve a run-out of less than 2–4 μm [7]. After profiling, the grits and the bond are at equal height (Fig. 4.59b). In the pre-sharpening stage the grit protrusion needed at the start of the grinding process is produced by electrolysis (Fig. 4.59b-1). At this stage the wheel surface has good conductivity, and hence the current is as great as that set at the power supply [70]. Pre-sharpening is performed at low speed and takes about 10–30 min [7]. Continuing the electrolysis leads to the forming of an oxide layer on the surface of the wheel, which acts as an insulating layer and therefore prevents further oxidation of the wheel. The oxide layer is removed relatively easily by the abrasive contact in grinding (Fig. 4.59b-3). If the oxide-layer reaches a certain thickness during the grinding operation, the electrolysis is started again automatically (Fig. 4.59b-4), as a result of which the blunt grits are removed from the grinding wheel and the sharp grits below are exposed. This cycle is called the ELID cycle, which occurs repeatedly throughout the grinding process [70]. The rate at which bond material is removed is set by the ELID parameters, namely voltage, current and the gap between the electrodes. The rate of bond removal should correspond to the development of wear on the abrasive grits [7].

ELID can be used with various methods such as surface grinding, side grinding, double-side grinding and lapping. With the ELID technique, so far both hard and brittle optical and non-optical materials have been ground, including ceramics, hard steels, ceramic glass, ceramic coatings, biomedical materials, etc., with different shapes (plane, cylindrical external and internal, spherical and aspherical lenses, etc.) and very different sizes [7]. With ELID, improvements can be achieved in terms of both dimensional accuracy and roughness. Electrolytic in-process dressing can however be used efficiently above all in the manufacture of optical and ultra-precision components [68]. For some applications ELID grinding can eliminate polishing and/or lapping operations. With ELID grinding, for example on brittle materials an extremely fine surface in the nanometre range (4–6 nm) can be produced [78].

4.8.2 Electro-Discharge Dressing and Electrocontact Discharge Dressing

Electro-discharge dressing (EDD) works by the principle of thermal material removal in electro-discharge machining (EDM). With this process the metal bond of the grinding wheel is removed by the extreme heat that is generated by the discharges of sparks between the bond and a counter-electrode (the tool). The grinding wheel and the tool are connected as electrodes to a direct current source with a pulsating voltage. With EDD the electric discharge takes place in a non-conductive liquid, the dielectric. As with EDM, the task of the dielectric is to bundle the discharges at the point where the gap width is smallest, where a discharge channel forms. Hence, with the electric discharge sparks of extremely high temperatures are produced in the discharge zone, as a result of which the bond material melts rapidly and some evaporates. The gap width between the wheel and the dressing counter-electrode has to be regulated precisely by a special control system, which makes the process a complex task [71]. For this reason, based upon EDD a new conditioning method has been developed, namely electrocontact discharge dressing (ECDD) [72–74]. The setup of the system and the mechanism of action of ECDD are shown in Fig. 4.60. Unlike with EDD, with ECDD no dielectric is injected into the gap between the electrodes. Therefore, with this process the bond material is not removed by the discharges of sparks, as is the case with EDD, but by arc discharges. Due to the counter-electrode continuously being f_{ed} into the grinding wheel, chips and particles of electrode become present in the cutting edge space of the abrasive layer. Because of the voltage applied, electric discharges are produced between these chips and the bond of the wheel, as a result of which the bond is removed by the action of heat. With this method not only can the wheel be sharpened in one process, but to a limited extent it can also be profiled [13]. Compared with conventional conditioning methods, the conditioning time with ECDD is shorter and it produces a clearly higher maximum profile peak height on the wheel (Fig. 4.61). ECDD is therefore an economically attractive alternative to other methods [75].

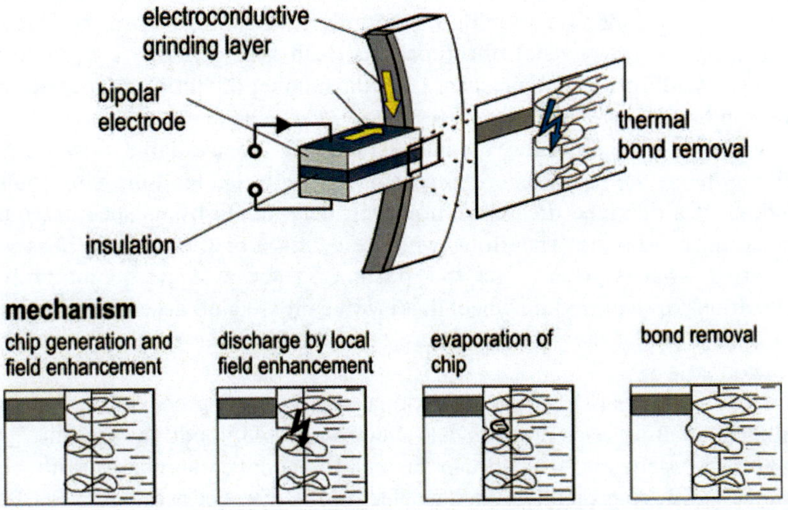

Fig. 4.60 Schematic of the ECDD grinding setup and mechanism [75] used with permission copyright Institute of Production Engineeringand Machine Tools (IFW), University Hannover

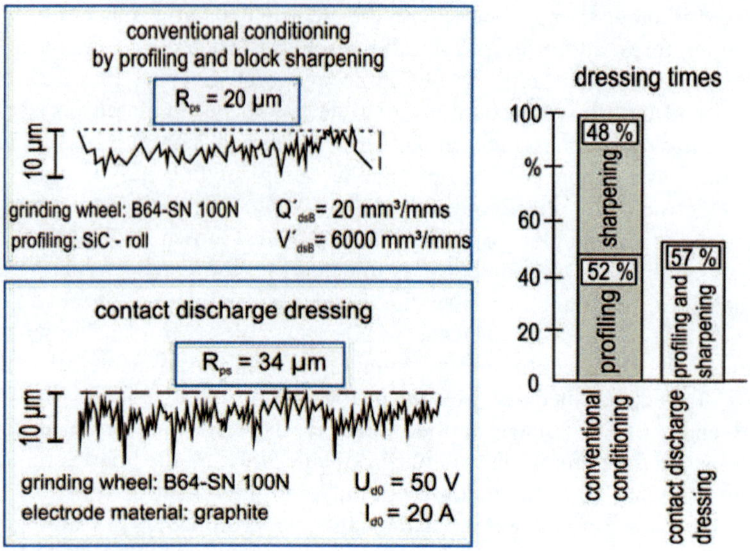

Fig. 4.61 Comparison of the process results achieved with conventional conditioning methods and ECDD [75] used with permission copyright Institute of Production Engineering and Machine Tools (IFW), University Hannover

4.9 Conclusions

The behaviour of a grinding wheel is influenced to a great extent by the conditioning method. With the conditioning of a grinding wheel, as well as producing the desired profile on the wheel, the process should also generate a suitable wheel topography with sufficient chip spaces and the number of cutting edges required for the task in hand.

A variety of conditioning methods and tools are available now. However, grinding wheels are conditioned predominantly with diamond dressing tools. The behaviour of a diamond dresser is, however, determined by its specifications and the dressing conditions. Therefore, when selecting a diamond dresser, as well as considering what is possible for the machine in terms of the geometry and the specifications of the grinding wheel that is to be dressed, all aspects of the diamond material used in the dresser and the dresser manufacturing process must be taken into account.

In recent decades new, non-conventional conditioning processes have also been developed, such as laser dressing and electro-assisted conditioning. Fine-grained superabrasive grinding wheels can be conditioned economically with electro-assisted conditioning methods such as electrolytic in-process dressing (ELID) and electrocontact discharge dressing (ECDD).

4.10 Nomenclature

Capital Letters

A	(–)	material absorbance
A_{US}	(μm)	ultrasonic amplitude
D	(mm)	diamond grit size
F'_n	(N/mm)	specific normal grinding force
F'_{rd}	(N/mm)	specific radial dressing force
F'_t	(N/mm)	specific tangential grinding force
G	(–)	grinding ratio
G_d	(–)	dressing ratio
H	(mm)	theoretic wheel profile height
I_{d0}	(A)	electrode current during electrocontact discharge dressing
K	(W/(cm°C))	thermal conductivity
K_D	(–)	diamond grit concentration
P	(W/cm^2)	laser beam intensity
P_m	(W)	laser average power
P_t	(μm)	total profile height
Q'_W	(mm^3/(mm s))	specific material removal rate
Q'_{Ws}	(mm^3/(mm s))	specific sharpening material removal rate
R_a	(μm)	average roughness height
R_{max}	(μm)	maximum peak-to-valley height
R_{ps}	(μm)	maximum peak height of the wheel
R_{ts}	(μm)	effective peak-to-valley roughness of the wheel
R_{sto}	(μm)	initial active peak-to-valley roughness of the wheel
R_z	(μm)	mean peak-to-valley height
T	(°C)	temperature
U_d	(–)	dressing overlap ratio
$U_{d(a)}$	(–)	dressing overlap ratio calculated in the axial direction
$U_{d(c)}$	(–)	dressing overlap ratio calculated along the wheel contour
$U_{d(bd)}$	(–)	dressing overlap ratio calculated from b_d
$U_{d(r)}$	(-)	dressing overlap ratio calculated in the radial direction
U_{d0}	(V)	electrode voltage during electrocontact discharge dressing
$\ddot{U}_P$	(–)	overlap ratio of the laser pulse
$\ddot{U}_S$	(–)	overlap ratio during the laser dressing
V'_W	(mm^3/mm)	specific material removal
V'_{Ws}	(mm^3/mm)	specific sharpening material removal
W_t	(μm)	total waviness height

Small Letters

$a_{d,real}$	(mm)	real total depth of dressing cut
a_{apd}	(mm)	axial engagement width

a_e	(mm)	grinding depth of cut
a_{ed}	(mm)	depth of dressing cut
a_{pd}	(mm)	engagement width
a_{rpd}	(mm)	radial engagement width
b_d	(mm)	active width of the dressing tool
d_b	(mm)	focus beam diameter
d_g	(μm)	grit size
d_R	(mm)	dressing roller diameter
d_s	(mm)	grinding wheel diameter
f	(mm)	focal length of the lens
f_{ad}	(mm)	axial dressing feed per grinding wheel revolution
f_d	(mm)	dressing feed
f_{rd}	(mm)	radial dressing feed per grinding wheel revolution
f_{US}	(kHz)	Ultrasonic frequency
k	(–)	model constant
q	(–)	grinding speed ratio
q_d	(–)	dressing speed ratio
n_R	(min^{-1})	number of dressing roller revolutions per minute
n_{rd}	(–)	number of roll-out roller revolutions
n_{sd}	(min^{-1})	number of dressing roller revolutions per minute
r_P	(mm)	tip radius of the form roller
r_R	(mm)	radius of the dressing roller
r_{RO}	(mm)	radius of the dressing roller at the beginning of the dressing process
$r_{R,US}$	(mm)	radius of the dressing roller during the ultrasonic assistance
r_S	(mm)	radius of the grinding wheel
r_{SO}	(mm)	radius of the grinding wheel at the beginning of the dressing process
t	(s)	time
t_a	(s)	spark-out time
t_e	(s)	moment of the dresser penetration in the wheel surface
t_0	(s)	laser pulse width
v_c	(m/s)	cutting speed
v_{cs}	(m/s)	cutting speed during sharpening
v_{fad}	(mm/min)	axial dressing feed rate
v_{fdres}	(mm/min)	dressing feed rate along the wheel contour
v_{frd}	(mm/min)	radial dressing feed rate
v_{ft}	(mm/min)	tangential feed rate
v_R	(m/s)	dressing roller peripheral speed
v_{sd}	(m/s)	grinding wheel peripheral during dressing

Greek Letters

α	($cm^2 s^{-1}$)	thermal difusivity
α_e	(°)	penetration angle

β	(°)	wheel contour inclination angle β
Δl_D	(μm)	wear length of diamond sticks
Δr_R	(μm)	radial dressing roller wear
Δr_s	(μm)	radial wheel wear
ΔV_D	(μm^3)	dresser wear in volume
ε_{Pd}	(°)	tip angle of the form roller
θ	(°)	full divergent angle of the laser beam
λ	(mm)	ultrasonic wavelength
μ	(–)	grinding force ratio
φ_{RO}	(rad)	starting angle of the point of the dressing roller to be looked
φ_S	(rad)	rotational angle of the grinding wheel

References

1. Saljé E, Harbs H (1990) Wirkungsweisen und anwendungen von konditionierungsverfahren. Annals of the CIRP 39: 337–340.
2. Klocke F, König W (2005) Fertigungsverfahren Band 2 – Schleifen, Honen, Läppen. 4. Auflage, Springer-Verlag, Berlin. ISBN: 3-5402-3496-9.
3. Tönshoff HK, Heuer W (1992) Konditionieren von CBN-Schleifscheiben. Industrial Diamond Review 26: 236–240.
4. Schulz A (1996) Das Abrichten von keramisch gebundenen CBN-Schleifscheiben mit Formrollen. Dissertation, RWTH Aachen.
5. Stuff D (1996) Einsatzvorbereitung keramisch gebundener CBN-Schleifscheiben. Dissertation, RWTH Aachen.
6. Wimmer J (1995) Konditionieren von hochharten Schleifscheiben zum Schleifen von Hochleistungskeramik. FBK-Produktionstechnische Berichte, Band 18, Hrsg.: Prof. Warnecke, Kaiserslautern.
7. Marinescu ID, Rowe WB, Dimitrov B, Inasaki I (2004) Tribology of abrasive machining processes, William Andrew, Inc., NY. ISBN: 0-8155-1490-5.
8. Marinescu ID, Hitchiner M, Uhlmann E, Rowe WB, Inasaki I (2007) Handbook of Machining with Grinding Wheels, Taylor & Francis Group, LLC, Boca Raton, FL. ISBN: 1-57444-671-1.
9. Völler N (2009) Dressing Technology. Technical guide, Winterthur Technology Group.
10. Friemuth T (2002) Herstellung spanender Werkzeuge. Fortschr.-Ber. VDI Reihe 2, Nr. 615, Düsseldorf, VDI-Verlag, zugl. Habilationsschrift Universität Hannover.
11. Liebe I (1996) Auswahl und Konditionierung von Werkzeugen für Außenrund-Profilschleifen technischer Keramiken. Dissertation Technische Universität Berlin, UNZE Verlaggesellschaft Potsdam, ISBN 3-8167-4509-1.
12. Holz R, Sauern J (1988) Schgleifen mit Diamant und CBN. Firmenschrift – Schleiftechnisches Handbuch, Ernst Winter & Sohn GmbH & Co. Norderstedt.
13. Hoffmeister HW, Hlavac M, Pekárek M (2008) Verfahren und Trends beim Konditionieren von hochharten und konventionellen Schleifscheiben. 7. Seminar, "Moderne Schleiftechnologie," Hrsg: Tawakoli T, ISBN: 978-3-00-024493-3.
14. Uhlmann E (1993) Tiefschleifen hochfester keramischer Werkstoffe. Dissertation TU Berlin.
15. Lauer Schmaltz H (1979) Zusetzung von Schleifscheiben, Dissertation, RWTH Aachen.
16. Zhang LC, Suto T, Noguchi H, Waida T (1995) A study of creep-feed grinding of metallic and ceramic materials. Journal of Materials Processing Technology 48: 267–274.

17. Cameron, et al. (2009) An investigation of the effects of wheel-cleaning parameters in creep-feed grinding. International Journal of Machine Tools and Manufacture, doi:10.1016/j.ijmachtools.2009.08.008
18. Sinot O, Chevrier P, Padilla P (2006) Experimental simulation of the efficiency of high speed grinding wheel cleaning. International Journal of Machine Tools and Manufacture 46: 170–175.
19. Neves AJ and Nazaré MH (2001) Properties, Growth and Applications of Diamond. Institution of Engineering and Technology, ISBN: 0852967853. 142–147.
20. Element Six TM Crystallography and Processing of Single Crystal (Monocrystalline) Diamond, Technical guide.
21. Bachmann PK and Messier R (1989) Diamond thin films. Chemical and Engineering News 67, No. 20: 24–39.
22. Buck Volker (2004) Diamonds are a Scientist's Best Friend. Forum Forschung 2003/2004, Universität Duisburg–Essen.
23. Linke B (2007) Wirkmechanismen beim Abrichten keramisch gebundener Schleifscheiben. Dissertation, RWTH Aachen.
24. Hessel D (2003) Punktcrushieren keramisch gebundener Diamantdiamantschleifscheiben. Dissertation, Universität Hannover.
25. Hessel D (2009) Dressing Technology, Technical guide, Dr. Kaiser.
26. Sen PK (2002) Synthetic diamond dresser logs. Serving the future needs of industry. Industrial Diamond Review 64, No. 594:194–202.
27. Graf W (1999) Diamant-Abrichtwerkzeuge. Informationsbroschüre der WST Winterthur Schleiftechnik AG, Winterthur.
28. Lierse T, Kaiser M (2002) Dressing of grinding wheels for gearwheels. Industrial Diamond Review 4: 273–281.
29. Weise C (2002) Was sind Industriediamanten? Eigenschaften, Anwendungen, Ausblicke. Diamond Business 2: 16–18.
30. Merz R (1994) Konzept zur Auswahl der Abrichtbedingungen bei der Einsatzzvorbereitung konventioneller Schleifscheiben mit Diamantprofilrollen. FBK-Produktionstechnische Berichte, Band 15. Hrsg.: Prof. Warnecke, Kaiserslautern.
31. Schmitt R (1968) Abrichten von Schleifscheiben mit diamantbestückten Rollen. Dissertation, Technische Universität Braunschweig.
32. Helletsberger H (2003) Grindology Paper C3 – Diamantabrichtrollen – Rollentypen und Verfahren. Tyrolit Schleifmittelwerke Swarovski K.G.
33. Minke E (1999) Handbuch zur Abrichttechnik, Herausgeber: Riegger Diamantwerkzeuge GmbH, 1. Auflage, Eislingen: Dischner Druck & Verlag.
34. Yegenoglu K, Jansen H, Janssens JP (1989) Diamant-Abrichtrollen zur wirtschaftlichen Serienfertigung, Werkstatt und Betrieb 122: 809–812.
35. Völler N und Gutsch V (2007) Richtiges Abrichten von Schleifscheiben sorgt für hohe Produktivität und Qualität. http://www.maschinenmarkt.vogel.de/.
36. Harbs U (1997) Beitrag zur Einsatzvorbereitung hochharter Schleifscheiben, Vulkan-Verlag, ISBN: 3-8027-8638-6, Zugl.: Dissertation, Universität Braunschweig.
37. Scheidermann H (1973) Einfluss der durch Abrichten mit zylindrischen und profilierten Diamantrollen erzeugten Schleifscheibenschneidfläche auf den Schleifbelag. Technische Universität Braunschweig.
38. Rohde G (1982) Abrichten von Korundschleifscheiben mit Diamantprofilrollen. Jahrbuch Schleifen, Honen, Läppen, Polieren, 51. Ausgabe, Vulkan Verlag.
39. Heuer W (1992) Außenrundschleifen mit kleinen keramisch gebundenen CBN-Schleifscheiben. Dissertation, Universität Hannover.
40. Tawakoli T (1993) High Efficiency Deep Grinding, Technology, process, planning and application. Mechanical Engineering Publication, London.
41. König W, Schulz A (1994) Abrichten von CBN-Schleifscheiben mit Diamantformrollen. Industrial Diamond Review 28, No. 4: 234–240.

42. Tawakoli T, Rasifard A, Vesali A (2009) Effect of the Coolant Lubricant Type and Dress Parameters on CBN Grinding Wheels Performance, ISSAT 2009, Australia, Gold cost. 27.09.–31.09.2009, 163/168
43. Çinar M (1995) Einsatzvorbereitung und Verschleißentwicklung keramisch gebundener CBN-Schleifscheiben. Dissertation, Universität Bremen.
44. Kaiser M (1997) Fortschrittliches Abrichten moderner Schleifscheiben. Jahrbuch Schleifen, Honen, Läppen und Polieren, Vulkan-Verlag, Essen, 324–339.
45. Uhlig U, Redeker W, Bleich R (1982) Profilschleifen mit kontinuierlichem Abrichten, Werkstatttechnik, 72, 6: 313–317.
46. König W, Arciszewski A (1987) Schleifen mit kontinuierlichem von schwer zerspanbaren Stählen. Industrie-Anzeiger, 110 61/62: 26–29.
47. Spur G and Holl SE (1995) "Ultraschallunterstütztes Schleifen von Hochleistungskeramik" in Jahrestagung, Kurzreferate. Deutsche Keramische Gesellschaft, Aachen, Germany.
48. Daus NA (2004) Ultraschallunterstütztes Quer-Seiten-Schleifen. Dissertation, Technische Universität Berlin.
49. Kappmeyer G (1999) Hochfrequenzhonen: Beitrag zum ultraschallunterstützten Dornhonen. Dissertation, TU Braunschweig.
50. Shamoto E, Moriwaki T (1999) ultrasonic diamond cutting of hardened steel by applying elliptical vibration cutting. Annals of the CIRP 48: 441–444.
51. Schmütz J (2002) Mikrobearbeitung duktiler Werkstoffe mit ultraschallangeregten Diamantwerkzeugen. Dissertation, Universität Bremen.
52. Thoe TB, Aspinwall DK, Wise MLH (1998) Review on ultrasonic machining. International Journal of Machine Tools and Manufacturing 38, No. 4: 239–255.
53. Ikuse Y et al. (1996) Development of new ultrasonic dressing equipment. International Journal of Japan Society for Precision Engineering 30, No. 3, 217–222.
54. Nomura M et al. (2005) Effects of ultrasonic vibration in truing and dressing of CBN grinding wheel used for internal grinding of small holes. Key Engineering Materials 291–292: 183–188.
55. Sroka F (2005) Konditionieren von Diamantschleifscheiben. Dissertation, Technische Universität Berlin.
56. Tawakoli T, Rasifard A, Azarhoushang B (2008) Dressing of CBN grinding wheels with ultrasonic assistance. International Journal of Mechatronics and Manufacturing Systems 1, No. 4: 321–331.
57. Tawakoli T, Rasifard A (2009) Ultraschallunterstütztes Abrichten von CBN-Schleifscheiben, ein neue Alternative zum konventionellen Abrichten, Abschlussbericht zum durch das BMBF im Rahmen des FH^3-Programmes geförderten Forschungsprojekt.
58. Jiao F et al. (2006) Ultrasonic dressing of grinding wheel and its influence on grinding quality. Key Engineering Materials 304–305: 62–65.
59. Hoffmeister HW, Maiz K (2002) Laserkonditionieren hochharter Schleifwerkzeuge. 10. Internationales Braunschweiger Feinbearbeitungskolloquium (10. FBK), 07-09. Braunschweig, Hrsg.: Hesselbach, J.
60. Xie XZ, Chen GY, Li LJ (2004) Dressing of resin bonded superabrasive grinding wheels by means of acousto-optic Q-switched pulsed Nd:YAG laser. Optics and Laser Technology 36: 409–419.
61. Jackson MJ et al. (2007) Laser cleaning and dressing of vitrified grinding wheels. Journal of Materials Processing Technology 185: 17–23.
62. Chen X, Pashby I (2004) Comparison of laser cleaning of Al2O3 and CBN grinding wheels. Key Engineering Materials 257/258: 365–370.
63. Wang XY, Wu YB, Wang J, Xu WJ, Kato M (2005) Absorbed energy in laser truing of a small vitrified CBN grinding wheel. Journal of Materials Processing Technology 164–165: 1128–1133.

64. Khangar A, Dahotre NB (2005) Morphological modification in laser-dressed alumina grinding wheel material for microscale grinding. Journal of Materials Processing Technology 170: 1–10.
65. Hosokawa A, Ueda T, Yunoki T (2006) Laser dressing of metal bonded diamond wheel. Annals of the CIRP 55 (1): 329–332.
66. Timmer JH (2000) Laserkonditionieren von CBN- und Diamantschleifscheiben. Jahrbuch Schleifen, Honen, Läppen und Polieren, 59 Ausgabe, Vulkan-Verlag, Essen, 324–339.
67. Zhang C, Shin YC (2002) A novel laser-assisted truing and dressing technique for vitrified cBN wheels. International Journal of Machine Tools and Manufacture 42: 825–835.
68. Klocke F, Demmer A, Klink A, Grüntzig, M (2006) ELID-Präzisionsschleifen von Keramikformwerkzeugen. Werkstatttechnik-online, Jahrgang 96: 48–352.
69. Ren CZ, Hua JH, Wu ZY, Wang TY, Jin WO, Jin XM (2006) Influence of the state of passivating films in electrolytic in-process dressing (ELID) grinding process. Key Engineering Materials 304–305: 24–28.
70. Fathima K et al. (2010) A knowledge based feed-back control system for precision ELID grinding. Precision Engineering 34 (1): 124–132.
71. Xie J, Tamaki J (2008) An experimental study on discharge mediums used for electro-contact discharge dressing of metal-bonded diamond grinding wheel. Journal of Materials Processing Technology 208: 239–244.
72. Tamaki J, Kondoh K, Iyama T (1999) Electro-contact discharge dressing of metal-bonded diamond grinding wheel utilizing a hybrid electrode. Journal of Japan Society for Precision Engineering 65 (11): 1628–1632.
73. Tonshoff, HK, Friemuth T (2000) In-process dressing of fine diamond wheels for tool grinding. Precision Engineering 24 (1): 58–61.
74. Xie J, Tamaki J, Kubo A, Iyama T (2001) Application of electro-contact discharge dressing to a fine-gained diamond grinding wheel. Journal of Japan Society for Precision Engineering 67 (11): 1628–1632.
75. Friemuth T, Becker JC, Hannebauer A (2000) Kontakterosives Konditionieren von metallisch gebundenen CBN-Schleifscheiben. Industrie Diamanten Rundschau (IDR) 34, No. 3: 248–253.

Chapter 5
Surface Integrity of Materials Induced by Grinding

L.C. Zhang

Abstract This chapter discusses some fundamentals in the surface integrity of workpiece materials generated by grinding, including surface roughness, microstructural changes and residual stresses induced by mechanical loading, thermal heating and phase transformation. The discussion concludes that different classes of materials have distinctive material removal mechanisms in grinding and hence have their individual surface integrity characteristics. The workpiece materials to discuss include metals, ceramics, semiconductors and composites.

Keywords Grinding · Surface integrity · Grinding wheels · Residual stress · Grinding mechanisms

5.1 Introduction

Grinding is a process that uses abrasives as the cutting edges to remove materials from a workpiece. Generally speaking, a grinding process involves four groups of variables that comprehensively affect the surface integrity of a ground component. These variables are:

(a) The performance variables of a grinding machine such as its loop stiffness, positioning accuracy, stability and thermal deformation.

L. Zhang (✉)
School of Mechanical and Manufacturing Engineering, The University of New South Wales, Sydney, NSW 2052, Australia
e-mail: Liangchi.Zhang@unsw.edu.au

M.J. Jackson and J.P. Davim (eds.), *Machining with Abrasives*,
DOI 10.1007/978-1-4419-7302-3_5,

(b) The grinding wheel variables such as the size of abrasives, properties of the abrasive material, wheel baking temperature, wheel structure, concentration, porosity, hardness and properties of bond materials.
(c) The properties of a workpiece material such as the yield stress, fracture toughness and phase transformation threshold.
(d) Grinding parameters such as the wheel speed, table speed, depth of cut, coolant type and its supply method, dressing conditions and properties of dressers.

Although the conventional grinding of metals is mainly a mechanical process, physical and chemical means have been extensively incorporated into the technology nowadays to improve the grindability and quality in dealing with difficult-to-grind materials, such as ceramics and silicon. In a sense, grinding is one of the most complex material removal processes. In addition to the complexity in processing, applications of grinding in various areas of high technology, such as optics and electronics have created further requirements for the quality of ground elements. Apart from the low surface roughness and high accuracy, which are the traditional requirements for grinding, the high technological applications of ground components often need tailored sub-surface properties, such as distribution of residual stresses and allowed variation of micro-structures. These imply that one must understand and consider the principal factors that determine the surface integrity of an element subjected to grinding [1].

Investigations on grinding can be grouped roughly into three categories, experimental study, theoretical modelling and the combination of the two, of which each has its merits and disadvantages. For instance, an adequate theoretical modelling enables one to understand the major roles of a grinding parameter, while an experimental study can often be done under limited conditions. Nevertheless, it is difficult for a theoretical model of grinding to give quantitative predictions because assumptions and simplifications are usually inevitable in modelling. On the other hand, real production conditions are much easier to achieve in experimental studies. This chapter will discuss the surface integrity generation mechanisms in the grinding of various workpiece materials – metals, ceramics, silicon and fiber-reinforced composites.

5.2 Residual Stresses and Subsurface Microstructures

5.2.1 Grinding of Metals

For a metal workpiece subjected to grinding, mechanical plastic deformation, thermal plastic deformation and irreversible deformation due to phase transformations are the major sources of residual stress generation [1]. Zhang and his co-workers have systematically analyzed grinding temperature and phase transformation [2–5] and have investigated the variation of residual stresses caused by surface mechanical loading [6], thermal plastic deformation [7], coupling of thermal plastic deformation and phase transformation [8], and by the combination of mechanical and thermal plastic deformation [9]. A conclusion of their research is that residual stress distributions in a ground component are sensitive to the combination of grinding-induced permanent deformation and that the change of the deformation sources can change the nature of residual stresses (i.e., compressive or tensile). Generally, mechanical plastic deformation produces compressive surface residual stresses, thermal plastic deformation generates tensile surface residual stresses, but phase transformation may lead to either compressive or tensile surface residual stresses depending on the volume increase or decrease of the transformed material phase. Under certain grinding conditions, plastic deformation by all the three sources can occur simultaneously in some metals. In the following, we will discuss surface grinding and the full coupling effect of all the sources of irreversible deformation on the variation of residual stresses [10, 11]. The reader who is interested in the mechanisms of residual stress generation by individual sources can find details in [2–9].

The deformation in a workpiece during surface grinding can be considered as a plane-strain problem, and the heat flux generated by grinding can be approximated by a triangular heat source, moving along the positive direction of x-axis on the workpiece surface, as shown in Fig. 5.1. The surface mechanical stress due to grinding can also be simulated by a moving triangular traction with its peak being coincident with that of the heat flux. In this way, the mechanical cutting of the grinding wheel in the finite element analysis can be simulated by equivalent nodal forces. The direction of the horizontal nodal forces depends on the type of grinding processes, being positive for an up-grinding and negative for a down-grinding. Meanwhile, the type of grinding processes alters the height of the heat flux, q_a, the height of the mechanical stress traction, p_a, and the location of the peaks of the surface mechanical traction and heat flux, $l_a = 2\zeta_a/L_C$, respectively, where L_C is the contact length between the grinding wheel and the workpiece, as shown in Fig. 5.1. In the figure, $T_\infty = 25°C$ is the ambient temperature.

The cooling effect of grinding fluid can be modelled by a variable coefficient of convection heat transfer, that is,

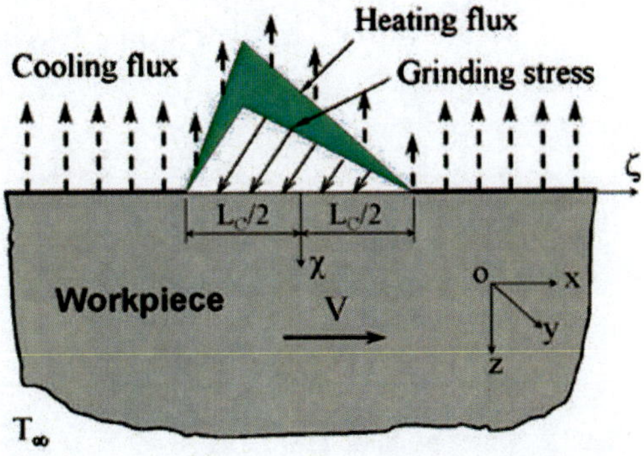

Fig. 5.1 A theoretical model of surface grinding with convection

$$q_c = \begin{cases} hT, & |2x/Lc| \geq 1 \\ whT, & |2x/Lc| < 1 \end{cases} \quad (5.1)$$

where q_c is the convection heat flux, h is the heat transfer coefficient of coolant, T is grinding generated temperature rise, w is an effective cooling factor over the workpiece surface, with $w = 1$ indicating a uniform convection over the whole workpiece surface, $0 \leq w < 1$ representing a less effective cooling, and $w > 1$ standing for a super-cooling inside the grinding zone. To account for the property change of work materials by phase transformation, a proper constitutive model was developed by Mahdi and Zhang [8] which was then integrated with the finite element method [10, 11]. It was found that when a workpiece experiences the critical temperature variation in grinding, phase change occurs at a certain distance away from the grinding zone, as shown in Fig. 5.2, where Taust is the austenite transition temperature of the work material (EN23 steel in this case). The phase change starts earlier if the coefficient of convection heat transfer, H, is higher, regardless of the variation of table speed. Moreover, less cooling in the grinding zone ($0 \leq w < 1$) accelerates the initiation of phase transformation if H is high, for instance, H = 1.

The variation of grinding conditions does not influence the two major surface residual stresses, σ_{xx} and σ_{yy}, when phase transformation occurs, as shown in Fig. 5.3. The residual stress σ_{yy} is nearly constant across the martensite zone (Fig. 5.3b). The longitudinal residual stress σ_{xx}, however, changes linearly, with a limit, within the martensite zone. At the boundary of the martensite zone, a rapid change of residual stresses occurs due to the sudden change of work material properties. Compared with σ_{yy}, σ_{xx} is more affected by the types of grinding operations, the ratio of horizontal to vertical forces and the fully coupled thermo-mechanical

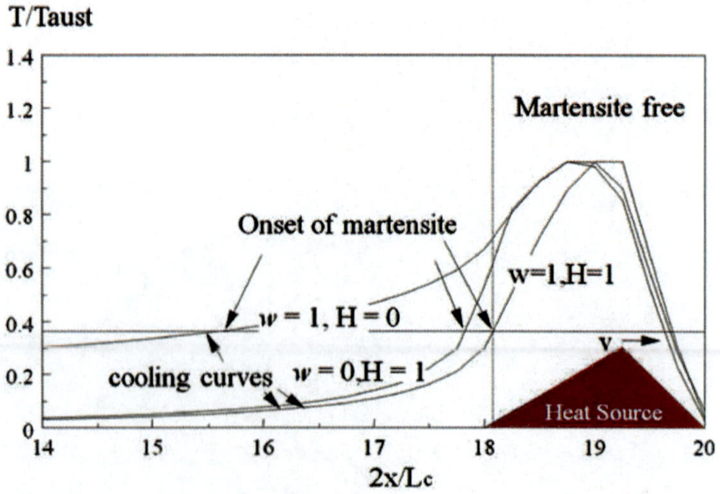

Fig. 5.2 Onset of phase change vs. heat source location

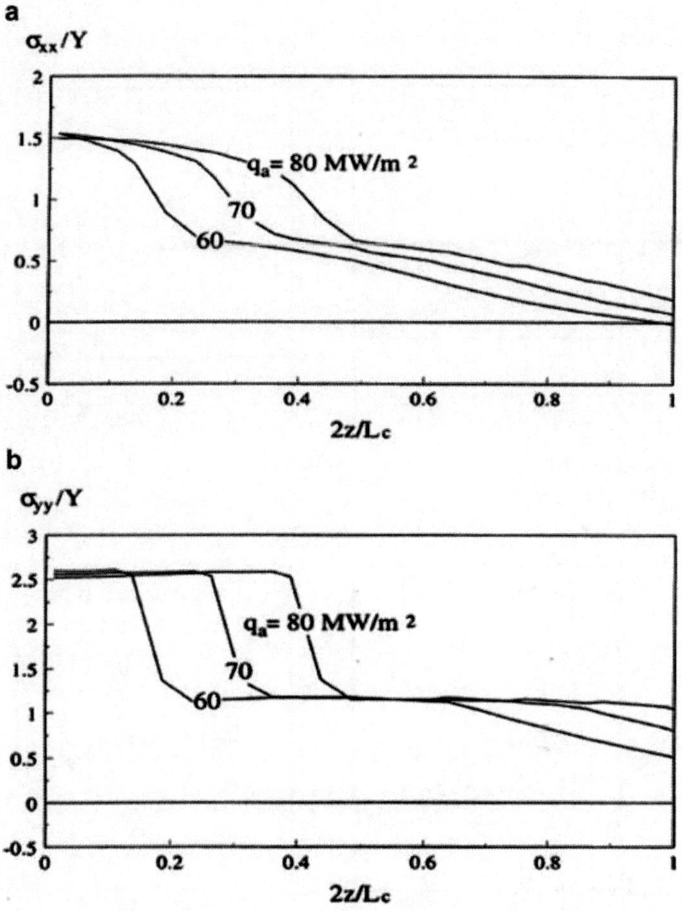

Fig. 5.3 Residual stresses without surface cooling ($H = 0$, $l_a = 0.25$, $VL_C/4\alpha = 1$where α is thermal diffusivity). (**a**) σ_{xx}/Y, and (**b**) σ_{yy}/Y, where Y is the yield stress of the workpiece material

grinding conditions with phase change. The maximum σ_{xx} and σ_{yy} are 1.5 and 2.6 times higher, respectively, than the initial yield stress of the work material, and they are tensile at the ground surface. These characteristics are similar to those of purely thermal residual stresses without a phase change [7]. However, the coupling of phase transformation with thermal and mechanical deformation brings about much higher tensile residual stresses at the workpiece surface.

The coupling of mechanical loading with thermal loading and phase transformation has a minor influence on residual stress distributions for a wide range of magnitude of surface traction, as shown in Fig. 5.4. Moreover, at a higher ratio of horizontal to vertical traction μ (e.g., $\mu = 0.3$ in the case of Fig. 5.4b), the increase of traction slightly decreases the longitudinal surface residual stress σ_{xx}. The reason for that is related to the reduction of the longitudinal strain within the grinding

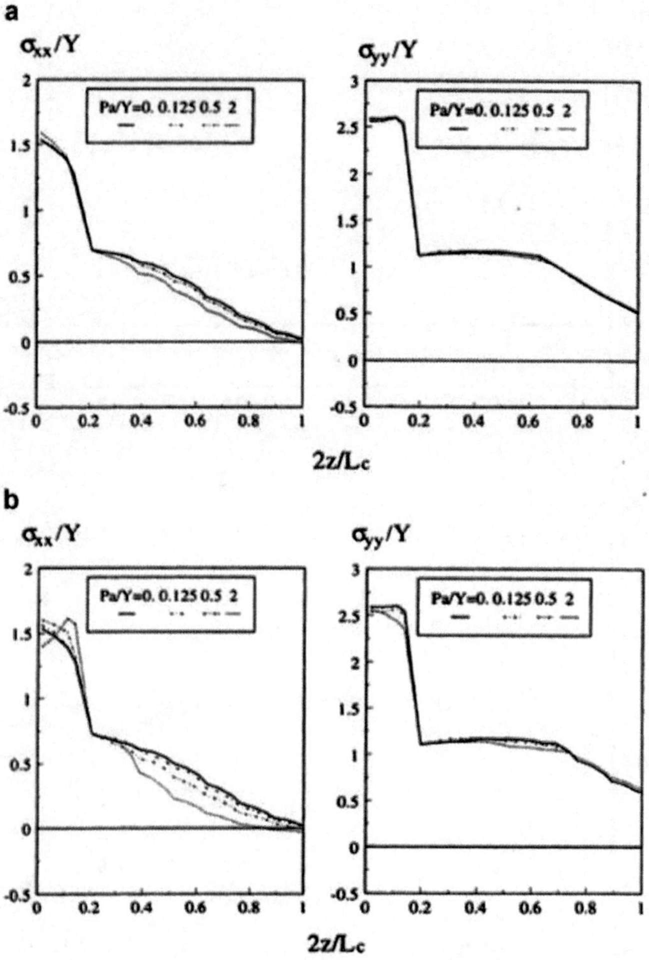

Fig. 5.4 The effect of mechanical traction on residual stresses (H = 0, $l_a = 0.25$, $VL_C/4\alpha = 1$ and $q_a = 60\ \mathrm{MW/m^2}$). **(a)** $\mu = 0.1$ and **(b)** $\mu = 0.3$

zone. Compared with thermal deformation and phase transformation, the effect of mechanical traction is minor. In the case with martensite phase transformation, the strains at the austenising grinding temperature play a central role in the formation of tensile residual stresses.

The effect of grinding conditions on the nature of residual stresses can be understood more deeply by comparing the influence of the individual causes and with different combinations. Figure 5.5 shows the role of each mechanism of different grinding conditions on the longitudinal residual stress σ_{xx}. Under sole mechanical conditions (Case 1), a very small, almost negligible residual stress is developed. Sole thermal grinding conditions without surface hardening (Case 2) lead to a tensile

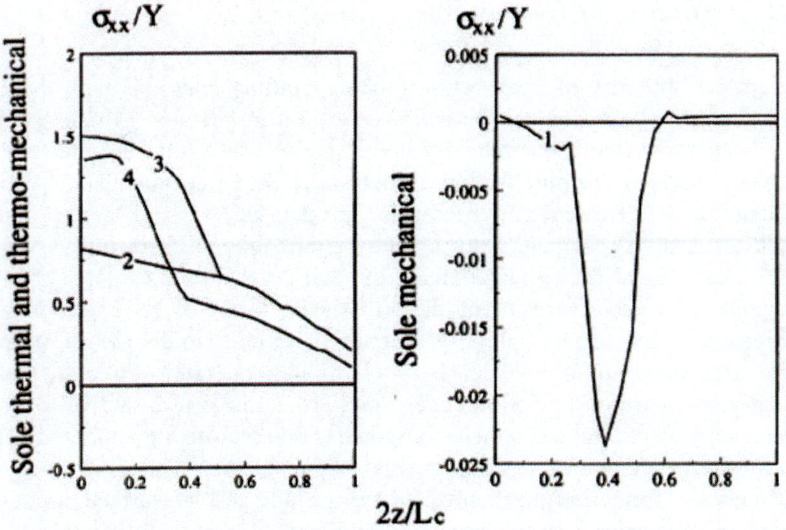

Fig. 5.5 Effect of individual grinding conditions. Case 1: $P_a/Y = 2.0$, $\mu = 0.1$; Case 2: $q_a = 80$ MW/m^2, Pe = 1, H = 0; Case 3: Case 2 coupled with phase transformation; Case 4: Case 3 coupled with Case 1

residual stress, which decreases gently with the subsurface depth. When the phase transformation (surface hardening) is coupled, see Case 3, a surface layer with a greater tensile residual stress is generated. The variation of σ_{xx} also becomes sharper compared with Case 2. By coupling the mechanical grinding conditions of Case 1 with those of Case 3, a considerable decrease in residual stress occurs, see Case 4. This means that mechanical grinding conditions may have stronger effects on residual stresses when combined with thermal conditions only, particularly at a lower value of $VL_C/4\alpha$.This is similar to the results of thermo-mechanical grinding conditions reported by [9] when the work material properties are temperature-independent.

The above study can be briefly summarized below: (1) The surface hardening associated with a phase change is slightly increased if cooling is more effective; (2) The influence of mechanical grinding conditions on residual stress distribution is minor when phase change takes place; (3) All the components of surface residual stresses become tensile when phase transformation occurs in the EN23 steel,which is independent of the cooling conditions and the type of grinding operations (up- or down-grinding); and (4) The distribution of the residual stress in the grinding direction is nearly linear in both the martensite and non-martensite zones, but the location of the maximum residual stress is related to the depth of martensite and surface mechanical loads. The understanding above, though from EN23 steel only, has been important to the late development of the grinding-hardening technology and its optimization [12–15].

5.2.2 *Grinding of Ceramics*

High surface integrity of hard, brittle monocrystalline materials with mirror-like and fracture-free surfaces can be achieved by ultra-precision grinding [16, 17]. It has been shown both theoretically and experimentally that the mechanism of material removal can be purely ductile when sufficient independent slip systems are activated [18, 19]. However, in the case of grinding polycrystals, certain fractured surface areas always appear, causing the degradation of their surface integrity [18–21]. Zarudi and Zhang [22] pointed out that defects such as pores or voids in a ceramic workpiece, a result of the sintering process of ceramics, contribute significantly to the surface integrity of ground ceramic components. Alumina is known as a material of low density with high temperature strength and high resistance to wear and corrosion. These properties make it attractive to industry for producing electronic and optical components that require high quality of surface and damage-free subsurface. Thus in this section of the chapter, we will aim to understand the formation mechanism of the surface and subsurface structures of alumina in ultra-precision grinding. The conclusions obtained, however, will be applicable to other ceramics of similar properties.

The materials used were polycrystalline alumina of 99.99% purity with average grain diameters of 1 and 25 μm (Kyocera, Japan). The grinding experiments were conducted on a precision surface grinder, Minini Junior CF CNC M286. A diamond grinding wheel of SD4000L75BPF (diameter 305 mm, peripheral speed 27 m/s) was used. The depth of cut was fixed at 100 nm but the table speed was varied from 0.02 to 1 m/min. A water based coolant Syntilo 3 (99% water, 1% mineral oil) was applied. The topography of the ground surfaces was explored by a high resolution scanning microscope (HRSEM), JSM-6000F. The surface roughness was measured by an atomic force microscope (AFM) and the subsurface structure of the specimens was studied by a transmission electron microscope (TEM), EM 430. Detailed procedures can be found in [18, 19, 22].

5.2.2.1 Surface Topography After Grinding

The surface topography of ground alumina is shown in Fig. 5.6. Mirror surfaces were obtained under all the table speeds used. Plastic grooves were clearly seen on the ground surface, suggesting the ductile mode of material removal. However, some pits were also observed. The variation of Rms roughness vs. table speed for both types of alumina is shown in Fig. 5.7. It is clear that Rms roughness increased with the increase of table speed. For the 1 μm-grained alumina, the Rms roughness increased from 30 to 50 nm with the increase of table speed from 0.02 to 1 m/min. For the 25 μm-grained alumina, the effect was more pronounced and the Rms roughness varied from 33 to 90 nm. It is worth noting that the Rms roughness was influenced by pits on the surface, as shown in Fig. 5.8 (determined by the technique of image analysis). It is obvious that the surface covered by pits decreases

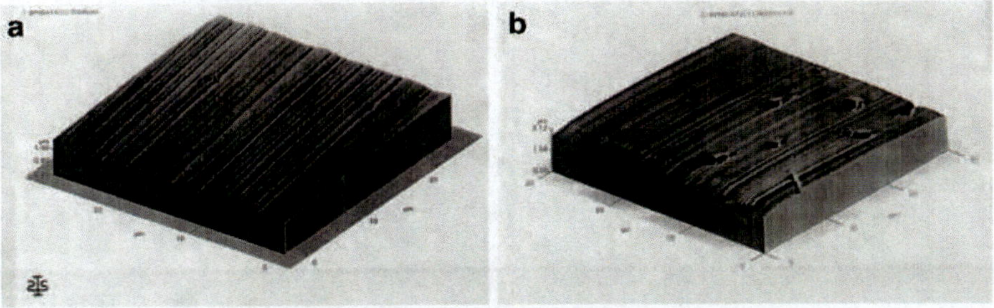

Fig. 5.6 Surface topography after grinding (Table speed 0.02 m/min): (**a**) the 1 μm-grained alumina; (**b**) the 25 μm-grained alumina

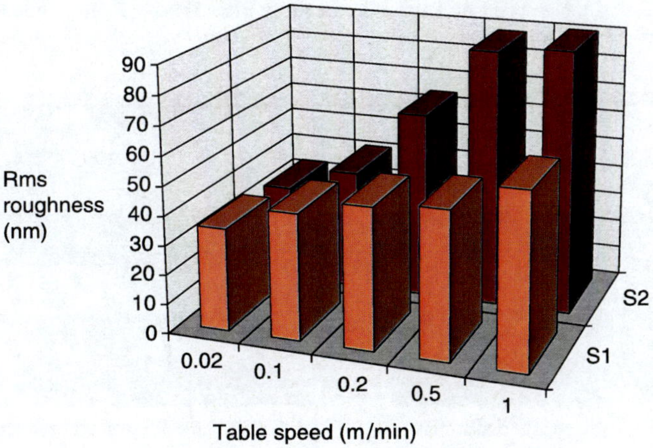

Fig. 5.7 Effect of table speed on Rms roughness. S1: the 1 μm-grained alumina, and S2: the 25 μm-grained alumina

significantly with the decrease of table speed, i.e., with the decrease of the uncut chip thickness. Furthermore, the pits in the 1 μm-grained alumina (1.5% under the table speed 0.02 m/min) is less than those in the 25 μm-grained alumina (5% under the same table speed).

The nature of pits can be explored by looking into their topography. As shown in Fig. 5.9, the central part of a pit resembles the cross-section of pores that were formed during the specimen processing (ceramic sintering). The peripheral edge of the pit has the characteristics of fracture. The fracture around the edge can be explained by the effect of pore on the grinding process. Briefly speaking, a pore in the material can act as a stress raiser and promote the microcracking near the pore edge during the abrasive cutting of the grinding wheel. Thus for a polycrystalline alumina with pores, it is impossible to obtain a perfect surface without any edge mirocracking around a surface pore. The reader can find more detailed discussions about the fracture mechanics analysis in reference [22].

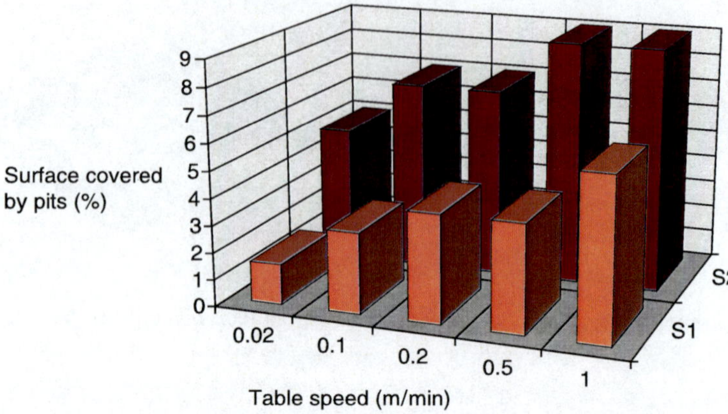

Fig. 5.8 Effect of table speed on surface covered by pits. S1: the 1 μm-grained alumina, and S2: the 25 μm-grained alumina

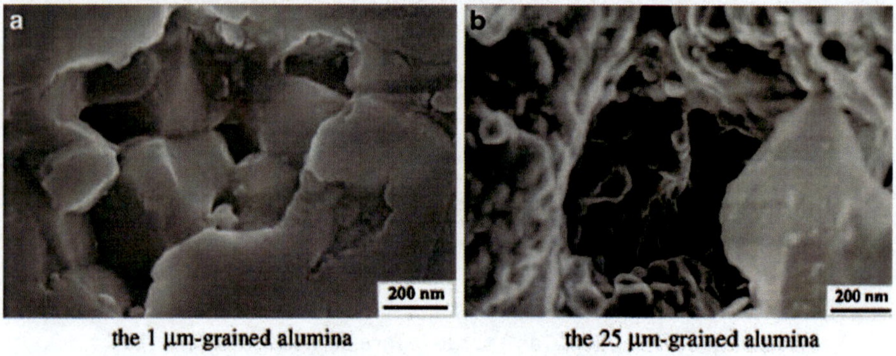

Fig. 5.9 The topography of pits (HRSEM) on ground surfaces

5.2.2.2 Subsurface Structure

Figure 5.10 shows that ductile-mode grinding created a layer with an extremely high density of dislocations, which was immediately beneath the ground surface. This layer spreads to the depth of 100–150 nm under the surface depending on the table speed. Following this, a zone with a much lower density of dislocations appears. Dislocations could be distinguished easily there. The depth of the second zone is also dependent on table speed and varies from 150 to 500 nm.

Microcracks were observed only in the vicinity of the pit edges, which are neither radial nor lateral cracks. The microcracking-free behaviour in the first layer can be explained by the nucleation of sufficient number of twin and slip systems that satisfy the von Mises criterion [23]. The second zone had less than five independent slip and twin systems and thus microcracking should be highly possible there. However, the above experimental observations do not provide any

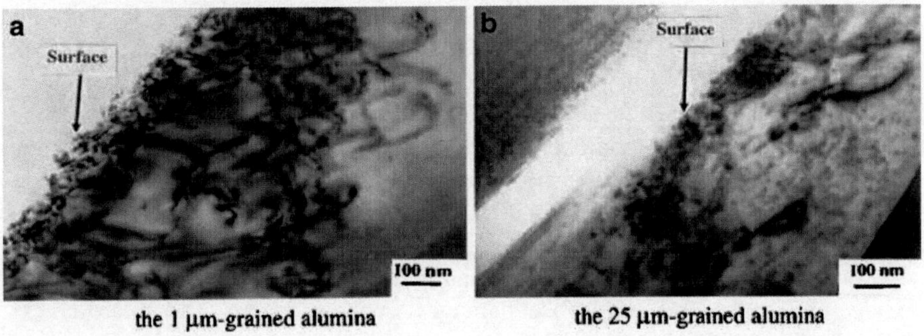

Fig. 5.10 Subsurface structure in alumina after ductile-mode grinding with the table speed=1 m/min

cracking evidence. This can be explained by the possibility of microcracking due to the pile-up of dislocations on an active slip plane. The threshold effective resolve shear stress (τ_s) for cracking could be determined by [24]:

$$\tau_s^2 = \frac{3\pi}{8}\left[\frac{\gamma\mu}{(1-\nu)L}\right] \tag{5.2}$$

where γ is fracture surface energy, μ is shear modulus, ν is Poisson's ratio, and L is the length of pile-up. Considering the length of pile-up of 150–500 nm (i.e., penetration depth of dislocation in grinding), we can see that for such a small pile-up length the nucleation of microcracks does not usually occur. It is also worth noting that the interaction of different slip systems was a rear event here and could not create microcracks.

In short, the investigation above on grinding alumina enables us to conclude that the ductile-mode of material removal is possible due to the initiation of more than five independent slip and twin systems, that the subsurface structure of alumina after ductile-mode grinding is composed of a layer with high dislocation density in the surface vicinity followed by a layer with a much lower dislocation density, and that the absence of microcracks in the immediate subsurface is the small length of pile-ups that cannot create sufficient stress for microcracking.

5.2.3 *Grinding of Composites*

Composites have been widely used in industry and are a class of most important materials across a broad range of disciplines. This is because composites have excellent properties, such as the low thermal expansion, superior damping capacity, and high specific strength and specific modulus.

Generally speaking, a composite contains at least two different materials whose properties are usually very different. Some typical examples include the carbon fiber reinforced polymer matrix composite [25] and the ceramic particle reinforced aluminium matrix composite [26].

To obtain a high quality surface, including surface integrity and dimensional accuracy, grinding is often an essential process in producing composite components. There have been a number of studies on the machinability of carbon fiber reinforced plastics (CFRP), focusing on the cutting, turning and drilling processes [27–34], as well as grinding [35–38]. As generally known, different machining processes have different effects on the surface integrity of a CFRP component. In a ground CFRP, surface imperfections can be the cracking origins. Furthermore, a variety of surface and subsurface damages, such as macro- and micro-cracks, de-laminations and fiber-matrix de-bonding, can occur.

In this section of the chapter, we will focus on the characterization of the subsurface damage of ground components and the mechanism of defect formation under various grinding conditions. The specimens used had a geometry of 45 × 15 × 4 mm, cut from the CFRP laminates with unidirectional carbon fibers of about 7–8 μm in diameter. The laminates were fabricated from a commercial resin system of F593 prepregs, cured under the constant pressure of 0.6 MPa at the temperature of 177°C for 2 h. The mechanical properties of the composites are as follows: tensile strength = 1,331 MPa, tensile modulus = 120 GPa, compressive strength = 1,655 MPa and compressive modulus = 115 GPa. All samples were pre-ground using a finer grinding condition to minimize the possible subsurface damage, so that the characterization of experimental results will be reliable. Figure 5.11 shows the cross-section image of a pre-ground specimen with the fiber orientation of 150°. It can be clearly seen that the specimen surface is smooth and the subsurface damage is negligible.

The grinding tests were conducted on a MININI M286 CN surface-grinding machine with an aluminium oxide grinding wheel BWA36HVAA. Down-grinding under a dry condition was employed as shown in Fig. 5.12. The grinding wheel was dressed regularly with a single point diamond dresser at the dressing

Fig. 5.11 Surface and subsurface quality of a specimen before grinding

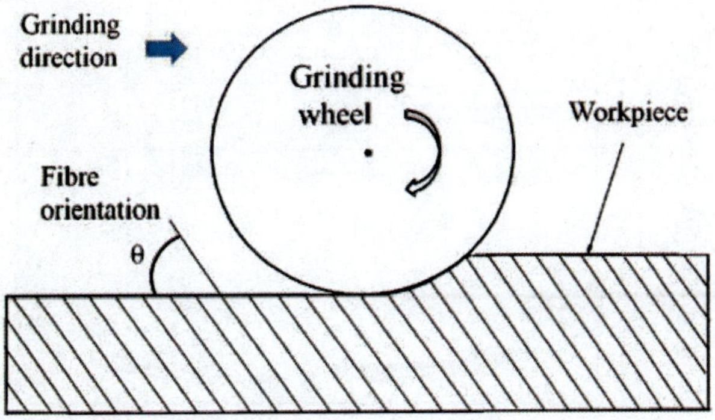

Fig. 5.12 Illustration of surface down-grinding

depth of 50 μm each time, dressing feed of 200 mm/min and wheel peripheral speed of 20 m/s. The fiber orientations of the specimens were $\theta = 0°, 30°, 60°, 90°, 120°$, 150° and 180° (equivalent to 0°), defined clockwise from the ground surface to the direction of the fibers as illustrated in Fig. 5.12. The wheel speeds used were 16, 19, 22, 25 and 27 m/s, respectively, and the table speeds were 1, 2, 3, 4 and 5 m/min, respectively. The surface integrity of a ground specimen was assessed in terms of surface roughness, surface damage and de-lamination in the subsurface, using a surface roughness tester, Surftest-402 and Surftest Analyzer, an optical microscope, Leica LEITZ DMRXE, and a scanning electronic microscope (SEM), Philips XL-30.

5.2.3.1 Influence of Fiber Orientation

Fiber orientation has a significant effect on the surface quality of a ground CFRP component. Figure 5.13 shows the relationship between the longitudinal surface roughness of the ground specimens and the fiber orientations under different grinding depths. The results indicate that the roughness is dependent on the fiber orientation remarkably. The ground components with the fiber orientation in the range of 120–180° have much rougher surfaces. Under relatively small grinding depths close to the diameter of the carbon fibers, such as 10 μm and 20 μm, the worst surface roughness occurs at $\theta = 0°$. The reason is that the rupture of the fibers and the fiber/matrix interface de-bonding occurred at the fiber orientation of 0°. Partial matrix in the machining zone was pulled off with the ruptured fibers during the grinding process, leaving some arc-shaped grooves where the fibers originally located. This makes the ground surface at $\theta = 0°$ rougher. However, when a large grinding depth (e.g., $\geq$50 μm) is used, the roughest surface appears at $\theta = 150°$, with the roughness about 2–3 times that at $\theta = 0°$ and at least ten times that at $\theta = 90°$. The results in the transverse direction are similar.

Figure 5.14 shows a typical subsurface defect, where some cracks are more than 100 μm in length. These defects manifest that the fracture of fibers and matrix is

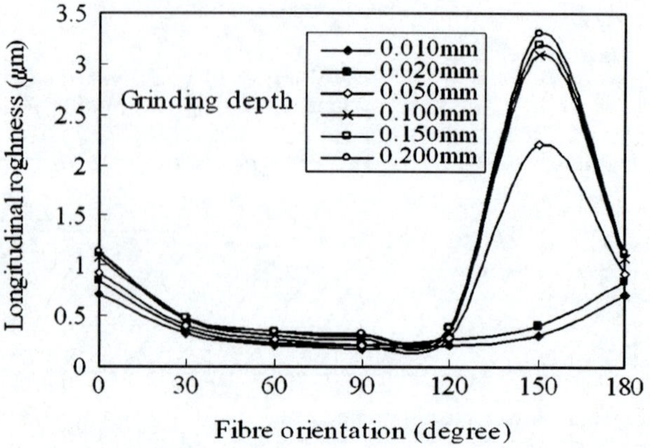

Fig. 5.13 Influence of fiber orientation on surface roughness

Fig. 5.14 Serious subsurface damage and cracks when $\theta = 150°$

below the cutting point of grits, which may be caused by fiber buckling and shearing. There is an inherent weakness of fiber-reinforced composites in the transverse direction due to the weak bonding of fiber/matrix interfaces [39]. At the fiber orientation of $\theta = 150°$, the serious buckling is prone to causing partial de-bonding and micro-cracks in fiber/matrix interfaces along fiber direction. As there are severe subsurface damages on ground CFRP, and these damages just resemble pre-service cracks and edge de-lamination on CFRP components, it would not be surprised that these components would have poor mechanical properties and low fracture toughness. Figure 5.15 shows typical cavitations caused by grinding.

Fig. 5.15 Damages on ground surface when θ = 150°

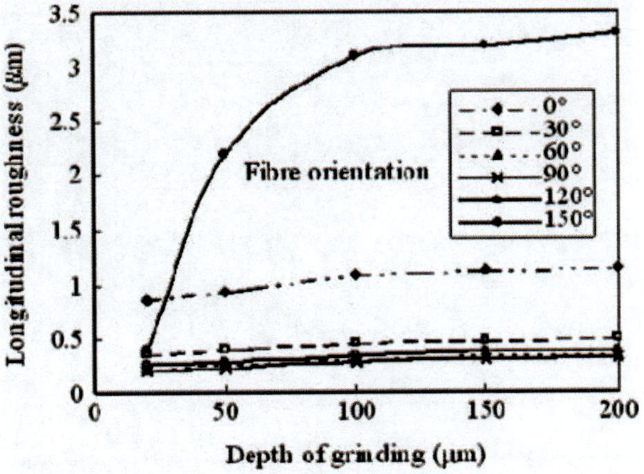

Fig. 5.16 Influence of grinding depth on surface roughness

5.2.3.2 Influence of Grinding Depth

The grinding depth has an obvious effect when θ = 150°. As shown in Fig. 5.16, at θ = 150° the increase in grinding depth results in a sharp rise of surface roughness until the depth reaches 100 µm. For other θ values, however, the effect is trivial. It is worth noting that the ground surface at θ = 150° shows a typical saw-toothed surface morphology as shown in Figs. 5.14, 5.17 and 5.18. This again implies that the chip fracture in grinding is often below the cutting edge of the active abrasive grits. The mechanism of material removal seems to be the same as that in single point cutting, which was investigated by Zhang et al. [40]. In other words,

Fig. 5.17 Subsurface damage at a shallow grinding depth (grinding depth = 20 μm, θ = 150°)

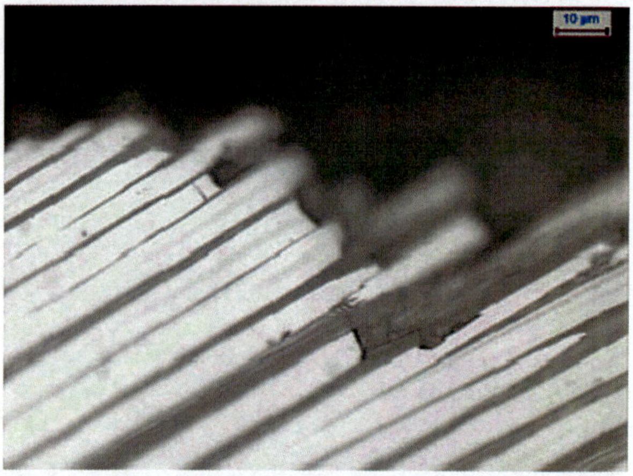

Fig. 5.18 Subsurface damage at a deep grinding depth (grinding depth = 100 μm, θ = 150°)

under the cutting of an abrasive grit, fiber/matrix de-bonding occurs first along the fiber/matrix interface and then fiber fractures when the tensile stress at the fiber surface becomes great enough to initiate the cracking.

By comparing the results in Figs. 5.17 and 5.18 corresponding to different grinding depths, we can see that the subsurface damage at a shallow grinding depth is less severe. This is because at a large grinding depth, an active grit would exert a greater axial force on a fiber, which shifts the cracking point of the fiber towards the deeper substrate, because fiber cracking needs a sufficiently large tensile stress. As a result, a deeper saw-toothed morphology is produced.

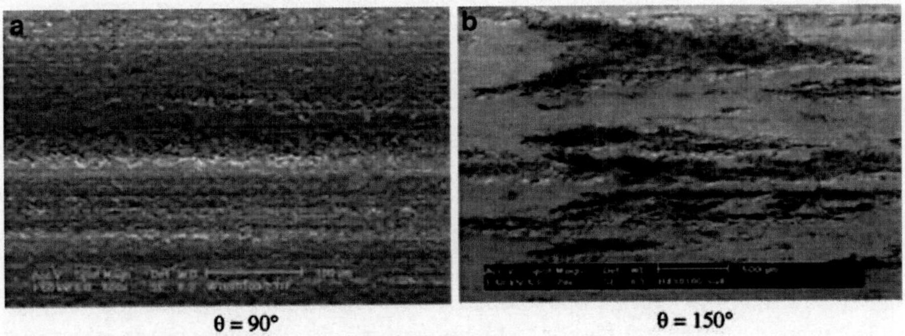

Fig. 5.19 Ground surfaces with matrix smearing

5.2.3.3 Influence of Grinding Wheel Speed and Table Speed

Grinding wheel speed and table speed are generally influential parameters. However, their effects are not obvious in the range of grinding conditions used in this section except the case of $\theta = 150°$. This may be due to the matrix smearing occurred in the grinding of a CFRP component. The smearing brings about a fictitiously small surface roughness that may conceal the effects of wheel speed and table speed. At $\theta = 150°$ the surface finish is much poor so that the smearing may not cover the surface so easily because of the much deeper surface damage. Figure 5.19 shows the comparison of the matrix smearing on the ground surfaces with different fiber orientations.

The above discussion brings about the following understanding:

1. A fiber orientation θ in the range of 120–180° has a greater effect on surface integrity. The most serious surface and subsurface damages occur at $\theta = 150°$.
2. Fiber fracture in grinding often takes place at a distance below the cutting edge. This becomes more severe when the depth of grinding becomes larger. However, the smearing of the matrix on a ground surface may conceal the effects of table and wheel speeds.
3. In order to obtain a satisfactory surface integrity in the grinding of CFRP components, one needs to take a smaller grinding depth or avoid the fiber orientation in the range of 120–180°.

5.2.4 Grinding of Monocrystalline Silicon

As the last example of materials in this chapter, this section discusses the grinding-induced surface integrity in a semiconductor, monocrystalline silicon. Monocrystalline silicon is an important material to the production of micro-electronics and micro-mechanical components. It has also been used widely as an etalon material. These applications require that the surface finishing must be high quality and subsurface structure be damage-free [41]. To achieve the requirements the mechanism of subsurface formation during surface processing, such as indentation [42–45], turning [46], grinding [47, 48] and polishing [41, 48–50], must be fully understood. These have led to extensive theoretical and experimental studies [51–56]. The theoretical analysis by means of molecular dynamic simulations has provided the evidence of amorphous phase change in indentation, sliding and polishing, and has established a criterion for predicting its initiation based on the octahedral shearing stress [41, 53, 55, 56]. Direct and indirect experimental studies have confirmed the amorphous transformation in ultra-precision grinding or turning. Some traces of the amorphous phase were also admitted inside the residual indentation marks [57–60]. In the following, we will discuss the subsurface damage in silicon due to grinding.

All experiments were conducted on precisely polished (100) surfaces of monocrystalline silicon with Rms roughness of 2 nm. Before a grinding experiment, the subsurface structure of specimens was examined to guarantee that a specimen was damage-free in its initial state. The grinding experiments were done on the same surface grinder as that used for ceramics, metals and composites (Minini Junior 90 CF CNC M286) discussed in the previous sections. The grinding parameters are listed in Table 5.1 below.

The subsurface structure of the specimens after grinding was investigated by means of TEM using CM12 and a scanning transmission electron microscope (STEM) using VG HB601. The cross-sectional view TEM samples were prepared using the method developed by Zarudi and Zhang [42].

The subsurface structure of silicon after grinding at a table speed of 1 m/min is shown in Fig. 5.20, featuring an amorphous transformation layer developed in the immediate vicinity of the subsurface region. Below this amorphous layer, some dislocations (two dislocation systems) appear. By recalling the experimental and theoretical studies aforementioned, it is clear that during the grinding of silicon, amorphous silicon was always the first subsurface damage to emerge, followed by the dislocations. This observation is further confirmed in Fig. 5.21 where the table speed was reduced to 0.02 m/min so that there is only a single dislocation, because at this lower table speed, the grinding force was much smaller.

The above results show that in the grinding of monocrystalline silicon, the most possible types of subsurface damages are the amorphous phase transformation and dislocations. To avoid the emergence of the amorphous phase, one must control the

Table 5.1 Grinding parameters

Grinding conditions		Depth of cut (nm)	100
Grinding wheel	SD4000L75BPF	Coolant	Syntilo 3
Wheel diameter (mm)	305	**Wheel dressing conditions**	
Wheel speed (m/s)	27	Type of dresser	Multipoint diamond
Grinding width (mm)	5	Wheel speed (m/s)	10
Table speed (m/min)	0.02, 0.1, 0.2, 0.5, 1	Dressing cross-feed rate (mm/rev)	0.01

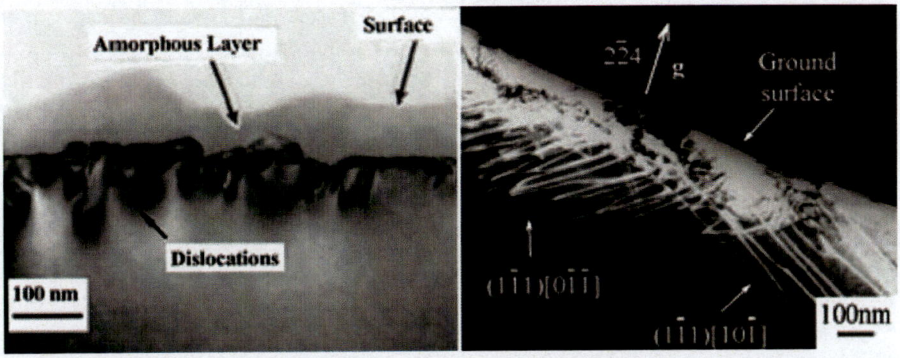

Fig. 5.20 Dislocations induced by grinding at a table speed of 1 m/min. Note that two dislocation systems appear

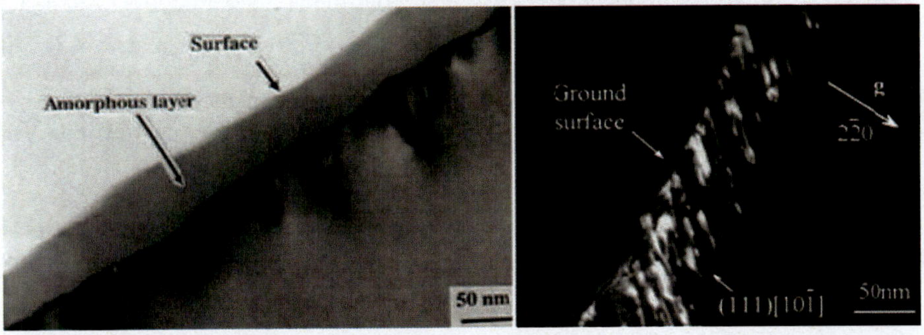

Fig. 5.21 The subsurface structure of silicon after grinding with a table speed of 0.02 m/min. Note that only one dislocation system appears

grinding parameters, such as the depth of cut and table speed, to make the grinding stresses be lower than the initiation stress (or threshold) of the phase transformation. The reader can find more details in [41, 50, 53].

5.3 Summary

The discussions in this chapter have demonstrated that in the grinding of different workpiece materials, the material removal mechanisms vary, which in turn leads to diverse types of subsurface damages.

With a metal workpiece, the material removal mechanism is via dislocations. Due to the ductility of the materials, cracking in a workpiece during grinding is usually not a problem. Tensile residual stresses can occur under certain combinations of thermal deformation and phase transformation.

In the case of a ceramics workpiece, to avoid macroscopic cracking, ductile regime grinding is necessary. The ductile-mode of material removal is achieved by the activation of sufficient number of independent slip systems. If a polycrystalline ceramics contains surface pores, it is impossible to eliminate the micro-cracking at the pore edge.

When grinding composites (fiber reinforced plastics in this chapter), fiber breakage, de-lamination and cavitations are the major sources of subsurface damages. To minimize these defects, smaller depth of abrasive cut is recommended. Meanwhile, it will be very helpful to obtain high surface integrity if the fiber orientation relative to the grinding direction can be controlled or adjusted.

In the grinding of monocrystalline silicon, amorphous phase transformation and dislocations are the most critical events to avoid. It is possible to obtain damage-free silicon surface (or perfect surface integrity) if critical stresses in grinding can be controlled [41, 50].

The discussion in this chapter has been limited to the surface integrity of workpiece materials under given grinding conditions. Nevertheless, there are many other influential factors as outlined in Sect. 5.1 of the chapter. A change of these will contribute to the achievable level of surface integrity. For a production aiming at an optimal grinding process, a comprehensive consideration is necessary to incorporate other factors such as properties of grinding wheels [61, 62] and variations of wheel-workpiece contact details [63, 64].

Acknowledgments The continuous support of Australian Research Council to this work is appreciated.

References

1. Zhang LC, Suto T, Noguchi H, Waida T (1992) An overview of applied mechanics in grinding, Manufacturing Review, 5, 261–273.
2. Zarudi I, Zhang LC (2002) A revisit to some fundamental wheel-workpiece interaction problems in surface grinding, International Journal of Machine Tools and Manufacture, 42, 905–913.
3. Mahdi M, Zhang LC (1994) Correlation between grinding conditions and phase transformation of an alloy steel, in: Advanced Computational Methods in Heat Transfer III, edited by LC Wrobel, et al, Computational Mechanics Publications, Southampton, pp. 193–200.
4. Mahdi M, Zhang LC (1995) The finite element thermal analysis of grinding processes by ADINA, Computer & Structure, 56, 313–320.
5. Zhang LC, Mahdi M (1995) Applied mechanics in grinding, Part IV: the mechanism of grinding induced phase transformation, International Journal of Machine Tools and Manufacture, 35, 1397–1409.
6. Mahdi M, Zhang LC (1994) A theoretical investigation on the mechanically induced residual stresses due to surface grinding, in: Progress of Cutting and Grinding, Vol. III, edited by N Narutaki, et al, Japan Society for Precision Engineering, Osaka, pp. 484–487.

7. Mahdi M, Zhang LC (1997) Applied mechanics in grinding, Part V: thermal residual stresses, International Journal of Machine Tools and Manufacture, 37, 619–633.
8. Mahdi M, Zhang LC (1998) Applied mechanics in grinding, Part VI: residual stresses and surface hardening by coupled thermo-plasticity and phase transformation, International Journal of Machine Tools and Manufacture, 38, 1289–1340.
9. Mahdi M, Zhang LC (1998) Residual stresses in ground components: effect of thermo-mechanical deformation, Proceedings of the 4th International Conference on Progress of Cutting and Grinding, Japan Society for Precision Engineering, Urumqi and Turpan, China, 5–9 October 1998, pp. 447–452.
10. Mahdi M, Zhang, LC (1999) Applied mechanics in grinding, Part VII: residual stresses induced by the full coupling of mechanical deformation, thermal deformation and phase transformation, International Journal of Machine Tools and Manufacture, 39, 1285–1298.
11. Mahdi M, Zhang, LC (2000) A numerical algorithm for the full coupling of mechanical deformation, thermal deformation and phase transformation in surface grinding, Computational Mechanics, 26, 157–165.
12. Zhang LC (2007) Grind-hardening of steel surfaces: a focused review, International Journal of Abrasive Technology, 1, 3–36.
13. Zarudi I, Zhang LC (2002) Modelling the structure changes in quenchable steel subjected to grinding, Journal of Materials Science, 37, 4333–4341.
14. Zarudi I, Zhang LC (2002) Mechanical property improvement of quenchable steel by grinding, Journal of Materials Science, 37, 3935–3943.
15. Nguyen T, Zhang LC, Zarudi I (2007) Grinding-hardening with liquid nitrogen: mechanisms and technology, International Journal of Machine Tools and Manufacture, 47, 97–106.
16. Zarudi, I, Zhang, LC (1997) Subsurface structure change of silicon after ultra-precision grinding, in: Advances in Abrasive Technology, edited by LC Zhang and N Yasunaga, World Scientific, Singapore, pp. 33–38.
17. Suzuki H, Wajima N, Zahmaty MS, Kuriyagava T, Syoji K (1997) Precision grinding of a spherical surface accuracy improving by on-machine measurement, in: Advances in Abrasive Technology, edited by LC Zhang and N Yasunaga, World Scientific, Singapore, pp. 116–121.
18. Zarudi I, Zhang LC, Mai YW (1996) Subsurface damage in alumina induced by single-point scratching, Journal of Materials Science, 31, 905–914.
19. Zarudi I, Zhang LC, Cockayne D (1998) Subsurface structure of alumina associated with single-point scratching, Journal of Materials Science, 33, 1639–1654.
20. Komanduri R (1996) On material removal mechanisms in finishing of advanced ceramics and glasses, Annals of the CIRP, 45, 509–514.
21. Bifano TG, Dow TA, Scattergood RO (1991) Ductile-regime grinding: a new technology for machining brittle materials, Transactions of the ASME Journal of Engineering Materials and Technology, 113, 184–189.
22. Zarudi I, Zhang LC (2000) On the limit of surface integrity of alumina by ductile-mode grinding, Transactions of the ASME Journal of Engineering Materials and Technology, 122, 129–134.
23. Zhang LC (2001), Solid Mechanics for Engineers, Paragrave Macmillan, Basingstoke, England.
24. Hagan JT (1979) Micromechanics of crack nucleation during indentations, Journal of Materials Science, 14, 2975–2980.
25. Zhang LC (2009) Mechanics and modelling of machining polymer matrix composites reinforced by long fibers, Chapter 1, in: Machining of Composite Materials, edited by JP Davim, ISTE-Wiley, New York, pp. 1–38.
26. Pramanik A, Zhang LC, Arsecularatne JA (2008) Machining of metal matrix composites: effect of ceramic particles on residual stress, surface roughness and chip formation, International Journal of Machining Tools and Manufacture, 48, 1613–1625.
27. Koenig W, Wulf Ch, Grass P, Willerscheid H (1985) Machining of fiber reinforced plastics, Annals of the CIRP, 34, 537–548.

28. Tagliaferri V, Caprino G, Diterlizzi A (1990) Effect of drilling parameters on the finish and mechanical properties of GFRP composites, International Journal of Machine Tools and Manufacture, 30, 77–84.
29. Kaneeda T (1991) CFRP cutting mechanism, Transactions of the North American Manufacturing Research Institute of SME, 19, 216–221.
30. Bhatnagar N, Ramakrishnan N, Naik NK, Komanduri R (1995) On the machining of fiber reinforced plastic (FRP) composite laminates, International Journal of Machine Tools and Manufacture, 35, 701–716.
31. Caprino G, Tagliaferri V (1995) Damage development in drilling glass fiber reinforced plastics, International Journal of Machine Tools and Manufacture, 35, 817–829.
32. Wang DH, Ramulu M, Arola D (1995) Orthogonal cutting mechanisms of graphite/epoxy composite, Part I: unidirectional laminate, International Journal of Machine Tools and Manufacture, 35, 1623–1638.
33. Zhang HJ, Chen WY, Chen DC, Zhang LC (2001) Assessment of the exit defects in carbon fiber-reinforced plastic plates caused by drilling, Precision Machining of Advanced Materials, Key Engineering Materials, 196, 43–52.
34. Mahdi M, Zhang LC (2001) A finite element model for the orthogonal cutting of fiber-reinforced composite materials, Journal of Materials Processing Technology, 113, 373–377.
35. Inoue H, Kawaguchi I (1990) Study on the grinding mechanism of glass fiber reinforced plastics, Journal of Engineering Materials and Technology, 112, 341–345.
36. Park KY, Lee DG, Nakagawa T (1995) Mirror surface grinding characteristics and mechanism of carbon fiber reinforced plastics. Journal of Materials Processing Technology, 52, 386–398.
37. Hu NS, Zhang LC (2003) A study on the grindability of multidirectional carbon fiber-reinforced plastics, Journal of Materials Processing Technology, 140, 152–156.
38. Hu NS, Zhang LC (2004) Some observations in grinding unidirectional carbon fiber-reinforced plastics, Journal of Materials Processing Technology, 152, 333–338.
39. Wang J, Karihallo BL (1994) Cracked composite laminates least prone to delamination, Proceedings: Mathematical and Physical Sciences, 444 (No. 1920), 17–35.
40. Zhang LC, Zhang HJ, Wang XM (2001) A force prediction model for cutting unidirectional fiber-reinforced plastics, Machining Science and Technology, 5, 293–305.
41. Biddut A, Zhang LC, Ali YM, Liu Z (2008) Damage-free polishing of monocrystalline silicon wafers without chemical additives, Scripta Materialia, 59, 1178–1181.
42. Zarudi I, Zhang LC (1999) Structural changes in mono-crystalline silicon subjected to indentation – experimental findings, Tribology International, 32, 701–712.
43. Zarudi I, Zhang LC, Cheong WCD, Yu TX (2005) The difference of phase distributions in silicon after indentation with Berkovich and spherical indenters, Acta Materiala, 53, 4795–4800.
44. Zarudi I, Nguyen T, Zhang LC (2005) Effect of temperature and stress on plastic deformation in monocrystalline silicon induced by scratching, Applied Physics Letters, 86, 011922.
45. Chang L, Zhang LC (2009) Deformation mechanisms at pop-out in monocrystalline silicon under nanoindentation, Acta Materialia, 57, 2148–2153.
46. Biddut A, Yan JW, Zhang LC, Ohta T, Kuriyagawa T, Shaun B (2009) Deformation in monocrystalline silicon caused by high speed single-point micro-cutting, Key Engineering Materials, 407–408, 347–350.
47. Zarudi I, Zhang LC (1998) Effect of ultra-precision grinding on the microstructural change in silicon monocrystals, Journal of Materials Processing Technology, 84, 148–158.
48. Zarudi I, Zhang LC (1996) Subsurface damage in single-crystal silicon due to grinding and polishing, Journal of Materials Science Letters, 15, 586–587.
49. Zhang LC, Zarudi I (1999) An understanding of the chemical effect on the nano-wear deformation in mono-crystalline silicon components, Wear, 225–229, 669–677.
50. Zhang LC, Zarudi I (2001) Towards a deeper understanding of plastic deformation in mono-crystalline silicon, International Journal of Mechanical Science, 43, 1985–1996.

51. Zhang LC (2008) Microstructural changes in silicon caused by indentation and machining, Chapter 4, in: Semiconductor Machining on the Micro-Nanoscale, edited by JW Yan and J Patten, Research Signpost, India, pp. 155–197.
52. Zhang LC (2006) Nano-characterisation of materials: silicon, copper, carbon nanotubes and diamond thin films, Chapter 8, in: Handbook of Theoretical and Computational Nanotechnology, Volume 8: Functional Nanomaterials, Nanoparticles, and Polymer Design, edited by M Rieth and W Schommers, American Scientific Publishers, Stevenson Ranch, CA, USA, pp. 395–456.
53. Zhang LC, Tanaka H (1999) On the mechanics and physics in the nano-indentation of silicon mono-crystals, JSME International Journal, Series A: Solid Mechanics & Material Engineering, 42, 546–559.
54. Zhang LC, Tanaka H (1998) Atomic scale deformation in silicon monocrystals induced by two-body and three-body contact sliding, Tribology International, 31, 425–433.
55. Vodenitcharova T, Zhang LC (2003) A mechanics prediction of the behaviour of mono-crystalline silicon under nano-indentation, International Journal of Solids and Structures, 40, 2989–2998.
56. Vodenitcharova T, Zhang LC (2004) A new constitutive model for the phase transformations in mono-crystalline silicon, International Journal of Solids and Structures, 41, 5411–5424.
57. Zarudi I, Cheong WCD, Zou J, Zhang LC (2004) Atomistic structure of monocrystalline silicon in surface nano-modification, Nanotechnology, 15, 104–107.
58. Zarudi I, Zhang LC, Zou J, Vodenitcharova T (2004) The R8-BC8 phases and crystal growth in monocrystalline silicon under microindentation with a spherical indenter, Journal of Materials Research, 19, 332–337.
59. Zarudi I, Zou J, McBride W, Zhang LC (2004) Amorphous structures induced in monocrystalline silicon by mechanical loading, Applied Physics Letters, 85, 932–934.
60. Zarudi I, Zhang LC, Swain MV (2003) Microstructure evolution in monocrystalline silicon in cyclic microindentation, Journal of Materials Research, 18, 758–761.
61. Zhang LC, Suto T, Noguchi H, Waida T (1993) Applied mechanics in grinding, Part II: modelling of elastic modulus of wheels and interface forces, International Journal of Machine Tools and Manufacture, 33, 245–255.
62. Lu G, Zhang LC (1994) Further remarks on the modelling of elastic modulus of grinding wheels, International Journal of Machine Tools and Manufacture, 34, 841–846.
63. Zhang LC, Suto T, Noguchi H, Waida T (1993) Applied mechanics in grinding, Part III: a new formula for contact length prediction and a comparison of available models, International Journal of Machine Tools and Manufacture, 33, 587–597.
64. Zhang LC, Suto T, Noguchi H, Waida T (1995) A study of creep-feed grinding of metallic and ceramic materials, Journal of Materials Processing Technology, 48, 267–274.

Chapter 6
Traditional and Non-traditional Control Techniques for Grinding Processes

Jian Liu, Chengying Xu and Mark Jackson

Abstract Owing to the highly demanding geometric accuracy and surface finish for many modern products, grinding processes have been extensively used in manufacturing industry. However, it is also well-accepted that grinding is one of the most complicated machining processes due to the high nonlinearities, intrinsic uncertainties and time-varying characteristics. Multiple challenging problems exist in the process that limits its overall quality and production in practice. With the increasing demands for higher part geometry accuracy, better surface integrity, more productivity, and other desired product parameters (e.g., minimization of subsurface micro-damage) with less operator intervention, various control methods have been studied and implemented to control position, velocity, force, power, temperature and the Material Removal Rate (MRR) during the grinding process, in order to achieve the desired system performance within certain cost/time. This paper reviews different control strategies in order to provide a guideline for academic researchers and industrial practitioners to improve the final product quality with increased possible process flexibility.

Keywords Grinding · Control systems · Mathematical modeling · Quality

6.1 Introduction

Grinding process is one of typical complicated nonlinear systems involving multiple input/output variables, such as position, velocity, force, power, temperature, etc.

C. Xu (✉)
Department of Mechanical, Materials and Aerospace Engineering, University of Central Florida, 4000 Central Florida Blvd, Orlando, FL 32816, USA
e-mail: cxu@mail.ucf.edu

M.J. Jackson and J.P. Davim (eds.), *Machining with Abrasives*,
DOI 10.1007/978-1-4419-7302-3_6, © Springer Science+Business Media, LLC 2011

Users usually have multiple goals for the process, such as final parts surface/subsurface quality, geometry accuracy, cost, cycle time. The process itself is disturbed in many ways, such as the varying abrasive property of the grinding wheel due to the wear, hardness variation of the workpiece, dimensional deviations due to elastic deformation of the workpiece-tool-machine system, unbalanced wheel due to the mounting error. The purpose of this survey paper is to review existing control methodologies employed in various grinding processes and to identify potential research directions as a future guideline.

6.2 Conventional Control Techniques

6.2.1 *Fixed-Parameter Control Technique*

The easiest way to control the grinding process is to design a fixed-parameter controller based on off-line system identification. Tönshoff et al. [1, 2] applied a fixed-gain force controller to an internal grinding process to increase the product quality and shorten the manufacturing cycle time by reducing the times of air grinding. In this particular application, the Proportional Integral Derivative (PID) control algorithm embedded in a microcomputer system is used to construct the digital closed control loop for internal grinding force control with fixed PID gains, as shown in Fig. 6.1.

Digital control is applied because the available analogue systems are often not suited to consider the limited conditions due to noncircular borings; however, the

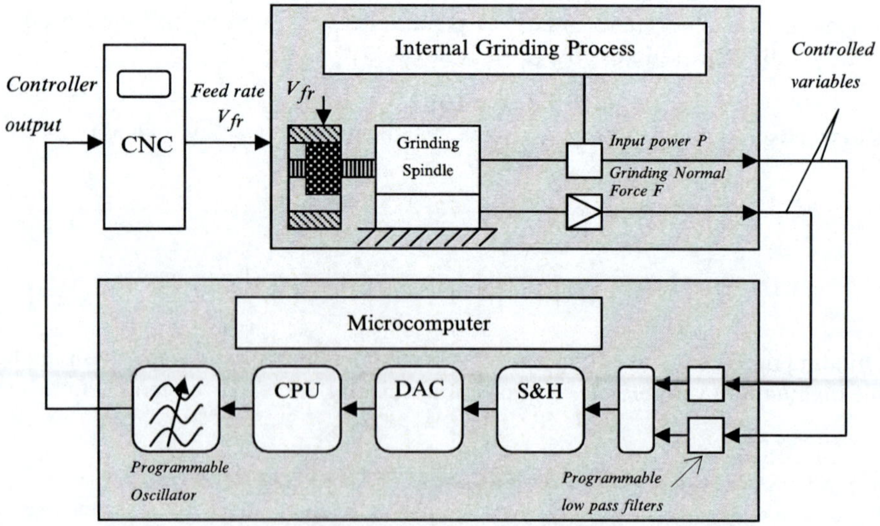

Fig. 6.1 Control loop for internal grinding

digital data processing allows the implementation of closed-loop controls, which at the same time increase the process stability, the efficiency and accuracy of machining and hence the finish surface quality. In Fig. 6.1, it also shows that the controlled variables, grinding force $F_n$and grinding spindle input power P, and the necessary hardware devices. The core part in this control system set up is a microcomputer, based on the 16bit-micro-processor MC 68010, which is used to execute process identification and digital control algorithm. PEARL (process and experimental automation real-time language) which was developed in Germany and standardized in DIN 66253 is used to program the microprocessor.

During the development of such a kind of controllers, prerequisites for control algorithm developing require a process identification for the following two goals: (1) enforcing the grinding force to reach the nominal value as soon as possible after the first cut; (2) Detecting dead times and use power measurement as controlled variable, namely carry out immediately a dead-beat-controller. Also, controller parameters tuning process has to be performed to follow the process variations. Therefore, the behavior of the controlled system could be characterized to some extent and the control law could be employed to this corresponding objective plant. Figure 6.2 illustrates the concept and connections of process identification and digital closed-loop control.

The process identification tries to exactly capture the real process using the data acquired and stored during the identification grinding cycle. Therefore, the order and the dead-time of the transfer function have to be known ahead. A model of a third order is supposed for the internal grinding cycle, and the Z-transformation of this process as well as its differential equation can be written as follows:

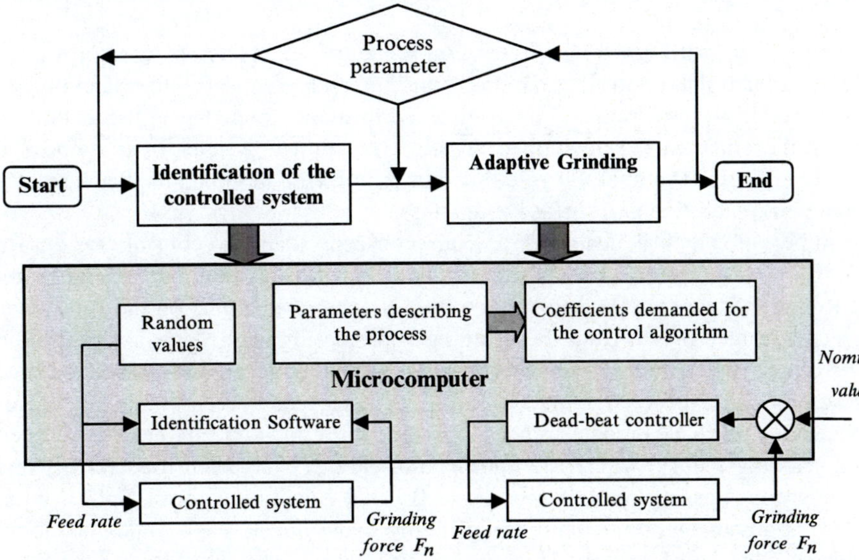

Fig. 6.2 Diagram of process identification and control

$$G_{(z)} = \frac{y_{(z)}}{u_{(z)}} = \frac{b_0 + b_1 z^{-1} + b_2 z^{-2} + b_3 z^{-3}}{1 + a_1 z^{-1} + a_2 z^{-2} + a_3 z^{-3}} z^{-d} \tag{6.1}$$

$$\begin{aligned} y_{(k)} = & -a_1 y_{(k-1)} - a_2 y_{(k-2)} - a_3 y_{(k-3)} + b_0 u_{(k-d)} + b_2 u_{(k-1-d)} \\ & + b_2 u_{(k-2-d)} + b_3 u_{(k-3-d)} \end{aligned} \tag{6.2}$$

where, k = 0, 1, 2 ...n, and parameters a_i and b_i are required to model the real process. By the random values of feed rate as input and the corresponding sampled grinding force values as output, both recorded during the identification grinding cycle, the process parameters a_i and b_i can be calculated by the Gaussian algorithm. Thereby, the six parameters will be used to calculate the coefficients of the controller. These coefficients are valid as long as the machining conditions are unchanged. When grinding condition has to be modified due to different feature requirements, the parameters should be identified again for the new grinding cycle.

In summary, the implementation of this kind of controller is usually simple. However, the fixed-gain controller does not have the adaptation ability and therefore cannot compensate for various process uncertainties and parameter variations. Hence, the overall system performance cannot be guaranteed in practical industry applications.

6.2.2 *Adaptive Control Technique*

In comparison with fixed-parameter control approach, adaptive controllers are able to adjust the control parameters on-line and are useful to maintain the system stability and improve the control performance over the entire operating time. Kaliszer et al. [3] developed an adaptive control system for a cylindrical grinding process. The overall cycle time was minimized based on feedback signals on the workpiece size and surface roughness.

Amitay et al. [4] designed a computerized adaptive control system to optimize both the grinding and dressing conditions for the maximum Metal/Material Removal Rate (MRR), subjected to two process constraints on the workpiece burn and surface finish. Both the grinding and dressing parameters are controlled based upon the data from on-line sensing of the grinding power and off-line measurement of surface roughness. On-line convergence to the optimal conditions proceeds along a predetermined optimization position trajectory derived from grinding theory, where the optimization strategy has been used for off-line grinding and dressing optimization as well. This Adaptive Control Optimization (ACO) is advantageous to optimize grinding cycle due to easy implementation, requiring a set up of a micro-computer, power sensor, and interfaces to the basic grinding machine.

The performance index of the grinding process is the volumetric material removal rate, which can be expressed in the below equation for external cylindrical grinding:

$$Z = \pi d_w v_f b \tag{6.3}$$

Since the workpiece diameter d_w is almost constant and the grinding width b is fixed for any particular optimization, the performance index can be simplified as the radial infeed velocity v_f, the optimization problem therefore can be formulated as:

$$\begin{array}{c} \text{Maximize} \quad v_f : \\ \text{Subject to:} \quad P \leqslant P_b \\ R_a \leqslant R_{ax} \end{array} \tag{6.4}$$

where P is the total grinding power, which can be sensed on-line during control process; P_b is the threshold grinding power; R_a is surface finish, which can be measured off-line, after several individual grinding cycle; R_{ax} is the desired surface finish limit.

The overall scheme for such a control system is shown in Fig. 6.3, which includes a grinding machine, a computer, and the operator interconnected. Control algorithm is executed by the computer in order to control the grinding process, based on some partially human-interfered expertise imported by the operator. Measurements of the grinding power P are fed from the grinding machine to the computer and the computer in turn controls the grinding parameters v_f and n_w to operate along the optimal locus (as seen in Fig. 6.4) and converge towards the optimal operating point. Moreover, the required surface finish limit R_{ax} and intermediate surface finish measurements R_a are inputs by the operator to the computer while off-line, and the computer suggests new dressing conditions to the operator which he sets on the grinding machine.

A pilot ACO system which constitutes a cylindrical grinder interfaced to an industrial computer was developed to demonstrate this control concept, as shown in Fig. 6.5. The pivot of the on-line control system here is an algorithm incorporating the optimal locus optimization skill, as illustrated in Fig. 6.4, with an ID-type

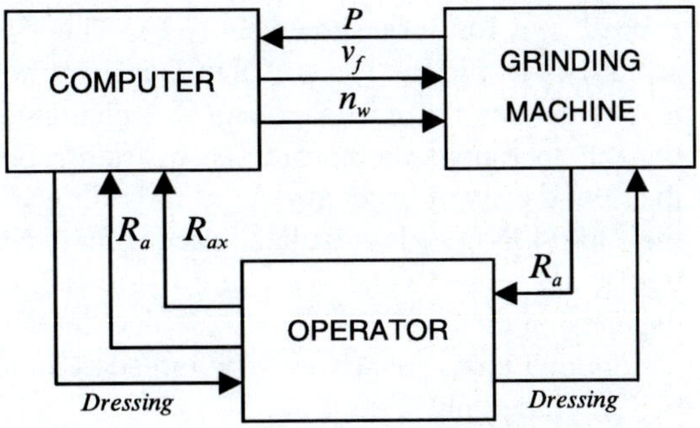

Fig. 6.3 Optimal adaptive control concept

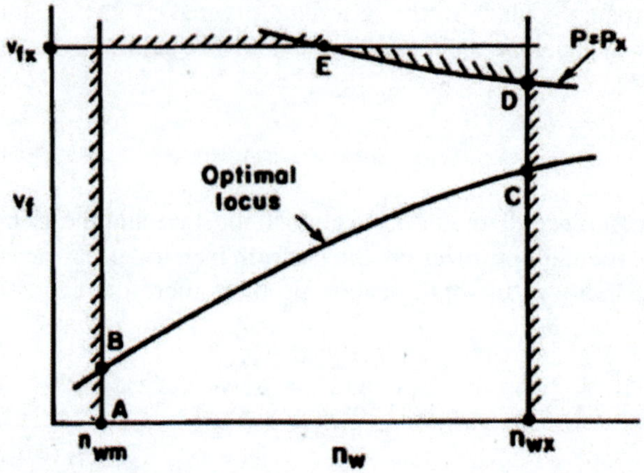

Fig. 6.4 Illustration of optimal locus together with machine tool limitations in $v_f - n_w$ plane

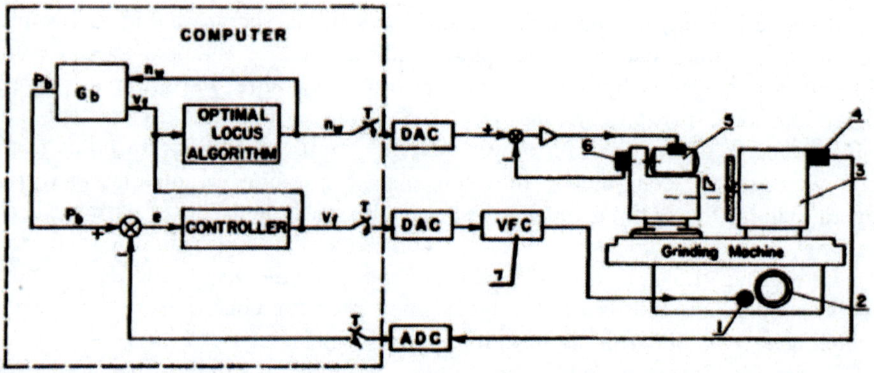

Fig. 6.5 On-line adaptive control system: 1 –stepping motor infeed drive; 2 – infeed control hand wheel; 3 – grinding wheel motor; 4 – power sensor; 5 – workpiece spindle DC motor; 6 – tacho-generator; 7 – voltage-to-frequency converter

control law for $v_f$as shown in (6.5). The grinding operation could start at an arbitrary point in the $v_f - n_w$ plan, but is immediately transferred by the computer to a point on the optimal locus by changing n_w. Thereafter, the trajectory of convergence towards the optimal operating point is along the optimal locus. For the rate of convergence, instead of providing an external reference, the reference to the control loop P_b is calculated according to Malkin's [5] model in the block G_b to which the control variables $v_f$and $n_w$are fed. Consequently, the convergence rate depends on the error $e = P_b - P$, which converges to zero when proceeding along the optimal locus adaptively. The infeed velocity is determined by the controller in Fig. 6.5 based on:

$$v_f(i) = v_f(i-1) + K_1 e(i) + K_2[e(i) - e(i-1)] \tag{6.5}$$

6.3 Adaptive Force Control Application

Other researchers such as Furukawa and Ohishi [6], Elbestawi et al. [7], Bhattacharyya et al. [8] also have applied adaptive control scheme in grinding processes. For instance, in Elbestawi's work, the focus is on predictive adaptive force control during grinding cycle. The configuration for the robotic disk grinding process is show in Fig. 6.6 and the grinding force model is summarized in the following equation, which combines a disk wear model, a metal removal rate model and the nonlinear stiffness of the robot according to a dynamic force model of Ulrich, B. J. [9] and Srivastava, A. K. [10].

$$\dot{F}_n = f_1\left\{V_f - \dot{H}f_2 \frac{\Lambda_w - \lambda F_n(F_n - F_{th})}{B(h - y_i)} + \frac{\Lambda_\theta(F_n - F'_{th})}{2\pi R^*(h - y_i)}\left[1 \frac{2F_n f_3}{(h - y_i)}\right] + \frac{F_n V_f f_4}{L_w}\right\} \tag{6.6}$$

where F_n is the normal cutting force; Λ_θ is the disk wear parameter; F'_{th} is the disk wear threshold force; L_w is the length of the workpiece; h and y_i are the heights of the initial and resulting workpiece surfaces from some arbitrary datum. $f_1 \sim f_4$ are four coefficients who are functions of (K_x, K_y), L_w and the distance travelled by the disk during a pass. The (K_{x1}, K_{x2}) and (K_{y1}, K_{y2}) are the horizontal and vertical robot stiffness values at the beginning and end points of the grinding pass, respectively. By analyzing the geometry of the grinding process, as show in Fig. 6.7, the coordinates $(x_i, y_i)$of the extreme edge could be expressed as:

$$x_i = x_0 - \frac{F_x}{K_x} - \left(h - H + R\sin\theta_0 - \frac{F_y}{K_y}\right)\tan(\theta_0 - \theta_i) \tag{6.7}$$

$$y_i = \frac{\sin\theta_i}{\sin\theta_0\cos(\theta_0 - \theta_i)}\left[H - R\sin\theta_0 + \frac{F_y}{K_y}\right] + h\left[1 - \frac{\sin\theta_i}{\sin\theta_0\cos(\theta_0 - \theta_i)}\right] \tag{6.8}$$

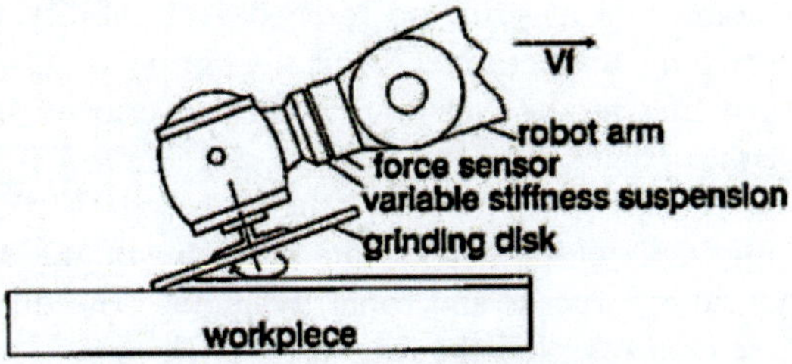

Fig. 6.6 Configurations for robotic disk grinding process

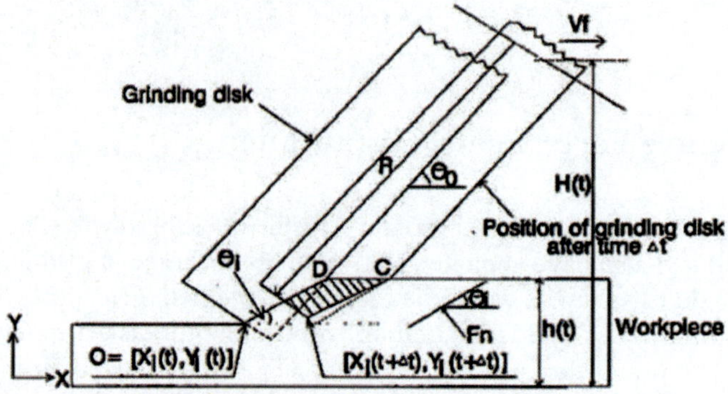

Fig. 6.7 Coordinate system and geometry of grinding disk

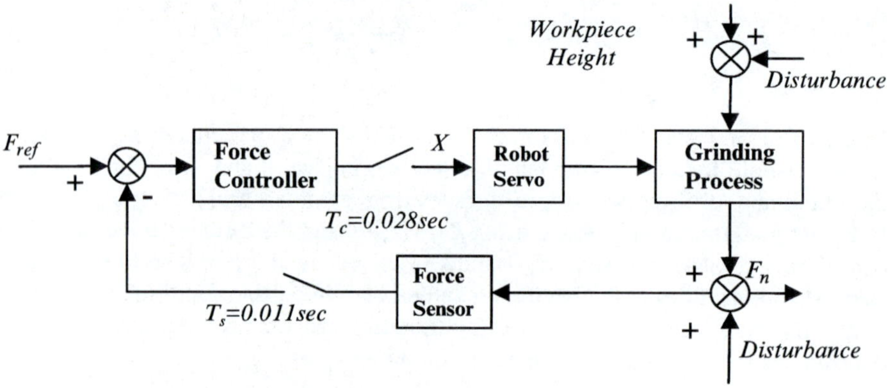

Fig. 6.8 Block diagram of the force feedback controller

Using force control (regulation) for grinding application can have several advantages: (1) It can be used to maintain specified MRR in order to either achieve the maximum productivity or save machining time, based on requirements for final product quality. (2) It can help reduce geometric errors in the finished workpiece dimensions due to variations in workpiece hardness, wheel wear, external disturbances, etc. (3) It can be used to detect and hence prevent workpiece burn since the monitoring of grinding force has the ability to indicate any severe changes in grinding temperature. Basically, grinding force control is quite significant for grinding process adaptive control, based on which more effective and efficient control algorithm.

A basic force control scheme based on PID feedback controller is shown in Fig. 6.8, which depicts the structure of the PID force controller including the grinding process and robot dynamics. The digital control signal implemented in the controller has the form as:

$$u(kT) = K_p\varepsilon(kT) + K_i[\xi(kT)] + K_d\left[\frac{\varepsilon(kT) - \varepsilon[(k-1)T]}{T}\right] \quad (6.9)$$

where, $u(kT)$ is the control signal at the current sample time, $\varepsilon(kT)$ is the force error signal ($\varepsilon(kT) = F_{ref}(kT) - F_n(kT)$) at the current sample time instant, $\xi(kT)$ is the cumulative sum of $\varepsilon(kT)$ at the current sample time, the three leading coefficients are the gains of proportional, integral, and derivative terms respectively.

Even though the above PID control structural is capable of regulating the grinding force in the presence of a step disturbance in the workpiece height, its performance highly depends on the robot arm location, metal removal and disk wear parameters. Therefore, adaptive control was found to be necessary to compensate for these variations in system parameters instead of the PID approach. The structure for this kind of adaptive controller is illustrated in Fig. 6.9.

Similar to the former example for Tönshoff's [11] work, the plant's model should be built firstly before the actual control works. The modeling parameters identification was proceeded under open-loop conditions, by commanding the robot end effecter to follow a pseudo-random trajectory signal while simultaneously measuring the normal force during grinding cycle.

Based on the identified plant model, two adaptive control algorithms, namely, Generalized Predictive Control (GPC) and Extended Horizon Control (EHC) are utilized to regulate the grinding force and evaluate the advantages adaptive control scheme shown above has over non-adaptive PID force controller. The objective of the GPC strategy is to drive the future plant output as close as possible to the future set point and the design objective of EHC is to bring the plant output to the set point within the control horizon with minimum effort.

Another example is Bhattacharyya's work [8]. In this work, the author focused on improving grinding surface integrity by avoid grinding burn and thermal damage, where phenomena arise from the point when a grinding wheel is in contact with

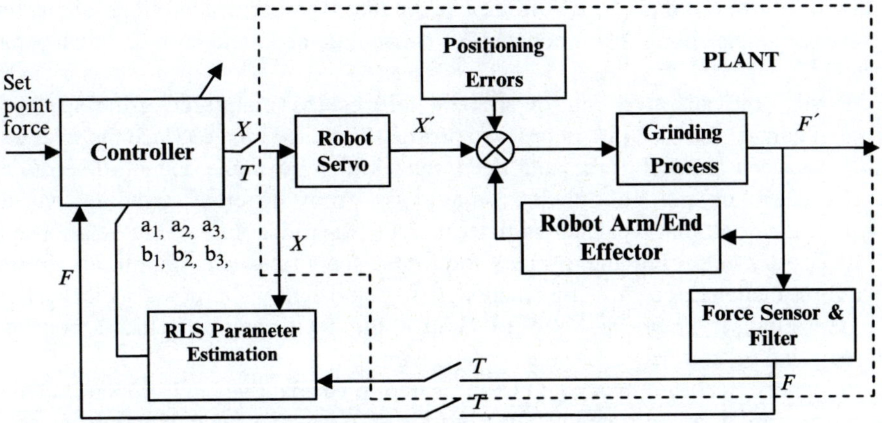

Fig. 6.9 Block diagram of the adaptive force control system

the workpiece, and considerable heat is generated and majority of the them enters the workpiece. This might be responsible for the undesirable surface finish and introduces residual stress. Therefore, by evaluating the surface finish and residual stress mainly, grinding surface integrity could be enhanced.

However, the surface integrity is only one aspect of optimum grinding conditions, which are defined when achieving the maximum metal removal rate while maintaining the required surface integrity, mainly in terms of surface finish and residual stress, so that the decision of optimum grinding conditions which has also been termed as "art of grinding" usually requires expensive and scarce skill and is a time consuming job. In this case, an adaptive control engine which is able to optimize automatically will be meaningful. It is important to know the model for predicting the onset of grinding burn based on which the adaptive control for grinding process proposed can detect and prevent thermal damage during machining and hence improve surface finish.

According to well-documented literature about the relationship between workpiece surface finish and grinding process variables, the surface finish is primarily affected by the normal force intensity and the wheel dressing conditions. Also, a proportional relationship was found between the surface finish and the metal removal rate. Surface finish is also affected by wheel regenerative chatter vibrations. Hahn and Lindsay [12, 13] proposed a correlation between thermal damage and normal force intensity. As for grinding burn, Cebals proposed that it was related to the grinding coefficient, defined as the ratio of the tangential grinding force to the normal grinding force. A coefficient of less than 0.4 indicates the production of thermally damaged parts. In general, based on the above investigation, both surface finish and grinding thermal damage/burn can be controlled by grinding at a reasonable work speed under controlled force intensities or ratio of the tangential grinding force to the normal grinding force with wheels, whose sharpness is maintained above safety level. Its advantages in requiring less grinding data and less computation load comparing with regression techniques facilitates the application in grinding adaptive control of workpiece finish surface integrity. However, when comes to the control procedure, more considerations about the process models should be conducted for thermal damage and surface finish separately.

Firstly, the adaptive control strategy utilizes the monitored grinding force data to control the feed rate in order to prevent thermal damage. Since the relationship between normal force and feed rate, which to some extent determines the success of the control law, may vary from different grinding wheel-workpiece combinations and hard to predict, identification of the relationship is necessary requiring the normal force and feed rates are monitored during grinding cycle. Once this crucial model is established via process identification in real-time, it could be used to control the feed rate to achieve constant force grinding and hence prevent thermal damage.

Secondly, surface finish could be measured in an adaptive control identification strategy similar to the strategy adopted for thermal damage, even though it can only be applied in off-line manner. When the model which connects normal force and vibration levels with workpiece surface finish is ready, the adaptive control system could use this model to execute surface finish control by varying the feed rate of the grinding wheel.

This control is an extension of the adaptive control strategy to achieve constant force grinding. Also, the identification stage could be re-initiated to produce a new model once the old one is invalidated by an alteration in dressing conditions, wheel, workpiece profile or material properties etc. The entire control algorithm development for thermal damage prevention is summarized as: firstly, an adaptive control identification strategy is used to compute a pre-burn value of the grinding coefficient from the monitored grinding force data. A control algorithm would then monitor the grinding forces, compute the current grinding coefficient and compare it with the pre-burn value. If a significant reduction is experienced, workpiece burn could be detected and the adaptive controller could reduce the wheel feed rate to produce less aggressive grinding conditions.

6.4 Non-linear Adaptive Control with Self-tuning Ability

By on-line identifying the parameters of the mathematical model, the grinding conditions could be calculated and adjusted in real time to achieve the optimum machining efficiency. Webster and Zhao [14] developed a time-optimizing adaptive control strategy for plunge grinding processes. An online identification of the system time constant from size measurements was incorporated in the controller design to make the method more robust in accelerating the spark-out grinding cycle. Guo et al. [15, 16] presented a nonlinear indirect adaptive control scheme based on the model following scheme and applied to the grinding force control of a form grinding process. During the process, a nonlinear adaptive observer was used to estimate the model parameter and therefore increase the robustness of the control system against parameter uncertainties. The stability of the whole closed-loop system is thus guaranteed with satisfactory system performance.

The mathematical force model of a grinding process belongs to the class of nonlinear systems with the form as below:

$$\dot{\underline{x}} = \underline{f}(\underline{x}) + \sum_{i=1}^{m} {}^{-}\varphi_{1i}(\underline{x})\vartheta_{1i} + \left[\underline{g}(\underline{x}) + \sum_{j=1}^{p} \underline{\varphi}_{2_j}(\underline{x})\vartheta_{2j} \right] u \tag{6.10}$$

$$y = h(\underline{x}) + \sum_{k=1}^{q} \underline{\varphi}_{3_k}(\underline{x})\vartheta_{3k} \tag{6.11}$$

where $\underline{x}(t)$, $u(t)$and $y(t)$ denote the state vector, the input and the output of the system at time instant t respectively. This model represents the uniform of nonlinear system with uncertain parameters and the purpose of the control work is to regulate $y(t)$ to be a desired value y_d, namely,

$$|y_d(t) - y(t)| < \varepsilon \quad \text{for a given } \varepsilon \text{ or } \lim_{t \to \infty} [y_d(t) - y(t)] = 0 \tag{6.12}$$

Adaptive controller becomes a suitable candidate for this kind of nonlinear control problem because it can dynamically compensate the variations of the uncertain system parameters in a real-time manner. Based on a reduced nonlinear system model, after deriving a nonlinear controller by extending the well known global linearization theory to the case where system parameters may be uncertain, a nonlinear adaptive observer was designed as below:

State estimation:

$$\begin{aligned} \dot{\hat{\underline{z}}}(t) &= \underline{R}\hat{\underline{z}}(t) + \underline{\Omega}(t)\hat{\underline{\theta}}(t) + \underline{g}(t) + \underline{k}(y - \hat{y}) + \begin{bmatrix} 0 \\ \underline{V} \end{bmatrix} \dot{\hat{\underline{\theta}}} \\ \hat{y} &= [1 \quad 0 \quad \ldots \quad 0]\hat{\underline{z}} \end{aligned} \tag{6.13}$$

Parameter adaptation:

$$\dot{\hat{\underline{\theta}}}(t) = \Gamma\underline{\phi}(t)(y - \hat{y}) \tag{6.14}$$

Auxiliary filter:

$$\dot{\underline{V}}(\mathrm{t}) = \underline{F}^*\underline{V}(\mathrm{t}) + \underline{\Omega}(\mathrm{t}), \; \underline{\phi}^T(t) = \underline{V}_1(t) \tag{6.15}$$

where $\hat{\underline{z}}$ the estimated state variable; $\hat{\underline{\theta}}$ the estimated parameter vector; $\hat{y}$ the estimated output; $\underline{k}$ constant vector that has to be chosen to guarantee the stability; $\underline{F}^*$ constant matrix that has to be chosen to guarantee the stability; After the parameter determination and stability analysis, in the real grinding control application, a reduced model for the grinding force control can be written as:

$$\dot{n}_w = a_{11}n_w + b_1 n_{wdesired}, \; \dot{\theta}_w = n_w \tag{6.16}$$

$$F_n' = K_f \left[\frac{n_w l_t(\theta_w)}{v_s}\right]^{e_1} [a]^{e_2} [D(\theta_w)]^{e_3} \tag{6.17}$$

where θ_w rotating angle of workpiece; F_n' normal grinding force per width of cut; $D(\theta_w)$ equivalent wheel diameter; a depth of cut; n_w workpiece speed; $l_t(\theta_w)$ lever of the tangent force; v_s wheel circumference speed; e_1, e_2, e_3 exponential coefficients; K_f Proportional factor (Fig. 6.10).

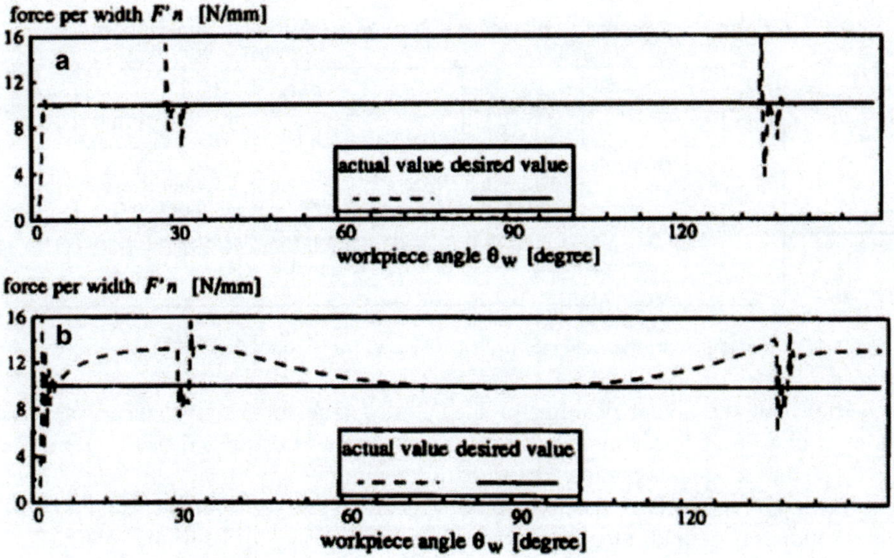

Fig. 6.10 Comparison for simulation results of (**a**) the adaptive control scheme with (**b**) PI-control scheme

By comparing the simulation results using this adaptive controller with PI-controller, it can be seen that the PI-controller has difficulties during shaping the unsymmetrical part of the features, which indicates that the uncertainty and time varying properties of a nonlinear system/process such as grinding requires a more adaptive control algorithm to regulate or optimize in order to improve the performance of the system/process.

Jenkins and Kurfess [17, 18] developed a self-tuning adaptive controller for the plunge grinding and traverse grinding processes. A pole-zero cancellation technique was developed and implemented to remove the process variations based on real-time model parameter estimation.

As shown in Fig. 6.11, the entire control system model consists of several functional blocks: $G_{FC}(s)$ – force controller; $G_{PC}(s)$ – the programmable multi-axis controller (PMAC) position control loop; $G_{TWP}(s)$ – the tool-workpiece and grinding process interaction, as well as estimation and adaptation functions. The gains of the adaptive force controller are updated in real-time style from the host computer that estimates the grinding process parameters and determines the force controller gains accordingly. The force controller is implemented on a PC-hosted PMAC using a high sampling rate, running continuously. Since force is being controlled via position, two control loops must be designed, a position loop and a force loop. The servo controller for the position loop is implemented using the PID servo algorithm in the PMAC. The gains are determined using Ziegler-Nichols PID tuning techniques. The resulting closed-loop PMAC position transfer function $G_{PC}(s)$ is represented by block between the commanded

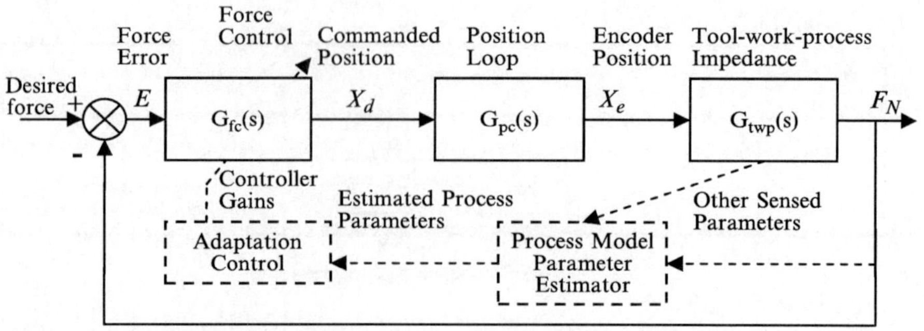

Fig. 6.11 Plant and controller scheme for PMAC-based force controller.

position and the actual position of the X-axis. Due to the high speed of servo loop a first-order linear model of position response is estimated using an autoregressive, least squares parameter estimation technique.

The adaptive approach taken in this work, namely the self-tuning regulator (STR) is an indirect method, also is a design procedure used to determine controller's parameters. In other words, the STR uses model parameter estimation to determine current states for use in the real-time design calculations to adjust the controller parameters. Different form the work done by Tönshoff who applied a fixed-gain controller to regulate the grinding force, the STR approach here performs real-time estimation of the grinding process and uses a pole-zero cancellation approach which will be shown briefly below to remove the process variation.

Based on the block diagram and grinding force model as below:

$$F_N - F_{TH} = \dot{x}_f \frac{A}{K_P V} \tag{6.18}$$

The derivation of the adaptive force controller was achieved via an important step which cancel the process dynamics pole by the zero of a PI-type controller, $G_{FC}(s)$, and the complete open-loop discrete transfer function is defined as:

$$G_{FC}G_{PC}G_{TWP}(z) = \frac{K_{PROP}}{1+\alpha} f_2(\eta)\left(\frac{1}{z - C_0}\right) \tag{6.19}$$

The implementation of this controller uses a windowed data handling scheme for recursive parameter estimation and the best window size was determined via experiments and simulations. This approach to grinding force control has demonstrated the potential for increasing productivity with the reduction of tarry time by use of the stable force control in plunge grinding, with higher fidelity in force regulation.

Brinksmeier and Popp [19] presented an adaptive control system to reduce the grinding time and ensure a good workpiece quality with a self-tuning force controller for external grinding processes. A system identification procedure was used to adapt the control parameters online in order to compensate for the time-varying grinding process behavior. The identification result was also used to monitor the wheel wear and compensate for the unknown system variations, such as the elastic deformation of the workpiece-tool-machine system.

The model for the grinding process taken in this work can be described with the p-transfer function as:

$$F_N(p) = \frac{1}{\frac{1}{\omega_d^2}p^2 + \frac{2D_d}{\omega_d}p + 1} \times \frac{\frac{1}{C}}{\frac{N}{C}p + 1} \tag{6.20}$$

where p – complex variable; ω_d, D_d – parameters of the feed drive behavior; N – compliance; C – abrasive property; F_N – grinding normal force; The disturbances mainly come from disturbances added to the measured plant output and those changing the system parameters.

Also, since digital control algorithm could effectively improve the controller accuracy, the above differential equation has to be transformed into a difference equation, represented by Z-transformation, similar to (6.1)

$$F_N(z) = \frac{b_1 z^{-1} + b_2 z^{-2} + b_3 z^{-3}}{1 + a_1 z^{-1} + a_2 z^{-2} + a_3 z^{-3}} \times z^{-d} \tag{6.21}$$

For the unknown parameters, as the process behavior can vary during the grinding cycles, the on-line process identification is needed, which is shown in Fig. 6.12. In general, a process variable which sufficiently represents the process state is very important. The normal grinding force is preferred here because it is convenient to monitor and allows additional utilization to reveal process mechanisms, namely

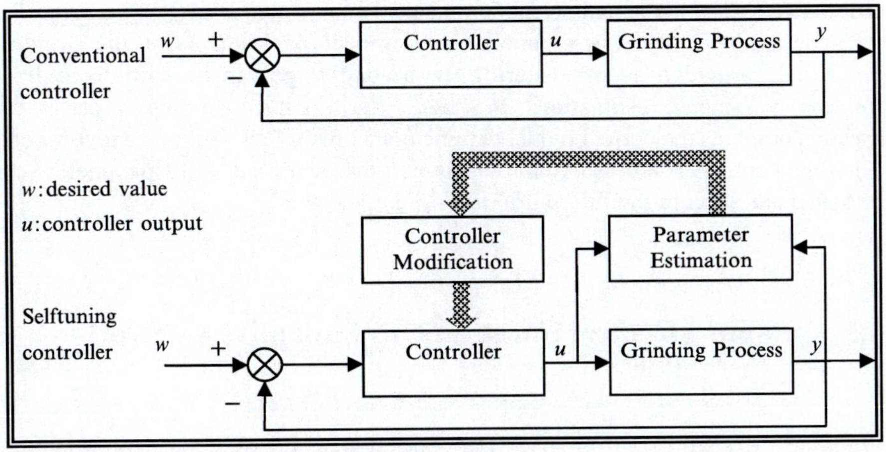

Fig. 6.12 The structure of a self tuning controller

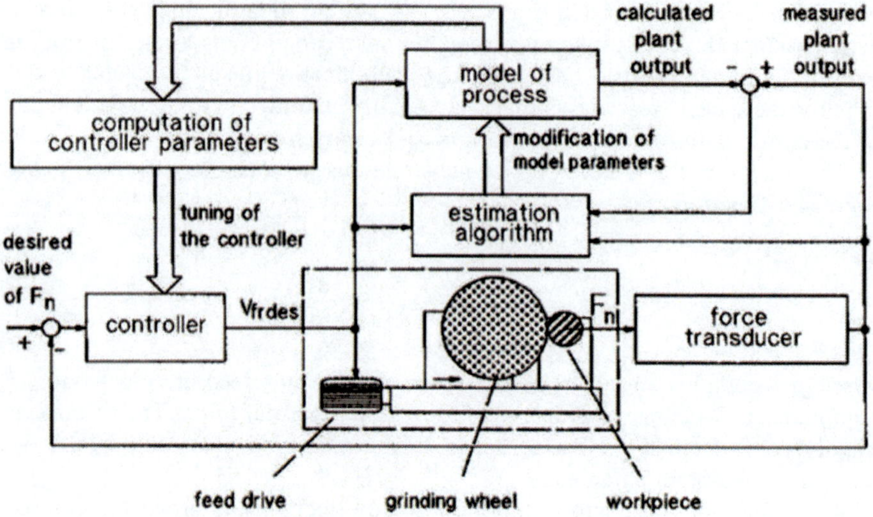

Fig. 6.13 Signal processing of the adaptive control system

more relative to most of the anomalies. The signal processing of the adaptive control system is illustrated in Fig. 6.13.

The hardware is based on an external VME-bus-based computer system, which contains a MC68020 CPU-board with coprocessor. The prototype of the adaptive control system consists of a studer S50-42 grinding machine with a bruetsch-CNC. The dead centers of the grinding machine contain piezoelectric force sensors measuring the grinding force in normal, tangential and axial directions, power meter detecting electrical power of the spindle motor etc.

For the adaptive strategy, two stages are applied. The task of the first stage is to detect the first cut while the second stage is the removal of workpiece eccentricities. After that, the actual selftuning adaptive controller is used. The controller is predetermined with an assumed set of model parameters for the grinding process. Information acquired during the second stage can be partially utilized for the parameter estimation. In short, during the grinding process the system computes the correct model parameters within a few workpiece revolutions. Then the controller is tuned continuously and the identified set of parameters can be saved and used in the future grinding.

6.5 Adaptive Control Constraint and Adaptive Control Optimization

Wada and Kodama [20] proposed the concept that the general adaptive control in grinding process can be classified into two categories depending on whether the controlled process is working within a constrained condition, or the ultimate objective function has a certain optimum value. The first classification is Adaptive Control Constraint (ACC) and the second one is Adaptive Control Optimization (ACO).

A typical Adaptive Control Constraint (ACC) scheme was developed by Hahn [12, 13] for grinding force control. In this work, a controlled-force grinding technique is presented to eliminate random size and taper fluctuations resulting from variable elastic deflections. Through this method, the affection coming from variations of stock, hardness and wheel sharpness could be eliminated from the sizing problem and allows the comprehensive grinding performance directly related to grinding force which could simplify practical grinding setups. Moreover, the control method is based on an analysis made of the rounding up dynamics of the controlled-force system and a critical work speed is found which must be exceeded in order to have a rounding-up action.

Adaptive Control with Optimization (ACO) system has been developed by multiple researchers, such as König and Werner [21], Kelly et al. [22], Xiao and Malkin [23–25], Li et al. [27], Dong et al. [28, 29], Srivastava et al. [30].

In Li's [27] work, based on the basic grinding models, the objective function and constraint functions for the multi-parameter optimum grinding process have been built and the non-linear optimum grinding control parameters have been obtained through computer simulation and the actual grinding process is controlled by these optimized parameters. According to the multi-parameter optimal theories, the objective function obtaining the shortest grinding time ϕ can be formulated with constraints as:

$$
\begin{gathered}
\text{Minimize :}\\
\phi = t_1 + t_2 + t_3\\
\text{Subject to the constraints :}\\
g_1 = P - P_b \leqslant 0 \ \text{(no - burn constraint)}\\
g_2 = z - q \leqslant 0 \ \text{(burning constraint)}\\
g_3 = R_a - R_{\max} \leqslant 0 \ \text{(roughness constraint)}\\
g_4 = RN - RN_{\max} \leqslant 0 \ \text{(roundness constraint)}\\
g_5 = r(t_1 + t_2 + t_3) - \Delta r = 0 \ \text{(size constraint)}\\
g_6 = u_l - u_i \leqslant 0, \ i = 1, 2 \ \text{(lower infeed constraint)}\\
g_7 = \mathrm{u_i} - \mathrm{u_u} \leqslant 0, \mathrm{i} = 1, 2 \ \text{(upper infeed constraint)}
\end{gathered}
\tag{6.22}
$$

where P is the power of the roughing stage and P_b the burning power limit of the workpiece; z is depth of the burning layer and q the depth of removal in the subsequent finishing stage. R_a is the actual surface roughness and $R_{\max}$ its maximum allowable value; RN is the actual out-of-roundness and $RN_{\max}$ its maximum allowable threshold. $r(t_1 + t_2 + t_3)$ is the total actual infeed for the whole grinding process and Δr the depth of removal denoting the radial workpiece allowance. u_l is the lower limit of the infeed rate and u_u the upper limit, whilst $i = 1$ denotes the roughing stage and $i = 2$ denotes the finishing stage.

The strategies of multi-parameter optimization include three main cases. The first one is for the conventional grinding process under the no-burn state, the grinding power being much lower than the burning limit in the roughing stage with the constrained models including $g_1 \sim g_7$; the second one, under the critical burning state, requires that the inequality should be $g_2 = 0$. The last one is under the burning state in the roughing stage, its constraint models including $g_2 \sim g_7$. These optimum strategies were implemented experimentally on an external semi-automatic grinder and the controlling system is illustrated in Fig. 6.14.

In Dong's [28, 29] work, an innovative technology, which allows for continuous variation of the infeed rate to further reduce the cycle time, is developed for optimal infeed control of cylindrical plunge grinding cycles. The controller is designed to identify the state of the cycle at each sampling instant from on-line measurements of power and size, and to then compute the infeed rate according to the optimal policy whereby the infeed rate is determined according to the active constraint at each segment of the cycle.

The objective of the grinding optimization is to find the optimal infeed rate $u(t)$ which minimize the cycle time while also satisfying constrains associated with the machine capability and workpiece quality as well as the parameters associated with

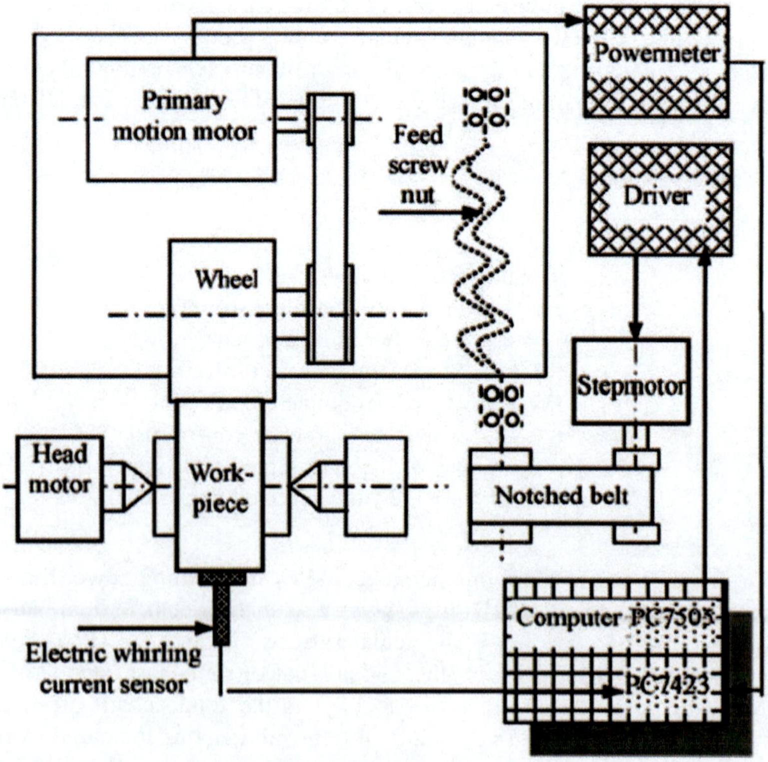

Fig. 6.14 Chart of the grinding controlling system

periodically dressing of the wheel. Namely, the magnitude of the grinding infeed rate which is related to productivity is constrained not only by machine limitations but also part finish quality requirements including thermal damage, surface roughness and dimensional tolerance. Consequently, the optimization problem can be formulated as:

$$\begin{gathered}\text{Minimize cycle time}: \\ t_c = \sum_{i=1}^{N} \delta t_i \\ \text{Subject to the constraints}: \\ g_1 : z(k) - \sum_{i=k+1}^{N} v(i)\delta t \leqslant 0 \\ g_2 : R_a(N) - R_{a,\max} \leqslant 0 \\ g_3 : r(N) - r_{\max} \leqslant 0 \\ g_4 : \left| \sum_{i=1}^{N} v(i)\delta t - Q \right| \leqslant tol \\ g_5 : u_l - u(k) \leqslant 0 \quad \textit{for all} \ \ k = 1, \dots, K \\ g_6 : u(k) - u_u \leqslant 0 \quad \textit{for all} \ \ k = 1, \dots, K\end{gathered} \tag{6.23}$$

where the inequality g_1 defines the depth of thermal damage limit above a critical temperature referred to as "workpiece burn"; inequality g_2 defines the surface roughness requirement; g_3 defines the out-of roundness at the end of the cycle and $r_{\max}$ is its maximum allowable value; g_4 defines the size requirement and the last two constraints represent limitations on the machine infeed rate where u_l is the lower bound and u_u is the upper bound. A schematic diagram of the corresponding control system is shown in Fig. 6.15. The core part is the block named "optimization" whose task is to identify the state of the cycle using measurement data of power and size and then compute the desired infeed rate v_d. The controller can be bypassed at first and last grinding sections, because the command infeed rate should

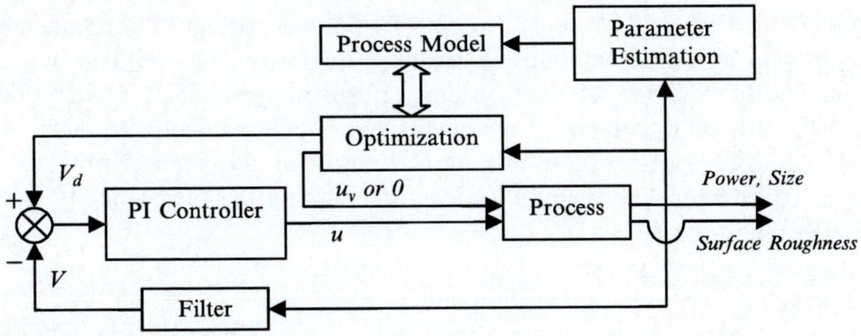

Fig. 6.15 Schematic diagram of the on-line control strategy

be at its maximum allowable limit for the first section of the cycle and at zero for the last. The actual infeed rate v is obtained on-line from size measurements, and parametric values of the time constant τ and the effective wear flat area A_{eff} can be estimated in the block "parameter Estimation" from measurements of power and size. Moreover, the control system uses feedback from power and size to compensate for modeling uncertainty caused by parameter variations and external disturbances.

This kind of control system seeks to adjust operating condition according to a predefined performance index in order to optimize the grinding process during real-time operation. The identified process models allow to predict workpiece burn and estimate the compliance of the workpiece-tool-machine system. The drawback of such control strategy is that some parameters cannot easily be measured online, e.g., surface roughness is often measured off-line, and hence is not controllable with such a scheme. In addition, the optimization calculation load is usually heavy for multiple process goals with multiple input/output constraints during online operation and the optimal solutions largely depend on a carefully selected initial condition, which needs to be determined by experienced operators.

Based on optimal control theory, Malkin and Koren [31] derived an accelerated spark-out method by reducing the time required to recover the accumulated elastic deflection in the system and therefore the cycle time was successfully minimized in cylindrical plunge grinding processes. And this optimal control policy is particularly advantageous for grinding systems having a long characteristic time constant. It has been proposed by Malkin [5] that this accelerated spark-out method can be beneficial to incorporate into the Adaptive Control Optimization (ACO) system in optimizing the grinding and dressing parameters for maximum Material Removal Rate (MRR) within constraints on workpiece burn and surface finish. Grinding processes are complex in nature with multiple cutting points, which are defined by a large number of grits possessing irregular shapes, sizes and spacing. Inevitable variations of the tool position, velocity and force fluctuation heavily affect the final part's geometry accuracy, surface finish, and the overall process Material Removal Rate (MRR). Moreover, various uncertainties and disturbances inherently exist in the process. It is therefore necessary to control the grinding process under various system variations. Jenkins et al. [17] presented a robust controller for a grinding system where the normal grinding force was decoupled from the tangential feed velocity of the system. Two separate control loops were designed as a standard feed velocity loop and a force loop, so that the two variables could be controlled simultaneously. A variety of other system performance specifications were successfully achieved, such as reduced overshoot value and transient settling time. The objective of the compensator design is to control the force and velocity to meet specific target while simultaneously decoupling the feed velocity from the normal grinding force.

The grinding system used in this work consists of a three-axis prismatic servo stage, which is controlled by a programmable multi-axis controller (PMAC). The controlled grinding system has two input trajectories, desired force and velocity, and two outputs, actual force and velocity. A least squares parameter estimation

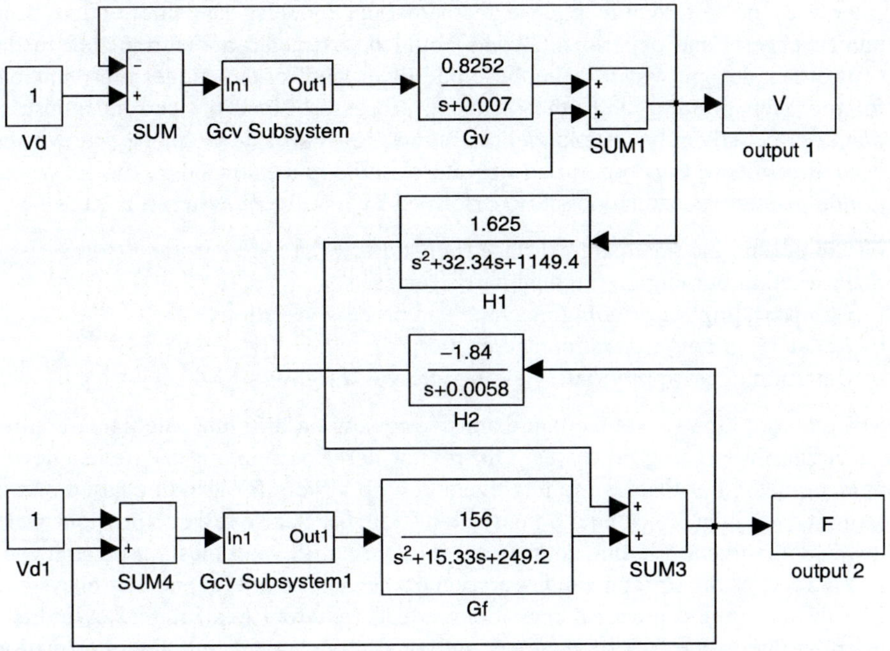

Fig. 6.16 System block diagram

technique (ARX, autoregressive external input, Ljung, L. [32]) is used to determine the transfer function between the input commands and the outputs, and analyze step commands in force and velocity. The block diagram of the closed-loop system with the appropriate transfer functions is shown in Fig. 6.16.

For the compensator design, a PID-type controller form is selected for implementation in both the velocity and force loops. In this case, by increasing the forward loop gain, cross coupling between force and velocity is easily decoupled. In general, many parameters needed for designing a new compensator in similar application circumstances can be gained by using the control design tools and procedures in this work.

6.6 Intelligent Control Techniques

Although various adaptive and robust control techniques have been studied to maintain the desired closed loop performance in the presence of parameter variations and process uncertainties, they often encounter difficulties in actual design when the system dynamics are not well known. In addition, in the control process, all the necessary coefficients of the controller need to be calculated and updated online, which makes the computation burden quite heavy. In contrast, intelligent controllers based on neural networks and fuzzy logic are attractive alternatives since they do not require precise mathematical models. The information from human experts and experimental data could be extracted and formulated in the control law design, which make these intelligent control techniques most suitable for precision grinding and other abrasive processes, since the current industrial practices heavily rely on experienced human operators in order to achieve the desired results. By incorporating operators' skills and knowledge, the following grinding activities can be possibly performed by intelligent systems, such as:

- Controlling the final part's surface roughness
- Preventing burning on the final part's surface
- Compensating for grinding machine and process variations
- Reducing grinding vibration and chatter
- Determining an appropriate dressing interval for grinding wheel

Rowe et al. [33, 34] provided an extensive review on different intelligent control techniques in grinding processes. This review describes the object-oriented development method of the generic intelligent control system for grinding based on the proposed modular conceptual framework (as Fig. 6.17), also reviews previous work towards intelligent grinding control and summarizes the previously used strategies and introduces the structure and operation of the generic intelligent control system. The most common practical operations are for external and internal cylindrical grinding due to the lack of accurate analytical models or incomplete information about the processes, where the intelligent control technologies emerged to be promising when the conventional methods often fail. The trend of increasing usage of machine intelligence in grinding systems and operations is clearer and more researchers are working in this area nowadays. A most significant reason is that the human specialist knowledge and lessons grained from previous operations can be incorporate in the controller design to ensure a better system performance for the future operations. A conceptual framework for a typical intelligent grinding machine is illustrated in Fig. 6.17, where all essential elements have been tested and integrated into practical grinding systems.

Above conceptual framework provides a general guide to the design of the intelligent control system. Of all the components, the executor plays the central role, which is basically a software drive, capable of selecting relevant software modules and integrating them to form an intelligent control system for a specific grinding process following predetermined rules. Besides, it consists of several other components, including I/O routines, process models and rules, a database, an intelligent parameter selection system, typical grinding cycles, safety strategies, adaptive strategies and learning strategies.

Nakajima et al. [35] presented a neuro & fuzzy in-process control technique for plunge grinding processes, where a back-propagation neural network was built up to predict the surface roughness during the process. The infeed rate to the speed ratio was controlled and the grinding efficiency was optimized with the desired final surface roughness, independent of the wheel surface condition. Xiao and Malkin [25] proposed an intelligent grinding system, where the system used power and the part size information (which was measured online) as the feedback signal to

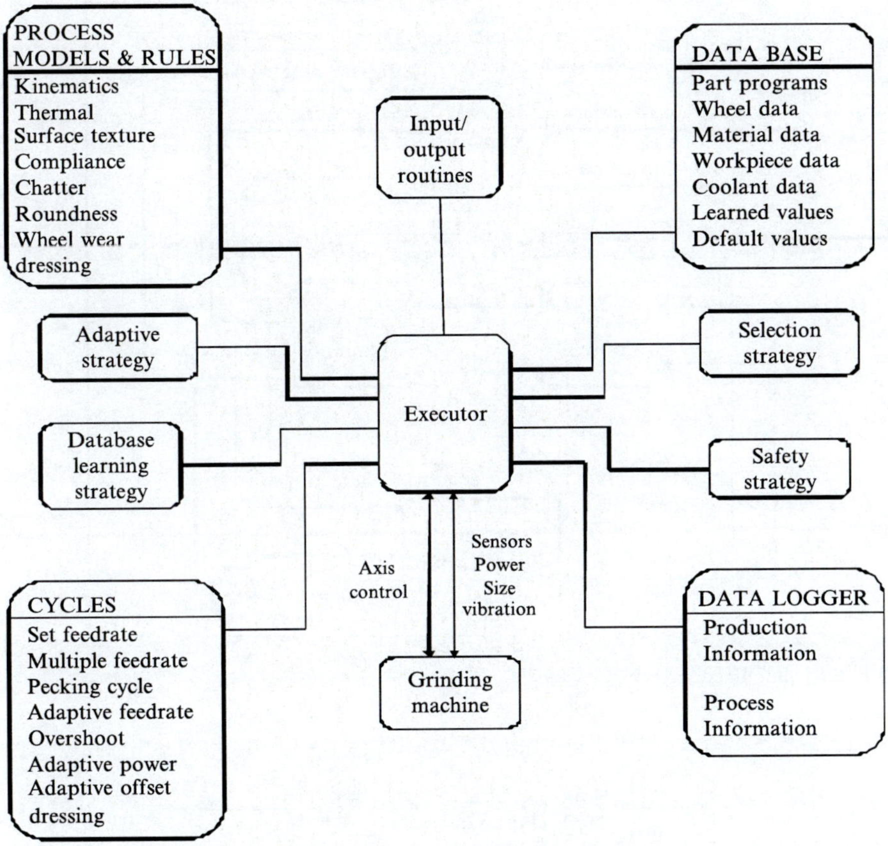

Fig. 6.17 Conceptual framework for a typical intelligent grinding machine

estimate and optimize the grinding performance for the next workpiece part. Additionally, the intelligent controller took the surface roughness and roundness measured during post-process to adjust the dressing feedrate and also identified the process to minimize the cycle time. The combined online system identification and the meta control ensured the resultant system intelligence. The integrated intelligent grinding system is illustrated in Fig. 6.18.

The fuzzy technique has been recognized and utilized to control complex grinding processes since early 1980s. The basic idea of a fuzzy inference system is to incorporate human's knowledge into a set of fuzzy IF-THEN rules, which involve operations on linguistic variables. The general fuzzy inference system can be shown in Fig. 6.19, which consists of four components as: a fuzzifier, a fuzzy rule base, a fuzzy inference engine and a defuzzifier.

Zhu et al. [26] did the first implementation of a fuzzy controller to obtain the desired surface finish by controlling the workpiece feedrate without any mathematical model. Zhao and Webster [14] applied fuzzy pattern recognition to the

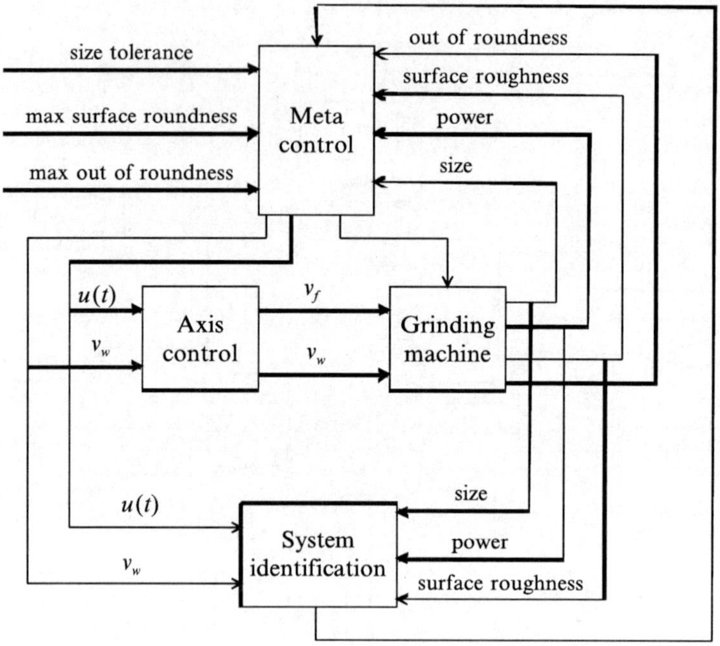

Fig. 6.18 Intelligent grinding machine

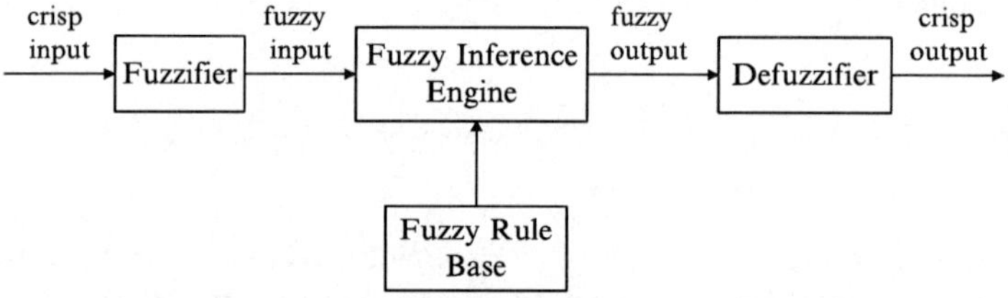

Fig. 6.19 General structure of a fuzzy inference system

automatic control of roller grinding processes. This adaptive control strategy could eliminate the highly skilled, time consuming and tedious task of manually adjustment of supporting steadies. The method is based on microcomputer controlled in-process size measurement techniques and enables faster and more accurate grinding, replacing the experienced operators by an expert system. The experimental results showed that this approach yielded a more reliable and more accurate result than an experienced operator could.

The hardware layout of the system is shown in Fig. 6.20, in which five studies are adopted, each of which is shown in Fig. 6.21, has a horizontal and a vertical pad driven by separate stepper motors.

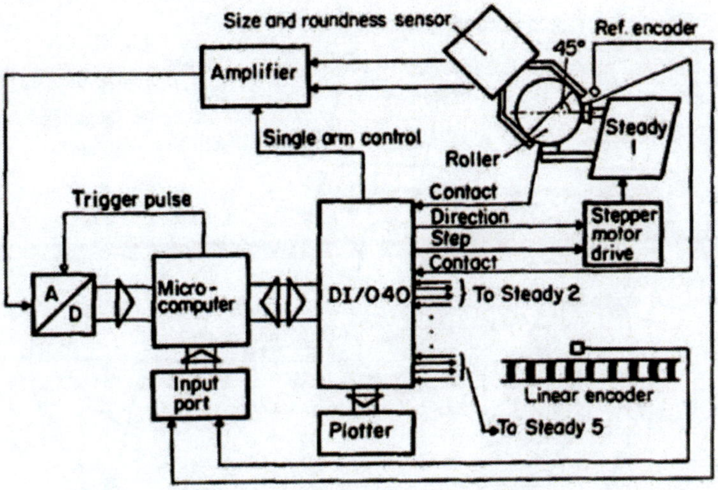

Fig. 6.20 Hardware layout of the computer control system for roller grinding

This computer control system using fuzzy pattern recognition has the following advantages:

(1) Reduces the dependence on operator's grinding experience and measuring skill by computer intelligence, thereby de-skilling the roller grinding operation
(2) It is more accurate and reliable to use the relative adjustments of the pads oriented by fuzzy pattern recognition than given by operators in accordance with their experience. Thus, it would be expected the roller form errors will be corrected in much fewer grinding passes, namely, sooner to achieve fine steady state error
(3) Reduces machining time as well as machining cost and make the operator's work easier and enables supervision of two or more machines simultaneously

Chen and Shin [36], Lee et al. [37] designed and implemented an intelligent and robust grinding process advisory system with fuzzy logic inferencing. The surface grinding process was automatically optimized using analytical grinding process models and heuristic knowledge without human intervention. The optimization procedure is presented using seven analytical models and heuristic knowledge base. These seven most representative and important models for surface grinding operation are: chip model, energy model, force model, surface finish model, stress model, temperature model and safety model, which are highly non-linear. In some cases, they are only useful for predicting approximate values due to the complexity of the underlying unknown physical mechanism. That's why the fuzzy heuristic rules and fuzzy inferencing are employed to represent the input and output relationships and to perform the optimization process. Therefore, the ambiguity of any physical quantity is described in terms of membership functions and fuzzy

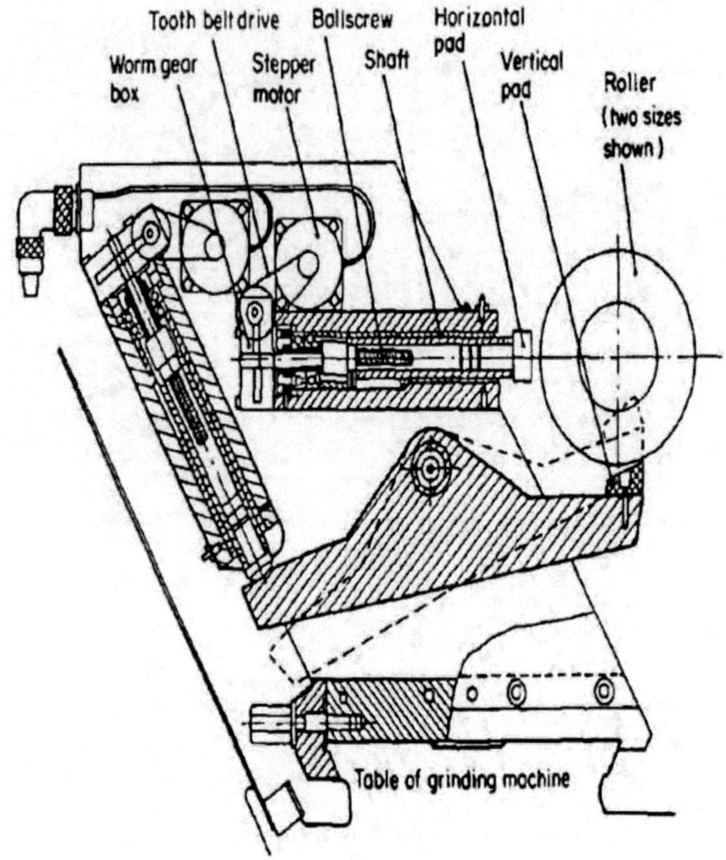

Fig. 6.21 Structure of steadies for roller grinding

inferencing, and approximate but robust solutions can be generated based on this fuzzy model. Moreover, a novel rule generation mechanism has been developed to convert the grinding models into fuzzy linguistic rules, which can be used generically with any grinding model in a wide operating range. This automatic rule generation mechanism also allows for the optimization of a grinding process without user intervention. Consequently, by autonomously selecting proper process conditions and operating parameters which can be monitored on-line, the system can be used as a tuning controller of a grinding process.

Lees's work is an implementation of the above model-based optimization scheme. In order to achieve, process models for grinding force, power, surface roughness, and residual stress are developed and grinding optimization is considered as constrained nonlinear optimization problems with mixed-integer variables and time-varying characteristics. The model-based optimization scheme has been applied to three different case study optimization problems for the surface grinding process, including minimization of grinding cost, minimization of cycle time, and process control.

The objective of optimization in process control is to find operating conditions to achieve desirable values of process outputs. In surface grinding, three aspects are supposed to be concerned at the same time which are desirable surface roughness, grinding power and G-ratio. In this case, the objective function can be given as follows:

$$J = \left|\frac{R_a^d - R_a}{R_a^d}\right| + \left|\frac{P^d - P}{P^d}\right| + \left|\frac{G^d - G}{G^d}\right| \tag{6.24}$$

where R_a^d, P^d, and G^d are the desirable process output values of surface roughness, grinding power and G-ratio. In practical situation, this idea can be extended by employing weighting factors $w_1 \sim w_3$ in the objective function:

$$\begin{gathered} \text{Minimize :} \\ J = w_1\left|\frac{R_a^d - R_a}{R_a^d}\right| + w_2\left|\frac{P^d - P}{P^d}\right| + w_3\left|\frac{G^d - G}{G^d}\right| \\ \text{Subject to :} \\ 0.10 \leqslant v_w \leqslant 0.30 \\ 0.75 \leqslant s_t \leqslant 2.0 \\ 12.7 \leqslant a_d \leqslant 50.8 \\ 0.015 \leqslant a \leqslant 0.03 \\ \sigma_r \leqslant 600 \end{gathered} \tag{6.25}$$

The constraints come out in terms of work speed v_w, the crossfeed s_t, the dressing depth a_d, the depth of cut $a$and maximum residual stress σ_r.

Xu and Shin [38, 39] developed a multi-level fuzzy controller and implemented to deal with the real-world nonlinear plants with intrinsic uncertainties and time-varying parameters, such as a creep-feed grinding process. The proposed fuzzy control strategy can be used to control a system with an input-output monotonic relationship or a piecewise monotonic relationship, which is efficient and cost-effective for practical control applications.

The multi-level fuzzy controller is a two-input single-output controller with a two-level, hierarchical structure with an adaptation mechanism embedded in the lower level to tune the output membership functions (MFs) of the first layer fuzzy controller as shown in Fig. 6.22. The first layer fuzzy control rules are generated based on the experience of human operators. In order to correct the impreciseness of the first layer fuzzy control rules and compensate for the time-varying behavior or system uncertainty, a self-organizing fuzzy controller is used to tune the output membership functions (MFs) of the first layer fuzzy controller based on the evaluation of system performance.

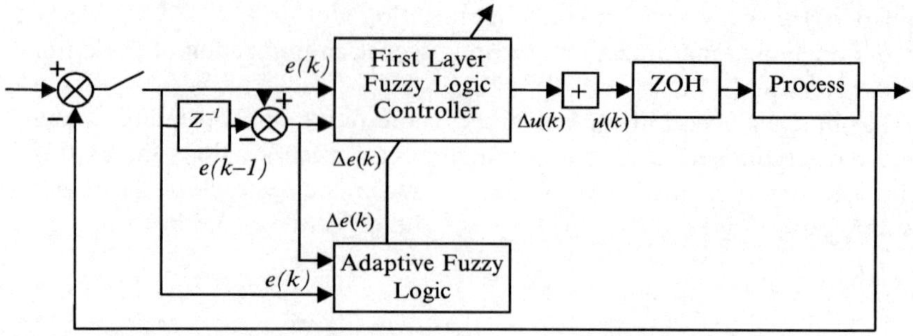

Fig. 6.22 Multilevel fuzzy control (MLFC) system

For the first layer fuzzy controller, the two inputs are the error $e(k)$ and the change of the error $\Delta e(k)$. The output is the change in the control action $\Delta u(k)$:

$$e(k) = y_d(k) - y(k) \tag{6.26}$$

$$\Delta e(k) = e(k) - e(k-1) \tag{6.27}$$

where $y_d(k)$ is the desired output at time instant k, $y(k)$is the actual output, $e(k)$ is the error and $\Delta e(k)$ is the change of the error. The change of control action is calculated based on the two inputs using a fuzzy inference mechanism, which is described in the fuzzy control rule as the following form:

$$R : \text{if } e_1 \text{ is } E_i \text{ and } e_2 \text{ is } E_j, \text{ then } u_1 \text{ is } U_{n(i,j)} \tag{6.28}$$

where e_1 is the error $e(k)$ at time instant k, e_2 is the change of the error $\Delta e(k)$, u_1 is the change of the control signal $\Delta u(k)$, E_i is the linguistic variable of e_1, E_j is the linguistic variable of e_2, and $U_{n(i,j)}$ is the linguistic variable of u_1. By employing Mamdani's minimum fuzzy implication and center average defuzzification, the fuzzy control law of the first layer is:

$$u_1 = \frac{\sum_{i,j} [(\mu_{E_i}(e_1) \cap \mu_{E_j}(e_2)) \cdot U_{n(i,j)}]}{\sum_{i,j} (\mu_{E_i}(e_1) \cap \mu_{E_j}(e_2))} \tag{6.29}$$

Similarly, the self-organizing fuzzy controller (the lower fuzzy layer) has the control rule in the form of:

$$R : \text{if } e_1 \text{ is } E_i \text{ and } e_2 \text{ is } E_j, \text{ then } u_1 \text{ is } C_{m(i,j)} \tag{6.30}$$

where $C_{m(i,j)}$ is the linguistic variable of C which is the change of the center of the output MF of the first layer fuzzy controller. Hence, using Mamdani's minimum implication and center average defuzzification, the center of output MF of the first layer controller should be tuned as:

$$U_{n(i,j)} = U_{n(i,j)} + \frac{\sum_{i,j} [(\mu_{E_i}(e_1) \cap \mu_{E_j}(e_2)) \cdot C_{m(i,j)}]}{\sum_{i,j} (\mu_{E_i}(e_1) \cap \mu_{E_j}(e_2))} \tag{6.31}$$

This proposed systematic procedure is proved to be capable of controlling nonlinear systems with uncertainties and time varying parameters to reduce tracking error and improve system performance by simulation results conducted on cargo ship steering and fuzzy cruise control, as well as another control work done on cutting force control for creep-feed grinding processes. The grinding force was maintained at the maximum allowable level for varying depth of cut of raw workpieces. Experimental results illustrate that the cycle time was reduced by up to 25% than that without any controller, and the effect of wheel wear was also compensated by the adaptation scheme. These indicated the effectiveness of the intelligent system in improving the productivity and enhancing the reliability of grinding processes in practical industrial applications.

Other researchers such as Mise et al. [40], Yuan et al. [41] have also used fuzzy logic technique in grinding processes. Mise's work proposed an autonomous control system which can compensate for the sizing points in an in-process control based on information from a total inspecting machine located in the post-processing end of an online-controlled measuring instrument.

6.7 Hybrid Control Schemes

In the case when the plant model is partially known, the fuzzy technique can be designed to work with conventional control methods to improve the system performance accuracy and efficiency. Xiao et al. [42] designed a fuzzy + conventional PI controller to reduce the multi-axis contour error and promote the dynamic performance of a grinding machine based on human knowledge and analytical models. The hybrid control system was implemented in an open architecture system for practical applications. This system is capable of optimizing the grinding and dressing parameters in response to in-process and post-process measurements which characterize the process and update the process model which has considered a more complete set of realistic constraints compared with previously proposed systems.

Orchard et al. [43, 44] presented a fuzzy predictive controller for a mineral concentration grinding plant in order to maximize the ore feed rate and to follow a predetermined particle size set-point. The process was a typical multi-input multi-output system, where the inputs were water and fresh ore flows, the outputs were percentage of solids, particle size and power demand. The proposed hybrid control scheme was based on a traditional linear multivariable ARX model, and a fuzzy characterization of the process objective functions and constraint functions. The process objective function was expressed as multiple fuzzy criteria based on operators' knowledge about the process itself and the corresponding requirements. The overall scheme of the fuzzy predictive control regulator is shown in Fig. 6.23. Compared with conventional predictive control method, simulation results illustrated a better system performance in minimizing the output errors and the input variations under multiple process input/output constraints and disturbances, such as different sizes and hardness variations of the in-feeding ore particle. Moreover, the computational time of the fuzzy predictive controller was reduced significantly, which ensured for a real-time operating environment.

Even though many fuzzy control schemes have been investigated and implemented in grinding processes, in most cases, the control rules are predefined and fixed during the entire operation process. The parameters of such controllers may not be optimal when the process is not accurately known or time-varying, and the desired system performance can not be guaranteed. Hence, adaptive fuzzy controllers would be desirable with the ability to tune the control parameters online. Tönshoff and Walter [11] designed a self-tuning fuzzy controller for an internal grinding process. The membership functions and the fuzzy rules were automatically adjusted without human intervention. Tang and Song [45] implemented a real-time fuzzy controller with online self-learning ability for a grinding process, which could provide better system performance than pure conventional linear PID controller under the system dynamic variations and process uncertainties, such as the variations of the infeed parts' materials, sizes and hardness. The block diagram of the closed-loop control system is illustrated in Fig. 6.24, where the fuzzy control rules are updated in real time continuously through online rule modification algorithm.

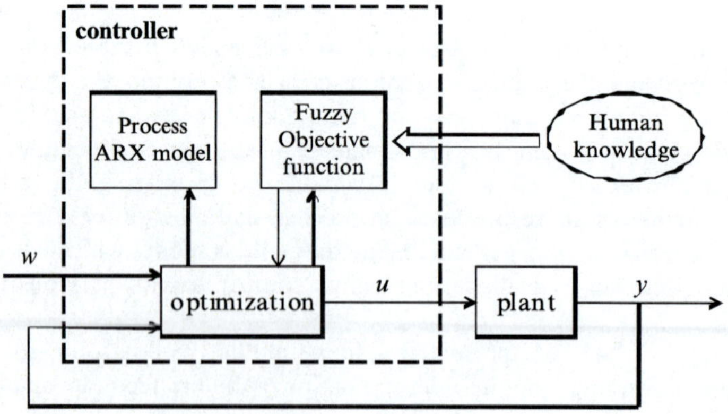

Fig. 6.23 Fuzzy predictive control scheme

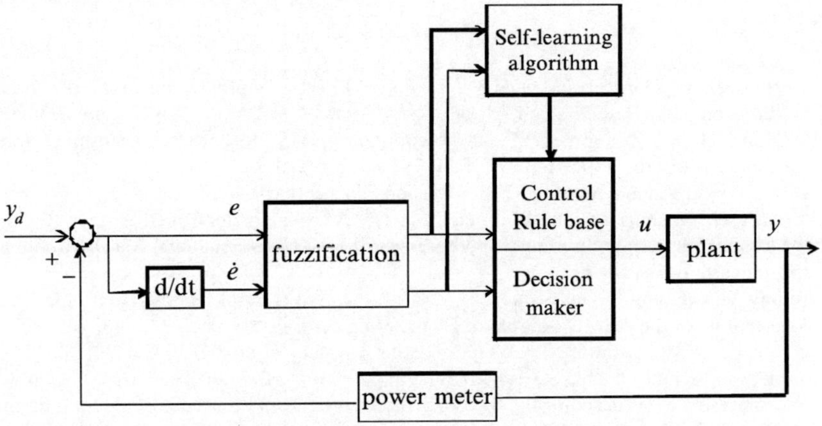

Fig. 6.24 A fuzzy controller with online self-learning scheme

6.8 Conclusions

It is well-accepted that model-based control systems can ensure the most satisfactory system performances as desired. The system mathematical models are usually set up either from physical laws or from experimental data. (Sometimes operators' knowledge can also be used to generate system model, but in often cases it is not accurate enough.) It is the conclusion of the authors that, in the future, the most efficient control systems will probably be hybrid in nature. Therefore, multiple possible information sources can be used to effectively control the entire grinding process. In addition, a particular control technique will be developed as a modular, which should be designed as an appropriate way for a specific grinding process.

General grinding processes are typical complicated nonlinear and time-varying multivariable systems, which have intrinsic interactions and inevitable uncertainties during the operation. Based on the models obtained from the system input/output data, hybrid control techniques possess the intelligence, which does not require a good mathematical model of the process. The optimal grinding conditions can be automatically determined within certain operation constraints, such as the grinding burning and chatter constraints. The appropriate control actions, wheel speed, table speed, depth of cut and width of cut, will be obtained by the intelligent inferencing without resorting to trial and error. Moreover, some process variables are often not measurable. Thus, an effective controller with an observer design will be beneficial to extend the current application range.

References

1. Tönshoff, H. K., Zinngrebe, M., and Kemmerling, M., 1986, "Optimization of Internal Grinding by Microcomputer-Based Force Control", Annals of the CIRP, Vol. 35, No. 1, pp. 293–296.
2. Tönshoff, H. K., Zinngrebe, M., and Kemmerling, M., 1991, "Optimization of Internal Grinding by Microcomputer-Based Force Control", Control of Manufacturing Processes, ASME Winter Annual Meeting, DSC Division, Vol. 28, pp. 67–77.
3. Kaliszer, H., Mishina, O., and Webster, J., 1979, "Adaptively Controlled Surface Roughness and Roundness during Grinding", Proceedings of the 20th International MTDR Conference, Birmingham, pp. 471–478.
4. Amitay, G., Malkin, S., and Koren, Y., 1981, "Adaptive Control Optimization of Grinding", Transactions of the ASME, Journal of Engineering for Industry, Vol. 103, pp. 103–108.
5. Malkin, S., 1981, "Grinding Cycle Optimization", Annals of the CIRP, Vol. 30, pp. 223–226.
6. Furukawa, Y. and Ohishi, S., 1984, "Adaptive Control of Creep Feed Grinding to Avoid Workpiece Burn", Proceedings of the Fifth International Conference on Production Engineering, Tokyo, Japan, pp. 64–69.
7. Elbestawi, M. A., Yuen, K. M., Srivastava, A. K., and Dai, H., 1991, "Adaptive Force Control for Robotic Disk Grinding", CIRP Annals – Manufacturing Technology, Vol. 401, pp. 391–394.
8. Bhattacharyya, S. K., Fowell, B., and Wallbank, J., 1984, "Control of Workpiece Surface Quality in Grinding", ASME Winter Annual Meeting on Production Engineering Division, New Orleans, LA, Vol. 12, pp. 425–444.
9. Ulrich, B. J., Srivastava, A. K., Elbestawi, M. A., and Veldhuis, S., 1989, "Force Modelling of the Robotic Disk Grinding Process", ASME Winter Annual Meeting, PED – Vol. 39, San Francisco, pp. 105–130
10. Srivastava, A. K., Rogers, D. B., and Elbestawi, M. A., 1993, "Optimal Planning of an Adaptively Controlled Robotic Disk Grinding Process", International Journal of Machine Tools & Manufacturing, Vol. 33, No. 6, pp. 809–825.
11. Tönshoff, H. K. and Walter, A., 1994, "Self-tuning Fuzzy Controller for Process Control in Internal Grinding", Fuzzy Sets and Systems, Vol. 63, No. 3, pp. 359–373.
12. Hahn, R., 1964, "Controlled-Force Grinding – A New Technique for Precision Internal Grinding", Transactions of the ASME, Journal of Engineering for Industry, pp. 287–293.
13. Hahn, R., 1965, "Some Characteristics of Controlled Force Grinding", Proceedings of the Sixth International Machine Tool Design Research Conference, pp. 597–609.
14. Zhao, Y. W. and Webster, J., 1990, "Fuzzy Pattern Recognition and Automatic Steady Control in Roller Grinding", Proceedings of Second IEEE International Conference on Computer Integrated Manufacturing, Troy, NY, pp. 395–401.
15. Guo, L., Schöne, A., and Ding, X., 1992, "A Comprehensive Approach to Nonlinear Adaptive Control and its Application to Form Grinding Processes", 31st IEEE Conference on Decision and Control, Tucson, AZ, Vol. 2, pp. 1267–1272.
16. Guo, L., Schöne, A., and Ding, X., 1993, "Grinding Force Control Using Nonlinear Adaptive Strategy", 12th World Congress IFAC, Sydney, Australia, Vol. 5, pp. 459–462.
17. Jenkins, H. E., Kurfess, T. R., and Dorf, R. C., 1996, "Design of A Robust Controller for A Grinding System", IEEE Transactions on Control Systems Technology, Vol. 4, No. 1, pp. 40–49.
18. Jenkins, H. E. and Kurfess, T. R., 1999, "Adaptive Pole-Zero Cancellation in Grinding Force Control", IEEE Transactions on Control Systems Technology, Vol. 7, No. 3, pp. 363–370.
19. Brinksmeier, E. and Popp, C., 1991, "A Self-tuning Adaptive Control System for Grinding Processes", Annuals of the CIRP, Vol. 40, No. 1, pp. 355–358.
20. Wada, R. and Kodama, H., 1977, "Adaptive Control in Grinding", Japanese Society of Precision Engineering, Vol. 11, No. 1, pp. 1–10.

21. König, W. and Werner, G., 1974, "Adaptive Control Optimization of High Efficiency External Grinding – Concept, Technological Basics and Application", Annals of the CIRP, Vol. 23, No. 1, pp. 101–102.
22. Kelly, S., Rowe, W. B., and Moruzzi, J. L., 1989, "Adaptive Grinding Control", Advanced Manufacturing Engineering, Vol. 1, No. 5, pp. 287–295.
23. Xiao, G., Malkin, S., and Danai, K., 1992, "Intelligent Control of Cylindrical Plunge Grinding", Proceedings of the 1992 American Control Conference, Chicago, IL, pp. 391–398.
24. Xiao, G., Malkin, S., and Danai, K., 1993, "An Autonomous System for Cylindrical Plunge Grinding", Proceedings of NSF Design and Manufacturing Systems Conference, pp. 399–403.
25. Xiao, G. and Malkin, S., 1996, "On-Line Optimization for Internal Plunge Grinding", Annals of the CIRP, Vol. 45, pp. 287–292.
26. Zhu, J. Y., Shumsheruddin, A. A., and Bollinger, J. G., 1982, "Control of Machine Tools Using the Fuzzy Control Technique", Annals of the CIRP, Vol. 31, No. 1, pp. 347–352.
27. Li, G. F., Wang, L. S., and Yang, L. B., 2002, "Multi-Parameter Optimization and Control of the Cylindrical Grinding Process", Journal of Materials Processing Technology, Vol. 129, pp. 232–236.
28. Dong, S., Danai, K., Malkin, S., and Deshmukh, A., 2004, "Continuous Optimal Infeed Control for Cylindrical Plunge Grinding, Part I: Methodology", Transactions of the ASME, Journal of Manufacturing Science and Engineering, Vol. 126, pp. 327–333.
29. Dong, S., Danai, K., Malkin, S., and Deshmukh, A., 2004, "Continuous Optimal Infeed Control for Cylindrical Plunge Grinding, Part I: Controller Design and Implementation", Transactions of the ASME, Journal of Manufacturing Science and Engineering, Vol. 126, pp. 334–340.
30. Srivastava, A. K., Ulrich, B. J., and Elbestawi, M. A., 1990, "Analysis of Rigid-Disk Wear During Robotic Grinding", International Journal of Machining Tool and Manufacturing, Vol. 30, No. 4, pp. 521–534.
31. Malkin, S. and Koren, Y., 1984, "Optimal Infeed Control for accelerated Spark-Out in Plunge Grinding", Transactions of the ASME, Journal of Engineering for Industry, Vol. 106, pp. 70–74.
32. Ljung, L., 1987, "System Identification: Theory for the user", Englewood Cliffs, NJ: Prentice-Hall.
33. Rowe, W. B., Li, Y., Mills, B., and Allanson, D. R., 1996, "Applications of Intelligent CNC in Grinding", Computers in Industry, Vol. 31, pp. 45–60.
34. Rowe, W. B., Chen, Y., Moruzzi, J. L., and Mills, B., 1997, "A Generic Intelligent Control System for Grinding", Computers Integrated Manufacturing Systems, Vol. 10, No. 3, pp. 231–241.
35. Nakajima, T., Tsukamoto, S., Murakami, D., and Yasuda, H., 1993, "Neuro & Fuzzy In-Process Control Grinding Techniques – Study on Intelligent Automation of Grinding Process", Journal of the Japan Society for Precision Engineering, Vol. 59, No. 8, pp. 1313–1318.
36. Chen, Y. T. and Shin, Y. C., 1991, "A Surface Grinding Process Advisory System With Fuzzy Logic", Control of Manufacturing Processes, ASME Winter Annual Meeting, DSC Division, Vol. 28, pp. 67–77.
37. Lee, C. W., Choi, T., and Shin, Y. C., 2003, "Intelligent Model-based Optimization of the Surface Grinding Process for Heat-Treated 4140 Steel Alloys with Aluminum Oxide Grinding Wheels", Transactions of the ASME, Journal of Manufacturing Science and Engineering, Vol. 125, pp. 65–76.
38. Xu, C. and Shin, Y. C., 2005, "Design of a Multi-level Fuzzy Controller for Nonlinear Systems and Stability Analysis", IEEE Transactions on Fuzzy Systems, Vol. 13, No. 6, pp. 761–778.
39. Xu, C. and Shin, Y. C., 2007, "Control of Cutting Force for Creep-feed Grinding Processes using a Multi-level Fuzzy Controller", ASME Transaction, Journal of Dynamic Systems, Measurement and Control. Vol. 129, No. 4, pp. 480–492.

40. Mise, R., Itoga, K., and Kato, C., 1994, "Applying Fuzzy Logic for Hybrid Control of Grinding Work", IEEE Proceedings of the Tenth Anniversary Advanced Technologies in Instrumentation & Measurement Technology, Hamamatsu, Japan, Vol. 2, pp. 615–618.
41. Yuan, L., Järvenpää, V. M., and Keskinen, E., 2004, "Design of Fuzzy Logic-Based Controller in Roll Grinding System with Double Regenerative Chatter", Proceedings of IMECE, ASME International Mechanical Engineering Congress and Exposition, Anaheim, CA, pp. 463–469.
42. Xiao, B. X., Xia, M., and Zhao C. M., 1996, "The Main Control Mode and Fuzzy Control Strategy of CNC System for Gear Hobbing and Grinding Machine", Proceedings of the IEEE International Conference on Industrial Technology, Shanghai, China, pp. 643–646.
43. Orchard, M., Flores, A., Muñoz, C., and Cipriano, A., 2001, "Predictive Control with Fuzzy Characterization of Percentage of Solids, Particle Size and Power Demand for Minerals Grinding", Proceedings of the 2001 IEEE International Conference on Control Applications, Mexico City, pp. 600–605.
44. Orchard, M., Flores, A., Muñoz, C., and Cipriano, A., 2001b, "Model-based Predictive Control with Fuzzy Characterization of Goals and Constraints, Applied to the Dynamic Optimization of Grinding Plants", The Tenth IEEE International Conference on Fuzzy Systems, Melbourne, Australia, Vol. 2, pp. 916–919.
45. Tang, Y.-G. and Song, G., 2002, "The Mill Load Control for Grinding Plant Based on Fuzzy Logic", Proceedings of the First International Conference on Machine Learning and Cybernetics, Beijing, China, pp. 416–419.

Chapter 7
Nanogrinding

Mark J. Jackson and J. Paulo Davim

Abstract Nanogrinding is an aspect of advanced manufacturing that has been growing rapidly in recent years. For instance, lenses and mirrors are manufactured to precise and ultraprecise standards using nanogrinding techniques. Most of these applications require a crack-free surface. Generally, hard and brittle materials such as glass, silicon, and germanium are commonly used for making these products. Traditionally, optical glasses require grinding and finishing by a polishing process in order to remove the damage caused by the previous machining operation and to obtain a flat surface. The conventional ground surface of glass results in a fine finish due to brittle fracture during the removal process and this surface needs to be polished to 20 nm R_{max} to reduce absorption and scattering of light on the glass surface. However, advances in advanced nanogrinding of brittle materials have led to the discovery of a ductile regime in which material removal is by performed by plastic deformation. Fracture mechanics predicts that even brittle solids can be machined by the action of plastic flow, as is the case in metal, leaving crack free surfaces when the removal process is performed at less than a critical depth of cut. This means that under certain controlled conditions, it is possible to machine brittle materials such as ceramics and glass using specially developed abrasive tools so that material is removed by plastic flow, leaving a smooth and crack-free surface. This chapter describes the recent advances in nanogrinding from computational analysis to the development of an ultra precise machine tool that is used to grind and polish flat, crack-free surfaces.

Keywords Nanogrinding · Microgrinding · Precision grinding · Brittle materials

M.J. Jackson (✉)
MET, College of Technology, Purdue University, 401 North Grant Street, West Lafayette, IN 47907, USA
e-mail: jacksomj@purdue.edu

M.J. Jackson and J.P. Davim (eds.), *Machining with Abrasives*,
DOI 10.1007/978-1-4419-7302-3_7, © Springer Science+Business Media, LLC 2011

7.1 Introduction

For the analysis of the molecular dynamic (MD) process simulations, micro mechanical state variables of all atoms and thermodynamic state variables of the system as a whole are accessible throughout the simulation. While some state variables are directly available, such as co-ordinates and velocities, others are available by statistical mechanics' analyses that need to be calculated as system quantities such as thermodynamic state variables: heat; temperature; and pressure. Some of the capabilities of MD for analyzing nanoscale cutting processes by simulation are demonstrated on the basis of application examples. For this purpose, results of orthogonal cutting process simulations of ductile, single-crystal copper is presented, although MD cutting and machining simulations results of brittle as well as ductile poly-crystalline materials can be found in the literature. The following figures show results of nanogrinding process simulations where hard tool tips move on {001} surfaces along <110> directions of fcc crystals. The following computational analysis is described in detail by Rentsch in the classic textbook on nano and micromachining edited by Jackson and Davim [1].

7.2 Analysis of Microstructural Deformation

The availability of the co-ordinates of the atoms in MD calculations allows one to visualize the instantaneous positions and to track the motion of the atoms individually as well in sequence. This information is not only useful for so-called snapshots of atomic arrangements, through which the progress of process simulations can be monitored, but it can be used for dynamic analyses on the basis of animations as well. For this purpose often the radii of the atoms are chosen in order to enhance a certain aspect or view of a specific arrangement than representing realistic atomic sizes. Using instantaneous co-ordinates rather than averaged positions, due to the vibration of the atoms, does not cause a significant error, because the maximum vibration amplitude of an atom in a stable MD calculation is less than 1% of the minimum bonding length.

The model in Fig. 7.1 was employed to study the chip formation process and the surface generation at a cutting edge in nanoscale cutting. By employing one-dimensional PBC along the y-axis, the orthogonal cutting condition reduced the model to a quasi-2D type with a small width, that allows one to correctly model the 3D fcc crystal structure of the copper workpiece. The model contained 71,000 work atoms and 11,000 tool atoms. Atoms at the bottom and to the left and right hand side of the model represent fixed boundaries, but were shifted with cutting speed to create the relative motion between workpiece and tool. Atoms in the layers next to these hard boundary atoms had thermostatic properties controlling the temperature at the workpiece boundary.

For diamond cutting of copper, no plastic and no significant elastic deformation of the tool was expected, calculating its internal structure was not of special

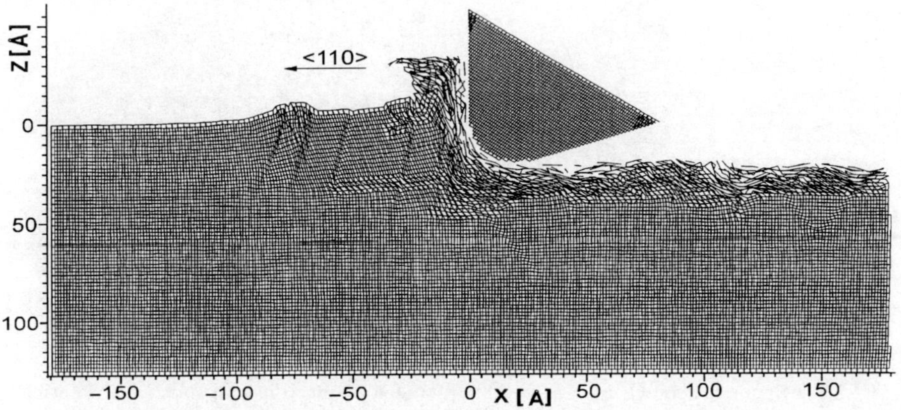

Fig. 7.1 Deformation graph, view <110> (after Rentsch in Davim and Jackson [1]). Used with permission copyright ISTE-Wiley (2009)

importance, but its shape and surface structure were for the tool-workpiece interactions. Its surfaces were formed by preferred diamond/fcc cleavage planes and were used as rake and clearance faces. The edge radius was not chosen to be atomically sharp (2 nm) in order to consider limited minimum edge radii because of surface stresses. To further provide a reasonable tool-workpiece contact model, the tool structure had the atomic density of diamond. Although the tool was modeled as a hard body with collectively moving atoms, i.e., no interaction within the tool, the interaction potential between tool and workpiece atoms needed to be specified. Data for the diamond/copper interactions based on a pair potential function were found. The cutting forces were calculated as reaction forces at the tool due to its feed motion. The work atom interactions are described by the EAM potential for copper. The cutting speed was restricted to 100 and 50 m/s. Lower speeds were not practical to simulate due to computational limitations.

Figure 7.1 shows a single-crystalline structure, that moved relative towards the cutting tool, whereupon material is deformed in front of the tool tip, the chip generation is initiated and dislocation loops can be identified at the generated workpiece surface. It shows areas of plastic deformation, dislocations and large elastic deformations in the sub-surface region. The method is based on horizontal and vertical connections between initial-neighbor atoms. Deformations show as sharp equilateral folds in neighboring layer lines within the otherwise rectangular structure or by narrowing mesh spacing as in case of strong elastic deformation. For large displacements between initial-neighbor atoms, the bond was considered to be broken and was not drawn anymore. In this way, highly deformed areas, like the chip and the newly generated surface, show few initial-neighbor lines. Deep running dislocations, observed in 2D MD cutting simulations if pair potentials are being used, could not be confirmed by employing this 3D model and the better EAM potential. This model predicts intensive plastic deformation at the generated surface with a thickness of only a few atom layers. At the same strain, a 2D model

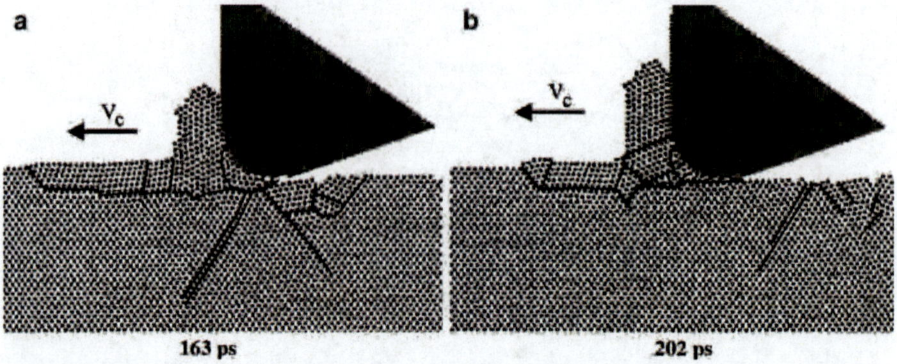

Fig. 7.2 Micro structural changes in the chip root area during chip formation (after Rentsch in Davim and Jackson [1]). Used with permission copyright ISTE-Wiley (2009)

always predicts larger dislocations than its 3D counterpart. The cutting process changes drastically when changing the ratio of depth of cut to cutting edge radius from 0.5 to 1.0, depending also on the crystalline orientation of the work. At a ratio of a/r = 1.0, the tool begins to utilize more its rake face for the chip formation. With the increase in depth of cut, the portion of twin dislocations in chip formation increases over dislocation slipping.

Such twinned areas can be seen ahead of the tool and the chip. The lower energy requirement for twinning makes the chip removal process at larger depth of cut more efficient and the cutting forces only increase under proportionally. Figure 7.2 shows two subsequent 2D views of microstructural snapshots of the same model as in Fig. 7.1, but at smaller depths of cut than before. Analyzing the changes in the microstructure deepens the understanding of the mechanisms of material removal for specific cutting conditions. In Fig. 7.2 areas with different crystal orientation are separated by lines and slip lines drawn for identified dislocations. Until state (a) was reached, the pre-deformation area in cutting direction (ahead of the chip) increased without an increase in chip length. Until state (b) was reached, this process had stopped, the pre-deformation area decreased while the chip grew in size. The micro structural plots show a change of the crystal orientation in the chip root area, that supports either the deformation away from the chip root (a) or the chip formation in the chip root area.

7.3 Cutting Forces, Stress and Temperature

By modeling systems of discrete particles and observing their progression over time, statistical mechanics provides a basis for the analysis and the description of the behavior of such systems. It has been demonstrated for a Gibbs micro canonical

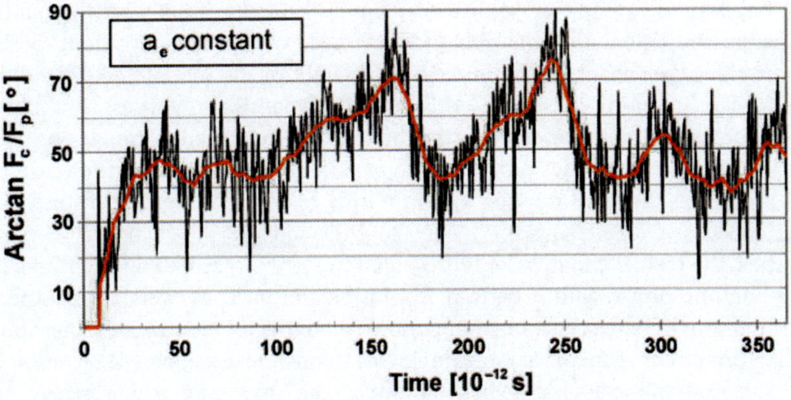

Fig. 7.3 Course of the cutting force ratio Fc/Fp (after Rentsch in Davim and Jackson [1]). Used with permission copyright ISTE-Wiley (2009)

ensemble, that taking time averages is in statistical agreement with taking phase-space averages and, that the numerical quality of results in MD can always be improved by longer calculations.

Cutting forces can be calculated as reaction forces at the tool due to its relative motion during contact with the workpiece atoms. For every time step Δt, which is usually in the range of a few femtoseconds (10^{-15} s), the force contributions of the workpiece atoms interacting with the tool are integrated. The dynamic character of such a system with a large number of degrees of freedom, i.e., all the freely moving workpiece atoms, emerge as fluctuations of derived, non-constant quantities. Figure 7.3 shows the course of the cutting force ratio for the simulation in Fig. 7.2.

The instantaneous tool forces, which will be newly calculated for every time step, fluctuate intensively. Calculating a moving average of the force ratio over, e.g., 1,000 time steps cancels out the fluctuations and leads to a smooth course. After overcoming the equilibration phase (no cutting forces), the first tool/workpiece contact was made and the force ratio changed to an average value of 1 (=arctan 45°) over a period of about 20 ps (10^{-12} s). Besides smaller maxima and minima during the observed total process simulation time, two gradually developing maxima in the course of the force ratio appear at about 165 and 245 ps. Figure 7.3 sheds light on the micro structural process that is related to the observed course of the force ratio.

Detailed information about the distribution of stresses and temperature in nanoscale cutting are of high interest for science and manufacturing industry. So far most of the MD results of cutting process simulations were presented as atomic, discontinuous sets of instantaneous data at individual atom sites, such as snap shot atomic positions, relative displacement and instantaneous atomic temperature. Besides the limited meaning of instantaneous atomic temperatures and stresses, looking at such large sets of 10,000, 100,000 or even millions of data is not practical from a point of view of efficient data analysis. Furthermore, it makes any attempt to compare MD results with, for instance, results from continuous mechanics difficult if not impossible. Taking advantage of the possibility of improving the quality of local values by calculating them as time averages over sufficiently long period of time provides the means to obtain a deeper insight of the model and the simulated process. Thus, aiming at macroscopic thermodynamic properties, suitable time intervals for averaging these properties have to be identified. Simulations showed that an average over about 1,000 time steps led in some cases to sufficiently stable mean properties, but still provide a certain time resolution in order to study details of the process. Considering the basics of MD and the physical nature of these quantities, the results can now be represented in form of gradual distributions as so-called contour plots, with a certain resolution in space as well as in time. The representation of stresses and temperature in terms of continuous distributions allows a direct comparison of continuous mechanics results and MD results.

In Fig. 7.4 the calculated distribution of the maximum shear stress of the orthogonal cutting process already shown in Figs. 7.1 and 7.2 is given. With the help of this distribution it is possible to determine where stress concentrations occur, how much the crystal structure influences the stress distribution, as well as the material removal process, and what the differences are in comparison to macroscopic, continuous mechanics processes.

This also allows the determination of how the process influences the motion of the workpiece and where new dislocations can occur, since areas of high shear stress are potential sources for formation or extension of dislocations. Similarly also temperature distributions can be calculated by adopting [7] to local volumes and calculating time averages. As it will be shown in the following chapter, MD cutting process simulations without the consideration of fluids lead to approximately concentric temperature distributions, in which the hottest area is the chip area and

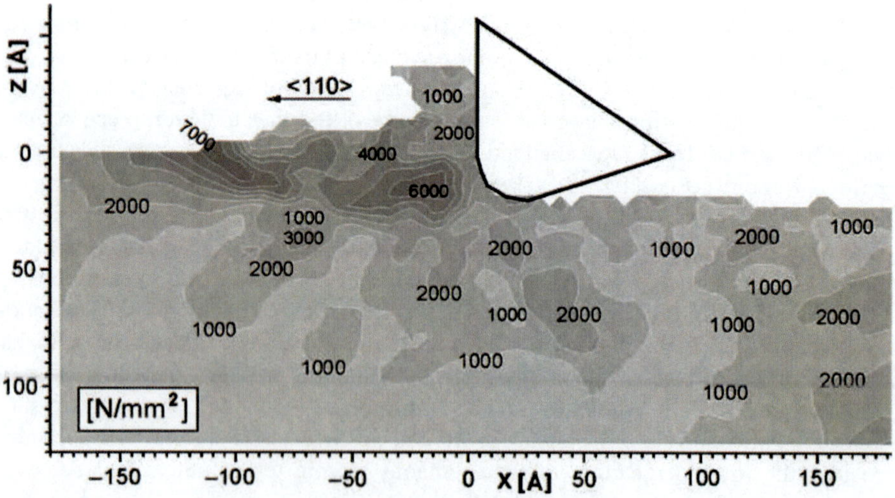

Fig. 7.4 Shear stress distribution in an orthogonal 3D MD machining model (after Rentsch in Davim and Jackson [1]). Used with permission copyright ISTE-Wiley (2009)

the chip root area. At the tip of the tool, the material is deformed at a high stress level, whereby a lot heat it generated. Hence, the high temperature areas extend from the chip under the tip of the tool, as one important source of heat generation, to the areas of shearing. It should be noted here, that modeling the tool by rigid, thermally not active atoms, does not enable the tool to conduct any heat. Therefore the tool acts like a thermal isolator, which further supports a concentration of heat in the chip.

Regarding the temperature distributions in MD cutting it should be further noted, that in most of the published work only the thermal conductivity through phonons is considered. The conductivity by electrons is neglected in such case, even though it is one order of magnitude larger than that of the phonons. Hence, the presented temperature levels as well as the local gradients would actually be lower than shown. New algorithms were developed to describe thermal conductivity more accurate by considering both, the electron and the phonon conductivity.

7.4 Three-Dimensional Machining Simulations

Two aspects of advances and recent developments in material removal process simulation using MD will be explained in some more detail, which are the possibility of carrying out complete 3D surface machining simulations and the consideration of fluids. For abrasive processes the model requirements are higher than for cutting processes, since orthogonal symmetry is not given and a quasi-2D model, cannot be applied. Besides the need to describe the geometry of abrasives, the model has to provide sufficient space for the deformation and chip formation of the three-dimensional material removal process. Figure 7.5a shows a snapshot of a molecular dynamics simulation to study material pile-up and chip formation in abrasive machining as a function of shape and orientation of the abrasives.

The simulation in Fig. 7.5 considered two pyramidal grains with diamond structure and two different orientations. The figure shows an advanced state of the 3D grinding simulation using a model with more than 100,000 copper atoms (the workpiece height was 6 nm). In several terms the simulation represents a high-end state-of-the-art MD simulation of the grain/workpiece contact as the interactions were based on an EAM potential function and the model considers two abrasives that cut at 100 m/s through a workpiece over its whole length. Hence, the periodic boundaries (for both directions of the horizontal plane) lead to complete groove formation by the grits in a cutting direction and describe a model setup with multiple grit/workpiece contacts that occurs in nanogrinding. By repeating the complete groove generation with relative-to-the-cutting-direction shifted abrasives, the machining of the whole surface can be realized. This provides the basis for 3D surface roughness and residual stress analyses of completely machined surfaces at realistic machining speeds (common grinding speeds range from about 5 to 80 m/s and high speed grinding up to about 250 m/s).

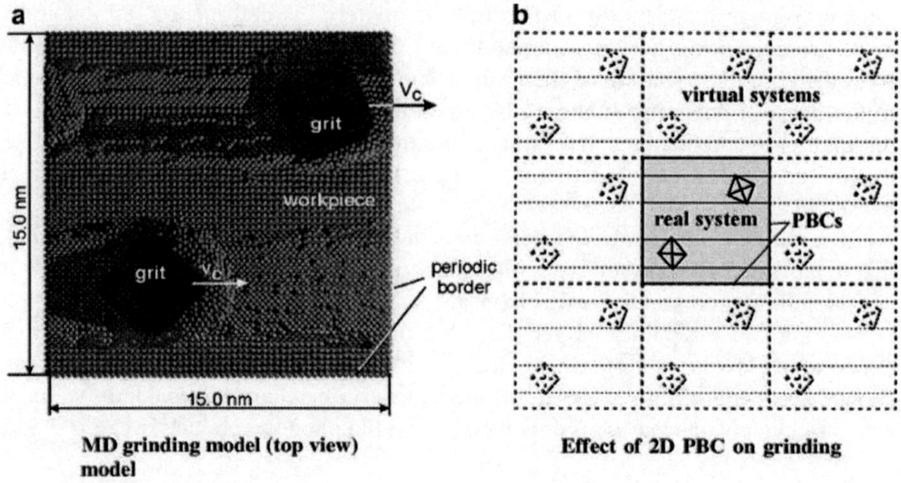

Fig. 7.5 Snapshot of a full model length scratching simulation with two hard abrasives (after 360,000 time steps, 144 ps) (after Rentsch in Davim and Jackson [1]). Used with permission copyright ISTE-Wiley (2009)

Thorough the analyses of the chip formation, the elastic and plastic response of the workpiece and process quantities in MD simulations have revealed clear and consistent effects. For example, the machining speed has a direct influence on the microscopic material removal process and the chip formation in MD simulations. The results suggest, that the sensitivity of the simulation results on the machining speed is less strong than observed in experimental investigations. A possible reason for this effect is, that the implemented boundary conditions and model settings have a strong impact on the dynamics of the finite process model. However, significant changes in magnitude of the machining speed lead to significant changes in chip shape and formation mechanism. Further quantities of the process are affected due to an increasing localization of the deformation process at high speeds. More direct effects on the process simulation results show the depth of cut, the grit orientation and the cutting edge radius. Hence, it is possible with MD to simulate the influence of grain shape and orientation on the efficiency of abrasive processes. On basis of bigger MD models, it will be possible to determine the energy dissipation by a direct analysis of elastic and plastic work and the microscopic mechanisms that determine the surface roughness in nanoscale machining processes.

7.5 Experimental Nanogrinding

The piezoelectric nanogrinding process is a process that relies on using a nickel-coated ceramic material with microscale diamond particles bonded to it that are cubo-octahedral in shape to machine nanoscale features in a variety of workpiece materials. The diamonds are bonded to the piezoelectric material by gaseous deposition, laser cladding, or directly bonding a porous tool to the material via an adhesive paste. The process is executed by applying a known sinusoidal frequency to the piezoelectric crystal in order to achieve a desired oscillatory displacement. Rapid vibration of the crystal will allow material removal rates to be increased, thus making it a nanomanufacturing process. The nanogrinding process is accompanied by wear of the diamond grains, and the rate of this wear plays an important role in determining the efficiency of the nanogrinding process and the quality of the nanomachined surface. Wear mechanisms in nanogrinding processes appear to be similar to that of single-point cutting tools, the only difference being the size of swarf particles generated. Figure 7.6 shows the arrangement of grinding swarf and abrasive grains that have been lost during the nanogrinding process.

Figure 7.6 shows abrasive grains with blunted cutting edges (wear flats), and abrasive grains with sharp cutting edges that are released from the surface of the piezoelectric crystal before they have chance to grind nanoscale chips from the surface of the workpiece. The process suffers with a loss of diamond grains even when the interfacial adhesion between diamond and piezoelectric material is very good. A more closely related process that has been reported widely is that of the wear of probes used in atomic force microscopy [2, 3]. However, these observations were purely experimental with no explanation of how to design probes that inhibit, or retard, wear. A performance index used to characterize diamond wear resistance is the grinding ratio, or G-ratio, and is expressed as the ratio of the change in volume of the workpiece removed, Δv_w, to the change in the volume of the diamond abrasive grain removed, Δv_s, and is shown in (7.1),

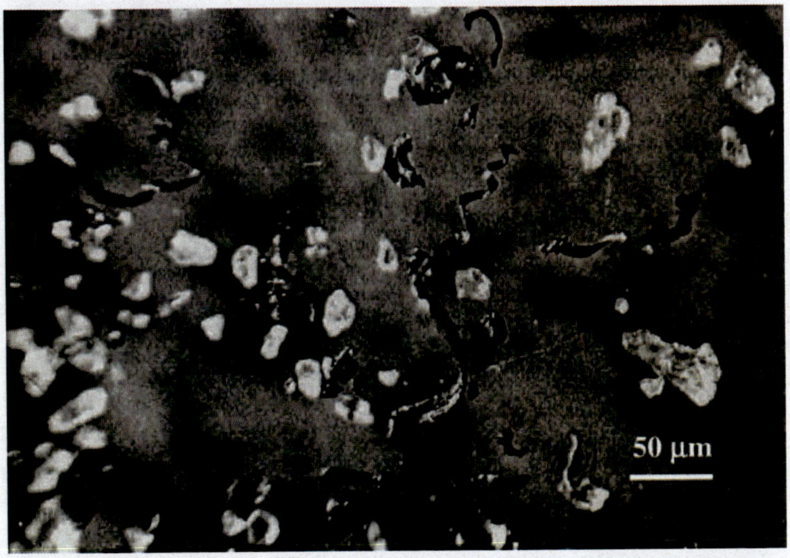

Fig. 7.6 Nanoscale grinding detritus showing blunt and sharp abrasive grains of diamond and metal chips. Used with permission copyright Springer (2007)

$$G = \Delta v_w / \Delta v_s \tag{7.1}$$

Grinding ratios for processes at the nanoscale have not yet been characterized. However, the complexities of wear of abrasive materials at any scale lead us to believe that the variety of different and interacting wear mechanisms involved, namely, plastic flow of abrasive, crumbling of the abrasive, chemical wear etc., makes the wear of diamond at the nanoscale too complicated to be explained using a single theoretical model [4]. The following analysis of diamond grains represented by loaded wedges assumes that grain fracture is the dominant wear mechanism when grinding at the nanoscale using the piezoelectric nanogrinding process.

7.5.1 Analysis of Nanogrinding Grains

Diamond grains are blocky in nature and possess sharp cutting points prior to nanogrinding workpiece materials. Figure 7.7 shows a collection of diamond abrasive grains that have well defined cutting points that form a wedge at their apex. When bonded into a strong matrix, these grains can be considered to be representative infinite wedges.

An infinite wedge represents the cutting point of an abrasive grain in contact with the workpiece material (Fig. 7.8).

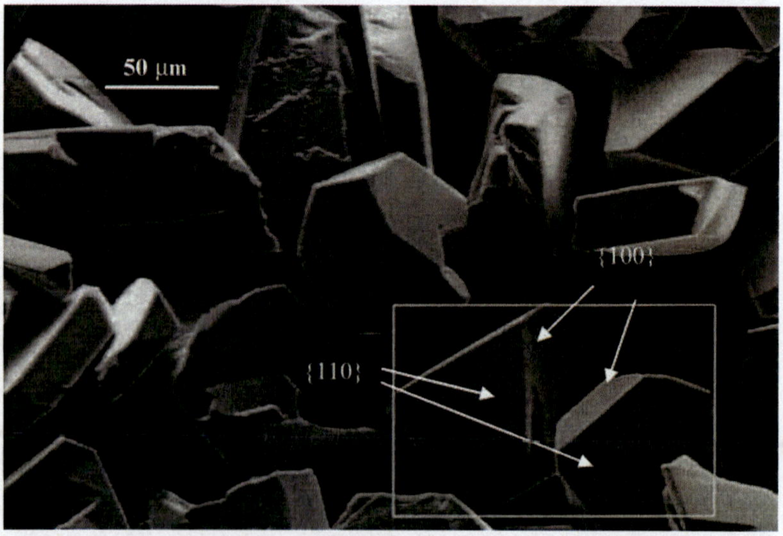

Fig. 7.7 A collection of diamond grains showing cutting points located at their apex, and locations of {100} and {110} planes. Diamond grains are approximately 60 µm in diameter. Used with permission copyright Springer (2007)

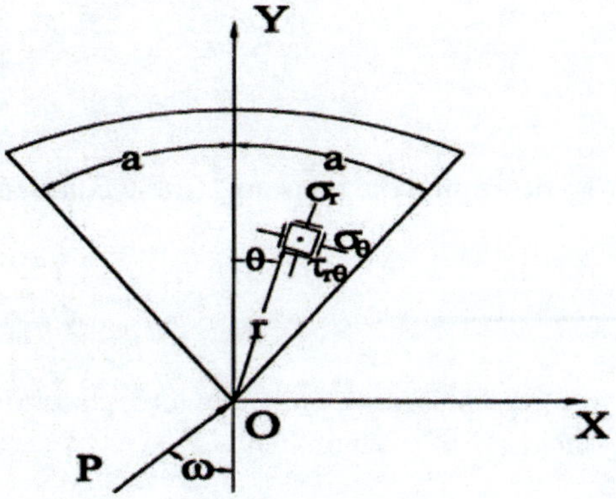

Fig. 7.8 The single-point, loaded infinite wedge. Used with permission copyright Springer (2007)

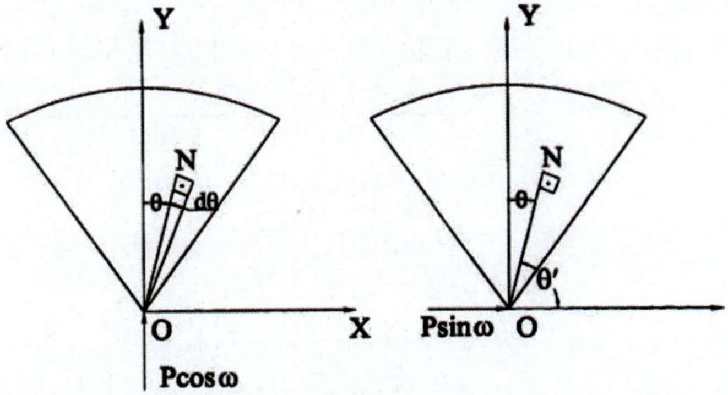

Fig. 7.9 The single-point, loaded infinite wedge showing force components, and the point *N* within the wedge at polar co-ordinates, r and θ. Used with permission copyright Springer (2007)

The wedge is loaded at the apex by a load P in an arbitrary direction at angle ω to the axis of symmetry of the wedge. Resolving the force into components P·cos ω in the direction of the axis, and P·sin ω perpendicular to that the stresses due to each of these forces can be evaluated from two-dimensional elastic theory [5]. The state of stress in the wedge, due to force P·cos ω, can be obtained from the stress function,

$$\varphi = C \cdot r \cdot \theta \cdot \sin\theta \tag{7.2}$$

where r and θ are polar coordinates, or the point N in Fig. 7.9, and C is a constant.

The stress function yields the following radial, tangential, and shear stress components,

$$\sigma_r = -2C\frac{\cos\theta}{r} \tag{7.3}$$

$$\sigma_\theta = 0 \tag{7.4}$$

$$\tau_{r\theta} = 0 \tag{7.5}$$

To determine the constant, C, the equilibrium of forces along the y-axis is:

$$P\cos\omega - \int_{-a}^{a} \sigma_r \cdot \cos\theta dA = 0 \tag{7.6}$$

where dA is an element of cross sectional area within the wedge. If, t, is the thickness of wedge, then,

$$\begin{aligned}\cos\omega P &= \int_{-a}^{a} 2C\frac{\cos\theta}{r} \cdot t \cdot r \cdot \cos\theta d\theta = 2Ct\int_{-a}^{a} \cos^2\theta d\theta \\ &= Ct[2a + \sin 2a]\end{aligned} \tag{7.7}$$

Therefore,

$$C = \frac{P\cos\omega}{t(2a + \sin 2a)} \tag{7.8}$$

And,

$$\sigma_r = -\frac{2P\cos\theta \cdot \cos\omega}{r \cdot t(2a + \sin 2a)} \tag{7.9}$$

Note that the negative sign denotes that the stress is compressive. The state of stress in the wedge, due to force P·sin ω, can be obtained from the stress function,

$$\varphi = C' \cdot r \cdot \theta' \cdot \sin\theta' \tag{7.10}$$

Therefore,

$$\sigma_r = -2C\frac{\cos\theta'}{r} \tag{7.11}$$

$$\sigma_\theta = 0 \tag{7.12}$$

$$\tau_{r\theta} = 0 \tag{7.13}$$

Equilibrium of forces along the x-axis (Fig. 7.9) yields the following solution for the constant, C,

$$P \sin \omega - \int_{\pi/2-a}^{\pi/2+a} \sigma_r \cdot t \cdot r \cdot \cos \theta' \mathrm{d}\theta' = 0 \tag{7.14}$$

$$P \sin \omega = -\int_{\pi/2-a}^{\pi/2+a} 2C \frac{\cos \theta'}{r} \cdot t \cdot r \cdot \cos \theta' \mathrm{d}\theta' = 2Ct \cdot \int_{\pi/2-a}^{\pi/2+a} \cos^2 \theta' \mathrm{d}\theta'$$
$$= -C \cdot t(2a - - \sin 2a) \tag{7.15}$$

$$C = \frac{P \sin \omega}{t(2a - \sin 2a)} \tag{7.16}$$

Thus,

$$\sigma_r = -\frac{2P \cos \theta' \cdot \sin \omega}{r \cdot t(2a - \sin 2a)} \tag{7.17}$$

Expressing in terms of the angle θ (where θ' is negative), yields,

$$\sigma_r = -\frac{2P \cos \theta \cdot \sin \omega}{r \cdot t(2a - \sin 2a)} \tag{7.18}$$

Therefore, the combined stresses are:

$$\sigma_r = -\frac{2P}{r \cdot t} \left[\frac{\cos \omega \cos \theta}{2a + \sin 2a} + \frac{\sin \omega \cos \theta}{2a - \sin 2a} \right] \tag{7.19}$$

It follows that σ_r vanishes for angle θ_o defined using the expression:

$$\tan \theta_o = \frac{1}{\tan \omega} \cdot \frac{2a - \sin 2a}{2a + \sin 2a} \tag{7.20}$$

This equation corresponds to a straight line through the apex as shown in Fig. 7.10. This natural axis separates the regions of compressive and tensile stresses. It can be seen that for values of angle ω which gives, $|\theta_o| > |a|$, the neutral axis lies outside the included angle of the wedge. This means that the whole area of the wedge will be under stresses of uniform sign. Expressing (7.19) in terms of the rake angle of the abrasive grain, β, and force components F_t and nF_t (Fig. 7.11), yields,

$$\sigma_r = -\frac{2F_t}{r \cdot t} \times \left\{ \frac{[n \cdot \cos(a - \beta) + \sin(a - \beta)] \cos \theta}{2a + \sin 2a} + \frac{[\{\cos(a - \beta) - n \cdot \sin(a - \beta)\} \cos \theta]}{2a - \sin 2a} \right\} \tag{7.21}$$

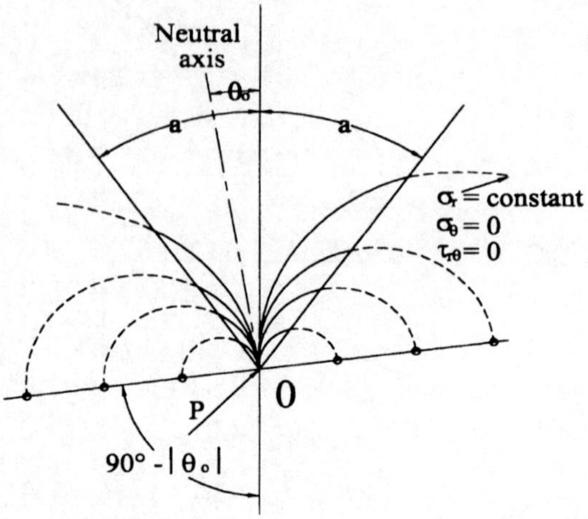

Fig. 7.10 Stress analysis of a single-point loaded wedge. Used with permission copyright Springer (2007)

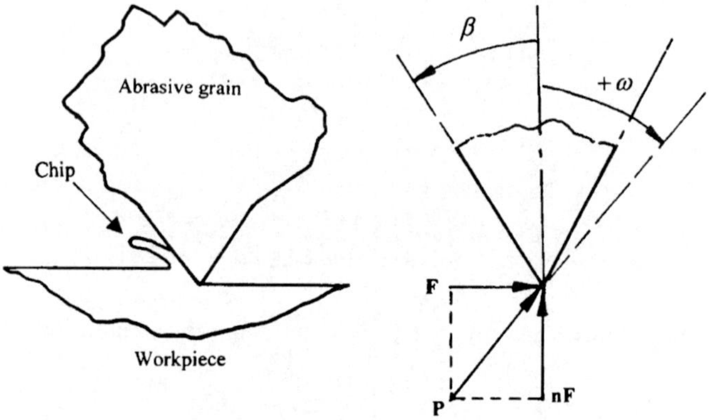

Fig. 7.11 Ideal wedge-shaped cutting point and grinding force diagram. Used with permission copyright Springer (2007)

It can be observed that:

$$\tan \omega = \frac{\cos(a - \beta) - n \cdot \sin(a - \beta)}{n \cdot \cos(a - \beta) + \sin(a - \beta)} \tag{7.22}$$

In the simple case of a wedge with the normal force nF_t along the wedge axis, a is equal to β, hence,

$$\tan \omega = 1/n \tag{7.23}$$

It is interesting to examine the radial stresses on the left-hand face of the wedge, which corresponds to the leading face of idealized wedge. Thus, for the left-hand face, θ is equal to −a, and from (7.19),

$$\sigma_r = -\frac{2P}{r \cdot t}\left[\frac{\cos\omega\cos a}{2a + \sin 2a} + \frac{-\cos a \cdot \sin\omega}{2a - \sin 2a}\right] \tag{7.24}$$

This stress is zero, i.e., the neutral axis coincides with the left-hand limit of the wedge, when,

$$\frac{1}{\tan\omega} = \frac{\sin a(2a + \sin 2a)}{\cos a(2a - \sin 2a)} \tag{7.25}$$

Thus if,

(a) a = β, then,

$$n = \frac{\sin a}{\cos a} \cdot \frac{2a + \sin 2a}{2a - \sin 2a} \tag{7.26}$$

(b) $a - \beta = \pi/2 - a$
(as is the case when F_t is parallel to the right-hand face of the wedge). From (7.23),

$$\tan\omega = \frac{\sin a - n \cdot \cos a}{n \cdot \sin a + \cos a} = \frac{\sin a - n\cos a}{n \cdot \sin a + \cos a} \tag{7.27}$$

And substituting in (7.25), yields,

$$\frac{n\sin a + \cos a}{\sin a - n\cos a} = \frac{\sin a}{\cos a} \cdot \frac{2a + \sin 2a}{2a - \sin 2a} \tag{7.28}$$

$$\frac{\frac{1}{2} \cdot n\sin 2a + \cos^2 a}{\sin^2 a - \frac{1}{2} \cdot n \cdot \sin 2a} = \frac{2a + \sin 2a}{2a - \sin 2a} \tag{7.29}$$

$$\frac{1}{2} \cdot n \cdot \sin 2a(2a - \sin 2a) + 2a\cos^2 a - \cos^2 a \cdot \sin 2a$$
$$= 2a \cdot \sin^2 a + \sin^2 a \cdot \sin 2a - \frac{1}{2} \cdot n \cdot \sin 2a(2a + \sin 2a)$$

$$\frac{1}{2} \cdot n \cdot \sin 2a(2a - \sin 2a + 2a\sin 2a) = 2a(\sin^2 a - \cos^2 a) + \sin 2a$$

$$n = \frac{1}{2a} - \cot 2a \tag{7.30}$$

Equation (7.25) expresses the condition for the whole of the wedge's cross-sectional area to be under compressive stress. It can be seen that this depends not only upon the rake angle, β, but also upon the ratio, n. In general the relative size of the region of compressive to the region of tensile stresses depends upon β and n as (7.20) and (7.22) indicate. Also, from (7.21), the magnitude of the stress on the left-hand face of the wedge is found to be dependent upon the tangential force component, F_t, and the force component ratio, n. Referring to (7.19), it can be seen that for constant stress, $\sigma_r = \text{constant}$,

$$r \cdot C_1 = C_2 \cdot \cos\theta + C_3 \cdot \sin\theta \tag{7.31}$$

where C_1, C_2, C_3 are constants. Equation (7.31) represents, in polar co-ordinates, the circumference of a circle tangent to the line,

$$0 = C_2 \cdot \cos\theta + C_3 \cdot \sin\theta \tag{7.32}$$

that is to the neutral axis at the point $r = 0$. However, the point $r = 0$ must be considered separately because the stress at that point approaches infinity, since by definition P is a point load. The central point of these circles are of constant radial stress, and so the point of constant maximum shear stress must lie on a line perpendicular to the neutral axis at the point where r is equal to zero. The radius of each of those circles depends upon the magnitude of the radial stress, σ_r.

7.5.2 *Fracture Dominated Wear Model*

Brittle materials exhibit high strength properties when loaded in compression than in tension. The ratio of rupture strengths is usually between 3:1 and 10:1. The existence of relatively low tensile stresses in the abrasive grains may cause failure by fracture to occur. To model the action of diamonds bonded to piezoelectric ceramics, one must consider a single active cutting grain to be classed as a wedge of constant width loaded at its inverted apex with point loads, F, and, nF, which represent the radial and tangential force components with reference to the grinding wheel in which the grain is supported, and P is the resultant force.

The stress distributions within point-loaded wedges can be determined analytically, and the results of such an analysis indicate that if tensile stresses exist within the wedge then it will occur at its maximum along the rake face. The existence of a tensile stress depends on the magnitude of the force ratio, n. If the ratio is especially small that a tensile stress exists in the wedge, then for a specific force ratio the tensile stress is proportional to the tangential grinding force, F. Stresses of this nature would extend to and beyond the abrasive grain-bonding interface. The fracture of abrasive grain, bonding phase, or the interface between the two, depends on the particular type of piezoelectric ceramic material used and the magnitude of the tensile stress induced during nanogrinding. Grains of diamond are ten times stronger in compression than in tension. The probability of grain fracture is likely to increase with an increase in tensile stress exerted in the grain although the magnitude of the stress may be slightly higher than one-fifth the magnitude of the maximum compressive stress in the grain. A significant barrier to the acceptance of stress patterns evaluated for such situations arises because point loads applied to perfectly sharp wedges produce infinitely high stresses at, and about, the point of contact. Loads must be applied over a finite area. It seems likely that higher tensile stresses are associated with higher grain fracture probability resulting in rapid loss of diamond grains and, consequently, lower grinding ratios. The wear model should incorporate the fact that the loads are spread over a finite area. This implies that point loads are applied along the rake face. The model should allow the examination between the wear rate of a diamond coated piezoelectric ceramic material and the general nature of stresses established in active cutting grains subjected to nanogrinding forces. This means that it is necessary to estimate the force components of grinding on each active cutting grain.

7.5.3 Nanogrinding Procedure

The experimental apparatus consisted of holding a polished specimen between the jaws of a vice so that a piezoelectric crystal oscillator traverses back-and-forth the specimen thus machining the specimen by creating a depth of cut between the diamonds adhered on the piezoelectric crystal and the workpiece material. Workpiece materials were polished with a 100 nm sized polishing compound. All samples were divided into four sections and each section was analyzed prior to machining and after machining occurred. The workpieces were mounted in a vice that was attached to a x-y-z linear slide in order to achieve accurate positioning of the workpiece. The piezoelectric crystal was mounted on a steel framework that was orthogonal to the workpiece. The whole unit was located within a tetrahedral space frame to dampen excess vibrations (Fig. 7.12). When the crystal and workpiece were aligned, the depth of cut was incremented in stages of 10 nm. The motion of the diamonds attached to the piezoelectric crystal generates a machining effect that is caused by the action of diamonds grinding into the workpiece material. Tracks or trenches are created by the diamond grain gouging the surface of the material when an electric current is applied. The material is removed until the end of the oscillating motion creates the material to plow.

The mechanism of oscillation can be described as a restricted bending mode that simulates a shear displacement of the crystal. At this point it is normal procedure to know how to estimate the number of grains contacting the surface of the workpiece. The estimation of the number of active cutting grains is estimated quite simply by driving the diamond coated piezoelectric ceramic at the prescribed specific metal removal rate into a piece of lead. The impression that the grinding wheel produces in the length of lead is equal to the number of cutting points that are active during the grinding stroke at that particular depth of cut. The effect of oscillating the diamond-coated crystal is shown in Fig. 7.13.

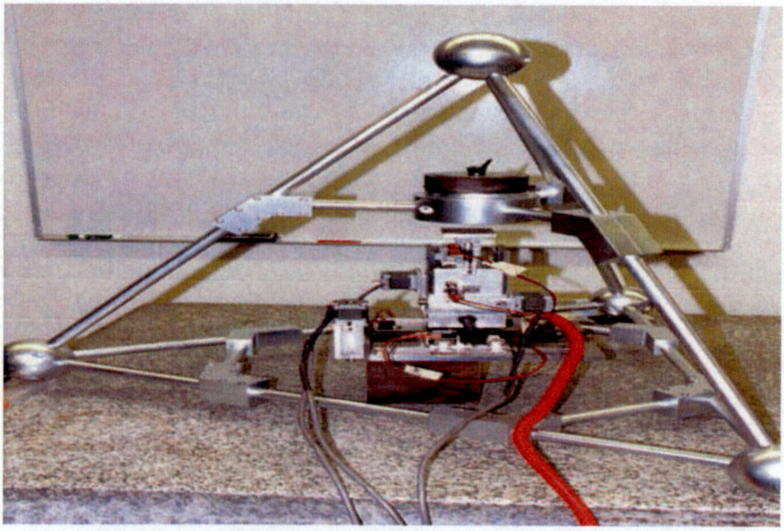

Fig. 7.12 Machining center capable of nanogrinding

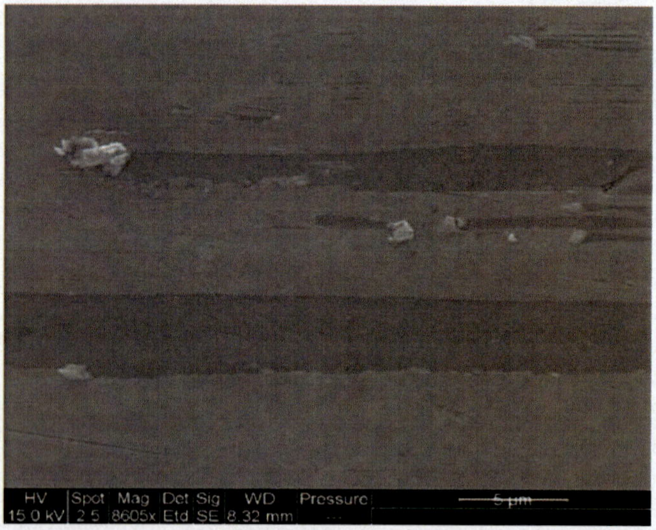

Fig. 7.13 Material removal by oscillating a diamond-coated piezoelectric ceramic crystal. Used with permission copyright Springer (2007)

The motion of the diamonds imparted by oscillating the crystal in the bending mode that causes a shear displacement to occur, which contributes to plowing of the material at the end of the nanogrinding stroke, is shown in Fig. 7.14.

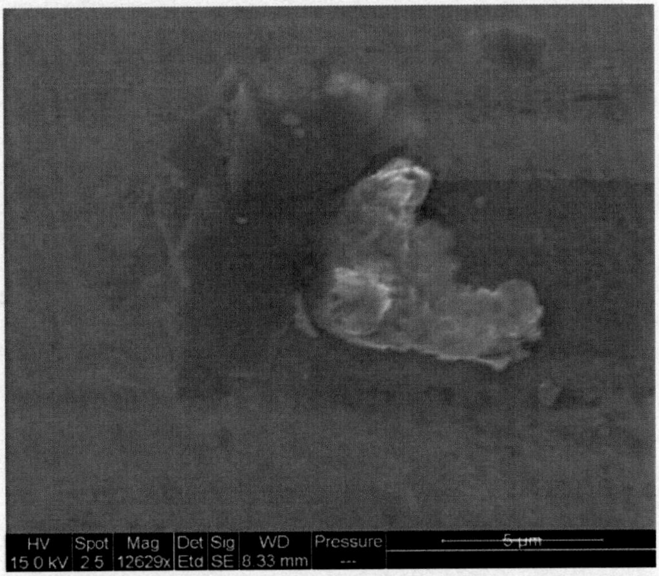

Fig. 7.14 Plowed material at the end of a nanomachined track. Used with permission copyright Springer (2007)

Fig. 7.15 Nanomachined tracks in steel. Used with permission copyright Springer (2007)

The effect of using the piezoelectric to machine tracks, or trenches, in engineering materials opens up the prospect of nanomanufacturing products that require geometric features such as channels so that fluids and mixed phase flows can be manipulated in devices such as micro-and nanofluidic "lab-on-a-chip" products. Figure 7.15 clearly shows such a channel produced using the piezoelectric nanogrinding process.

The measured force components of the nanogrinding operation are measured using a dynamometer. These components of force are then applied to a model abrasive grain by dividing the grinding force data into the number of active cutting grains over an area that simulates the abrasive grain-workpiece contact area. Stresses established in this area are calculated using finite elements. The wear of the piezoelectric material by diamond loss, expressed in terms of a grinding ratio, and its relationship to the stress levels set up in the model grain is investigated using a stress analysis method.

7.5.4 Stress Analysis

The assumed geometry of an ideal grain in the vicinity of its cutting edge is a simple symmetrical wedge of constant width with an included angle of 70° that results in a rake angle of −35°. There is no wear flat on the model cutting grain. In order that a finite element method is used to evaluate stresses in the wedge, the wedge was subdivided into 210 diamond-shaped elements with a total of 251 nodes. Forty-one nodes were constrained at the boundary of the wedge and the leading five nodes on the rake face were loaded (Fig. 7.16). The tangential and normal grinding forces were replaced by equivalent forces acting perpendicular to (normal load) and along (shear load) the rake face of the wedge.

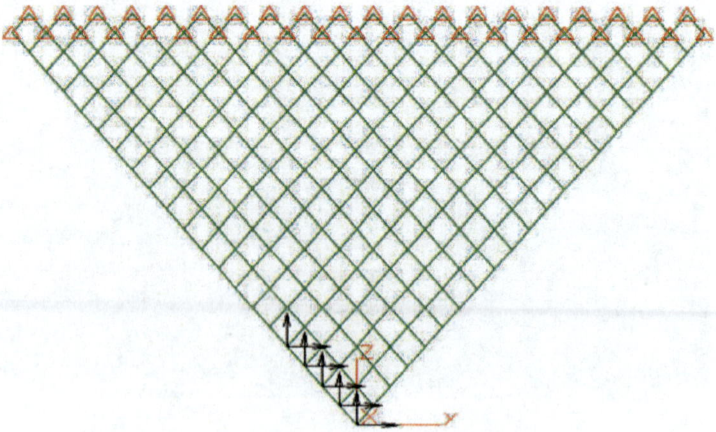

Fig. 7.16 Finite element assemblage with grinding loads applied at the rake face nodes. Used with permission copyright Springer (2007)

The concentrated loads at the five nodes are representative of the distributed and normal loads acting on the rake face over the abrasive grain-chip contact length. The normal force distribution on the rake face was taken to be maximum at the cutting edge and decreasing linearly to zero at the end of the abrasive grain-chip contact length. The shear force was taken to be constant over the first half of the contact length, decreasing linearly to zero over contact length.

Grinding loads were also applied directly to the rake face and at the tip of the grain without calculating equivalent forces. This was performed in order to compare and contrast the effect of different force distributions on the stresses generated within the wedge.

To measure the value of using the maximum tensile stress as a way to estimate grain fracture tendency, the correlation between the two sets of data were calculated for each set of data. The region of fracture initiation was also located using Griffith's criterion of fracture. For,

$$\frac{\sigma_c}{\sigma_t} \cdot \sigma_1 + \sigma_3 > 0 \tag{7.33}$$

Then,

$$\sigma_1 = \sigma_t \tag{7.34}$$

But for,

$$\frac{\sigma_c}{\sigma_t} \cdot \sigma_1 + \sigma_3 < 0 \tag{7.35}$$

Then,

$$(|\sigma_1| - |\sigma_3|)^2 + 8\sigma_t(|\sigma_1| - |\sigma_3|) = 0 \tag{7.36}$$

where σ_1 and σ_3 are the principal stresses, assuming that $\sigma_1 > \sigma_3$, σ_t is the ultimate tensile strength of the abrasive grain, and σ_c is the ultimate compressive strength. For diamond grain material, the ratio of σ_t and σ_c is 0.1.

The results of the two-dimensional stress analyses were consistent with the experimentally determined stress distribution obtained by Loladze [6] when cutting soft metal with photoelastic tools. The maximum tensile stress always occurs at the rake face at a distance from the cutting edge ranging from 1.5 to four times the abrasive grain-chip contact length, the exact magnitude of the coefficient depends on the loading conditions for a particular machining event. For a given value of the tangential force component, F, the higher the force ratio, F/nF, the greater the distance the maximum tensile stress is away from the cutting edge.

These results indicate that mechanically induced fracture occurs at a finite distance away from the cutting edge. When using Griffith's criterion, the influence

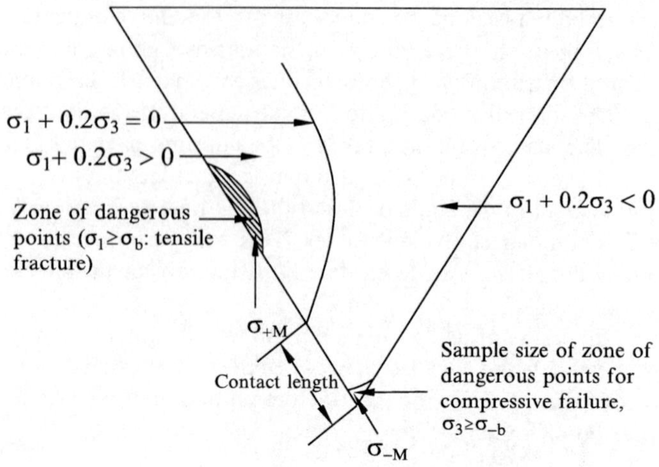

Fig. 7.17 Griffith's criterion applied to the idealized wedge showing tensile and compressive fracture initiation zones. Used with permission copyright Springer (2007)

of mechanically induced stresses indicate that fracture initiation zones are established. Figure 7.17 shows the occurrence of such zones in an idealized wedge. The first zone is located around the point of maximum tensile stress and is always at the rake face.

Failure in this zone is tensile in nature and would initiate fracture at a point on the rake face of the order of two-to-three times the abrasive grain-chip contact length away from the cutting edge. This type of fracture is consistent with fracture on scale comparable with the chip thickness. The second much smaller zone is located at the immediate vicinity of the cutting edge. Failure is compressive in this region and results in small-scale crumbling of the cutting edge leading to the formation of a wear flat on the abrasive grain. The correlation between the magnitude of the maximum tensile stress in the model abrasive grains and the appropriate grinding ratio (Table 7.1) is high and is dependent on the way the forces are applied to the grains. It would be expected that the higher the tensile stress, the greater is the rate of diamond wear and consequently the corresponding grinding ratio. Perfect linear correlation in accordance with this would result in a correlation coefficient of -1.

The correlation coefficient between the maximum tensile stress and the grinding ratio is significant. This is to be expected as the force ratio may vary slightly. However, if the tangential component of the grinding force changes significantly without a change in force ratio, then it is expected that the maximum tensile stress will change significantly and reduce the grinding ratio. The calculation and application of equivalent grinding loads produce a lower correlation coefficient compared to directly applied grinding loads. This implies that grinding loads are simply not point loads acting act the tip of the inverted apex and along the abrasive grain-chip contact length of the diamond grain. In fact, directly applied grinding forces

Table 7.1 Correlation coefficient between maximum tensile stress and grinding ratio for an idealized wedge using experimental data. Comparison is also made between the methods of applying loads to the idealized wedge models

Workpiece material	Exact wedge model with point loads applied to apex of wedge	Approximate finite element model: equivalent grinding forces applied to rake face of wedge	Approximate finite element model: grinding forces applied directly to the rake face of the wedge
Diamond on steel	−0.8	−0.9	−0.94
Diamond on MgO	−0.6	−0.7	−0.8
Diamond on copper alloy	−0.55	−0.65	−0.78
Diamond on aluminum alloy	−0.7	−0.85	−0.95
Diamond on silicon	−0.85	−0.87	−0.9

produce better correlation coefficients. This means that for perfectly sharp diamond grains, one must apply the component grinding loads directly to the rake face.

It can be seen from Table 7.1 that induced tensile stresses account for the loss of grain material from the diamond coated piezoelectric ceramic material. Therefore, the maximum tensile stress is the best indicator of diamond performance, in terms of grinding ratio, during a nanogrinding operation. The analysis performed on perfectly sharp diamond grains has provided a strong correlation between maximum tensile stress induced in the grain material and the wear parameter, grinding ratio, for the experimental data used in this chapter.

Correlations with other data sets have not proved so fruitful. From this, we can safely assume that the mechanism of grain fracture is not the dominant mechanism, which implies that other mechanisms are operating. The correlation coefficient demonstrates that a tougher grain material must be used in order to limit the effects of abrasive wear and the formation of wear flats, or a stronger bond, and possibly a higher volume of bond between diamond and piezoelectric crystal, is required to nanogrind under the current experimental conditions. Therefore, the present method of calculating the correlation coefficient between the maximum tensile stress and the grinding ratio demonstrates its potential application to the wider problem of selecting abrasive grains based on specific metal removal rates and the nature of the nanogrinding operation.

When porous tools are used to embed diamonds or any other abrasive material, the same analysis can be used but account of the properties of the bonding bridge must be made. The bonding bridge can be made of a variety of different materials but the most common one used for dressable applications is the vitrified type, which is made from a mixture of clays, glasses, and minerals. The emphasis on using dressable types for nanogrinding is based on their ability to be re-sharpened by dislodging worn grains and by microstructural phase transformations by focusing optical energy on the bonding bridges that hold the grains in place.

7.5.5 Porous Nanogrinding Tools

Porous nanogrinding tools are composed of abrasive particles (sub micron size) embedded in a vitrified bond with porosity interspersed between grinding grains and bonding bridges. The porosity level is approximately 15–21%. Figure 7.18 shows the image of a nanogrinding tool. The vitrified bonds are specially engineered to promote the formation of texture that creates ridges of cutting planes and nanogrinding "peaks" of α-Al_2O_3 in the preferred (012), (104), and (110) planes. The peaks created due to laser modification of the surface aid the nanogrinding process. Vitrified bonds are composed of glasses that are formed when clays, ground glass frits, mineral fluxes such as feldspars, and chemical fluxes such as borax melt when the grinding wheel is fired at temperatures in the range, 1,000–1,200°C. With reference to raw material nomenclature, a "frit" is a pre-ground glass with a pre-determined oxide content, a "flux" is a low melting point siliceous clay that reduces surface tension at the bond bridge-abrasive grain interface, a "pre-fritted" bond is a bond that contains no clay minerals (i.e., clays and fluxes), and "firing" refers to vitrification heat treatment that consolidates the individual bond constituents together [7]. Considering

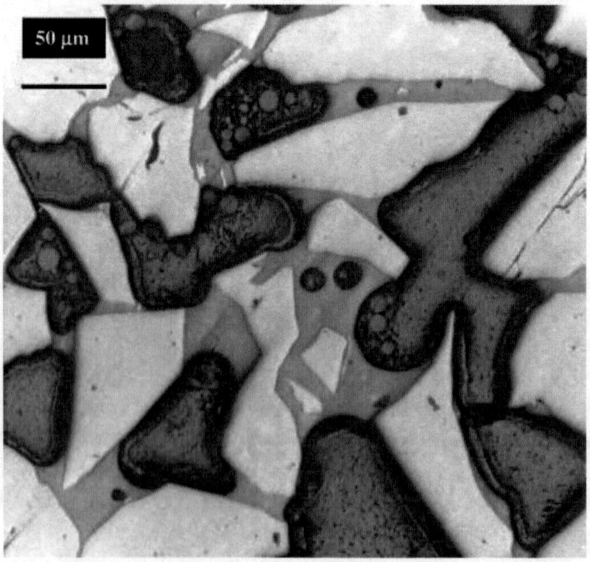

Fig. 7.18 Structure of the porous tool used for nanogrinding. Used with permission copyright Springer (2007)

Fig. 7.19 A collection of quartz particles in a vitrified bonding system. The quartz particle on the left has a circumferential crack extending into the dissolution rim. Used with permission copyright Springer (2007)

individual bond constituents, mineral fluxes and ground glass frits have little direct effect on the ability to manufacture grinding wheels. However, most clays develop some plasticity in the presence of water (from the binder), which improves the ability to mould the mixture so that the wheel, in its green state, can be mechanically handled.

Clays and clay-based fluxes contain an amount of free quartz that has a detrimental effect on the development of strength during vitrification heat treatment. Clays are used to provide vitrified grinding wheels with green strength during the heat treatment process. However, when the glass material solidifies around the particles of clay and quartz, the displacive transformation of quartz during the cooling stage of vitrification leads to the formation of cracks in the glass around the quartz particle (Fig. 7.19). The strength of the bonding bridge is impaired and leads to the early release of the abrasive particle during the cutting of metal.

The basic wear mechanisms that affect vitrified grinding wheels are concerned with grain fracture during metal cutting, fracture of bond bridges, mechanical fracture of abrasive grains due to spalling, and fracture at the interface between abrasive grain and bond bridge. Failure in vitrified silicon carbide grinding wheels is more probable due to the lack of a well-developed bonding layer between abrasive grain and glass bond-bridge. The bonding layer is approximately a few micrometers in thickness, and is caused by the use of a high clay content bonding system. High glass content bonding systems tend to aggressively decompose the surface of silicon carbide abrasive grains. In vitrified corundum grinding wheels, high glass content bonding systems are used extensively and lead to bonding layers in excess of 100 μm in thickness. In addition to the formation of very thin bonding layers in vitrified silicon carbide grinding wheels, the use of high clay content bonding systems implies that there is an increase in the amount of quartz in the bond bridges between abrasive grains. Although the likelihood of decomposition of silicon carbide surfaces is reduced, the probability of bond bridge failure is increased due to the increased quartz content. Therefore, the dissolution of quartz is of paramount importance in order to compensate for thinner interfacial bonding layers.

The dissolution of quartz in a liquid phase does not require a nucleation step. One process that determines the rate of the overall reaction is the phase-boundary reaction rate that is fixed by the movement of ions across the interface. However, reaction at the phase boundary leads to an increased concentration at the interface. Ions must diffuse away from the reaction interface so that the reaction can continue. The rate of material transfer and the diffusion rate are controlled by molecular diffusion in the presence of a high-viscosity liquid phase. For a stationary solid in an unstirred liquid, or in a liquid with no fluid flow produced by hydrodynamic instabilities, the rate of dissolution is governed by molecular diffusion. The effective diffusion length over which mass is transported is proportional to $\sqrt{Dt}$, where D is the diffusion coefficient and t is time, and therefore the change in thickness of the solid, which is proportional to the mass dissolved, varies with $\sqrt{t}$. Natural, or free, convection occurs because of hydrodynamic instabilities in the liquid which gives rise to fluid flow over the solid. This enhances the kinetics of dissolution. Generally, a partially submerged solid undergoes more dissolution near to the solid-liquid interface. Below this interface the kinetics of dissolution of the solid can be analyzed using the principles of free convection.

The boundary layer thickness is determined by the hydrodynamic conditions of fluid flow. Viscous liquids form much thicker boundary layers that tend to impede material transfer. Higher liquid velocities promote the formation of thinner boundary layers and permit more rapid material transfer. Considering the dissolution of quartz in glass materials, the high viscosity and slow fluid flows combine to give thick boundary layers. Also, the diffusion rate is much slower in viscous silicate liquids than in aqueous solutions, thus giving a tendency for the reaction process to be controlled by material-transfer phenomena rather than by interface reactions.

Difficulties encountered when developing a dissolution model arise from the fact that the phase boundary between quartz particle and molten glass moves during the diffusion process. The problem of a fixed boundary can be solved without difficulty although this is not equivalent to the conditions associated with a moving boundary between quartz particle and a highly viscous glass melt. The development of dissolution models is required to determine the magnitude of quartz remaining in the bonding system after a period of heat treatment. The models are then compared with experimentally determined quartz content of the bonding systems using X-ray diffraction techniques.

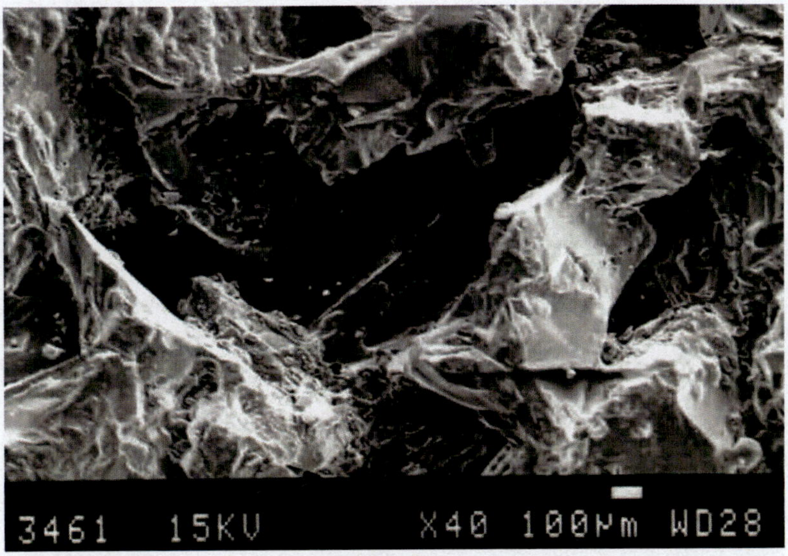

Fig. 7.20 Bonding bridge failure in a vitrified grinding wheel caused by the displacive transformation of quartz at high temperature during heat treatment. Used with permission copyright Springer (2007)

7.5.5.1 Dissolution Models for Quartz in Bonding Bridges

When densification occurs in a vitrified grinding wheel, the cooling rate is reduced to prevent thermal stress cracking in the bonding layer between abrasive particles. Cooling rates are reduced when crystalline inversions occur that involve volume changes. The inversion range for quartz and cristobalite are 550–580°C and 200–300°C, respectively. Since the formation of cristobalite is rare in most vitrified bonding systems used for grinding wheels, the rapid displacive transformation of quartz tends to promote the formation of cracks in bonding bridges (Fig. 7.20). Once the grinding grain is lost the remaining bonding bridges can be modified using a high power laser to create an oriented texture that forms "peaks" of α-Al_2O_3 in the preferred (012), (104), and (110) planes.

When quartz-containing bonds begin to cool form the soaking, or vitrification, temperature it is thought that the liquid phase relieves stresses resulting from thermal expansion mismatch between itself and the phases, β-quartz, β-cristobalite, and mullite, to at least 800°C. At 800°C, stresses will develop in quartz particles and the matrix that causes micro-cracking to occur. The shrinkage behaviour of quartz and the glass phase has been described by Storch et al. [8]. Between the temperature range, 573 and 800°C, the glass phase shrinks more than the quartz phase that causes tangential tensile stresses to form cracks in the matrix. At 573°C,

β-quartz transforms to α-quartz that causes residual stresses to produce circumferential cracking around quartz particles (Fig. 7.19). Some of these cracks have been seen to propagate into the glass phase [9]. Similar observations occur in the cristobalite phase. Spontaneous cracking of quartz has been found to occur over a temperature range that depends on the size of the quartz particles [10]. Particles larger than 600 μm' diameter cracked spontaneously at 640°C, whereas smaller particles of less than 40 μm diameter cracked at 573°C. This observation agrees with temperature-dependent micro-cracking reported by Kirchhoff et al. [11]. To maintain the integrity of the bond bridges containing coarse quartz particles, the grinding wheel must remain at the vitrification temperature until the quartz particles have dissolved. The dissolution model assumes that at a constant absolute temperature, T, a particle of quartz melts in the surrounding viscous glass melt, and that the rate of change of the volume of quartz present in the melt at a particular instant in time is proportional to the residual volume of quartz. The above assumption is based on the fact that alkali ions diffuse from the viscous glass melt to the boundary of the quartz particle, thus producing a dissolution rim around each quartz particle. A high reaction rate will initially occur which continuously decreases as the quartz particle is converted to a viscous melt.

Jackson and Mills [12] derived a mathematical relationship that accounts for the change in density when β-quartz transforms to α-quartz on cooling from the vitrification temperature, thus,

$$m_{T,t} = M\gamma \exp\left(-At^{1/2} \exp\left[\frac{-B}{T}\right]\right) \tag{7.37}$$

where, $m_{T,t}$, is the residual mass fraction of quartz at a constant time and temperature couple, M is the original mass fraction of quartz prior to heat treatment, γ is the ratio of densities of β-quartz and α-quartz, A and B are constants, t is time, and T is absolute temperature. The model was compared with experimental data determined using the powder X-ray diffraction method. The experimental work was divided into two parts. The first part concentrates on comparing the dissolution model with X-ray diffraction data using "sintering" bond compositions that are used in vitrified silicon carbide nanogrinding tools, while the second part focuses on comparing the model with "fusible" bond compositions that are used in high-performance vitrified corundum nanogrinding tools.

7.5.5.2 Preparation of Nanogrinding Wheel Structure

The raw materials used in the experimental study were Hymod Prima ball clay, standard porcelain China clay, potash feldspar, and synthetic quartz (supplied as silica flour). The chemical analysis of the raw materials is shown in Table 7.2. Rational analysis of the raw materials was performed to reveal the mineralogical composition of the raw materials. The rational analysis appears in Table 7.3.

Table 7.2 Chemical analyses of raw materials

Oxide (wt.%)	China clay	Ball clay	Potash feldspar	Quartz
Al_2O_3	37	31	18.01	0.65
SiO_2	48	52	66.6	98.4
K_2O	1.65	1.8	11.01	0.35
Na_2O	0.1	0.2	3.2	0.04
CaO	0.07	0.2	0.09	0.00
MgO	0.03	0.3	0.09	0.00
TiO_2	0.02	0.9	0.00	0.07
Fe_2O_3	0.68	1.1	0.11	0.03
Loss on ignition	12.5	16.5	0.89	0.20

Table 7.3 Mineralogical analyses of raw materials

Compound (wt.%)	China clay	Ball clay	Potash feldspar	Quartz
Quartz	4.05	12.77	4.93	98.40
Orthoclase	0.00	15.23	64.96	0.00
Kaolinite	79.70	62.71	2.17	0.00
Mica	13.94	0.00	0.00	0.00
Soda feldspar	0.8	1.69	27.07	0.00
Miscellaneous oxides/losses	1.51	7.60	0.87	1.60

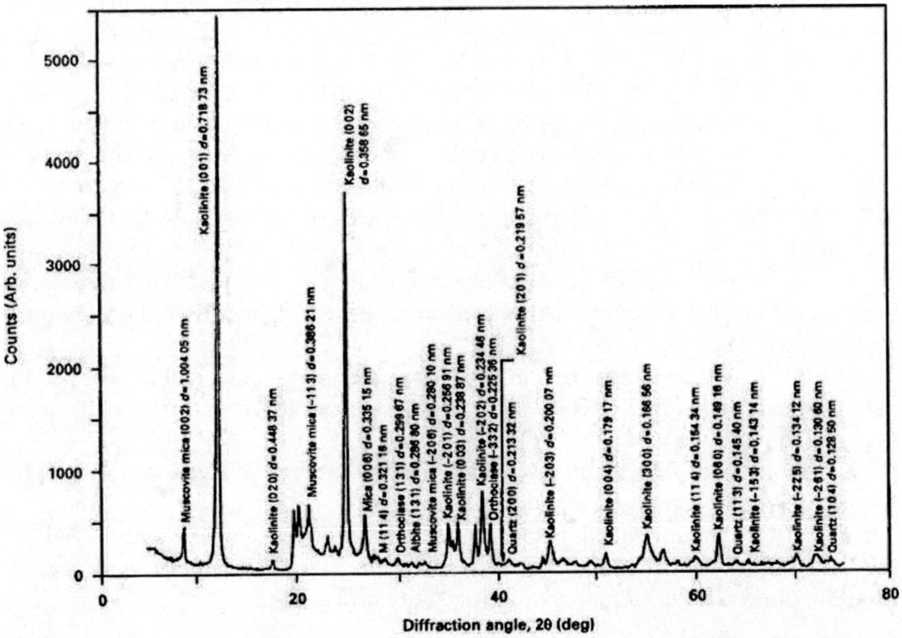

Fig. 7.21 X-ray diffraction spectrum of China clay showing crystallographic planes and interplanar distances of various mineral phases in the clay. Used with permission copyright Springer (2007)

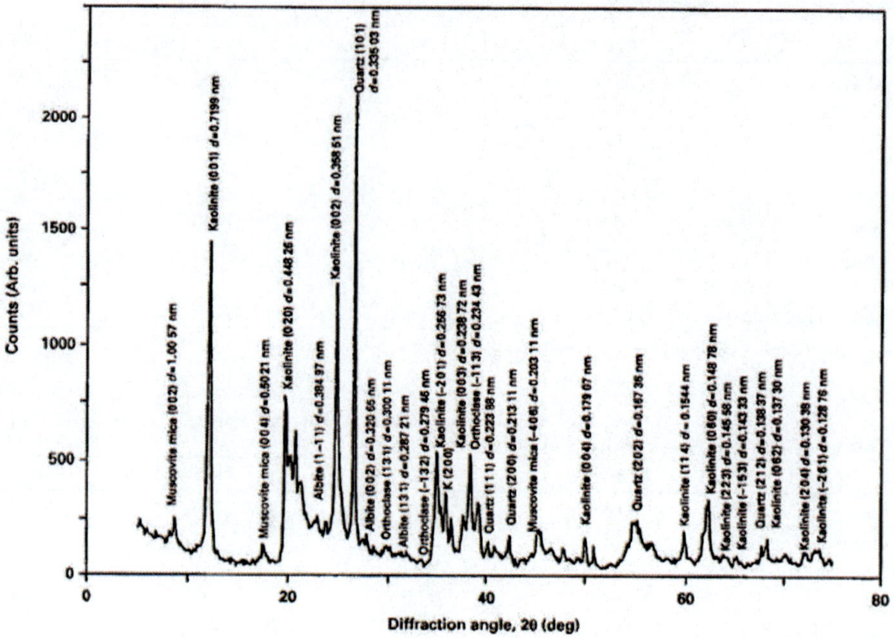

Fig. 7.22 X-ray diffraction spectrum of ball clay showing crystallographic planes and interplanar distances of various mineral phases in the clay. Used with permission copyright Springer (2007)

The characteristic X-ray diffraction spectra for ball clay and China clay are shown in Figs. 7.21 and 7.22. The bond mixture described is one typically used in vitrified silicon carbide grinding wheels where the erosion of the abrasive grain is reduced by using high clay content bonding systems.

Fusible bonding systems using a mixture of ball clay and potassium-rich feldspar were made to test the model developed by Jackson and Mills [12]. The ball clay used contained 12.77 wt.% quartz, and the feldspar contained 4.93 wt.% quartz. The bonding system was composed of 66 wt.% ball clay, and 34% feldspar.

The initial quartz content, M, of the bond mixture was 10.1 wt.%. The bond mixture described is one typically used in high-performance vitrified corundum grinding wheels.

The raw materials were mixed in a mortar, pressed in a mold, and fired at various temperatures. A heating rate of 2.9°C min^{-1} was employed until the vitrification temperature was reached.

The typical soaking temperature was varied between 1,200 and 1,400°C for "sintering" bond compositions, and 950 and 1,050°C for "fusible" bond compositions in order to simulate industrial firing conditions. The samples were cooled at a rate of 1.8°C min^{-1} to avoid thermal stress fracture. The fired samples were crushed to form a fine powder in preparation for X-ray diffraction.

7.5.5.3 X-Ray Diffraction of Bonding Systems

The dissolution model was compared with experimental data using the X-ray powder diffraction method. X-ray diffraction of the raw materials was performed on a Phillips 1710 X-ray generator with a 40 kV tube voltage and a 30 mA current. Monochromatic Cu kα radiation, $\lambda = 0.154060$ nm, was employed. A scanning speed of 2°/min for diffraction angles of 2θ was used between 2θ angles of 10° and 80°, and the X-ray intensity was recorded using a computer. The spectrum was then analyzed and compared with known spectra. Powder specimens were prepared by crushing in a mortar and pestle in preparation for quantitative X-ray diffraction. To eliminate the requirement of knowing mass absorption coefficients of ceramic samples for quantitative X-ray diffraction, Alexander and Klug [13] introduced the use of an internal standard. First, the ceramic sample is crushed to form a powder – the sizes of particles should be small enough to make extinction and micro-absorption effects negligible. Second, the internal standard to be added should have a mass absorption coefficient at a radiation wavelength such that intensity peaks from the phase(s) being measured are not diminished or amplified. It should be noted that the powder diffraction mixture should be homogeneous on a scale of size smaller than the amount of material exposed to the X-ray beam, and should be free from preferred orientation. The powder bed that is subjected to "X-rays" should be deep enough to give maximum diffracted intensity. The expected equilibrium phases from the fired mixtures are quartz (unreacted and partially dissolved), mullite, cristobalite and glass. However, from the samples tested, the compounds quartz, mullite and glass were successfully detected. A calibration curve was constructed using a suitable internal standard (calcium fluoride), a diluent (glass made by melting potash feldspar), and a synthetic form of the phase(s) to be measured. Synthetic mullite had a purity greater than 99.8%, whilst powdered quartz had a purity greater than 99.84% SiO_2. The method used for quantitative analysis of ceramic powders was developed by Khandelwal and Cook [14]. The internal standard gave a fairly intense (111) reflection ($d = 0.1354$ nm) lying between the (100) reflection for quartz ($d = 0.4257$ nm) and the (200) reflection for mullite ($d = 0.3773$ nm). Using copper kα radiation ($\lambda = 0.15405$ nm), the corresponding values of diffraction angle 2θ are: (100) quartz = 20.82°; (111) calcium fluoride = 28.3°; and (200) mullite = 32.26°. Figure 7.23 shows the calibration curve generated by varying proportions of calcium fluoride, synthetic quartz and mullite. Mass fractions of the crystalline phases in the mixture can be read from the calibration lines by measuring the intensity ratio of the phase(s) to the internal standard.

Figure 7.24 shows the diffraction peaks of interest for quantitative analysis lying between 15° and 40° of the diffraction angle 2θ. The figure shows the reflections of the (111) plane of calcium fluoride, (200) plane of mullite, and the (100) plane of quartz. In order to calculate the mass fractions of quartz and mullite in the mixture, the height of the chosen diffraction peak and its width at half-height were measured from the diffraction spectrum. The product of these two measures were then compared with that of the internal standard, and the resultant intensity ratio was

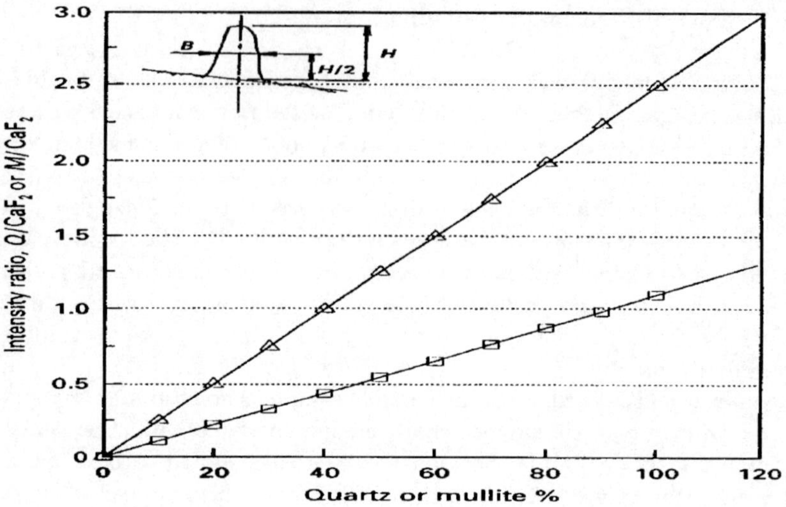

Fig. 7.23 Calibration curve for quantitative analysis of X-ray determined quartz and mullite using the CaF_2 (111) plane generated by the internal standard. Used with permission copyright Springer (2007)

used to find the exact mass fraction of the phase(s) measured in the glass that has been X-rayed.

7.5.5.4 Refractory Bonding Systems

In addition to comparing the experimental results to the dissolution model, results published in the literature were also used to test the accuracy of the model. The composition of the experimental mixtures was matched to those specified by Lundin [7]. Lundin's experimental mixtures were composed of 25 wt.% quartz (13.2 μm particle size), 50 wt.% clay (kaolin), and 25 wt.% flux (potassium feldspar – 25 μm particle size).

The constants A and B for the sintering bonding system were calculated,

$$A = 5.62 \times 10^8 \tag{7.38}$$

$$B = 33,374 \tag{7.39}$$

From which the experimental activation energy, Q, is 132.65 kcal/mole. The residual quartz content for the sintering bonding system is,

$$m_{T,t} = 26.25 \cdot \exp\left[-5.62 \times 10^8 \cdot t^{1/2} \cdot e^{-33,374/T}\right] \tag{7.40}$$

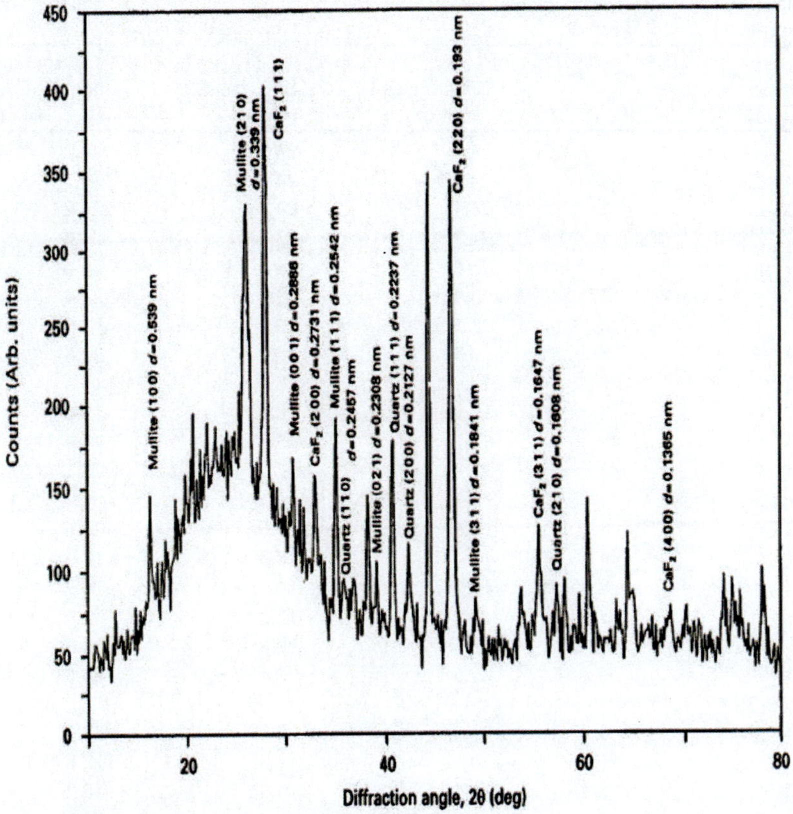

Fig. 7.24 X-ray diffraction spectrum of a vitrified bonding system showing the interplanar distances of crystallographic planes of mullite, quartz and calcium fluoride. Scan rate was 2°/min. Used with permission copyright Springer (2007)

The data comparing Lundin's experimental results, the author's experimental results, and the dissolution model are shown in Table 7.4. When the data are plotted as the logarithm of $(-\ln[m/M]/t^{1/2})$ versus the reciprocal of absolute temperature, 1/T, then all data fits a straight-line relationship. The gradient was calculated to be 33,374, the constant B, using two data points. Lundin's experimental gradient gave a value of 32,962 using the least squares method, and 34,000 for the present work. The corresponding activation energies for both systems are 131 kcal/mole for Lundin's work [7], and 135 kcal/mole for the present work, respectively. Figures 7.25 and 7.26 show the effects of time on residual quartz content at different temperatures according to (7.40) together with comparative experimental data. A comparison was made with dissolution models published in the literature. One of the earliest models was derived by Jander [15]. The equation can be expressed:

Table 7.4 Residual quartz content of a sintering bonding system at various vitrification temperatures

Temp. (°C)	Time (h)	Lundin's exp. result (wt.%)	Exp. result (wt.%)	Jackson and Mills' [12] result (wt.%)
1,200 (1,473 K)	1	24.1	24.2	24.2
1,200	1	24.7	24.3	24.2
1,200	1	26.1	24.8	24.2
1,200	2	23.7	23.8	23.4
1,200	2	23.6	23.9	23.4
1,200*	2	23.4	23.4	23.4
1,200	4	21.3	22.2	22.3
1,200	8	20.3	20.9	20.8
1,200	18	19.0	18.5	18.6
1,200	18	18.9	18.6	18.6
1,200	48	15.2	15.1	14.9
1,250 (1,523K)	1	22.7	22	22.1
1,250*	2	20.6	20.6	20.6
1,250	4	18	18.5	18.6
1,250	8	15.5	16	16.2
1,250	18	12.6	12.5	12.6
1,250	48	8.3	7.8	8.0
1,300 (1,573K)	0.5	22.6	20.4	20.6
1,300	0.5	21	20.9	20.6
1,300	1	20	18.3	18.6
1,300	2	16.1	15.9	16.2
1,300	4	13.4	12.8	13.2
1,300	8	10	9.7	9.9
1,300	18	5.9	5.8	6.1
1,300	50	1.6	1.8	2.3
1,300	120	0.3	0.2	0.6

Lundin's [7] experimental data are compared with the author's experimental data and the model [12]

The asterisk indicates values used for deriving the constants used in the theoretical model

$$\left(1 - \sqrt[3]{1 - Z}\right)^2 = \left\{\frac{C_1 \cdot D}{r^2}\right\} \cdot t \tag{7.41}$$

where Z is the volume of quartz that has been dissolved, r is the original particle radius, and D is the diffusion coefficient for the diffusing species. This equation can be transformed into mass fractions using Archimedes' law, thus,

$$\left(1 - \sqrt[3]{\frac{m}{M}}\right)^2 = C_2 \cdot t \tag{7.42}$$

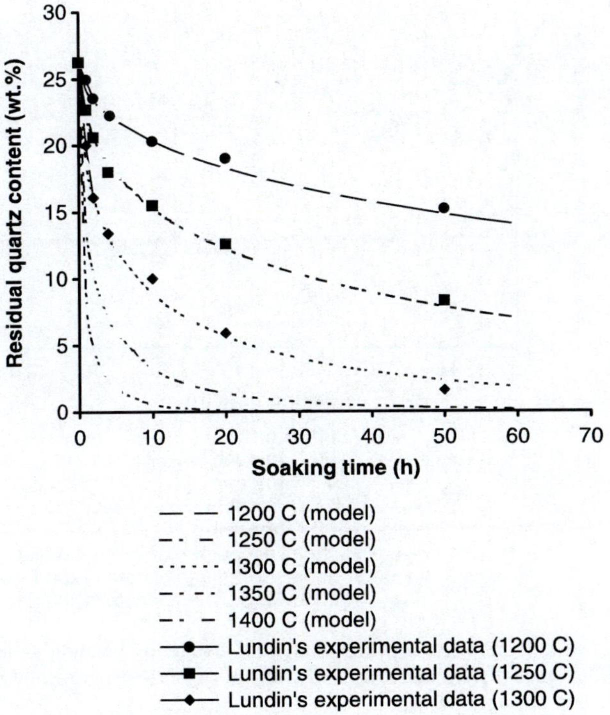

Fig. 7.25 Effect of time on residual quartz content of a sintering bonding system according to Jackson and Mills's model [12], and compared with Lundin's experimental data [7]. Used with permission copyright Springer (2007)

where C is a constant dependent on soaking temperature and initial particle size of quartz. Krause and Keetman [16] expressed the dissolution of quartz as a function of isothermal firing time, viz,

$$M - m = C_3 \cdot \ln t \tag{7.43}$$

where M is the initial quartz content, and m is the residual quartz content after time,t. The unit of time here is seconds such that after 1 s of firing the residual quartz content is equal to the initial quartz content. Monshi's dissolution model [17] can be transformed into the following equation assuming isothermal firing conditions:

$$\ln\left\{\frac{m}{M}\right\} = -C_6\sqrt{t} \tag{7.44}$$

Jackson and Mills' model [12] for isothermal firing conditions is transformed into:

$$\ln\left\{\frac{m}{\gamma \cdot M}\right\} = -C_7\sqrt{t} \tag{7.45}$$

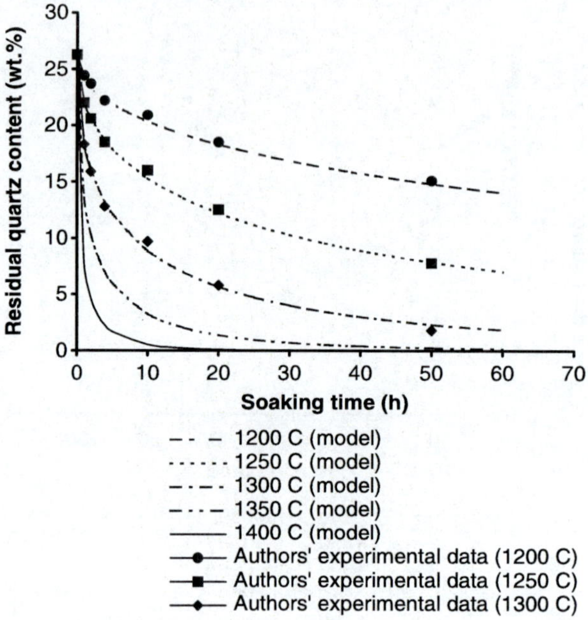

Fig. 7.26 Effect of time on residual quartz content of a sintering bonding system according to Jackson and Mills's model [12] and compared with the authors' experimental data. Used with permission copyright Springer (2007)

where γ is the ratio of densities of β- and α-quartz. Constants for all the equations presented here are calculated using quartz mass fraction data after 18 h' firing. The constants are dimensioned in seconds. The equations shown were compared with experimental data generated by Lundin [7] for a clay-based material containing 40 wt.% kaolin, 40 wt.% quartz, and 20 wt.% feldspar. According to the transformed equations, the mass fraction of quartz can be calculated as follows,

Jander's model [15]

$$m = 41.9 \cdot \left(1 - \left\{1.55 \times 10^{-6} \cdot t\right\}\right)^{3/2} \tag{7.46}$$

Krause and Keetman's model [16]

$$m = 41.9 - (2.58 \cdot \ln t) \tag{7.47}$$

Monshi's model [17]

$$m = 41.9 \cdot e^{-4.5 \times 10^{-3} \sqrt{t}} \tag{7.48}$$

Jackson and Mills' model [12]

Table. 7.5 Residual quartz content for different soaking times at 1,300°C for a sintering bonding system composed of 40 wt.% kaolin, 40 wt.% quartz, and 20 wt.% feldspar (Lundin's mixture number M21 [7]) compared with other dissolution models

Time (h)	Lundin's exp. data [7]	Jander [15]	Krause and Keetman [16]	Monshi [17]	Jackson and Mills [12]
0	41.9	41.9	0.00	41.9	41.9
0.5	35.9	41.72	22.55	34.61	34.76
1	32.8	41.54	20.76	31.97	32.12
2	29.2	41.19	18.97	28.58	28.72
4	23.2	40.49	17.18	24.39	24.51
8	19.5	39.11	15.39	19.49	19.59
18	13.3	35.72	13.30	13.30	13.36
24	10.7	33.74	12.56	11.13	11.19
48	6.9	26.18	10.77	6.43	6.51
120	3.6	7.85	8.96	2.17	2.17
190	2.7	0.00	7.22	1.00	1.01
258	2.0	0.00	6.43	0.54	0.55

$$m = 41.73 \cdot e^{-4.5\times 10^{-3}\sqrt{t}} \tag{7.49}$$

The transformed equations are then tested using data provided by Lundin [7]. Referring to Table 7.5, it can be shown that the mass fraction of quartz obtained using the equations derived by Jander [15] and Krause and Keetman [16] did not agree with Lundin's experimental results [7].

The results obtained using Monshi's model [16] are in much better agreement compared to Lundin's data. However, the results obtained using Jackson and Mills' model [11] is more accurate at predicting the mass fraction of quartz remaining owing to the differences in the density of quartz. After long periods of heat treatment, the model predicts lower magnitudes of mass fractions of quartz when compared to Lundin's experimental results [6].

7.5.5.5 Fusible Bonding Systems

The constants, A and B, for the fusible bonding system were determined using time and temperature couples at 2 and 10 h and were calculated to be, -5.2×10^8 and $-33{,}205$, respectively. The dissolution equation then becomes,

$$m_{T,t} = 10.06 \exp\left[-5.2 \times 10^8 t^{1/2} \cdot e^{-33{,}205/T}\right] \tag{7.50}$$

Equation (7.50) is used to compare the experimentally determined mass fraction of quartz remaining after heat treatment with the predicted values. The calculated mass fraction of quartz remaining after a period of heat treatment is calculated using the

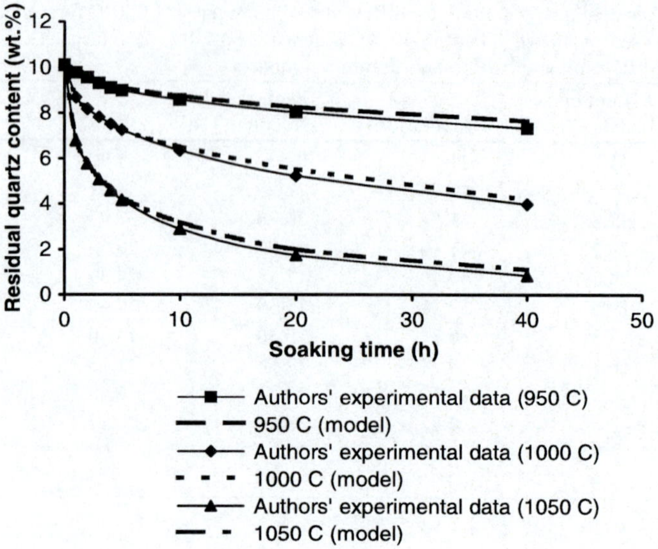

Fig. 7.27 Effect of time on residual quartz content of a fusible bonding system according to Jackson and Mills's model [12] and compared with the author's experimental data. Used with permission copyright Springer (2007)

equation derived by Jackson and Mills [12]. The results of the dissolution model compare well with the experimental data over short periods of time. However, over longer periods of heat treatment the model tends to become less accurate (Fig. 7.22). A comparison was made with published dissolution models. The equations shown were compared with experimental data at 1,050°C. The equations shown were compared with experimental data at 1,050°C. According to the transformed equations, the mass fraction of quartz can be calculated as follows (Fig. 7.27):

Jander's model [15]

$$m = 10.1.\left(1 - \left\{3.44 \times 10^{-6} \cdot t\right\}\right)^{3/2} \tag{7.51}$$

Krause and Keetman's model [16]

$$m = 10.1 - (0.59 \cdot \ln t) \tag{7.52}$$

Monshi's model [17]

$$m = 10.1 \cdot e^{-6.4 \times 10^{-3} \sqrt{t}} \tag{7.53}$$

Jackson and Mills' model [12]

Table 7.6 Residual quartz content for different soaking times at 1,050°C for a fusible bonding system compared with other dissolution models

Time (h)	Exp. data	Jander model [14]	Krause and Keetman model [15]	Monshi's model [16]	Jackson and Mills' model [11]
0	10.1	10.1	0	10.1	10.1
1	6.84	9.91	5.23	6.88	6.86
2	5.79	9.72	4.82	5.87	5.86
3	5.13	9.54	4.58	5.21	5.19
4	4.7	9.36	4.41	4.7	4.68
5	4.28	9.18	4.28	4.28	4.28
10	3.2	8.28	3.87	2.99	3
20	2	6.6	3.46	1.81	1.82
40	1.1	3.62	3.04	0.89	0.89

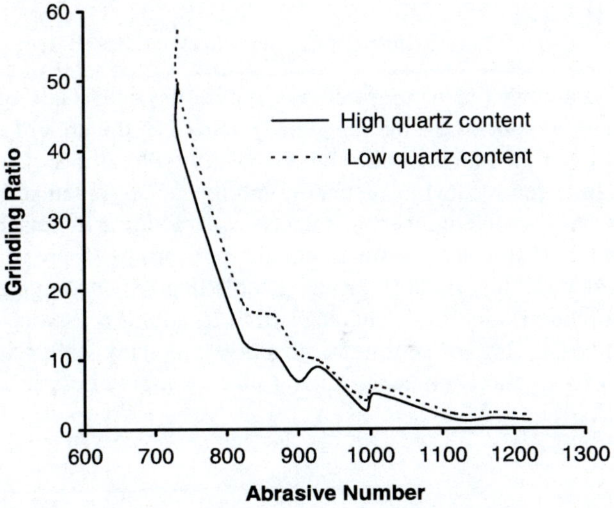

Fig. 7.28 Effect of the abrasive number on the grinding ratio for a high-quartz content and a low quartz-content bonding system. Used with permission copyright Springer (2007)

$$m = 10.06 \cdot e^{-6.37\times10^{-3}\sqrt{t}} \tag{7.54}$$

With reference to Table 7.6, it can be shown that the mass fraction of quartz obtained using the equations derived by Jander [15] and Krause and Keetman [16], did not agree with the experimental results at 1,050°C. The results obtained using Monshi's model [17] are in much better agreement compared to the experimental data. However, the results obtained from Jackson and Mills's model are more accurate at predicting the mass fraction of quartz remaining owing to the differences in the density of quartz. After long periods of heat treatment, the model predicts slightly lower magnitudes of mass fractions of quartz when compared to the experimental results.

The dissolution of quartz during heat treatment has a significant effect on the wear of vitrified grinding wheels. Figure 7.28 shows the effect of using a high and a low quartz content bonding system on the wear of vitrified corundum grinding wheels grinding a large number of tool steel materials [18]. The classification of tool steels is in the form of an abrasive hardness number, which is a weighted average of the number of carbides contained within the tool material. As shown in Fig. 7.28, the grinding ratio, or G-ratio, is a measure of the efficiency of the grinding wheel. It is the quotient of the volume of workpiece material removed and the volume of the wheel material removed. The figure demonstrates the effectiveness of reducing the quartz content of the bonding system of porous nanogrinding tools.

7.6 Conclusions

This chapter has shown that nanogrinding is in its embryonic stages of development and a fundamental understanding has already started in the area of computational analysis. However, the real breakthrough discovery will in the area of the development of machine tools capable of nanogrinding and its practical application to machining a variety of engineering materials in addition to machining optical materials and materials used in semiconductor applications. There is also a requirement to manufacture high quality grinding tools that will be required to continuously maintain form and shape characteristics in addition to very small surface roughness values. A form of continuously dressed, or sharpened, processes will be required in order to make the nanogrinding process highly desirable to micro and nanomachinists.

References

1. Rentsch R (2009) 'Nanoscale Cutting', published in 'Nano and Micromachining', Edited by Davim, J. P., and Jackson, M. J., ISTE-Wiley Publishers, London, pp. 1–26.
2. Zhao Q, Dong S, and Sun T (2001) High Technol. Lett., **7**, (3), 84–89.
3. Ahao QL, Dong S, and Sun T (2001) Key Eng. Mater., **202–203**, 315–350.
4. Jackson MJ (2001) J. Manuf. Process., **3**, 17–28.
5. Timoshenko SP, and Goodier JN (1970) Theory of Elasticity, 3rd Edition – International Student Edition, McGraw-Hill Kogakusha Ltd., Tokyo, pp. 109–113, and pp. 139–144.
6. Loladze TN (1967) Requirements of tool materials, Proceedings of the Eighth International Machine Tool Design and Research Conference, Pergamon Press, pp. 821–842.
7. Lundin ST (1959) Studies on Triaxial Whiteware Bodies, Almqvist and Wiksell, Stockholm, Sweden.
8. Storch W, Ruf H, and Scholze H (1984) Ber. Deut. Keram. Ges., **61**, 325.
9. Binns E (1962) Sci. Ceram., **1**, 315.
10. Ford WF, and White J (1951) Trans. J. Br. Ceram. Soc., **50**, 461.

11. Kirchoff G, Pompe W, and Bahr HA (1982) J. Mater. Sci., **17**, 2809.
12. Jackson MJ, and Mills B (1997) J. Mater. Sci., **32**, 5295–5304.
13. Alexander IE, and Klug HP (1948) Anal. Chem., **20**, 886.
14. Khandelwal SK, and Cook RL (1970) Am. Ceram. Soc. Bull., **49**, 522–526.
15. Jander W (1927) Z. Anorg. U. Allgem. Chem., **163**, 1–30.
16. Krause P, and Keetman E (1936) Sprechsaal, **69**, 45–47.
17. Monshi A (1990) Investigation into the strength of whiteware bodies, Ph.D. Thesis, University of Sheffield, United Kingdom.
18. Jackson MJ (December 1995) A study of vitreous-bonded abrasive materials, Ph.D. Thesis, Liverpool John Moores University, United Kingdom.

Chapter 8
Polishing Using Flexible Abrasive Tools and Loose Abrasives

Han Huang, Libo Zhou, and Ling Yin

Abstract This chapter presents the recent developments of polishing technologies which are relevant to practical applications, including robotic polishing of aerospace parts, mirror surface finishing of brittle solids, and microbore and free-form surface polishing. The polishing processes investigated involve the mechanical interactions between different work surfaces and flexible abrasive tools, such as belts and films, free abrasives and abrasive slurry flow. The polished products are of very different geometries as well, varying from three-dimensional free-form surfaces to inner wall surfaces of bores of several hundred microns in diameter. A variety of materials are concerned, including difficult-to-machine super alloys, hard and brittle solids and soft and ductile metals. The common characteristics of these polishing processes is that they are all used for the finishing of industrial products without making significant change in the product forms or profiles.

Keywords Abrasives · Polishing · Grinding.

8.1 Introduction

Polishing is one of the finishing technologies that have been extensively used in manufacturing industries [1–5]. It is probably the oldest manufacturing technology, with applications dating back to Neolithic man [6]. Polishing processes involve the mechanical interaction of the workpiece surface and the polishing media (mainly consisting of abrasive particles). In most of the cases, polishing also involves a third

H. Huang (✉)
School of Mechanical and Mining Engineering, The University of Queensland, St Lucia, QLD 4072, Australia
e-mail: han.huang@uq.edu.au

M.J. Jackson and J.P. Davim (eds.), *Machining with Abrasives*,
DOI 10.1007/978-1-4419-7302-3_8, © Springer Science+Business Media, LLC 2011

element: the surface of polishing pad. Surface finish and subsurface damage after polishing are influenced by the process parameters, such as applied pressure or load, relative speed between the workpiece surface and the polishing media, work material, pad shape and mechanical properties and polishing environment. Polished surfaces often have apparent merits, such as isotropic surface characteristics and strength, low subsurface damage, high bearing ratio due to uniform roughness profile and low residual stress. In the extensive literature published, polishing has always been used for the production of low surface roughnesses without the generation of product form [5].

This chapter provides a summary of our recent developments of polishing technologies relevant to practical applications. The polishing processes developed were used for the finishing of industrial products without the significant change of the product forms or profiles, but not necessarily producing extremely low surface roughness. So hopefully the topics presented in this chapter can extend our conventional knowledge of polishing. In this chapter, we first report the robotic polishing of backing strips of honeycomb seals and free-form turbine airfoils using abrasive belts for aerospace industries. We then introduce the development of an efficient polishing process for fabricating optic fibre end faces using a combination of abrasive films and polishing suspensions, which requires to produce very low surface roughness. The final section will describe the polishing processes with free abrasives and abrasive slurries for polishing small-diameter holes and a free-form industrial component.

8.2 Polishing with Flexible Abrasive Tools

8.2.1 Robotic Polishing of Aerospace Components Using Abrasive Belts

High-pressure turbine vanes and honeycomb seals are two of the core components in a turbine engine, as shown in Fig. 8.1. After a certain period of service at high-temperature and high-pressure environment, the components are severely worn and distorted due to heat fatigue. These components thus need to be repaired as replacement is far more expensive. During overhauling of a turbine vane, the defective areas on airfoil are covered with braze material (Fig. 8.1a) and skilled workers are required to remove the excessive braze material on the airfoil using abrasive belt polishing to restore the required airfoil profile, as shown in Fig. 8.2. Similarly, in the overhauling of honeycomb seals, worn honeycomb must be chiseled off from a backing strip (see Fig. 8.1b) and belt polishing is subsequently used to remove the honeycomb residues on the back strip to achieve the required surface finish for welding new honeycomb parts. The polishing operations using abrasive belts are both labor intensive, time consuming and quality inconsistent. Therefore, the automation of polishing process is of great importance to reduce the cost and to improve the quality.

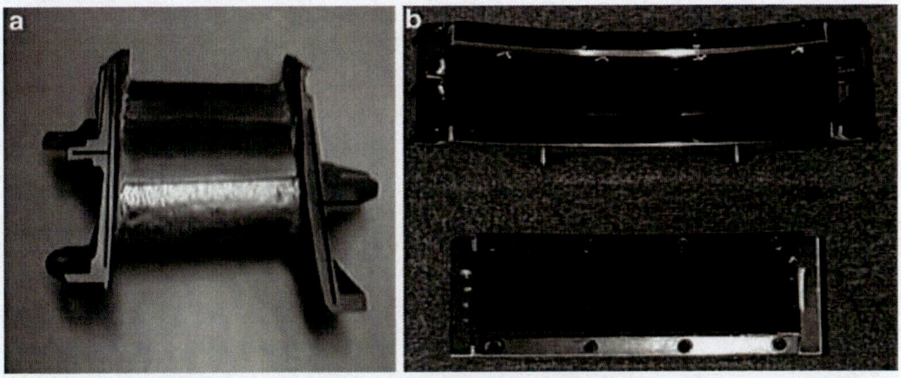

Fig. 8.1 (**a**) High-pressure turbine vanes after brazing; (**b**) backing strips of honeycomb seals with honeycomb residues on

Fig. 8.2 Manual belt polishing of turbine vane

Industrial robots have been extensively used to replace human operators in machining processes. The applications of the robots include deburring [7–11], chamfering [11, 12] and finishing [13–15] of components that have constant geometry. As compared with conventional robotic applications, such as material handling and welding, robotic machining systems for chamfering, deburring and finishing need force control with the aid of position control, which is often named as hybrid or compliance control [7, 9]. In hybrid position/force control systems, a constant cutting force must be maintained by adjusting the position of the robot handling mechanism normal to the part edge, or by adjusting the feed rate of the cutting tool tangential to the edge. The positioning accuracy can also be independently controlled by an active end-effector that is used to perform the position or feed rate adjustments [9, 13]. The hybrid position/force control method leads to significant improvement of robustness and position accuracy for deburring and chamfering robotic systems, in comparison to the non-adaptive systems. In the case of turbine vane overhauling [16], a consistent contact force is no longer the guarantee for avoiding over-cutting or under-cutting as various material removal rates are required to machine different areas. As compared with machining of components with well defined geometries, a major difficulty that is encountered in the turbine vane overhauling is that the part profile is severely distorted. This means that the geometry of vanes differs significantly from each other. Conventional methods like teaching-and-playing and programming-and-cutting no longer meet the positioning requirements. Automated task planning becomes essential. The 3D profile of the vane airfoil makes the position control even more difficult. The solution to this problem could be to use a 3D shape measurement sensor for compensating the part tolerances [17]. In this section, the robotic polishing processes developed for 3D turbine vanes [16, 17] and 2D honeycomb backing strips (which can be considered to have well defined geometry) [15, 16] will be introduced.

8.2.1.1 Work Materials and Polishing Requirements

The components in turbine engines use superalloys to meet the normal aerospace engineering requirements, i.e. high strength-to-weight ratio, high fatigue resistance, high corrosion resistance and superior high-temperature strength. Turbine vanes and honeycomb seals are often made of Inconel materials. These materials have poor machinability that is long recognized by manufacturers. The chemical composition of the base and braze materials of the vanes is shown in Table 8.1. The material mainly consists of cobalt, chromium and nickel. As mentioned in the last section, the turbine vane airfoil has a three-dimensional hollow structure, as shown in Fig. 8.3. The parts to be repaired are also distorted, compared with their original design. This makes a great challenge for robot machining as tool path must be flexible or varied part by part. Table 8.2 shows the quality requirements for the vane overhaul. A crucial requirement is that the over-cutting and under-cutting must be smaller than 0.1 mm, almost reaching the accuracy limit of the finishing robotic systems. It should be noted that the wall thickness ranges from 0.8 to 2 mm, which is quite small compared with the over- and under-cutting requirements. The complex 3D form of the airfoil also makes the processing extremely difficult.

Table 8.1 Compositions of main elements in a jet engine vane

Element	Cobalt	Chromium	Nickel	Tungsten	Tantalum	Carbon
Base (wt.%)	55.7	23.0	10.0	7.0	3.5	0.6
Braze (wt.%)	45.6	23.8	25.0	3.5	1.8	0.3

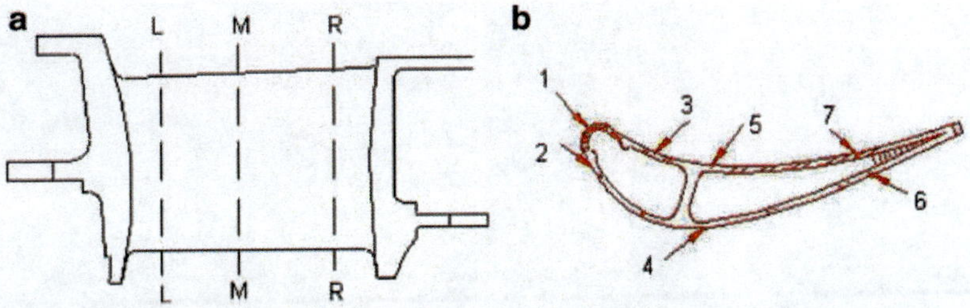

Fig. 8.3 (**a**) Turbine vane and (**b**) cross-sectional view of vane airfoil

Table 8.2 Quality requirements of jet engine vane repair [18]

Profile smoothness	Polished airfoil surface must be visually smooth and free of abrupt variations, transition lines and stubble dimples
Surface roughness	Average surface roughness, R_a, must not be greater than 1.6 μm
Dimension accuracy	Wall thickness of the airfoil should meet the minimum wall thickness requirements at the check point in three sections (LL, MM and RR), as shown in Fig. 8.3 and over-cutting and under-cutting must be smaller than 0.1 mm

8.2.1.2 Systems for Robotic Polishing

Robotic systems for polishing vanes and honeycomb backing strips consist of a six-axis finishing robot, a robot controller, an index table and passive compliance polishing stations, as shown in Fig. 8.4. The finishing robot, the passive compliant tools and the index table are classified as the device category. The system also has various sensors, such as an in situ profile measurement sensor for measuring airfoil profiles, pneumatic sensors for detecting part jamming in a self-aligned end effector, a laser through-beam sensor for confirming proper gripping, inter-lock and belt breaking detectors. The robot controller implements the device control functions, which controls all hardware devices and sensors. For the system for polishing 3D vane airfoils in Fig. 8.4b, host personal computers are needed as the robot controller is not suitable for the fast development of tool paths for polishing. The vane polishing systems is thus more complicated, requiring different levels of control, including device (robot) control and supervisory (host computer) control. For the PC-based host controller provides supervisory control for the entire system, housing all intelligent automation software modules. An interface between the host computer and the robot controller is also needed. The host computer is interfaced to the robot controller through digital I/Os. The data communication between the host computer and the robot controller is relayed by the interface computer that is running MS-DOS through RS 232 serial line in the manufacture's proprietary protocol. The host and interface computers are connected using an Ethernet LAN (local area networking).

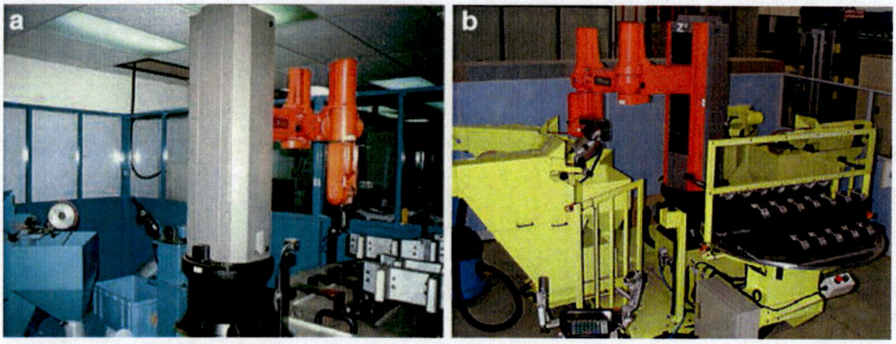

Fig. 8.4 Robotic polishing system for (**a**) honeycomb backing strips and (**b**) turbine engine vanes

During a typical working cycle, the robot hand picks up a part from the index table and moves to the station for polishing. The tool path for polishing the backing strips is pre-determined according to the original template. The strips are mounted firmly on a holding template with the same circular curvature to minimize the effect of part distortion, as can be seen from Fig. 8.1b. The polishing of the strips is thus two dimensional. The development of tool path for the turbine vane airfoil is fairly challenging as the airfoil has a three-dimensional free form. As shown in Fig. 8.3a, the geometries of the airfoil sections at LL, MM and RR are different. Due to the severe part distortion, the tool path must be varied part by part. In this work, a LVDT station is used to measure airfoil profile. The measured airfoil profile data are transported to the host computer for tool path planning. After achieving the tool path, the robot hand carries the turbine vane to the two tool stations for rough and fine polishing, respectively. All the operations work under a complicated multi-tasking schedule, which allows the robot running at a full strength. For example, when the host computer is planning the tool path for the first vane, the robot hand will be carrying the second vane for profile measurement. The processes for vane airfoil and backing strip both include rough and fine polishing procedures. Rough polishing aims to remove the excess braze material or honeycomb residue to reach the dimensional requirements. Fine polishing is to achieve the required surface roughness.

8.2.1.3 Part Gripper and Polishing Head

As the honeycomb backing strips are distorted, it is difficult to use one end-effector to grip all the parts. The fixtures for holding backing strips are thus separated from the gripper. Using this concept design, any types of backing strips can be machined using the robotic system. As the gripper and fixtures are precisely matched, the gripping and positioning are thus error-free. Its repeatability was tested over 300 times of loading and unloading of components [14].

For the turbine vane polishing, the end-effector and gripper is not separated [19]. However, its open-and-close mechanism is specially tailored for the vane. As shown

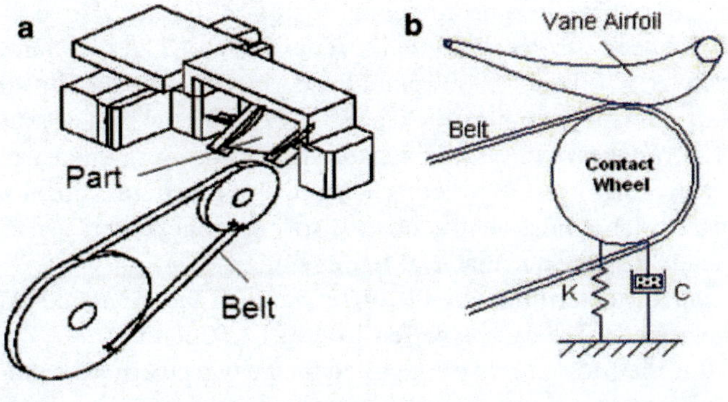

Fig. 8.5 (**a**) Part gripper and (**b**) polishing head with passive force control [16]

in Fig. 8.5a, the vane is gripped by a rigid semi-open rectangular frame and the part can be rotated during machining. The design optimizes the flexibility and operationality and demonstrates to have high reliability on workpiece loading and unloading. The polishing heads used in the two robotic systems are similar. A typical polishing head is schematically shown in Fig. 8.5b, where the contact wheel is suspended on a spring-damper system. The contact force between the belt and the vane can be expressed as follows

$$F \approx k(\Delta l_0 + \Delta l) + c\frac{\Delta l}{\Delta t} \tag{8.1}$$

where F is the force, k the spring stiffness, Δl_0 the pre-set compressed spring length, Δl the compressed or extended length of spring due to the variation in removal amount, c the viscous damping coefficient and Δt the time step.

During polishing, when the layer of the strip or vane to be removed at the contact location is thicker than the desired, the contact wheel and the spring will be pressed down, resulting in a greater force, and hence a higher removal amount. The sensitivity of the contact force to the removal amount can be adjusted by changing the spring stiffness and the pre-load on the spring. By adjusting the damping coefficient, the abrupt changes on the polished profile and the vibration generated during polishing can be attenuated.

8.2.1.4 Robotic Polishing Processes

Process Parameters

Abrasive belt polishing is a complicated material removal process, involving a number of process parameters. The surface quality required on the machined parts and the cycle time defined for the robotic polishing process can only be achieved with the comprehensive understanding of effects of the process parameters. The important process parameters for robotic polishing are summarised below.

- *Belt speed.* It is proportional to the speed of the motor that drives the belt. A faster belt speed results in a greater removal rate. The recommended motor speed is between 700 and 1,800 rpm considering the belt stabilization.
- *Feed rate.* It refers to the tangential speed at which the workpiece moves against the belt. The robot provides an adjustable factor for users to vary the feed rate. Different feed rates were applied when polishing different local areas of the vanes, but no such adjustment is needed for the honeycomb seals. The slower feed rate leads to a greater material removal rate.
- *Depth of cut.* It is determined by the offset value of the robot end-effector along z-axis. The offset value can be varied from 0.1 to 0.5 mm.
- *Pre-load.* It is the spring force pre-applied on the workpiece. The pre-load can be adjusted by changing the spring coefficient and its pre-elongated length. A greater load produces a greater removal rate.
- *Abrasive belt.* The commonly used abrasive belts for polishing superalloys are manufactured by Norton. Nortan structural abrasive belts normally have a better performance and longer tool life than the Nortan conventional belts. However, for the cost saving purpose, the Norton conventional abrasive belts were chosen for polishing backing strips and turbine vanes. The abrasives of Norton belts are made from ceramic aluminum oxide, specially designed for aerospace alloys. The mesh size of abrasives used ranges from 40 to 120, normally.

Preliminary polishing experiments designed using the "Taguchi" method [20] were carried out to explore the level of importance of the polishing parameters. Figure 8.6 shows the effect of process parameters on the polishing performance. It is seen that the material volume required to be removed has the most important effect on the polishing process, followed by the belt wear and pre-loaded contact force. Removal volume is determined by the status of part repair and polishing time depends on the cycle time requirements, which are not controllable. So is belt wear. Among the three adjustable or controllable parameters, i.e. pre-load, belt speed and feed rate, the feed rate appears to be least influenced process parameter.

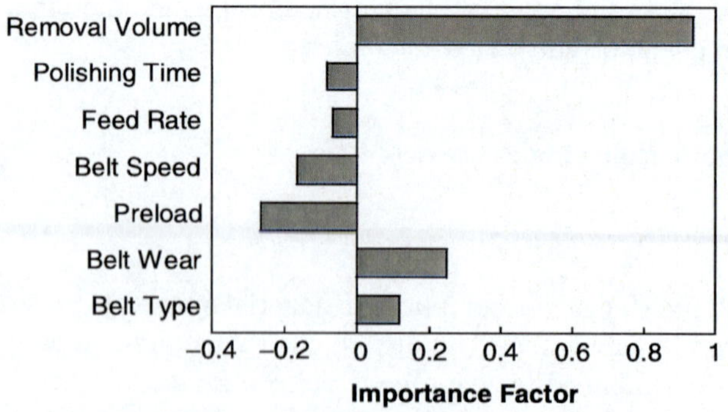

Fig. 8.6 Importance level of process parameters in robotic polishing

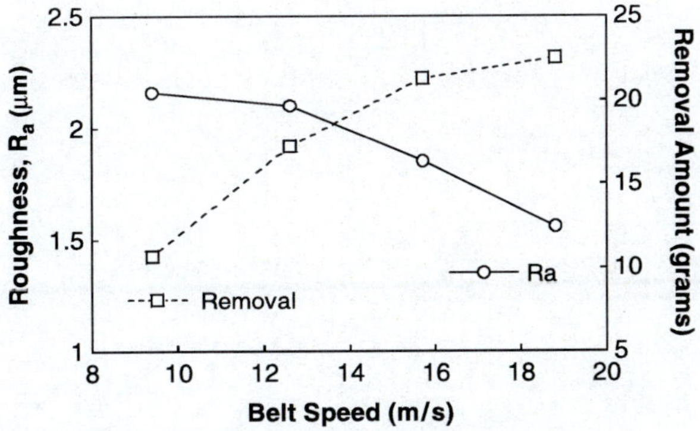

Fig. 8.7 Roughness and removal amount per pass plotted as a function of belt speed in vane polishing. Feed rate = 1.2 m/min, preload = 19 kg

Material Removal and Surface Finish

The effects of the three adjustable polishing parameters on the material removal and surface quality were experimentally obtained. Figure 8.7 shows the effect of belt speed on the material removal amount and surface roughness when polishing turbine vanes. It is seen that the removal amount increases with the increase in belt speed, but the surface roughness decreases with the rising speed. Similar to a conventional grinding process, increasing the speed of abrasive belt will increase the number of cutting points in a unit time, and thus increase the removal rate. The surface finish is improved too as higher speed reduces chip thickness. However, it should be noted that the increase in belt speed can generate more heat. Excessive heat would damage the airfoil surface. In addition, the high belt speed can cause some extents of vibration, which in return influences the surface quality.

Figure 8.8 shows the effect of feed rate on the removal amount and surface roughness. Different from the machining process with position control, in a belt polishing process, the smaller the feed rate is, the more material amount is removed. A faster feed rate decreases the surface roughness, or produces parts with better surface finish.

The effect of pre-load on removal rate and roughness is shown in Fig. 8.9. As expected, the greater force is applied on the polishing head, the more material is removed. The roughness of the polished surface is increased with the increasing pre-load. Apparently, with the greater load applied on the belt, the depth of cut is increased, resulting in higher removal and thus worse surface finish. Again, the increase of the pre-load for pursuing a higher productivity is restricted because excessive heat will be produced.

Abrasive belts with mesh sizes of 40, 60 and 80 were tested. The greater grit size increases the removal rate, but worsen the surface finish. As a result, a longer polishing time is required to meet the roughness requirement, leading to an increased cycle time. The optimal selection is the mesh size 80 for vane airfoils and mesh size 60 for backing strips in terms of surface finish and removal rate.

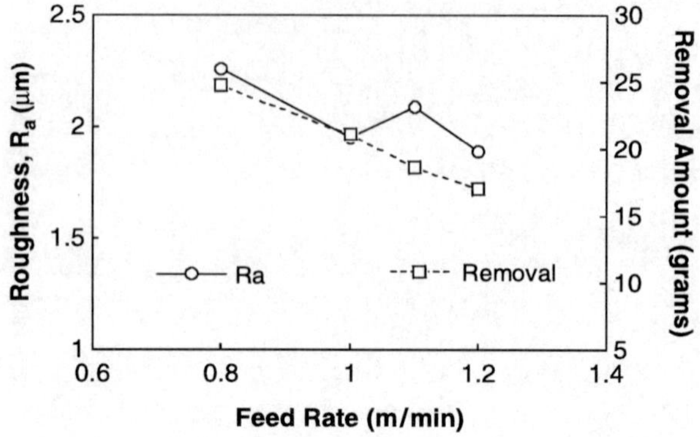

Fig. 8.8 Roughness and removal amount per pass plotted as a function of feed rate in vane polishing. Belt speed = 12.6 m/s, preload = 19 kg

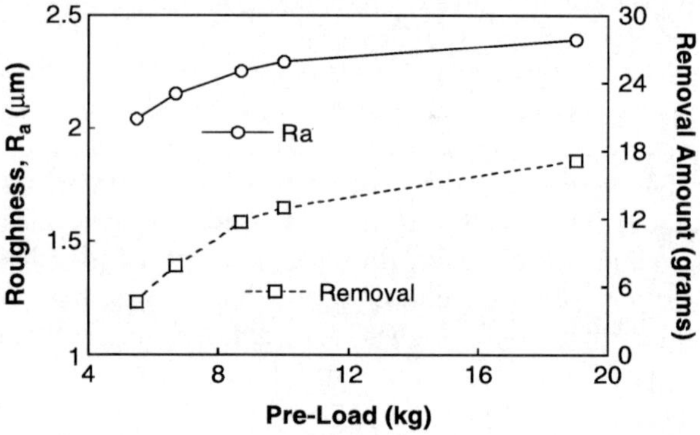

Fig. 8.9 Roughness and removal amount per pass plotted as a function of preload in vane polishing. Belt speed = 12.6 m/s, feed rate = 1.2 m/min

There always exists contradiction between removal rate and surface quality. In order to achieve a higher removal rate (implying a shorter cycle time), greater depth of cut, feed rate, pre-load and reasonably larger grit size should be used. This, at the same time, increases surface roughness. Reducing the mesh size, the depth of cut and the applied load can improve the surface roughness, but will significantly affect the removal rate or cycle time.

Tool Wear and Compensation

The wear status of an abrasive belt (Norton R981 #80) on the removal rate is shown in Fig. 8.10. The new belt with fresh sharp cutting edges achieves a much greater removal rate than the used one. After machining three workpieces, the cutting performance of the belt is significantly reduced. It is not surprising to see from

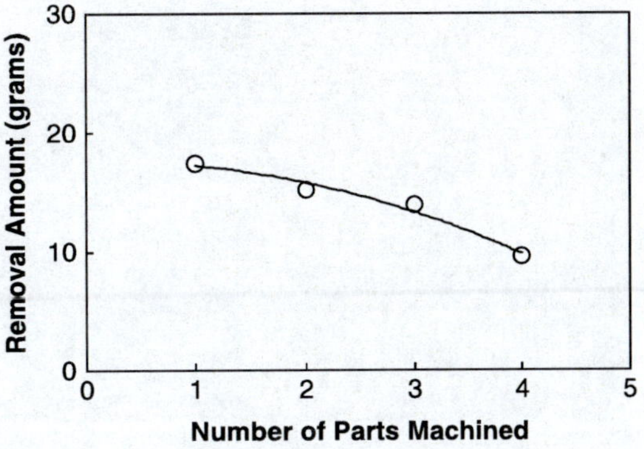

Fig. 8.10 Effect of number of vanes machined an abrasive belt on removal amount per pass. Belt speed = 12.6 m/s, feed rate = 1.2 m/min, preload = 19 kg

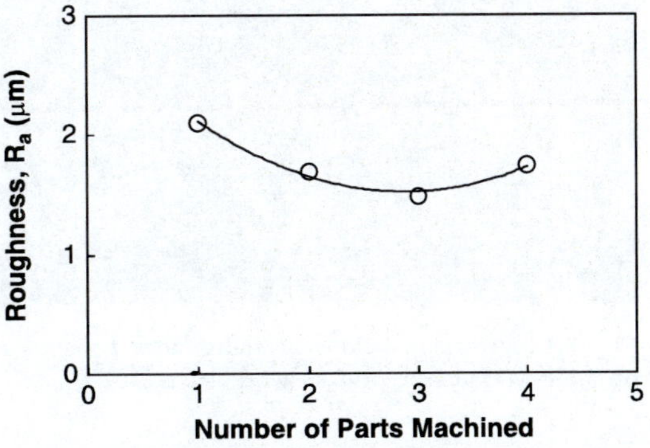

Fig. 8.11 Effect of number of vanes machined an abrasive belt on roughness. Belt speed = 12.6 m/s, feed rate = 1.2 m/min, preload = 19 kg

Fig. 8.11 that the surface finish is improved when the number of vanes machined by an abrasive belt is increased or the belt is getting duller. When the sharp edges of abrasives disappears, the depth of cut is decreased, leading to a lower value of R_a.

The belt wear was also examined using scanning electron microscope (Philips 430). Figure 8.12 shows the topographies of the abrasive belt that has polished one to four pieces. The wear of the abrasive grain was developed after machined one vane (Fig. 8.12a) and increased with the further polishing. After machined three vanes the plateau was built up. Grain pulling-off was observed on the belt after polishing of four workpieces. This corresponds to the drop of the material removal in Fig. 8.10 as a result of losing some effective cutting grains. At the same time, excessive heat may be generated. The belt should be replaced at this situation.

To save the machining cost and machine down-time, an abrasive belt should be used to polish workpieces as more as possible, but this is compromised by the requirement of a constant removal rate. In other words, a constant material removal rate has to be maintained during the course of polishing. As a consequence, the belt wear must be compensated. The compensation can be made by varying abrasive

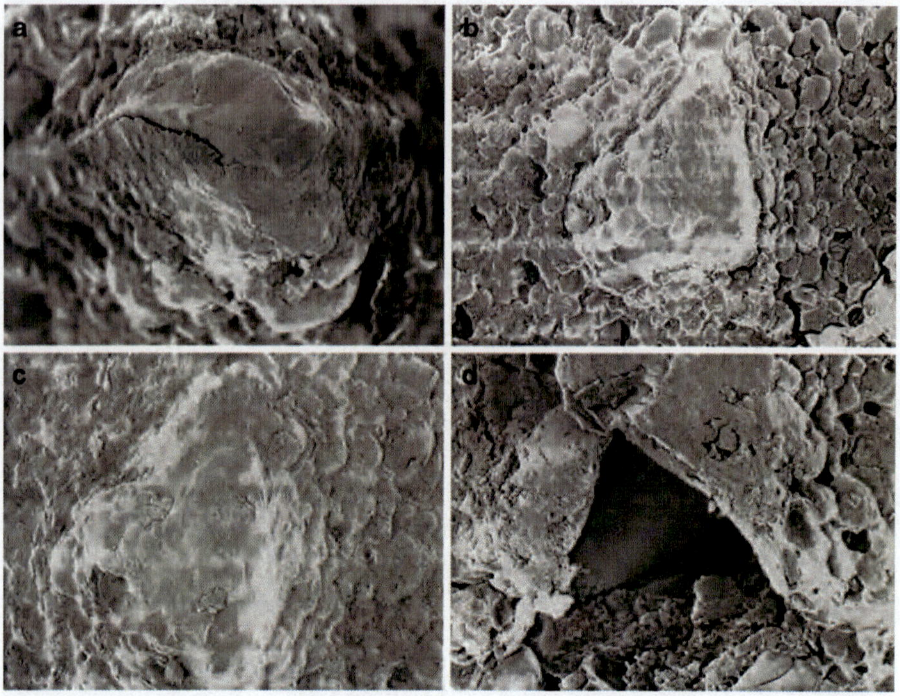

Fig. 8.12 Abrasive belt wear status after polished (**a**) one workpiece, (**b**) two workpieces, (**c**) three workpieces, (**d**) four workpieces [16]

Table 8.3 Optimized process parameters for polishing engine components.

	Part no.	1	2	3	4	5	6
Backing strip	Belt speed (m/s)	9.42	11.51	14.13	14.13	15.70	15.70
(Norton #60)	Feed rate (m/min)	0.72	0.61	0.58	0.54	0.50	0.50
Vane airfoil	Belt speed (m/s)	9.84	14.45	16.13	10.47	12.57	
(Norton #80)	Feed rate (m/min)	1.20	1.13	1.08	1.20	1.20	

belt speed and feed rate, according to the results shown in Figs. 8.7 and 8.8, as these two parameters can be adjusted in the robot programing. Pre-load can not be adjusted because its change is manually manipulated. Table 8.3 summarises the optimal belt speed and feed rate used for compensating tool wear in the polishing of backing strips and vane airfoils. The belt speed is increased and the feed rate is decreased to compensate the belt wear, aiming at maintaining a constant removal rate. It should be noted that to meet the surface finish requirement, a fine polishing process is arranged after the rough polishing of vane airfoils. The same passive compliance tool and compensation method are applied. The values underlined in Table 8.3 are for the fine polishing procedure, which has a much smaller removal rate than that for rough polishing. Figure 8.13 shows the results of surface

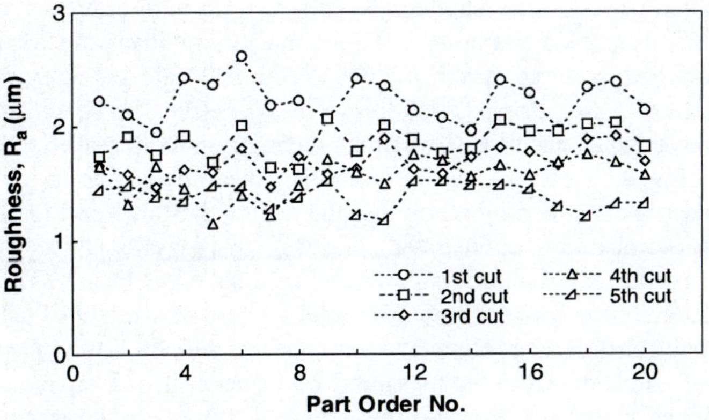

Fig. 8.13 Effect of polishing pass (or cut) of an abrasive belt on surface roughness of vane airfoils. A new belt is used for each vane

roughness for each abrasive belt under different cuts using the conditions in Table 8.3. Though the removal rate is kept unchanged for the first three cuts, but the surface roughness decreases with the increased cutting times.

8.2.1.5 Process Optimization and Quality Assurance

The selection of suitable abrasive belt for fine polishing of vane airfoil is critical for achieving the required surface finish. The mesh size for polishing is supposed to be finer than that for rough grinding in order to obtain shallower scratching marks. In manual operation, fine polishing is conducted using a "scotch" belt, which uses very fine abrasives and a different bond structure from the belt for rough polishing. The "scotch" belt is much more expensive than the abrasive belt used for rough polishing. Moreover, the use of different belts in fine and rough polishing processes adds difficulty for tooling management in automation systems.

As shown in Fig. 8.12, the SEM topographies of the used belts have indicated that after polished three vanes, the sharp edges of abrasive grains are worn. The cutting edges become dull and not suitable for material removal in rough polishing. Also indicated in Fig. 8.13, the roughness values are below 1.6 μm (which meets the final roughness requirement of the polished vane airfoils), when the belt is used to polish the fourth and fifth vanes. Therefore, the belt disposed after rough polishing can be re-used for fine polishing. To make sure that the surface roughness is below 1.6 μm, a small pre-load with soft springs is applied for fine polishing in contrast to the use of a large pre-load for rough grinding. The belt replacement plan in the robotic system for vanes is that both abrasive belts for rough and fine polishing are replaced after machined three vanes. The used belt in rough polishing is then re-used for fine polishing. The idea using used belts for polishing is of great significance because the production cost is reduced and the tool management is convenient.

In the robotic machining system for 3D vane airfoils, various process optimization strategies are used. For example, different machining plans are associated with different tool path sections by varying belt speed, feed rate and approaching angle. Repeated tool paths are also used to remove the local transition boundaries between the brazed layers and vane base. This enables visually smooth airfoils to be achieved, as shown in Fig. 8.14a. Except the surface finish, the dimensional accuracy of airfoil wall thickness is examined by sectioning the airfoil, as shown in Fig. 8.14b, where satisfactory profiles are demonstrated. The wall thickness of the polished airfoils must also meet the requirement. As shown in Fig. 8.3b, seven locations are selected to examine the minimum wall thickness. Figure 8.15 shows the relative thickness values for the polished airfoils. Here the relative thickness is defined as the subtraction of the nominal wall thickness from the measured wall thickness of a repaired airfoil. The three machined parts are all within the tolerance values, as the relative thickness values are positive.

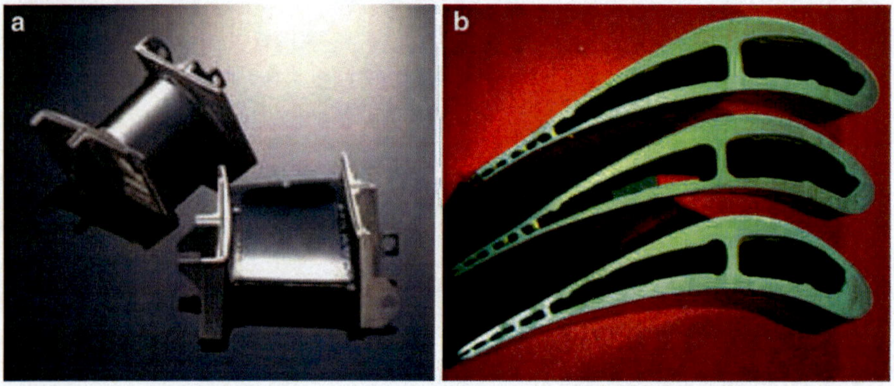

Fig. 8.14 (**a**) Polished turbine vanes [19] and (**b**) sectioned pieces of vane airfoils

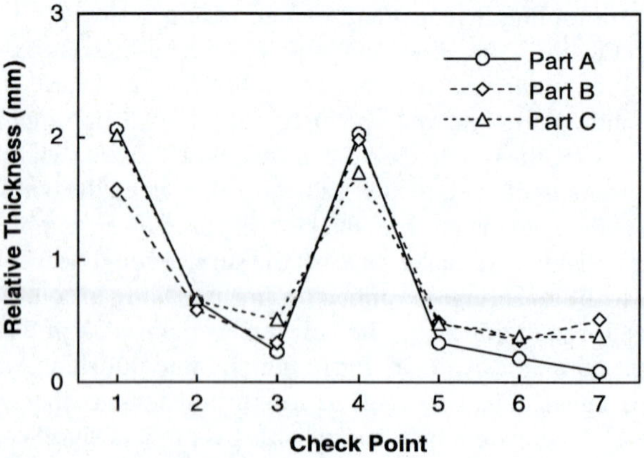

Fig. 8.15 Relative wall thickness of polished airfoil at defined checking points

8.2.2 Polishing of Fibre Optic end Faces Using Abrasive Films

Fibre optic connectors consist of a ferrule of diameter of 2.5 mm with a glass fibre of 125 μm in diameter centered in the ferrule [21]. When assembling the fibre into the ferrule, it exits at the ferrule's end face. The end face thus needs to be polished to achieve a smooth surface [22]. Ferrules are often made of ceramics, such as zirconia. In the current production line manual polishing is required to remove the excess fibre and epoxy bead first and several stages of machine polishing are then followed [23]. After polishing, optical inspection, such as insertion and return loss, are conducted to examine the optical quality of the polished connectors. The commonly encountered problem in the production line is the difficulty for maintaining flush of fibre and ferrule heights after polishing [24, 25]. Several extra polishing procedures are needed to meet the required relative height between fibre and ferrule faces, and thus the machining cycle time is long. In recent years, a great effort [26–33] has been directed towards developing high efficiency grinding and polishing processes for fiber connectors to meet the increasingly high demand from optic communication industries. In this section, we report the effect of polishing on the surface integrity and geometric measures of connector end faces, including surface roughness, fiber height or undercut, apex offset and radius of curvature. The relationships between geometrical parameters and optical quality, in terms of return and insertion losses, are discussed.

8.2.2.1 Fibre Optic Connector and Polishing Set-Up

Fibre optic connectors being studied consist of a glass fiber of 125 μm in diameter and a ferrule housing of 2.5 mm in diameter. The fibre is centered in the ferrule and exited from its end face. The ferrule is made of yttria partially stabilized zirconia. The ferrule is keyed in a mechanical assembly to align and hold the fibre rigidly. The end face of connectors is convex, spherically curved with a radius of curvature of about 20 mm. The protruded fiber end needs to be cleaved and the epoxies are removed by manual polishing prior to machine polishing.

The functionality of fibre optic connectors is to couple two fiber cables for optic communication. Thus, the geometric quality of fibre connectors will directly affect the quality of light transmission, including the reflection loss caused by the roughness of fiber end faces, the fiber separation, the lateral misalignment, the angular misalignment, the core and cladding diameter mismatch, the refractive index profile difference, and the contamination on fibre end faces. The geometrical parameters include fiber and ferrule surface roughness of R_a (arithmetic mean value), fiber height or undercut, apex offset and radius curvature. "Fibre height" and "fibre undercut" are defined to characterize the relative distances between the fiber end face and the best fit of spherical surface of the average ferrule end face [22]. The fiber undercut refers to the situation where a fiber is positioned inside a ferrule, while the fiber height indicates that the fibre protrudes above a ferrule. "Radius of

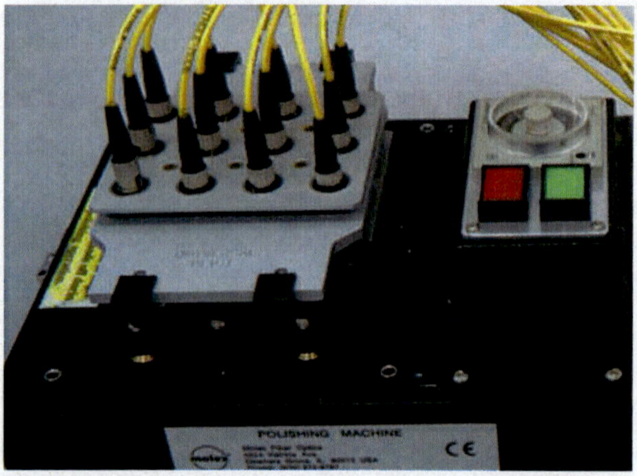

Fig. 8.16 The machine for polishing optic fibre connectors

curvature" is the radius of best fit of a spherical ferrule end face and "apex offset" is the linear distance from the center of a fiber to the apex or highest point on the best fit of its spherical end face [22]. Geometrical quality for the polished connector end faces was evaluated using an automated noncontact interferometer system for array-type fiber-optic connectors (AC-3005, Norland).

A polishing machine (8671X-6100 Series, Molex) was used for finishing the fibre connectors. The machine, as shown in Fig. 8.16, has a timer and a pressurized holder. The universal holder can contain 12 connectors. The machine provides a circular orbital oscillation for polishing. During polishing, individual connector rotates along its own circular orbit of 17.5 mm in diameter at a speed of 15 rpm, while polishing media rotates at a speed of 285 rpm.

8.2.2.2 Effects of Abrasive and Polishing Protocol

The polishing of fibre connectors includes rough and fine processes. Disposable and self-adhesive films were used for polishing. SiC films (see Fig. 8.17a) of grit sizes of 5 and 3 μm were used in rough polishing for material removal. The films were put onto an air-cushion metal pad with 12 circle orbits. Diamond (see Fig. 8.17b) or alumina films of grit sizes of 0.5, 0.1 and 0.05 μm were used in fine polishing, which were laid on a flat metal pad. Lubricants were sprayed on the abrasive films during polishing.

Fig. 8.18 shows the optical images of the fiber end faces taken prior to and after polishing with propylene glycol diamond suspension of grit size of 50 nm. Figure 8.18a is the typical surface for a cleaved and epoxy-removed fiber connector end face. Figure 8.18b shows the surface of a connector polished using a 5 μm SiC film, where polishing marks, scratches and chippings are observed. Almost all the damages occurred on the fibre material, i.e. fused silica of lower hardness and

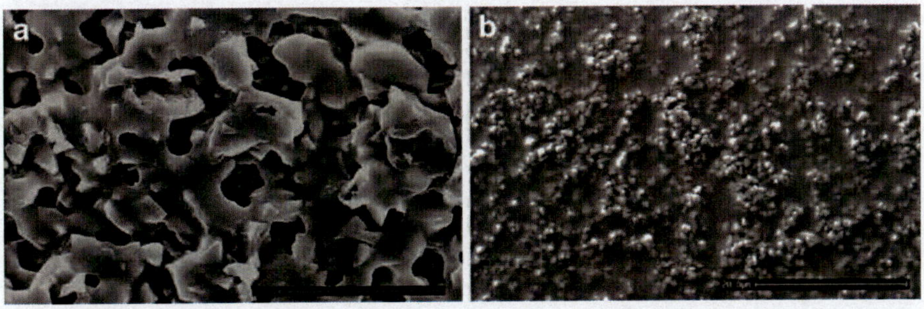

Fig. 8.17 SEM micrographs of polishing films for fibre end faces. (**a**) SiC abrasive film for rough polishing and (**b**) diamond film for fine polishing

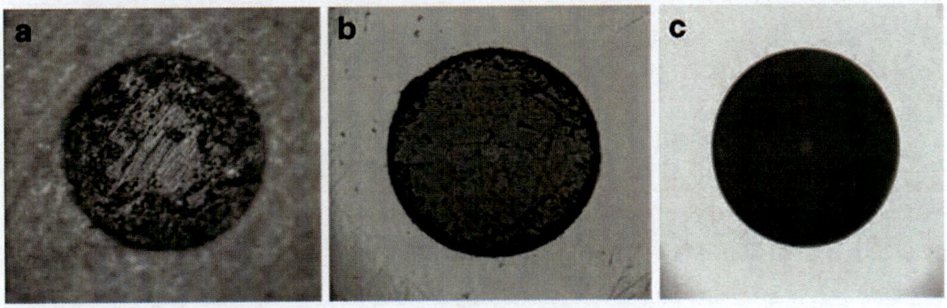

Fig. 8.18 Fibre end faces. (**a**) As-received, (**b**) polished for 30 s using 5 μm SiC film, and (**c**) further polished for 30 s using 0.1 μm diamond film

toughness than zirconia, include chipping and deep scratches. In contrast, only very shallow scratches are observed on the ferrule surface, which has higher toughness and hardness then silica. The polishing-induced damage using a 3 μm SiC film is less severe than that using a 5 μm SiC film, as expected. Figure 8.18c is the second-step polished end face using a 0.1 μm diamond film. No visible scratches and damage are observed. Polishing with fine diamond abrasives also produced better fibre height. As shown in Fig. 8.19, the fibre height of 53 nm left after polishing using the fine diamond film is much smaller than the height of ~2 μm produced by the 5 μm SiC film.

Combinations using different abrasives with different grit sizes for rough and fine polishing were tested. The polishing step had the same cycle time of 30 s. The polishing protocols shown in Table 8.4 consists of a rough polishing procedure using coarse SiC films, aiming to remove the protruded fiber height after cleaving and epoxy removal with high efficiency, and one or two fine polishing procedures with diamond or alumina abrasives, aiming to obtain good surface finish. Six connectors were simultaneously polished to examine the repeatability of the results. Each protocol shown in Table 8.4 can produce satisfactory surface quality. It is found that by judicious selection of abrasive types and grit sizes used for the rough

Table 8.4 Parameters for rough and fine polishing processes

Process	Step 1 Film on air cushion metal pad (30 s)	Step 2 Film on flat metal pad (30 s)	Step 3 Film on flat rubber pad (30 s)	Polishing quality
A3	3 μm SiC	0.5 μm alumina	0.05 μm alumina	Satisfactory
B3	3 μm SiC	0.1 μm diamond	0.05 μm alumina	Satisfactory
A2	5 μm SiC	0.1 μm diamond		Satisfactory
B2	5 μm SiC	0.1 μm diamond		Satisfactory
C2	3 μm SiC	0.5 μm diamond		Satisfactory

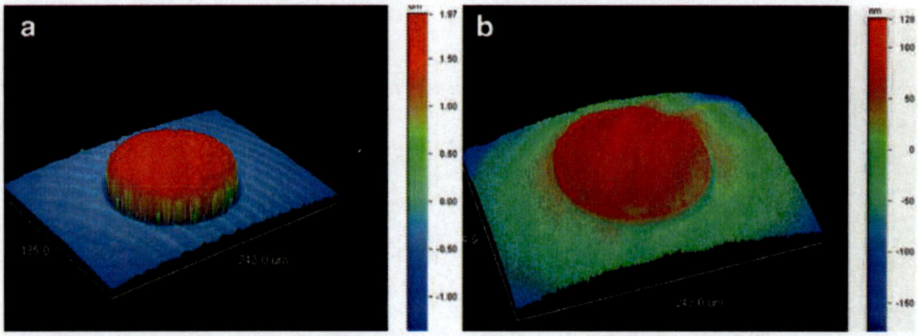

Fig. 8.19 3D optical interference images showing the fibre heights after polishing using (**a**) 5 μm SiC film, and (**b**) 0.1 μm diamond film [29]

and fine polishing, the two-step polishing protocols (i.e. A2, B2 and C2) with a total circle time of 60 s can produce satisfactory outcome, which normally needs a three-step polishing (i.e. A3 and B3) with a circle time of 90 s, in terms of ferrule surface roughness, fiber height/undercut, radius of curvature and apex offsets.

8.2.2.3 Effect of Suspensions on Surface Quality and Optic Performance

Five lubricants containing different suspensions were used. Table 8.5 summarizes the details of lubricants used. Sign "A" stands for 91% volume isopropyl alcohol. Sign "B" refers to a suspension with colloidal silica of 50 nm. Sign "C" and "D" represent propylene glycol and water-based diamond suspensions with particle size of 50 nm, respectively. Sigh "E" is a water-based alumina suspension of particle size of 50 nm.

The effect of different suspensions on the surface roughness of fibre connectors is shown in Fig. 8.20. The colloidal silicon suspension ("B") produced the smoothest surfaces for both fiber and ferrule. The surface roughness values of fiber polished using alcohol and water-based alumina suspensions are almost the same. However, the fiber roughness values using both propylene glycol- and water-based

Table 8.5 Polishing suspensions [32]

Suspension	Description
A	Isopropyl alcohol (91 vol.%)
B	Colloidal silica suspension of grit size of 50 nm
C	Propylene glycol diamond suspension of grit size of 50 nm
D	Water-based diamond suspension of grit size of 50 nm
E	Water-based alumina suspension of grit size of 50 nm

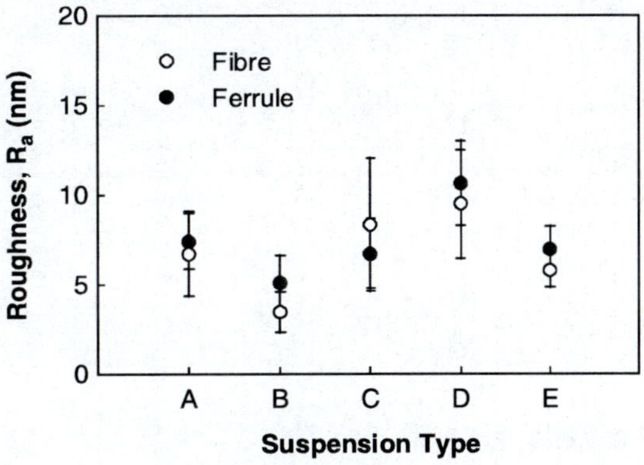

Fig. 8.20 Effect of suspension on surface roughness of fibre and ferrule

diamond suspensions ("C" and "D") are considerably higher, two times that of the silica suspension and one third those of the alcohol and alumina suspensions. The surface roughness values of ferrule using alcohol, propylene glycol diamond suspension and alumina suspensions are at the same level. The ferrule roughness for the water-based diamond suspension is two times greater than that using the silica suspension and about one third greater than that using alcohol or propylene glycol diamond or alumina suspension.

Figure 8.21 shows the effect of suspension type used on the fiber height of connectors. The fiber height resulted from using silica suspension is the smallest. In contrast, the fiber heights for water-based diamond and alumina suspensions are more than two times higher than that with silica suspension and 20–30% higher than those with alcohol and propylene glycol diamond suspensions.

Figure 8.22 demonstrates the effect of suspension on the radius of curvature. The curvature of fibre end faces polished using alcohol is convex curved, whose radius is smaller than that using silica suspension. Polishing using propylene glycol-, water-based diamond and alumina suspensions generated concave end faces, with negative radii of curvature.

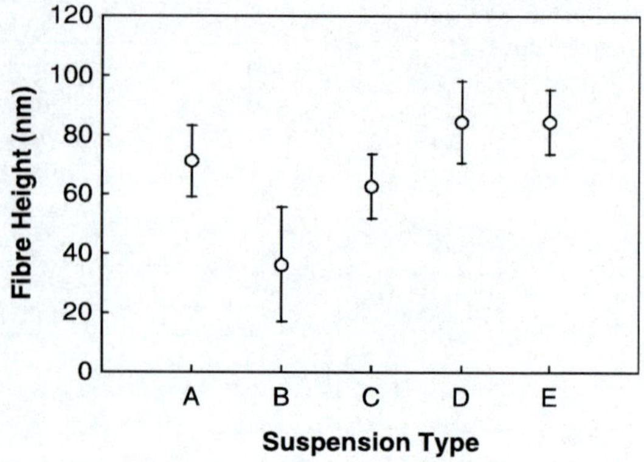

Fig. 8.21 Effect of suspension on fibre height

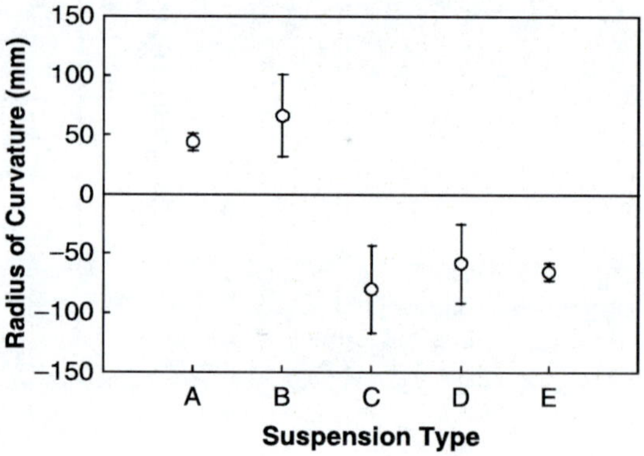

Fig. 8.22 Effect of suspension on the radius of curvature of polished end faces

Values of apex offset of the fibre connectors polished using the five suspensions are shown in Fig. 8.23. The alcohol generated the smallest apex offset. The silica suspension gave a slightly larger apex offset than that of the alcohol, whereas the propylene glycol diamond and alumina suspensions gave the largest offsets. The water-based diamond suspension produced a higher offset than the alcohol and the silica suspension but lower than the propylene glycol diamond and alumina suspensions.

After polishing, the connector end faces must be carefully cleaned with alcohol and optical tissues for optical performance testing. Optical performance for each suspension is plotted in Fig. 8.24. The minimal value of return loss required for acceptable physical contact transmission is 45 dB. The return losses for alcohol and silica suspension are both better than 45 dB. However, the return losses for propylene glycol and water-based diamond suspensions are lower than 45 dB. The return loss for alumina suspension just meets the requirement. The required insertion loss for a fibre connector is 0.3 dB or below. Silica suspension produced a minimum

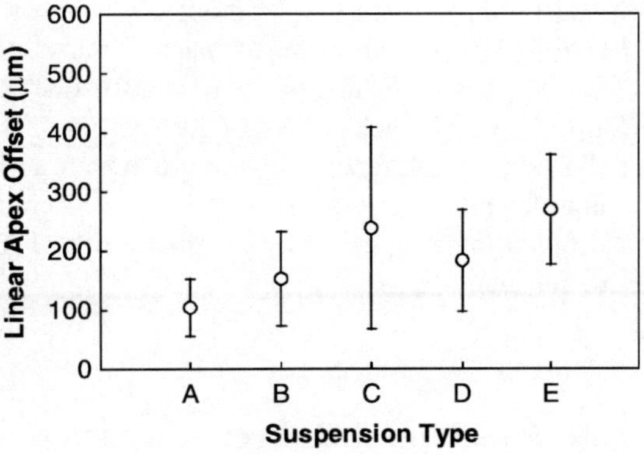

Fig. 8.23 Effect of suspension on the linear apex offset of polished fibres

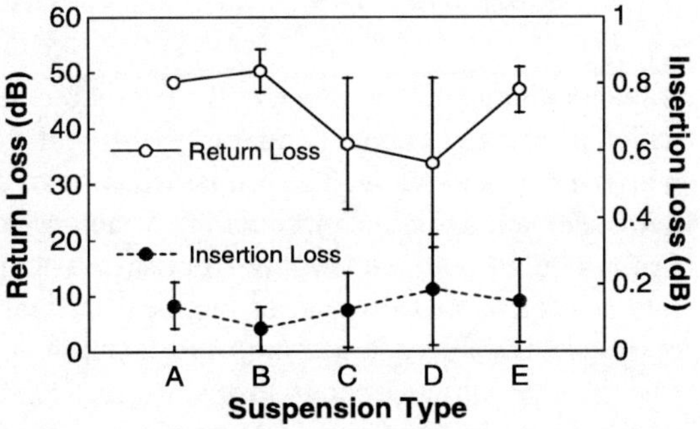

Fig. 8.24 Effect of suspension on optic performance

insertion loss of 0.07 dB on average. Alcohol, propylene glycol diamond and water-based alumina suspensions produced acceptable insertion losses, which are twice of that for silica suspension. Water-based diamond suspension resulted in the highest insertion loss, with the maximum value higher than 0.3 dB. The results show that suspensions have a significant effect on the geometrical quality and the optical performance of the polished connectors. Using silica suspension gave the best geometrical quality, including the finest fiber and ferrule surface finish, the smallest fiber height and the comparable radius of curvature and apex offset, and hence the best optical performance of the acceptable return loss and the significantly small insertion loss. Alumina suspension did not offer an advantage over alcohol in terms of both geometrical and optical qualities. Diamond suspensions did not provide better results either in terms of geometrical parameters and optical losses.

Better optical performance of a fiber-optic connector results from the well-polished connector end face. This is because better fiber and ferrule surface finish, smaller fiber height and apex offset, and reasonable radius of curvature of polished connect end faces could provide a better physical contact between the two mated connectors for optical communication.

In summary, with the judicious selection of polishing parameters and protocol, the efficiency for polishing physical contact connectors can be improved at least 30%. The best polishing process involves only two-step polishing with a total cycle time of 1 min. To achieve return loss of < -45 dB and insertion loss of <0.3 dB, the polished fiber and ferrule surface much have a roughness, R_a, of smaller than 50 nm, a fiber height or undercut of -25 or 75 nm, a radius of curvature ranging from 10 to 75 mm, a linear apex offset of smaller than 150 μm and an angular apex offset of smaller than 0.2°.

8.3 Polishing with Free Abrasives

8.3.1 Polishing of Microbores Using Liquid Suspended Abrasive Flow

Device miniaturization is one of the frontier technologies in the twenty-first century, which has brought about completely different ways in which human beings interact with the physical world. The miniaturization of devices associated with a number of fields demands for the production of micro-scale mechanical components with manufactured features ranging from several to several hundred microns with tight tolerances in a broad range of engineering materials. These fields mainly include optics, electronics, medicine, bio-technology, communications and avionics. The rapidly growing needs in making products from macroscopic to nanoscopic scales from photonic, microelectronic and biomedical industries echoes the international research direction in modern manufacturing. Micromachining technologies have attracted great attention of research in the fabrication of miniaturized industrial products that are extensively applied into our society [34–45].

In the making of the miniaturized products, a common challenge is from the fabrication of microbores of diameters ranging from several tens to several hundreds of microns. The microbore are a typical feature in various products, such as fluidic filters, grids, bio-medical filters, ink-jet printer nozzles, fuel injection nozzles, optical ferrules, high-pressure orifices, standard defects for testing materials, micropipettes, pneumatic sensors and manipulators, guides for wire-bonders and spinning nozzles, and fuel injection nozzles [34]. Great research efforts have been directed towards developing the micromachining technologies for microbores in the past two decades. Among them, micro electrical discharge machining (EDM) [40–42] and laser machining [43–45] have been widely used. Other technologies for such applications include microcutting [39, 46], ultrasonic machining [47], microforming and micromolding [48], micro electro-chemical-machining (ECM) [34] and micropunching [49]. There exist certain limitations in those technologies in terms of capability of handling various materials and surface finish of microbores. For example, micro-EDM is only capable for making holes on conductive or semiconductive materials. Laser machining has limitation to obtain microholes with large (length/diameter) aspect ratios. For majority of the technologies mentioned above, bore surface quality is not satisfactory and a final finishing process is still required to meet a stringent quality requirement on microbores.

The challenges for polishing microbores are obvious. Bores are so small that conventional polishing tools cannot be manipulated inside the holes, so does the characterization of machined surface quality. Consequently, non-conventional methods for tooling, machining and characterization must all be developed. Polishing using a slurry flow is one of the useful methods for polishing microbores [50–52]. In this section, we introduce an abrasive flow technology developed for polishing microbores. The bore diameter investigated varied from 260 to 500 μm and the length-diameter ratio ranged from 25 to 50. Both metal and ceramic materials were investigated. Three-dimensional characterization methods involving stylus profilometry and optical interferometry were also employed to investigate the effects of flow parameters on inner wall surface roughness and topography.

8.3.1.1 Experimental Apparatus and Working Principle

The experimental apparatus consists of a motor-driven hydraulic pump that is capable of generating pressure up to 40 MPa, a direction control valve, a hybrid cylinder, a work unit, two pressure gauges and a slurry tray. In this work, hydraulic oil 32 is pressurized in the drive power unit and supplied to the hybrid cylinder through a manually operated directional control valve, which pushes abrasive slurry going through a microbore to be polished. As illustrated in Fig. 8.25, the workpiece is precisely aligned in the work unit. The pre-filled abrasive slurry is forced to flow through the microbore, which is defined as one polishing pass. When one polishing pass is finished, the used slurry flows to a tray and fresh slurry is added to the slurry portion of a hybrid cylinder for the next pass of polishing. O-rings are doubly inserted between the two ends of the workpiece in order to avoid leaking of slurry under high flow pressure.

Abrasive flow polishing is in fact a slurry flow in turbulent flow regime through a microhole [53]. The critical Reynolds number for turbulent flow was 2,300 in this

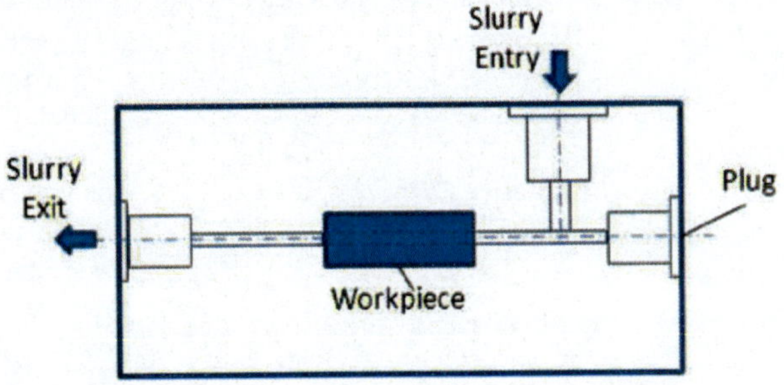

Fig. 8.25 Illustration of slurry flowing route

case, which was calculated using the equation [54]: $R_e = UD/v$, where R_e is the Reynolds number, U the mean velocity in m/s, v the kinematic viscosity in m^2/s and D the hydraulic diameter in meters. In the apparatus developed, U varied from 12 to 29 m/s, v was kept at 1.1×10^{-6} m^2/s, and D ranged from 260 to 500 μm. So the Reynolds number varied from 8,390 to 9,090, much higher than the critical number of 2,300 for turbulent flow.

The test materials selected include 304 stainless steel, S45C steel and partially stabilized zirconia. All the specimens have a cylindrical shape with a length of 13.6 mm and outer diameter of 6 mm. Diameters of bores are 500 μm for stainless steel specimens, 400 and 500 μm for S45C and 260 μm for zirconia. The bores in specimens made of stainless steel and steels were drilled, but the holes in zirconia specimens were formed by micro injection molding. Polishing conditions are summarized in Table 8.6. The pressure in abrasive flow polishing was kept constant at 10 MPa. Polishing slurry used alumina abrasives of grit size of 17.5 μm with a concentration of 3.44 vol.%. Flow speeds applied were 19.3 m/s for the 500 μm holes, 25 m/s for the 400 μm holes and 35.5 m/s for the 260 μm holes. Polishing passes of 5, 10, 15 and 20 were used.

8.3.1.2 Surface Characteristics and Roughness of Polished Bores

Figure 8.26 show the inner wall surfaces for the stainless steel specimens. Figure 8.26a is the typical surface produced by drilling as the drilling marks are observed and the surface has long grooves parallel to the drilling direction. When pressurized abrasive slurry passes the hole, the plastic grooves are worn away due to impact of abrasive particles. The drilling marks are gradually diminished when the polishing pass is increased, as shown in Fig. 8.26b–d.

Figure 8.27 shows the inner wall surface changes of S45C steel with bores of 500 μm in diameter. Similar to stainless steel specimens, the bores were first drilled. As shown in Fig. 8.27a, the drilled surface is rough, but the drilling grooves are not apparent. The surface finish after five polishing passes appears to be improved

Table 8.6 Polishing conditions

Workpiece (13.6 mm in length, 6 mm in outer diameter)	304 stainless steel, 500 μm inner diameter S45C steel, 400 and 500 μm inner diameters Zirconia, 260 μm inner diameter
Abrasive flow	Alumina of grit size of 17.5 μm Concentration: 3.44 vol.% Pressure: 10 MPa Solution: water

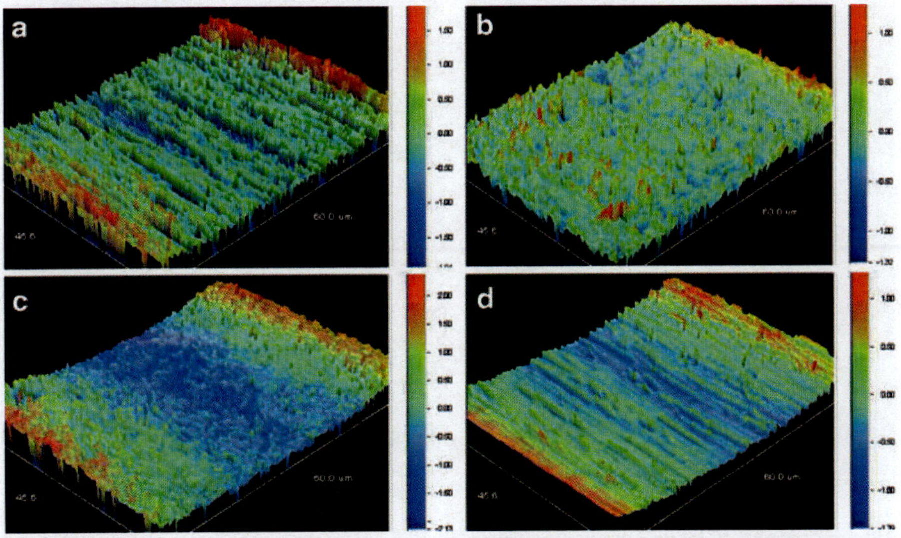

Fig. 8.26 White light interferometer images of inner wall surfaces of 304 stainless steel specimen (500 μm in diameter) (**a**) prior to and after different abrasive polishing of (**b**) 5, (**c**) 10 and (**d**)15 passes

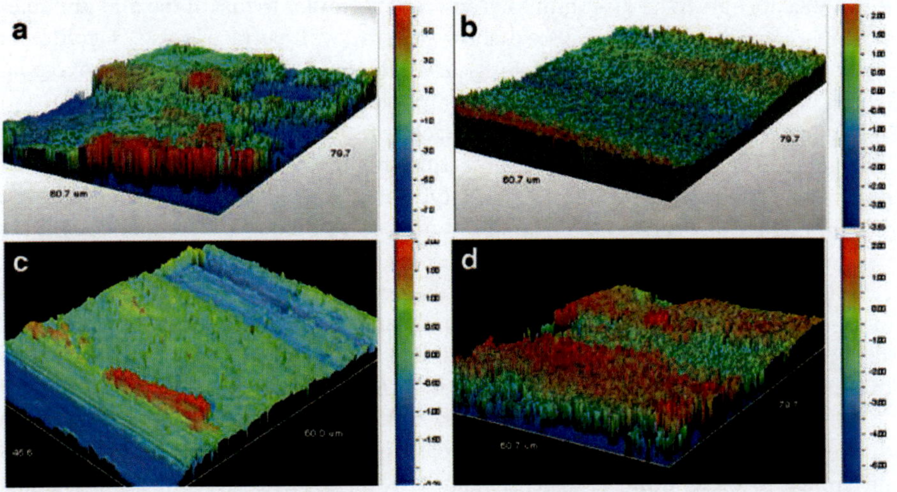

Fig. 8.27 White light interferometer images of inner wall surfaces of S45C steel specimen (500 μm in diameter) (**a**) prior to and after different abrasive polishing of (**b**) 5, (**c**) 10 and (**d**) 15 passes

significantly, as can be seen in Fig. 8.27b. After that, the surface finish improves continuously and gradually with the progress of polishing, as shown in Fig. 8.27c, d.

For comparison, the inner wall surfaces of S45C steel with bores of 400 μm in diameter is shown in Fig. 8.28. The initial surface of inner wall of the bore in

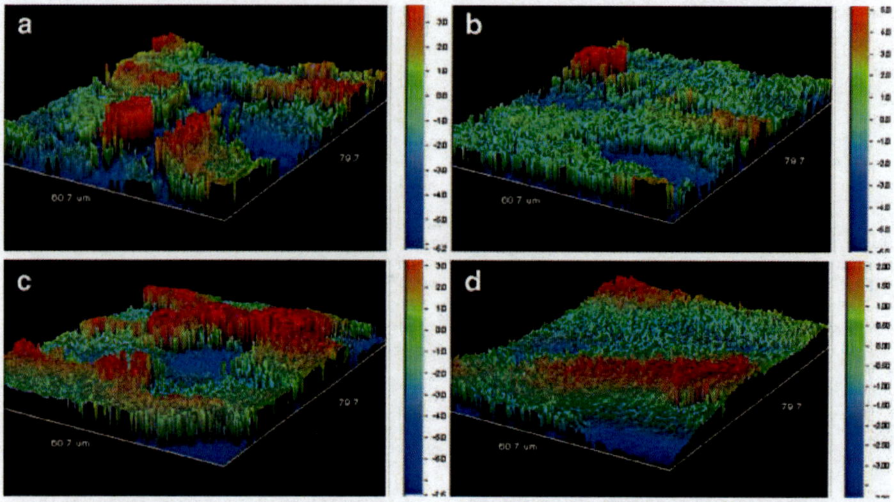

Fig. 8.28 White light interferometer images of inner wall surfaces of S45C steel specimen (400 μm in diameter) (**a**) prior to and after different abrasive polishing of (**b**) 5, (**c**) 10 and (**d**) 15 passes

Fig. 8.28a appears to be discontinuous and rough, similar to that of the 500 μm inner wall in Fig. 8.27a. Figure 8.28b–d show that the wall surface finish is gradually improved with the increasing polishing pass. However, the roughness improvement is slower than that in the polishing of microbores with greater diameters.

Figure 8.29 shows the inner wall surfaces of zirconia specimens with bores of 260 μm. In Fig. 8.29a, the powder-injection moulded bore surface after sintering exhibits to have defects, such as porosity. Pores are randomly distributed in the zirconia material. Apparently, porosity would contribute to the surface roughness measured. It is difficult to remove the pores via polishing. As shown in Fig. 8.29b, a relatively large pore of about 20 μm in diameter is observed on the wall surface after polished five passes. However, there are no pores observed on the bore surfaces polished 10 and 20 passes, as shown in Fig. 8.29c and d. It is also clear that an increase in polishing pass results in a much smoother surfaces.

Values of surface roughness of stainless and S45C steels with bores of 500 μm in diameter are plotted against polishing pass in Fig. 8.30. There is a significant difference in their initial surface roughness, with the S45C steel having a much rougher surface than the stainless steel after drilling. After five polishing passes, their surfaces become quite similar in terms of roughness. Further polishing improves the surface finish, but with very limited improvement. It is apparent that at the late stage of polishing, the trends of surface finish improvement for the two materials become similar. The surface roughness of stainless steel bores drops from 1.58 to 0.48 μm, with 70% of improvement after 20 polishing passes. However, the roughness of S45C steel bores reduces from 4.02 to 0.78 μm after 20 passes, with a more significantly improvement rate.

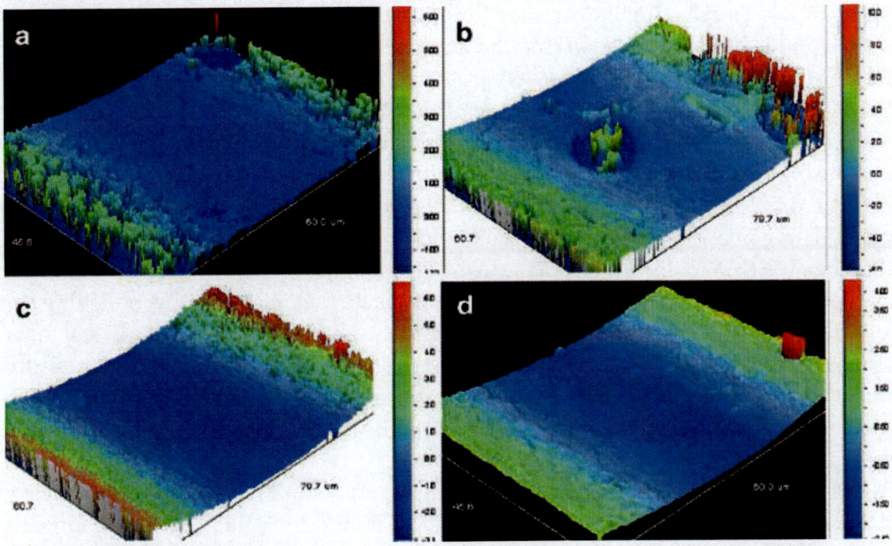

Fig. 8.29 White light interferometer images of inner wall surfaces of zirconia specimen (260 μm in diameter) (**a**) prior to and after different abrasive polishing of (**b**) 5, (**c**) 10 and (**d**) 20 passes

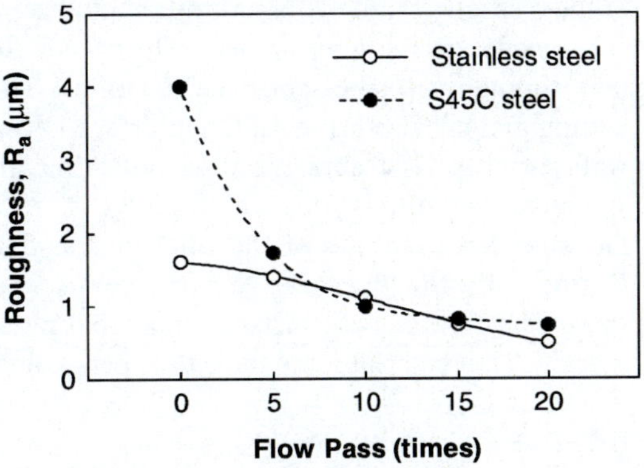

Fig. 8.30 Average surface roughness plotted as a function of polishing pass for different materials. Bore diameter = 500 μm

The values of surface roughness of the S45C steel bores of 400 and 500 μm in diameters are compared and shown in Fig. 8.31, together with the roughness values of zirconia bores. It is seen that the starting surface roughness for the larger S45C bore is much greater than that of the smaller bore. However, after polishing the surface of the larger bore is more significantly improved than the smaller bore with the progress of polishing. This suggests that the bore diameter plays important role during polishing. In the polishing processes studied, both the flow pressure and the flow volume per pass were kept constant, leading to different abrasive flow speeds of 19.3 and 25 m/s for 500 and 400 μm bores, respectively. Though the average velocity was greater for the smaller bore, however, its velocity gradient dropped more significantly. In other words, the abrasive flow velocity near the wall surface in the smaller bore could still be lower. This means that the polishing of a smaller bores is more difficult. The improvement of surface roughness of zirconia bores is significant too, decreasing from 3.5 to 0.5 μm after 20 polishing passes. The decreasing rate of roughness is similar to that of the S45C of 500 μm bore though its bore diameter is only 260 μm. This suggests that a different material removal rate existing in S45C steel and zirconia, with the later more easily to be removed in the abrasive flow polishing process.

The results shown in Figs. 8.30 and 8.31 indicate that abrasive flow polishing can significantly improve the inner wall surfaces of stainless steel, S45C steel and zirconia bores of less 500 μm inner diameters. The surface roughness for all the materials investigated is reduced with the progress of polishing, yielding at least 60% improvement in surface finish. The roughness decreasing rate appears to be relevant to both bore size and material.

Based on the results from the topographic analyses in Figs. 8.26–8.29, it is clear that the material removal in the polishing of stainless steel, S45C steel and zirconia is mainly due to abrasion. The abrasive flow polishing process developed can be considered as a lapping operation where abrasive slurry becomes a self-forming grinding stone as it proceeds through the inner wall of the microbores [54–56]. The maximum load that can be applied to the abrasive slurry is the cross sectional area of the restricting inner area multiplied by the applied pressure. Some of this force is consumed in pushing the abrasive slurry onto the inner wall and in getting the slurry flow through the inner wall (i.e. friction). The remaining force is applied to the cutting grains of abrasives at the interface between the abrasive slurry and the inner wall surfaces. The abrasive flow polishing finishes surfaces by forcing abrasive slurry through microbores with high pressure. The pressurized slurry interacts with the inner wall surface of the microbores, thus eroding bore material gradually. During polishing, the abrasive particles suspended in the slurry first erode away the sharp spots on the wall surface, thus yielding a more uniform, reliable and smooth surface. The material removal is thus gentle and uniform, like honing the inner wall

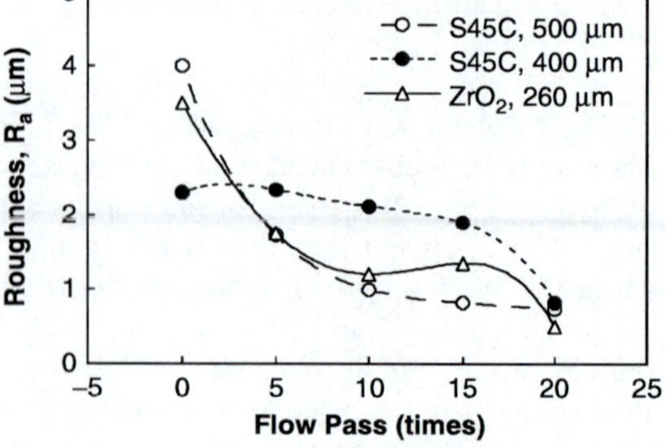

Fig. 8.31 Average surface roughness plotted as a function of polishing pass for different bore diameters

surfaces. Bores of various materials, such as soft steel, stainless steel and ceramics, can be successfully polished using this technology.

In summary, the abrasive flow polishing demonstrates to be a feasible technique for polishing microbores made of stainless steel, mild steel and ceramic material. It can be certainly used to polish microbores of 260 μm or larger. The surface roughness of the polished microbores decreases with the increase in pass of slurry flow, but the improvement is gradually slowed down. The polishing efficiency appears to be relevant to bore material and bore diameter, but less dependent on their initial surface status.

8.3.2 Polishing of Free-Form Component with Free Abrasives

In manufacturing industries, numerous components of free forms that need to be machined and polished. Those components normally have a complex shape, which is not well defined. There are stringent quality requirement on surface finish, however, almost no demand in terms of geometric accuracies. A copper plated egg component is a typical component in this category. The copper plated eggs are formed using an electroplating technique, which first forms a layer of copper with a thickness of 1–2 mm surrounding real egg shells. After forming, abrasive belt polishing and buffing are required prior to the final gold plating. In the production, the polishing/buffing processes of copper eggs are manually performed. A highly skilled worker can only polish 60 components or buff 30 components each day. In addition, as the belt polishing and buffing requires very skillful operators, the maintaining of consist quality is of great challenge. The automation of the grinding and buffing processes has thus become significant.

In this section, we report a study of the finishing processes using tumbling polishing with free abrasives for free-form copper components. The study focuses on the development of tumbling methods, which includes a detailed investigation of the effects of polishing variables, such as mill type, abrasive type and mesh size and the rotation speed, on the polishing efficiency and surface finish.

8.3.2.1 Tumbling Method and Apparatus

Tumbling has been extensively used in mineral processing for refining mines and particles. The process involves numerous collisions among the mines or particles charged in a tumbling container or tumbler, which refines or breaks the mines and particles. In the polishing process that uses the tumbling method, abrasives are charged into a container together with components to be machined. The collisions between the abrasives and the components will erode the unnecessary material of the components away and polish the component surface. Apparently, this method can handle workpieces with irregular shapes. The energy level of collisions can be controlled by adjusting the tumbling conditions.

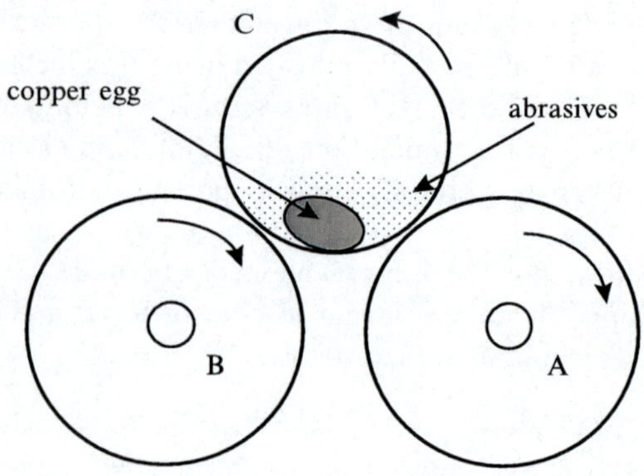

Fig. 8.32 Illustration of tumbling method for polishing free-form component

The tumbling equipment used here include laboratory ball mills, which is slightly different from the conventional one in the mining industry. Smaller and softer (plastic) containers are used to avoid the damage caused by the collision between the components and container wall. Three types of tumbling mills are used for polishing free-form copper eggs, which are described as follows.

The schematic illustration of a typical tumbling mill is shown in Fig. 8.32. The mill consists of two rollers and a cylindrical container. The roller A is driven by an electric motor with adjustable rotational speeds and the roller B is free for rotation along the axis. The container C is freely put on the gap between the two rollers and it is driven to rotate by the roller A. The free-form components and abrasives charged in the container thus rotate or move alongside their own trajectories, being dependent on the rotation speed of the container. The workpiece material can be removed by micro-cutting, micro-ploughing, micro-grooving or fatigue cracking, which is determined by the component material and energy level of collision due to the relative movements of the components and abrasives, resulting in different polishing effect.

A planetary mill is another type of tumbling mill. It involves the rotational movement of a container about two separate parallel axes, analogous to the motion of the earth around the sun. As illustrated in Fig. 8.33, the planetary container that is charged with the components and abrasives rotates about the point A, at the same time, also rotates about the axis O of itself. This results in a complex movement to both component and abrasives. As a very high rotational speed of the container can be operated, the energy level of the planetary mill is much higher than that of the tumbling mill. Thus the planetary mill can be used for the applications that requires relatively larger material removal.

The vibratory mill has the simplest structure for tumbling. As shown in Fig. 8.34, a container charged with components and abrasives as is loaded on a vibrating platform, which is sinusoidally driven by a slider-crank mechanism with an electric motor. The components and the abrasives also vibrate inside the container. The relative motions between them produce the polishing effect. Polishing efficiency and production capacity can be controlled by adjusting the frequency and amplitude of the moving platform.

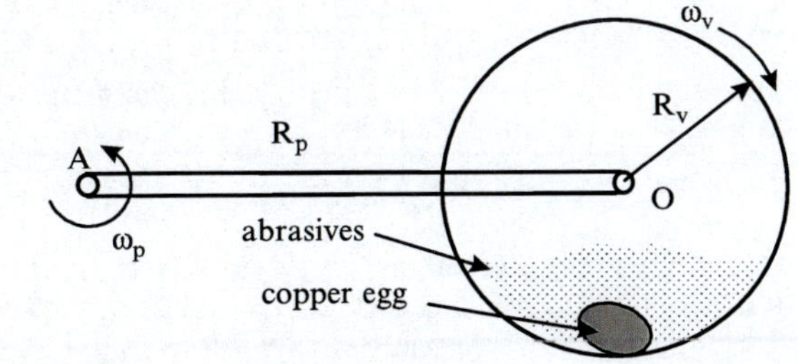

Fig. 8.33 Schematic illustration of a planetary mill

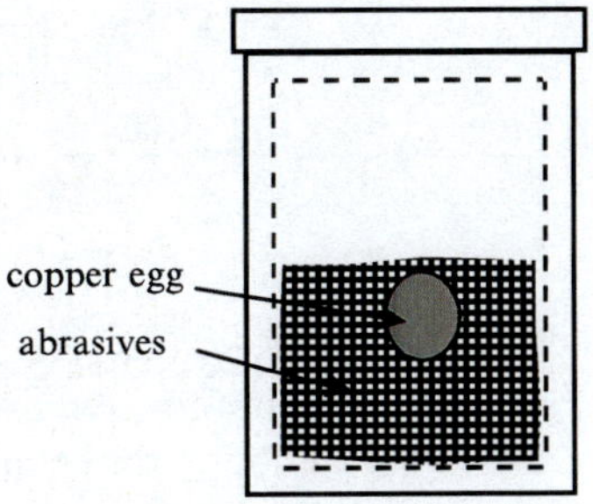

Fig. 8.34 Illustration of motion in a vibratory mill

8.3.2.2 Polishing Media

Three types of abrasives, silicon carbide (SiC), alumina (Al_2O_3) and garnet with mesh sizes of 100, 200 and 400 were used to examine the effect of abrasive type and mesh size on the material removal and surface finish. Focusing on the use of conventional abrasives is based on the consideration of the production cost. The effects of fine diamond powders are also examined, particularly for fine polishing. The diamond powder has particle sizes of 40–80 μm and particles in diamond suspension have a size of 90 μm.

8.3.2.3 Effects of Polishing Conditions

As shown in Table 8.7, the normalized removal weight for the planetary mill using SiC abrasive of mesh size of 100 is several times higher than that obtained using the tumbling mill, and more than 20 times greater that with the vibratory mill. Here the normalized removal weight is defined as the removal amount divided by the component weight before polishing. The planetary mill demonstrated to be a more efficient polishing than the tumbling mill due to its higher energy level of collisions. The tumbling and vibratory mills though have much smaller removal rates, but both have potential for mass production because the collisions between components during polishing can be avoided due to their movement styles.

The removal effect involved in the planetary mill is not uniformly distributed. The removal rate at the egg ends is smaller than that in the middle. However, though

Table 8.7 Effect of milling conditions on removal weight and surface roughness

Mill type	Abrasive type	Polishing time (h)	Normalized removal weight	R_a (μm)	R_{max} (μm)
Tumbling	Alumina #100	48	0.24	0.97	9.07
Tumbling	SiC #100	48	0.08	0.88	6.44
Planetary	SiC #100	48	0.55	1.02	7.23
Planetary	Alumina #400 + Diamond (40–80 μm)	24	0.98	1.59	11.02
Planetary	Alumina #400 + Diamond (40–80 μm), wet	24	0.82	1.01	10.97
Vibrotary	SiC #100	48	0.02	0.95	8.74

the tumbling mill is not as efficient as the planetary, it is easy to control the removal rate and has a uniform removal around the free-form component. A higher efficiency with tumbling mill could be achieved by increasing the rotation speed and size of the tumbler as the current speed is only 136 m/min (for a container diameter of 90 mm).

R_a and R_{max} are used to characterize the surface finish of the polished egg samples, measured using a Talor-Hobson (Talysurf) Profilometer. The roughness values of R_a and R_{max} are 1.87 and 15.85 μm, respectively for the as-received egg components after belt polishing. In Table 8.7, the values of R_a and R_{max} are much improved after tumbling 1 or 2 days. The effect of mill type on the surface roughness is not significant.

The abrasives used also plays an important role in the polishing of free-form components. As can be seen from Table 8.7, in the tumbling mill for the same mesh size, alumina has obvious advantage over silicon carbide in terms of removal rate, though silicon carbide does produce a slightly better surface finish. The combination of alumina and diamond appears to give a much improved material removal rate, in comparison to alumina alone. However, during dry polishing in the planetary mill, the surface finish is not much improved. This could be due to aggressive removal caused by diamond abrasives. The addition of isopropyl alcohol (80 vol.%) into the combination polishing media (20 wt.%) has significantly improved the surface finish (whose value of 1.01 is almost the same as that produced by SiC#100), while still maintaining a reasonably high removal rate.

Free abrasives of silicon carbide and alumina as the polishing media in the tumbling mill are compared. Figure 8.35 shows the normalized removal weights of SiC and Al_2O_3 abrasives. It is apparent that the removal rate with Al_2O_3 abrasives is much higher than that with SiC.

In Fig. 8.36, values of average surface roughness polished using SiC and alumina abrasives are plotted as a function of the polishing time. For both the abrasives used surface roughness is improved with the progress of polishing. When polished for

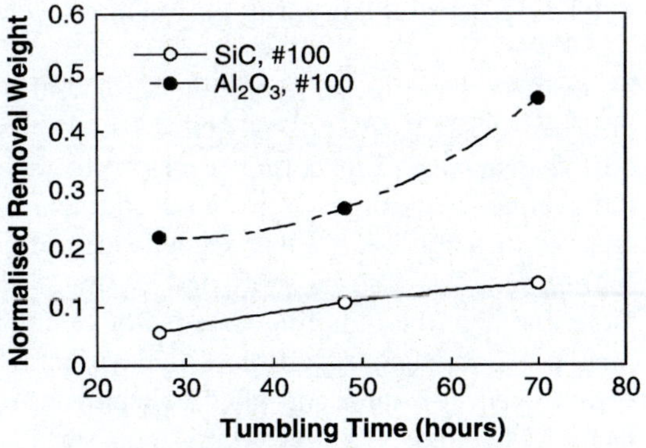

Fig. 8.35 Removal weight plotted as a function of polishing time during tumbling (container speed 136 m/min, container diameter 90 mm and filling ratio 0.8 kg/l volume)

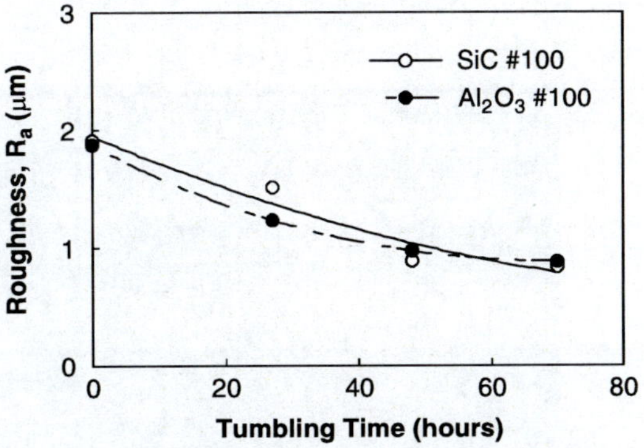

Fig. 8.36 Surface roughness of egg components plotted as a function of tumbling time for Al_2O_3 #100 and SiC #100 abrasives (container speed 136 m/min, container diameter 90 mm and filling ratio 0.8 kg/l volume)

27 h, the roughness produced by alumina is lower than that by SiC. However, further polishing of another 21 h demonstrated that SiC can produce a slightly better surface finish than alumina. This is because that after 27 h of polishing, uneven spots on the workpiece surface was still not fully removed. As alumina abrasive produced a greater removal, the surface finish of the component was thus slightly better. Once the uneven spots were removed fully, SiC abrasives gives a better surface finish as they produced a less aggressive removal than alumina.

For all the abrasives the profile of the polished free-form components is much improved. The polished components clearly have a much smoother surface than that prior to polishing, as illustrated in Fig. 8.37. The type of polishing machines determines the polishing efficiency, which is dependent on the relative movements between the components and the polishing media. The tumbling method is feasible to replace the complete manual polishing and partial buffing. Nevertheless, a comprehensive study has to be carried out to find out the optimum combination of abrasives. In addition, under current tumbling conditions using the tested abrasives, mirror surface finish won't be achieved. The development of suitable polishing media are needed.

8.3.2.4 Effect of Tumbling on Buffing Cycle Time

To evaluate the effect of tumbling on the buffing cycle time, five egg components polished using the tumbling method with alumina #100 abrasives were further buffed manually. The buffing times were then recorded and are compared with the average cycle times from the production line (shaded numbers) in Table 8.8. It is seen from Table 8.8 that second stage of belt polishing can be spared and the buffing time can also be shortened, saving about 30% of the total polishing time. Note that here the tumbling time is not calculated as tumbling can be operated in mass production, not requiring skilled workers. Figure 8.38 shows the surfaces after manual belt polishing and after polishing using tumbling method. The tumbling using Al_2O_3 abrasives improves both the surface finish and egg profile, thus, it is not surprising that the buffing time had been shortened. It should be noted that the fine buffing time for Part 5 was significantly longer than the industrial cycle time because the rough buffing was removed from the processing.

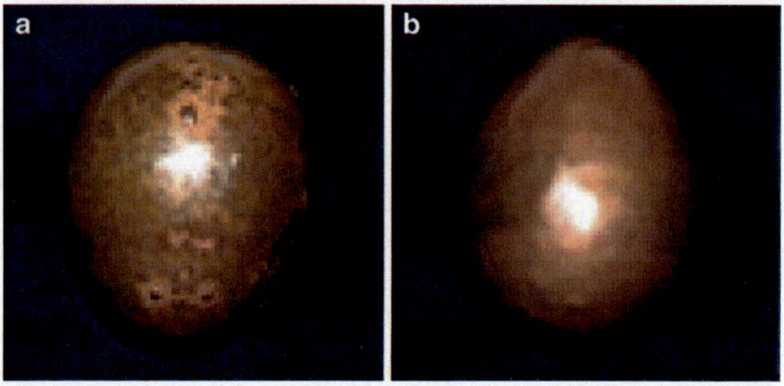

Fig. 8.37 Surfaces of egg component (**a**) prior to machining and (**b**) after polishing using the tumbling method with Al_2O_3 #100

Table 8.8 Comparison of cycle times (in minutes) with and without tumbling[a]

Part number	Belt polishing (#150)	Belt polishing (#320)	Rough buffing[+]	Fine buffing[+]	Time improved (%)
Manual	3	7	3	12.5	
1	3	–	3	10	37.3
2	3	–	3	12	29.5
3	3	–	3	12	29.5
4	3	–	3	13	25.5
5	3	–	–	15	29.5

+ sign – indicates that the process was NOT applied

[a]Buffing was carried out by a skilled worker in the production line

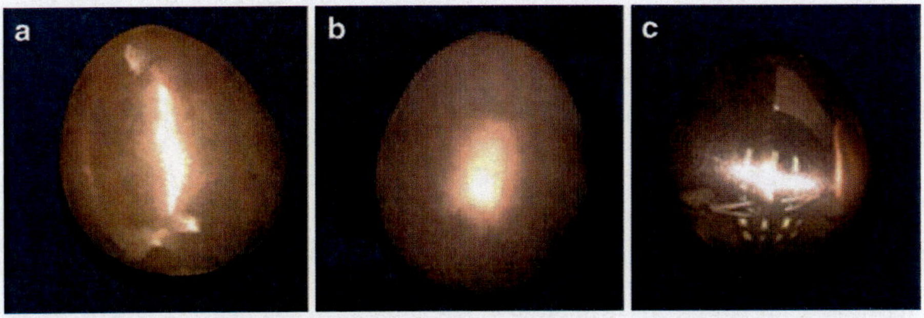

Fig. 8.38 Egg component (**a**) after manual belt polishing, (**b**) after tumbling using Al_2O_3 abrasives, and (**c**) after manual buffing

The results of tumbling experiments can be summarized as follows

- The large lumps on the egg surface could not be removed using the tumbling method due to its uniform polishing effect. So a rough polishing with abrasive belt was still required as the lumps had to be removed manually.
- The egg components polished using the tumbling method had much smoother profiles than those before tumbling. However, the surface finish was not significantly improved. It is apparent that the surface finish strongly depends on the polishing media used. Optimum combinations of polishing media, including the combination of different abrasives or mesh sizes, should be further studied to achieve a satisfactory surface finish.
- Using Al_2O_3 abrasives produced a more efficient polishing than SiC abrasives, but using SiC abrasives resulted in a slightly better surface finish.
- The planetary mill is the most efficient polishing machine, followed by the tumbling mill, and then vibratory mill. But the tumbling mill enables the most uniform material removal across the exposed surface.
- The tumbling method is feasible to replace fine belt polishing and rough buffing. Without inclusion of the tumbling time, the total machining cycle time was about 30% reduced.

8.3.2.5 Barrel Polishing

The results from the tumbling experiments indicated that polishing media is an important factor which controls the surface finish of polished products. Polishing was further conducted using different polishing media, which are commercially available from a barrel polishing specialist in Japan, UJIDEN. Two types of polishing media were tested, which were specially developed for polishing copper egg components. The barrel polishing experiment was carried out using a UJIDEN centrifugal barrel, whose working principle is similar to the tumbling or planetary mill, but with high energy level. The barrel finishing had two steps: rough and fine polishing using two different types of the barrels, respectively. As can be seen in

Table 8.9, TOSALIT PC-6-15 was used for rough polishing, which combined conventional abrasives with a cleaning/cooling compound (NF-44). The NF-44 compound was used to reduce the scratching from the work material being removed and prevent from the over-heating of egg components during polishing. For fine polishing two different combinations, i.e. WL#46 + AM-50 and WL#36 + AM-55, were used to examine the effect of mesh size of walnut particles on the surface roughness. Note WL#36 and WL#46 are polishing materials made of walnut shells, a non-conventional polishing media. Again, AM-55 is a cooling media. In the fine polishing, rubber containers were used to avoid collisions, which had the similar function to the plastic container used in our tumbling experiments. Detailed polishing conditions are listed in Table 8.9 (Fig. 8.39).

The polishing results were satisfactory after a total of 3 h (rough + fine) polishing. Figure 8.38 shows two copper eggs polished using the barrel finishing method with a mirror surface finish. It was difficult to visually distinguish the difference of surface roughness produced by the barrel polishing and manual buffing (Fig. 8.37c).

8.4 Concluding Remarks

The polishing processes with flexible abrasive tools, including belts and films, free abrasives and abrasive slurries have been investigated. The polished products are of very different geometries, varied from three-dimensional free-form surfaces to inner wall surfaces of microbores. Very different work materials were also studied, including difficult-to-machine superalloys, hard and brittle solids and soft and ductile metals.

The development of the robotic polishing processes for automation of the manual operations of aerospace engine components, which are labour intensive and requires extensive process knowledge and polishing skills, involved a great deal of investigations of polishing hardware and optimal process parameters. Aerospace engine components are made of superalloys, which fall into difficult-to-machine category. Benchmark tests showed that the systems achieved better

Table 8.9 Conditions and results from barrel polishing of egg components

Process		Rough polishing	Fine Polishing	
	Type	Planetary	Planetary	
Barrel machine	Capacity	4 rooms (6 L/room) 2–3 eggs/room	160 rubber rooms (2.6 L/room) 1 egg/room	
	Rotation speed	140 rpm	120 rpm	
Polishing media		TOSALIT PC-6-15	TOSALIT WL #46	TOSALIT WL #36
Filling ratio (work/media) (%)		50	50	50
Cleaning/cooling compound		NF-44	AM-50	AM-55
Polishing time (h)		2	1	1
Roughness, R_a (nm)			~22	~30
Egg profile			Perfect	Perfect

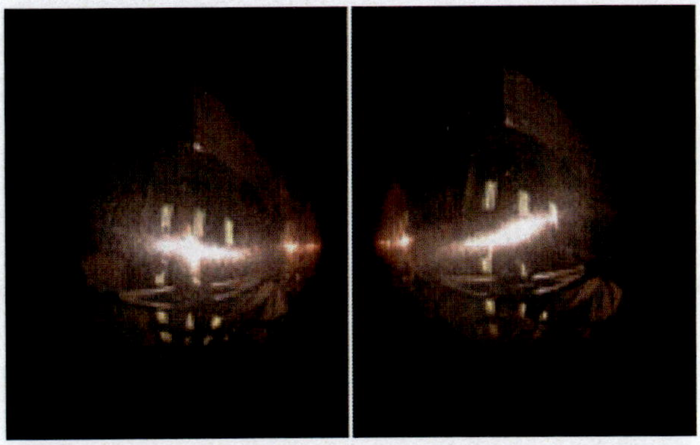

Fig. 8.39 Copper egg components after barrel polishing

and more consistent quality and considerably reduced the cycle times, in comparison to the respective manual operations. In particular, for the polishing system for free-form turbine airfoils, it demands for the synergistic combination of process and system solutions to meet the stringent quality requirement and 3D design criteria, such as profile smoothness and airfoil wall thickness. Though the surface roughness required for the components is not so high as the conventional polishing deals with, the technologies developed has great implications to the automation of finishing processes for precision mechanical components, such as three-dimensional moulds or dies.

The polishing of optic fibre connectors produced mirror surface finish and stringent geometric accuracies. Since the connectors consist of two different materials, i.e. zirconia and silica, the development of the polishing process for them thus involved a great deal of investigations of optimal selection of polishing abrasives and suspensions. There is a distinct difference in material response to polishing between glass fibre and zirconia ferrule, especially with coarser abrasives in which brittle chipping occurs in silica fiber, but plastic deformation dominates in the removal of zirconia material. For coarse polishing, diamond abrasives are not recommended because they induce severe surface and subsurface damage in silica. Silicon carbide abrasives can be applied to have efficient material removal to reduce the fiber height with only moderate surface and subsurface damage in fiber. For fine polishing, fine diamond abrasives produced damage-free and smooth end faces, scaling average surface roughness smaller than 20 and 30 nm for fiber and ferrule, respectively. The results also showed that the suspensions have a significant effect on the geometrical quality and the optical performance of the polished connectors. The colloidal silica suspension resulted in a markedly improved geometrical quality. The optimal combination of polishing media and suspension enables the improvement in cycle time, while maintaining the satisfactory quality requirements.

In the abrasive flow polishing of microbores, the results showed that the polishing surface was significantly influenced by the bore size being polished when both the abrasive flow pressure and the flow volume passing the bore were maintained unchanged. The average surface finishes of the as received microbores made on stainless steel, mild steel and zirconia, varied initially, but after several passes of polishing they reached almost the same value. This indicates that the polishing with abrasive flow has an efficient removal of uneven parts, such as spots and scratches, of the work material, which were left from the previous machining procedure, but is difficult to remove the bulk material. For the polishing conditions used in this investigation, the final roughness of wall surfaces for those materials reached 0.6 μm. This suggests that the abrasive flow polishing has great potential for finishing of both metal and ceramic microbores.

The surface quality and efficiency of polishing free-form copper components with free abrasives are strongly influenced by the abrasives used and the energy level of the tumbling barrels, less relevant to the barrel types. A couple of important factors require great attention. First, during polishing, abrasives in the polishing media are needed to remove the uneven areas of components, and additives are also required to wrap the removed chips to avoid their damage of polished surface. In the fine polishing procedure, additives also plays a role in cooling the polished areas. Second, great care must be taken to avoid the collisions between the components themselves and the component and the barrel container. While traditional abrasives, such as alumina and diamond, still play important role in free abrasive polishing, non-conventional materials, such as walnut shell, can also be used for fine polishing.

Acknowledgements The authors are grateful to experimental assistances and valuable discussion from ZM Gong, XQ Chen, K Ramesh, QF Li, ZJ Xiong, FZ Fang, YC Liu and PL Teo. This work was financially supported by the Australian Research Council (ARC), the Department of Education, Science and Technology (DEST) Australia and the Agency for Science, Technology and Research (A-STAR), Singapore.

References

1. Kasai T et al (1990) Improvement of conventional polishing conditions for obtaining super smooth surfaces of glass and metal works. Ann CIRP 39/1:321–324
2. Weule H and Timmermann S (1990) Automation of the surface finishing in the manufacturing of dies and molds. Ann CIRP 39/1:299–303
3. Evans CJ et al (1998) Rapidly renewable lap: theory and practice. Ann CIRP 47:239–244
4. Cai GO et al (1998) Analysis on lapping and polishing pressure distribution. Ann CIRP 47/1:235–238
5. Spur G and Engel H (1999) Lapping and polishing of glass and ceramics. In: Jahnamir S et al (ed) Machining of Ceramics and Composites. Marcel Dekker, New York. 177–207
6. Woodbury RS (1958) History of the grinding machine. Technology Press, Boston
7. Stephen TM et al (1987) Control of tool/workpiece contact force with application to robotic deburring. IEEE J Robot Autom RA-3:7–18
8. Proctor FM and Murphy KN (1989) Advanced deburring system technology. ASME PED 38:1–12

9. Bone GM et al (1991) Force control for robotic deburring. ASME J Dyn Syst Meas Control 113:395–400
10. Elbestawi MA et al (1992) An automated planning, control and inspection system for robotic deburring. Ann CIRP 41/1:397–401
11. Hollowell R and Guile R (1987) An analysis of robotic chamfering and deburring, ASME DSC 6:73–79
12. Ge DF at al (1995) Dexterous polishing of overhanging sculptured surfaces with a 6-axis control robot. Proceedings of the IEEE International Conference on Robotics Automation. Austin, Texas, USA. 2090–2095
13. Ozaki F et al (1995) A force controlled finishing robot system with a task-directed robot language. J Robot Mechatron 7:383–388
14. Zhou L and Huang H (2002) Automatic robotic system for jet engine overhaul – system design and development for honeycomb repair. Int J Adv Manuf Technol 19:370–376
15. Zhou L et al (2000) Automatic robotic system for jet engine overhaul – process development and enhancement for honeycomb repair. J Jpn Soc Precis Eng 66:1895–1900
16. Huang H et al (2002) Robotic grinding/polishing for turbine vane overhaul. J Mater Process Technol 127:140–145
17. Huang H et al (2003) SMART robotic system for 3D profile turbine vane repair. J Adv Manuf Technol 21: 275–283
18. Chen XQ et al (2002) Process development and approach for 3D profile grinding/polishing. In Chen XQ et al (ed) Advanced Automation Techniques in Adaptive Materials Processing. World Scientific, Singapore. 19–54
19. Chen XQ et al (2002) Adaptive robotic system for 3D profile grinding/polishing. In Chen XQ et al (ed) Advanced Automation Techniques in Adaptive Materials Processing. World Scientific, Singapore. 55–90
20. Taguchi G (1986) Introduction to Quality Engineering. Asian Production Organization, Tokyo
21. Derickson D (1998) Fiber optic test and measurement. Prentice-Hall, New Jersey. 621–638
22. TIA Standard. Telecommunications Industry Association, Arlington
23. Ultra Tec Manufacturing Inc. (2001) Fiber Optic Connector Polishing. Santa Ana
24. Kanda T et al (1995) New micro-finish surface technology for the fabrication of optical device endfaces. Proc SPIE 2576:84–91
25. Karaki-Doy T et al (1998) Development of a new automatic processing machine for optic-fiber connector ends. Bull Jpn Soc Precis Eng 22:216–222
26. Huang H et al (2004) Micro/meso ultra precision grinding of fiber optic connectors and its influence on optic performance. Precis Eng 28:95–105
27. Huang H et al (2004) Ultraprecision abrasive machining of fiber optic connectors. Key Eng Mater 257/258:171–176
28. Huang H and Yin L (2006) An experimental investigation of fabrication mechanisms of optic fibre end faces using nano/micro-indentation and nanogrinding. Int J Nanomanuf 1:47–61
29. Yin L et al (2004) Polishing of fiber connectors. Int J Mach Tools Manuf 44:659–668
30. Lin Y and Huang H (2008) Brittle materials in nano-abrasive fabrication of optical mirror-surfaces. Precis Eng 32:336–341
31. Chen WK et al (2005) Loose abrasive truing and dressing of resin bond diamond wheels for grinding fibre optic connectors. J Mater Process Technol 159:229–239
32. Yin L et al (2005) Influences of nanoscale abrasive suspensions on the polishing of fibre optic connectors. Int J Adv Manuf Technol 25:685–690
33. Yin L et al (2004) High-quality grinding of polycrystalline silicon carbide spherical surfaces. Wear 256:197–207
34. Masuzawa T (2000) State of art of micromachining. Ann CIRP 49:473–488
35. Van Brussel H et al (2000) Assembly of Microsystems. Ann CIRP 49:451–472
36. Huang H and Kuriyagawa T (2007) Nanometric grinding of axisymmetric aspherical mould inserts for optic/photonic applications. Int J Machining Machinability Mater 2:71–84

37. Chen WK et al (2005) Machining of micro aspherical mould inserts. Precis Eng 29:315–323
38. Ramesh K et al (2004) Microgrinding of deep micro grooves with high table reversal speed. Int J Mach Tools Manuf 44:39–49
39. Masuzawa T and Tonshoff HK (1997) Three-dimensional micromachining by machine tools. Ann CIRP 46/2:621–628
40. Huang H et al (2003) Ultrasonic vibration assisted electro-discharge machining of microholes in Nitinol. J Micromech Microeng 13:693–700
41. Odom B (2001) Making microholes. Manuf Eng 126:1–8
42. Rajurkar KP and Yu ZY (2000) Micro EDM can produce micro parts. Manuf Eng 125:1–4
43. Zheng HY and Huang H (2007) Ultrasonic vibration assisted femtosecond laser machining of microholes. J Micromech Microeng 17:N58–N61
44. Tonshoff HK et al (1997) New machining concept for manufacturing microstructures with excimer lasers. Mach Sci Technol 1:153–162
45. Gower MC (2000) Industrial application of laser micromachining. Opt Express 7:56–67
46. Weule H et al (2001) Micro-cutting of steel to met new requirements in miniaturization. Ann CIRP 50/1:61–64
47. Egashira K and Masuzawa T (1999) Microultrasonic machining by the application of workpiece vibration. Ann CIRP 48/1:131–134
48. Geiger M et al (2001) Microforming. Ann CIRP 50/2:445–462
49. Joo BY et al (2001) Development of micro punching system. Ann CIRP 50/1:191–194
50. Kurobe T et al (2001) Development of high speed slurry flow finishing of the inner wall of stainless steel capillary – polishing and gas flow characteristics of various size of capillaries. Precis Eng 25:100–106
51. Kurobe T et al (2002) Application of high speed slurry flow finishing method for finishing of inner wall of fine hole die - effects of the hardness of die materials on the polishing characteristics. Precis Eng 26:155–161
52. Sakuyama T et al (1999) Advances in Abrasive Jet Machining. Development of a new abrasive jet machining device. In Wang J et al (ed) World Scientific, Singapore. 291–298
53. Yin L et al (2004) Abrasive flow polishing of micro bores. Mater Manuf Process 19:187–207
54. Giles RV and Pollard D (1997) Schaum's outline of theory and problems of fluid mechanics and hydraulics. McGraw-Hill, New York
55. Rhoades LJ (1987) Abrasive flow machining. Soc Manuf Eng MR87-163:1–15
56. Rhoades LJ (1991) Abrasive flow machining in a new era of manufacturing. Soc Manuf Eng MR91-523:1–16

Chapter 9
Impact Abrasive Machining

Yasser M. Ali and Jun Wang

Abstract Impact abrasive machining is a broad class of manufacturing processes that include surface cleaning, descaling, deburring, sand-blasting, shot-peening, abrasive water or air jet micromachining, and abrasive waterjet (AWJ) cutting, milling, turning, and drilling. Starting from the basic principles of fluid and solid mechanics, this chapter describes the underlying physical processes and mechanisms affecting jet formation and material removal, and leads to a practical discussion of factors affecting process performance. Adaptations of the basic cutting process to operations such as milling, turning, micromachining, surface cleaning, or surface treatment are explained; and various means for process improvement are discussed. Throughout, focus is on AWJ cutting as the common generic process, from which all other processes are treated as special cases. This chapter is suitable for practicing engineers and researchers, and includes an extensive list of references to guide further study.

Keywords Impact machining · Abrasives · Micromachining · Abrasive water jets · Process modeling

9.1 Introduction

Impact abrasive machining uses a mixture of fluid and solid particles that is accelerated, then impacts the surface of a target workpiece; thus, causing permanent deformation or material removal. This definition encompasses a broad field of processes, with similar physical nature:

J. Wang (✉)
School of Mechanical and Manufacturing Engineering,
The University of New South Wales, Sydney, NSW 2052, Australia
e-mail: j.wang@unsw.edu.au

M.J. Jackson and J. P. Davim (eds.), *Machining with Abrasives*,
DOI 10.1007/978-1-4419-7302-3_9, © Springer Science+Business Media, LLC 2011

- When material removal is unfavourable, it is called *wear*, while favourable material removal is called *machining*. Wear processes (e.g. damage in pipes carrying abrasive slurry, damage to turbine blades due to particles in the gas stream, surface erosion by rainfall, etc) were among the early topics studied in tribology; they provide basic knowledge about material removal mechanisms that may be valid to machining [1–4].
- In explosive forming and welding processes [5, 6], a fluid-solid mixture is accelerated by radial expansion of gases, presenting conditions of extreme impact velocity and pressure. However, they do not necessarily form a *jet*: an almost unidirectional pencil of narrow flow stream.
- When a jet removes material from a target surface by impacting at near-normal angle, the process is called *erosion*. When impact is at near-tangent angle, it is called *abrasion*. Both erosion [7–12] and abrasion [13, 14] may take place during abrasive jet machining, where the term *abrasive* refers to the nature of the solid particles used (being harder than the target material), irrespective of the material removal mechanisms.
- Other abrasive machining processes, such as grinding and polishing [15–17], involve abrasive particles scratching a surface at relatively low velocity (tens of m/s), whereas processes studied here involve *impacting* on the surface at hundreds of m/s – thus, delivering much greater energy.

A common form of *impact abrasive machining* is the use of a narrow jet of fluid and solid-particle mixture, which is called *Abrasive Jet Machining*. The jet can be composed of any fluid. However, for economic and environmental reasons, it is usually a mixture of water and air. Other gases, steam, oils, or super-water are used in special cases. Water has higher density and is incompressible, compared with air; hence, it is capable of applying greater impact pressure. Water usually contains some air, and air usually contains some moisture; when referred to as *waterjet* or *airjet*, it is according to the dominant fluid in the jet. Solid particles can be any material, shape, or size. However, they should be harder than the target workpiece, so that they remove material from the target at minimum damage to their own. Particles having irregular shapes with sharp edges can penetrate and cut more into the target, whereas blunt particles induce more plastic deformation and compressive residual stresses. The proportion of liquid to solid in the jet can vary significantly. In one extreme, solid particles are reduced to zero, to have a pure fluid jet. Because of its low mechanical strength, a pure waterjet can only induce elastic deformation on a solid target. Prolonged application of a pure jet, or application at near-tangent angle, may result in fatigue and delamination of a thin surface layer.

There are several manufacturing processes, within the ambit of Abrasive Jet Machining, that meet the above definitions and limits, as given in Table 9.1. They share many physical mechanisms and operational attributes, and can be studied as special cases of one common process:

Table 9.1 Scope of abrasive jet processes

	Water (super-water, oils, etc.)	Air (nitrogen, steam, etc.)
No particles (oblique impact)	Descaling, paint stripping, etc.	Cleaning
Blunt or round particles (near-normal impact)	Peening	Blasting
Irregular sharp particles (normal or oblique impact)	Cutting, machining	Micro-machining

- Surface cleaning, descaling, deburring, sand blasting, etc are operated using water, air, or steam jets, with or without abrasives, and are usually applied at oblique impact.
- Hydro-honing or shot-peening, uses spherical cast-iron, hard-steel, or glass beads as particles, at near-normal impact and relatively low velocity, such that only plastic deformation – no material removal – is induced.
- Abrasive airjet micromachining utilizes rough particles, but at low pressure, and produces only shallow cuts in the target material.
- Abrasive waterjet milling, turning, drilling, etc, as the more general case.

This chapter focuses on AWJ cutting as the most *common* process, and other processes, as identified above, are special cases. We study AWJ in greater detail, and refer to other processes when applicable. Section 9.2 explains how an abrasive jet is generated, while Sect. 9.3 studies material removal due to abrasive particles impact. Sections 9.4 and 9.5 discuss various approaches to improve the process and various machining operations.

9.2 Generation of Abrasive Jet

Potential energy can be stored in any fluid by compressing it to a high pressure. However, cost and environmental factors limit the practical options. Oils and other chemicals are not "green" enough. Steam can be pressurized to 35 MPa, but the added cost of the boiler is discouraging. In abrasive jet machining, only water and air seem viable. Air is highly compressible, but moisture condensation is a serious limitation above 5 MPa. Typical Abrasive Air Jet (AAJ) systems operate ~1 MPa, and are suitable for micromachining or cleaning applications. Alternatively, AWJ systems typically operate at pressures up to 400 MPa, and systems up to 600–700 MPa are available [18–21]. They are suitable for deep cutting, even on hard materials. When pressurized fluid is released through a small opening (orifice), potential energy is converted into kinetic energy, and a high-speed jet is produced. Waterjet speeds in excess of 1,000 m/s are available in production.

Such high pressures are achievable using crankshaft-driven pumps or fluid-driven intensifiers. Yet, capital, maintenance, and operating costs are excessive at ultra-high pressures, and there is an economic limit on the maximum usable pressure. Figure 9.1 shows a schematic of a typical AWJ machine. Naturally, numerous commercial variations exist, and there are websites [22, 23] and associations [24, 25] for waterjet systems manufacturers, where product specifications and commercial information can be found. This schematic is the general case, treating pure waterjet and abrasive airjet as special cases of AWJ.

When solid particles are injected into a fluid stream, drag forces result in their acceleration, and a substantial amount of kinetic energy can be transferred to the particles. If particle size is relatively large, blunt, and impact velocity is low, it can only induce on the workpiece residual compressive stresses by plastic deformation, e.g. in *peening* processes [26–30]. When particles are small, irregular with sharp corners, and impact velocity is high, they can penetrate into target material, removing it, leaving a permanent crater. Typical abrasive particles include garnet, varieties of sand, corundum (aluminum oxide, Fig. 9.2), silicon carbide, and other super abrasives such as CBN and synthetic diamond.

The selection of the abrasive material is decided by production economics, and the requirement that particle hardness must be greater than workpiece hardness (a 1.6 ratio is recommended [31] and 1.25 is absolute minimum [1]) for efficient material removal. Lower hardness ratios yield excessive particle damage and lower material removal. Abrasive particle sizes range from tens of microns to several hundred microns, according to a mesh number – or an average characteristic dimension of the range of particles included. Abrasive particles, their properties, standards, and suppliers can be found from several associations [32–35].

9.2.1 Pure Fluid Jet

For an ideal potential (incompressible, irrotational, inviscid, and isentropic) fluid [36], under one-dimensional steady-state flow conditions, energy is conserved, and

$$\frac{P_i}{\rho_i} + \frac{v_i^2}{2} = \frac{P_o}{\rho_o} + \frac{v_o^2}{2}, \tag{9.1}$$

where P is hydrostatic pressure, v is average velocity of the stream, and ρ is fluid density. The subscripts i and o refer to inlet and outlet of the flow, respectively.

Since, by assumption, $\rho_i = \rho_o$, $P_i \gg P_o$, and $v_o \gg v_i$, (9.1) reduces to

$$v_o \approx \sqrt{\frac{2P_i}{\rho}}. \tag{9.2}$$

This is the upper limit on the maximum velocity of a pure fluid jet as it exits the nozzle (Fig. 9.3.) For a waterjet, $\rho_w \approx 1{,}000\ \text{kg/m}^3$, and for $200 \leq P_i \leq 600$ MPa, the maximum jet velocity is $630 \leq v_o \leq 1{,}100$ m/s.

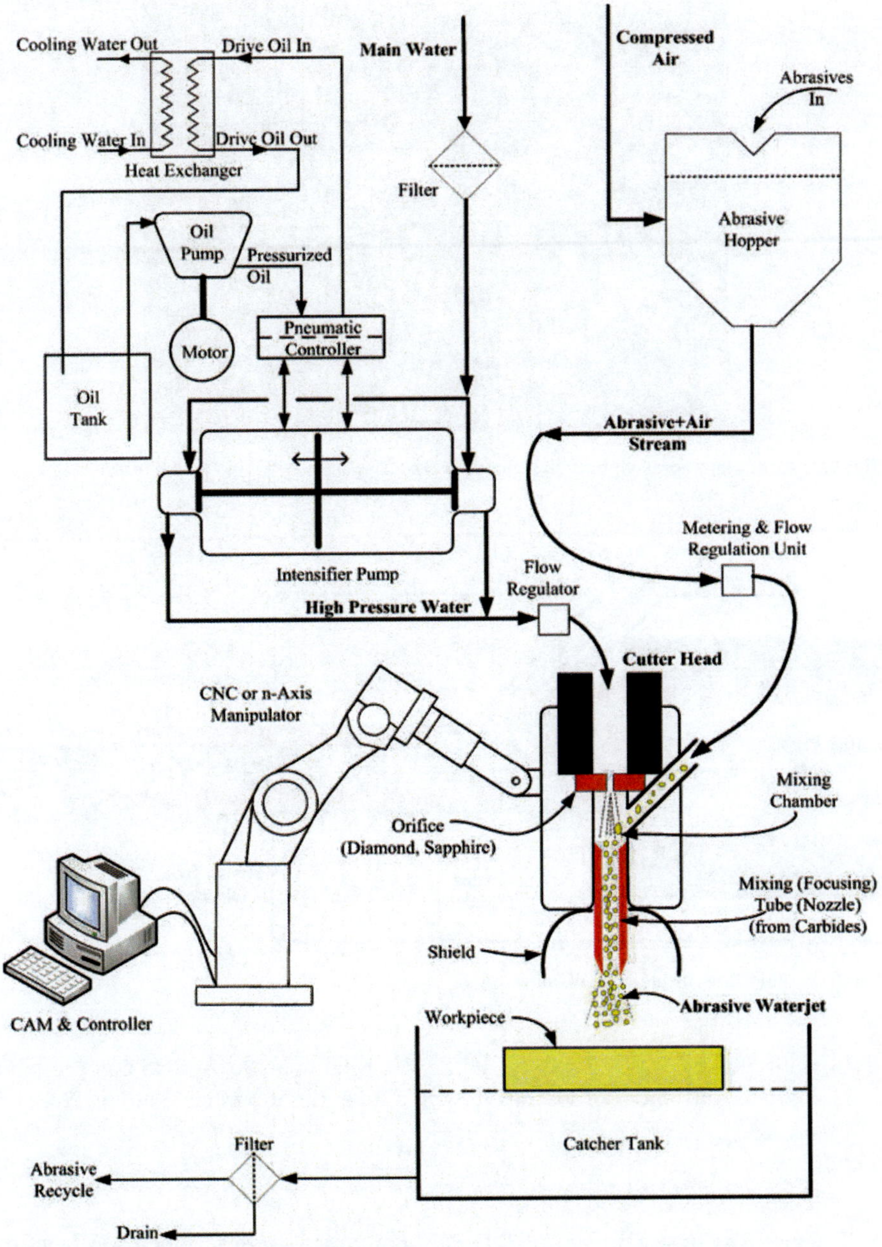

Fig. 9.1 Schematic of an abrasive waterjet (AWJ) cutting system

Fig. 9.2 Corundum abrasive particles, nominal size 27 μm

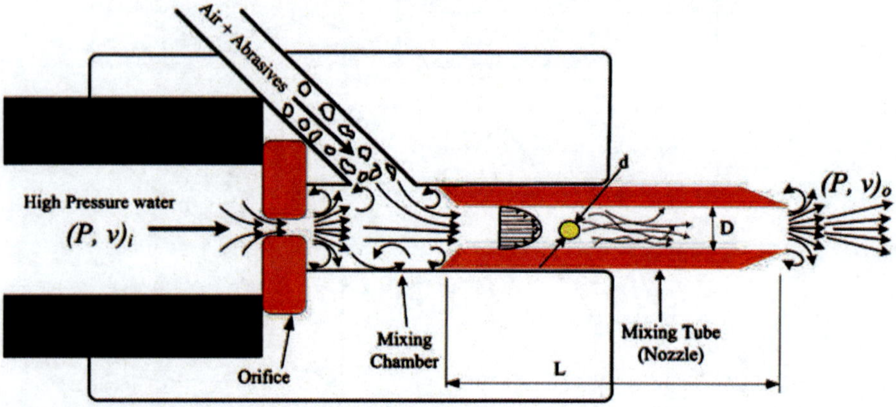

Fig. 9.3 Fluid flow inside an AWJ nozzle

In reality, water is viscous fluid and flow is rotational. Air is also compressible and its flow is entropic. For viscous flow in a tube, the Reynolds number, *Re*, is

$$Re_{tube} = \frac{\rho v D}{\mu}. \tag{9.3}$$

The viscosity of water is $\mu \approx 0.001$ Pa.s. For typical well-designed AWJ cutting heads, the orifice opening in Fig. 9.3 is 0.25–0.5 mm. Inner diameter of the mixing tube is typically ~3× orifice diameter, i.e. $0.75 \leq D \leq 1.5$ mm. The mixing tube length, *L*, is typically $L/D \approx 100$. Hence, typical AWJ conditions ($10^5 \leq Re \leq 10^6$) are well into the turbulent flow regime. For turbulent flow, fluid motion is no longer one-dimensional, nor laminar streamlines, and there is a significant amount of energy loses due to vortices and secondary flows. The real exit velocity of a pure waterjet is less than (9.2). Pressure loses occur at entry and exit across the orifice, entry and exit across the mixing tube, in the mixing chamber, and due to friction inside the mixing tube. The pressure drop is given by

$$\Delta P = k\frac{\rho v^2}{2}, \tag{9.4}$$

where k is a loss coefficient that is found experimentally and tabulated in fluid mechanics textbooks [36]. For well-designed and unworn nozzles, $k \approx 0.05$, for each loss component. Adding all the losses, the pressure drop from (9.4) is ≈20% of initial pressure, P_i. The real exit velocity of a pure jet is $v_{0r} \approx 0.9\ v_0$. This quick calculation formula matches experimental measurement within ±5% [37].

This estimate is for the *average* velocity of the jet. Yet, point velocity within the jet varies. It is maximum at the centerline, and almost zero at the turbulent boundary layer near the wall. The variation of point velocity, u, as a fraction of maximum center velocity, U, for turbulent flow, follows a 1/7 power law [36]:

$$\frac{u(r)}{U} = \left(1 - \frac{2r}{D}\right)^{\frac{1}{7}}, \tag{9.5}$$

where r is radial distance from center of the tube, $0 \leq r \leq D/2$, and $U/v_{0r} = 8/7$.

9.2.2 *Abrasive Jet*

It is common to consider that abrasive particles are entrained by the fluid stream and gain acceleration, with negligible initial velocity of the particles and perfectly smooth spheres inside the mixing tube. The accelerating motion of a sphere relative to a turbulent viscous fluid has no analytical solution, some asymptotic approximations [38–41] are valid only during the entry stage or very slow (creeping) laminar flow [36]. An approximate expression for the drag force, F_D, applied by the fluid on the particle is

$$F_D = C_D\frac{\pi\rho}{8}d^2w^2 + \frac{\pi\rho}{12}d^3\dot{w}, \tag{9.6}$$

where C_D is drag coefficient, d is diameter of the spherical particle, w is relative velocity between the particle and fluid, and the dot indicates time derivative. The first term on the right-hand side is due to viscous drag between the sphere and fluid under *steady flow* motion, while the second term is the inertia component due to relative acceleration, which is estimated (albeit for ideal potential flow) from the concept of *virtual mass* [42–44]. The drag force is also given by

$$F_D = \frac{-\pi}{6}\rho_s d^3\dot{w}, \tag{9.7}$$

where ρ_s is the density of the sphere. Equating (9.6) and (9.7), simplifying, and solving the differential equation, with initial condition $w(0) = v_{0r}$, gives

$$\frac{w(s)}{v_{0r}} = \exp^{\left(\frac{-C_D \rho s}{2d(2\rho_s+\rho)}\right)}, \tag{9.8}$$

where s is the distance along the length of the mixing tube. The relative velocity decays exponentially with the distance along the mixing tube. At nozzle exit, $s = L$. It can be seen that particle exit velocity is always less than v_{0r}, but can be maximized (i.e. $w(L) \to 0$) by increasing the drag coefficient or mixing tube length, or by decreasing the size or density of the sphere (but this also reduces the kinetic energy stored in the particle). Increasing mixing tube length improves exit particle velocity; but for economic (e.g. nozzle wear and cost) and operating (e.g. pressure losses along the tube) reasons, it cannot be made too long. The drag coefficient can be approximated as [45]:

$C_D = 0.5$	$400 \le Re < 3 \times 10^5$
$C_D = 0.08$	$Re = 3 \times 10^5$
$C_D = 0.000366\, Re^{0.4275}$	$3 \times 10^5 < Re \le 2 \times 10^6$
$C_D = 0.18$	$Re > 2 \times 10^6$

where, for a sphere,

$$Re_{sphere} = \frac{\rho w d}{\mu}, \quad w \to v_{0r}. \tag{9.9}$$

Hence, there is a critical value, $Re = 3 \times 10^5$, beyond which C_D drops sharply, and any advantage gained from using higher pump pressure is lost. Increasing water viscosity, μ, reduces turbulent flow and enables more efficient transfer of energy. It was found experimentally [46] that polymeric additives (e.g. super-water) result in bigger compact length and better performance of AWJ.

In AWJ, a commonly used abrasive material is garnet, whose density varies depending on the supplier, but on average $\rho_s \approx 4\ \rho_w$. At nozzle exit, $s = L \approx 100\ D \approx 75$ mm for a 0.75 mm nozzle, and from (9.9) and (9.8), it can be found that

$$\frac{w}{v_{0r}} = e^{\left(\frac{-1}{480d}\right)} \ \&\ v_{0r} d \leqslant 0.3. \tag{9.10}$$

Finally, utilizing (9.2) an approximate expression can be obtained for the average garnet particle velocity as it exits the nozzle of an AWJ system, i.e.

$$v_p \approx 0.04\sqrt{P_i}(1 - e^{-1/480d}) \ \&\ d\sqrt{P_i} \leqslant 7.4. \tag{9.11}$$

Real abrasive particles are not spherical. Their irregular shapes result in greater turbulence and lower drag forces. Some energy is consumed in rigid body rotation of the particles. Further, the above analysis assumes only a single particle. Real AWJ streams involve numerous particles, and the energy they absorb reduces energy available to the fluid. Hence, v_{0r} tends to decreases along the length of the mixing tube. Moreover, drag forces on one particle are lower when it is moving within the wake of another particle. The random motion of particles leads to radial velocity, frequent inter-particle collisions, and collisions with (and wear of) the wall of the mixing tube. Furthermore, inter-particle collisions cause particle fragmentation, so that particle size distribution at nozzle exit may be significantly different from as-supplied [47]. Finally, (9.11) applies to *average* particle velocity. From (9.5), fluid velocity varies along the radial direction, and not all particles are accelerated equally. However, due to random particles sizes and shapes, turbulent mixing, and inter-particle collisions, those variations even out and it is a good estimation to consider that abrasive particles exit the nozzle at almost uniform velocity distribution.

As the abrasive waterjet is released from the nozzle, constraints applied by the tube are removed, and the jet is free to expand radially at the expense of further decrease in axial velocity [37]. In AWJ cutting, there is no advantage in placing the workpiece at large standoff distance from the nozzle. Standoff is only a clearance, to avoid collision between the workpiece and nozzle, and to allow an escape path for rebounding abrasive particles. AWJ machine manufacturers recommend that standoff distance should be maintained $\leq 3D$, i.e. $\approx$2 mm. At such small standoff, velocity drop after exit from the nozzle is negligible [48].

All the above uncertainties and inefficiencies reduce the real particle velocity, v_{pr}, as it hits the workpiece surface. It was found from the balance of linear momentum, and experimentally [49], that

$$v_{pr} = \frac{\alpha v_p}{1 + \dot{m}_p / \dot{m}_w}, \tag{9.12}$$

where $\dot{m}_p$and $\dot{m}_w$are mass flow rates of the abrasives and waterjet, respectively, and α is a mixing efficiency coefficient. For well-designed systems, α is typically $0.57 \leq \alpha \leq 0.71$. AWJ manufacturers recommend an optimum abrasive flow rate$\dot{m}_w / \dot{m}_p \approx 50$. Notice that (9.12) applies only to the axial component of particle velocity. Both the radial velocity component and rigid body rotation are ignored, because of their lesser effect on material removal.

9.2.3 Design Rules

If pressurized water at 400 MPa is released through a 0.25 mm orifice, expands inside a 0.75 mm mixing tube and hits a target surface spot ~1 mm diameter, even assuming no loses inside the nozzle, impact pressure at the surface is about one order of magnitude less. Such low pressure cannot induce plastic deformation or material removal, except in very soft materials, e.g. plastics, leather, timber, etc.

[50] A pure waterjet is not suitable for machining metals, ceramics, or other hard materials. At a large standoff distance, a waterjet may disintegrate into droplets, whose impact pressure may be high enough to induce residual stresses, e.g. waterjet peening [30]. Prolonged exposure to waterjet may cause material failure by fatigue, and consequent delamination of a thin surface layer [10, 11, 51, 52], e.g. waterjet cleaning, descaling, paint stripping, etc. [53, 54]

In AWJ and AAJ machining of hard materials, the fluid has no direct effect on material removal, and the impact pressure by the fluid can be ignored. However, water (and to lesser extent, air) has an important indirect role in machining: it

- Stores and transfers the energy required to accelerate abrasive particles,
- Guides the abrasive particles and focuses impacts within a small spot,
- Flushes debris and abrasive particles away from the working zone, and ensures that fresh surface material is always exposed,
- Provides some lubrication between the particle and workpiece, reduces frictional heating, and
- Provides an effective convection cooling medium, where heat generated during deformation is immediately extracted from the workpiece. Indeed, this is the major advantage of AWJ, which is often viewed as a "cold" machining process: a first choice when cutting heat-sensitive materials.

Notwithstanding all the above contributions of the waterjet, Sect. 9.3 below focuses only on material removal due to impact by the abrasive particle and ignores the effect of water on the process.

From the above considerations, it seems appropriate to model the average real abrasive particle velocity at the time of impact on the workpiece surface as

$$v_{pr} = A \frac{\dot{m}_f \sqrt{\frac{P_i}{\rho_f}}}{\dot{m}_p + \dot{m}_f} \left\{ 1 - e^{\left(\frac{-BL}{d}\right)\left(\frac{\rho_f}{2\rho_p + \rho_f}\right)} \right\} \quad \& \quad \frac{Cd}{\mu_f} \sqrt{P_i \rho_f} \leqslant 10^5, \tag{9.13}$$

where the subscripts p and f refer to the abrasive particles and jet fluid, respectively, A, B, and C are parameters that depend on the design of a particular nozzle, vary with nozzle wear [55], and are best determined experimentally for the specific configuration, and all other symbols are as defined earlier. In the absence of better knowledge, for AWJ and for v_{pr} within $\pm 15\%$, assume $\alpha \approx 0.65$, $A \approx 0.8$, $B \approx 0.25$, and $C \approx 0.65$. For pure airjet or AAJ, due to compressibility of air, the factor of 2 in (9.2) is reduced to 1.4. A more thorough analysis of AAJ can be found in [56].

Equation (9.13) captures all the essential features of AWJ generation and the effect of all major variables:

- Increasing pressure, P_i, increases particle velocity, v_{pr}, to a limit, beyond which a sharp decline in drag, B, prevents additional velocity gains.
- Reducing particle size, d, or density, ρ_p, or increasing mixing tube length, L, can yield limited increase in particle velocity. When the exponential term in (9.13) almost vanishes, they become ineffective.

- Increasing fluid density, ρ_f, reduces particle velocity.
- Increasing fluid viscosity, μ_f, is beneficial, provided pressure losses inside the mixing tube are not excessive.
- For AWJ, maximizing the abrasive flow rate is always beneficial, as long as the fluid to abrasive flow ratio $\dot{m}_f : \dot{m}_p$ is maintained under 50:1.
- Changing abrasive particle size, within the limits in (9.13), at a *constant abrasive flow rate*, alters the number of abrasive particles per second, but does not change the total energy stored, nor average particle velocity, v_{pr}.
- Only very short mixing tubes lead to reduced particle velocity. At the typical $L/d \geq 300$, the effects of tube length and particle size cancel out.

A typical commercial AWJ system, similar to Fig. 9.1, uses a 0.25 mm orifice, $D = 0.75$ mm and $L = 75$ mm mixing tube, filtered water, and garnet abrasives. For such a system, (9.13) becomes

$$v_{pr} \approx 0.025\sqrt{P_i}\left(1 - e^{-1/480d}\right) \;\&\; d\sqrt{P_i} \leqslant 4.8, \tag{9.14}$$

which is a more accurate estimator $\dot{m}_w/\dot{m}_p \leqslant 50$than (9.11). At 400 MPa, and provided that nominal particle size $d < 0.24$ mm and, then the average particle velocity at the instant of impact is $v_{pr} \approx 500$ m/s. This result and the closed-form solution in (9.13) are in agreement with the numerical estimation in [57] and other analytical and experimental results [58], but it is simpler and more intuitive to use.

9.2.4 Research Directions

A proper analysis of AWJ or AAJ would require the solution of a number of coupled partial differential equations for the three-phase fluid dynamics of a mixed flow of water, air bubbles (for AWJ; or air with moisture content, for AAJ) and solid abrasive particles; along with the contact and impact solid dynamics among the particles, and between particles and the mixing tube wall; plus thermodynamics and heat transfer, multi-body kinematics, and statistical mechanics. The fluid dynamics is viscous compressible, rotational, entropic, turbulent, mixed-phase with moving, deformable, discontinuous, and randomly-moving boundaries. The solid dynamics involves elastic-plastic multi-body collisions, fracture, fragmentation, and all sorts of constitutive materials behaviors; and subject to random body motions and weakly defined boundary conditions. All those behaviors are coupled in an exceedingly nonlinear manner that precludes the possibility of any closed-form solution, even in some statistical sense. It is safe to say that a problem like AWJ is not solvable, from first principles of continuum and statistical mechanics. Even with current advances in computational methods, and the exponential growth in computing power and high-performance computers, this problem remains a challenge, and there may be decades before it can be simulated numerically with satisfactory realism.

Efforts towards more advanced analysis of jet generation and characteristics are a huge volume of research, as reviewed in [37, 59], and include:

- Development of high-pressure equipment and systems;
- Optimization of orifice and nozzle design and alignment;
- Characterization of abrasives size, shape, and velocity distributions;
- Estimation and minimization of pressure drop, energy and momentum losses, and other inefficiencies during the mixing process;
- Particle acceleration models, similar to that in (9.13);
- Particle fragmentation and size reduction; and
- Wear of the mixing (focusing) tube and its effect on jet characteristics; to name but some of the research topics. Most of those issues are relevant to machine and nozzle design. Once a specific machine or cutting head is selected, those parameters cannot be altered. Hence, for production and process planning of AWJ, (9.13) remains a more useful model of controllable process variables.

There are significant efforts to simulate the jet flow using Computational Fluid Dynamics (CFD). These include [59–61] solving the 2D axi-symmetric problem for the axial and radial velocity distributions of the fluid jet, then fitting the numerical solution to an equation, which is then used to estimate particle velocity. Similar treatments were reported in [62, 63]. In [64] random position and velocity of particles were simulated by mean of a fractal point set, without physical justification why abrasive jets may have a fractal dimension, or why their motions follow any specific statistical distribution. Further, at the current state of computational abilities, it is still difficult to simulate the multi-physics of the coupled motion of both fluid and particles, as they interfere with the behaviour of each other in a highly nonlinear manner.

There have been extensive efforts in developing techniques for experimental measurement of pure waterjet and abrasive particle velocity inside the AWJ [37] and AAJ [56]. Techniques include magnetic and inductive methods, reverse estimation from impact crater depth or from impact force measurement, various types of laser (interference, time of flight, or Doppler) velocity meters, or using ultra-high speed photography. All techniques include measurement uncertainties, difficulties in discriminating between fluid and particle, and they only detect average behaviour outside the nozzle, i.e. not the axial and radial distributions inside the nozzle. Results from experimental measurements are in agreement with (9.13).

9.3 Material Removal by Impact

In a typical AWJ machine, abrasive particles mass flow rate could be ~10 g/s. For garnet particles, and nominal average size ~0.15 mm, this amounts to ~10^6 particles per second; which may seem a huge number, but it is only one particle per microsecond. At impact velocity ~500 m/s, the full process of particle indentation and rebound lasts a fraction of μs. The jet spreads over ~1 mm spot size, ~40× the area for a single particle impact. Traverse (workpiece displacement) speed is typically 1–100 mm/s, which is negligible compared to impact velocity, and only fraction of a μm/μs – which is negligible compared to particle size.

From this quick calculation, it is reasonable to treat AWJ as a sequence of single-particle impacts: each particle indents, removes material, and rebounds away from the target as an independent event, without interference from other particles or from workpiece motion. The final effect of an AWJ cut is the geometric superposition of craters produced by all particles. This may not be simple to calculate due to the random variation in particles sizes, shapes, and times of their arrival at particular spots on the target. There may also be interference among incoming and rebounding particles, removed debris, or the waterjet. However, it should be possible, in principle, to develop an average model to account for all those perturbations, *if* the effect from single particle impact is known with reasonable confidence.

In AWJ, all forms of irregular particle shapes can be found in the abrasive jet. Rigid body rotation of the particles inside the jet stream, sliding and rolling motion during impact on the target, and localized deformation and fragmentation of the particles during impact, all provide additional uncertainty. It is not possible to estimate the distribution of real particle geometries at the instant of impact on the surface. It is assumed that all those random effects balance, and some "average" abrasive particle behavior is modeled by spherical impact. The main advantage of such assumption is that the contact mechanics of a sphere is one of the most studied and understood [65–69].

In general, material response to impact is described either as *ductile* or *brittle* erosion, as shown in Fig. 9.4. Ductile erosion is relevant to metals and similar materials that are capable of significant plastic deformation. Brittle erosion applies to ceramics, glasses, and hard metals that crack and fragment under impact.

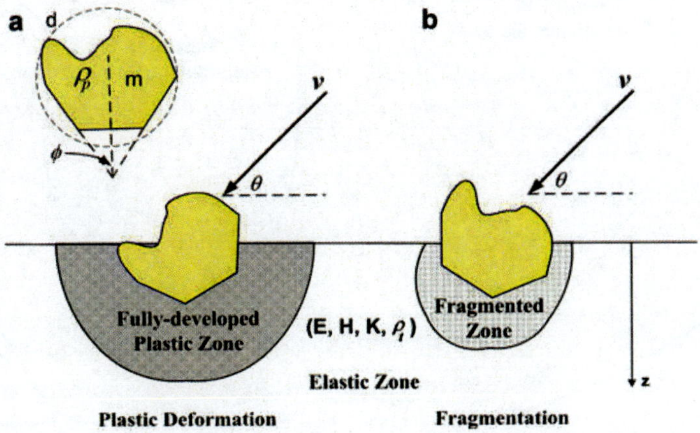

Fig. 9.4 Ductile and brittle erosion modes

9.3.1 Ductile Erosion

Considering a rigid particle hitting a smooth flat surface, as shown in Fig. 9.4, the general oblique impact condition can be decomposed into *normal* (indentation-like) and *tangential* (scratching-like) components. For normal impact, $\theta = 90°$, the particle pushes the surface down, and material beneath is compressed. As indentation continues, material in the high compression zone escapes pressure, by moving sideways, then upward towards the free surface. After the particle rebounds, a permanent crater is left on the surface, and excess lips (elongated flakes) are formed. The successive application of numerous particles results in cantilever formations that are easy to separate from the mother material (Fig. 9.5).

If an abrasive particle has mass, m_p, and initial impact velocity, v, its stored kinetic energy is ½ $m_p\, v^2$. Upon impact on the target surface, some fraction of the kinetic energy is consumed in displacing the target surface and forming a permanent crater, such that [1]

$$m_p(v\sin\theta)^2 \propto \int_0^h HA(x)dx, \tag{9.15}$$

where H is target material hardness, and h is indentation or crater depth. The projected area function, $A(x)$, is unknown, except for some regular shapes. However, exact expression for that function is not required, by noting that crater volume, V, is

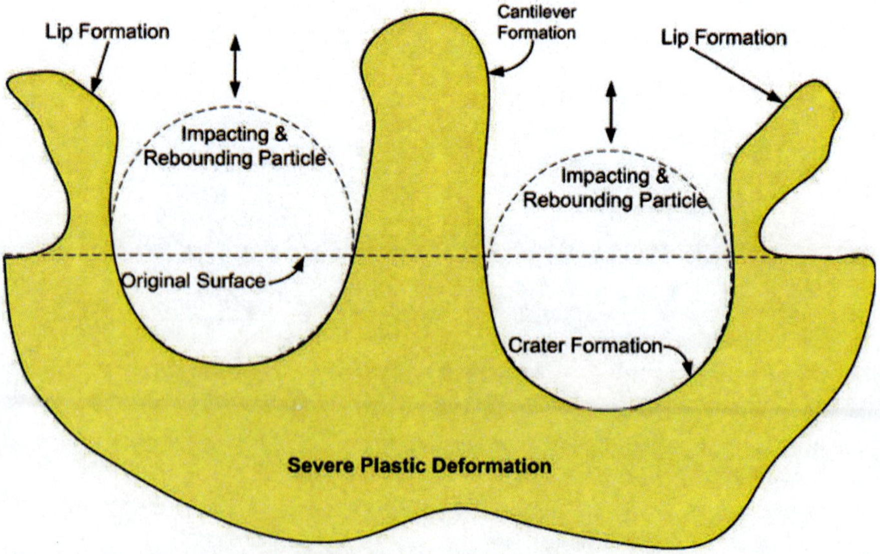

Fig. 9.5 Material displacement and removal by successive particles impact

$$V = \int_0^h A(x)dx. \tag{9.16}$$

Consequently, the volume removed by impact, V_r, is

$$V_r \propto \frac{m_p (v \sin \theta)^2}{H}, \tag{9.17}$$

which is a relative estimate of material removal in ductile materials by normal impact.

During tangential (scratching) motion, as shown in Fig. 9.6, shear stresses increase at the expense of compressive stresses. Stresses become tensile on the trailing edge and more compressive on the leading edge of the particle. Crater formation includes the ploughing effect from sliding and friction. The volume removed in this case, as in (9.17), is also proportional to the kinetic energy of the particle:

$$V_r \propto \frac{m_p (v \cos \theta)^2}{H}. \tag{9.18}$$

The combined volume removed due to oblique impact is obtained by addition of the normal and tangential components:

$$\frac{V_r}{m_p} = k_n \frac{(v \sin \theta - v_n)^2}{H} + k_t \frac{(v \cos \theta - v_t)^2}{H} \tag{9.19}$$

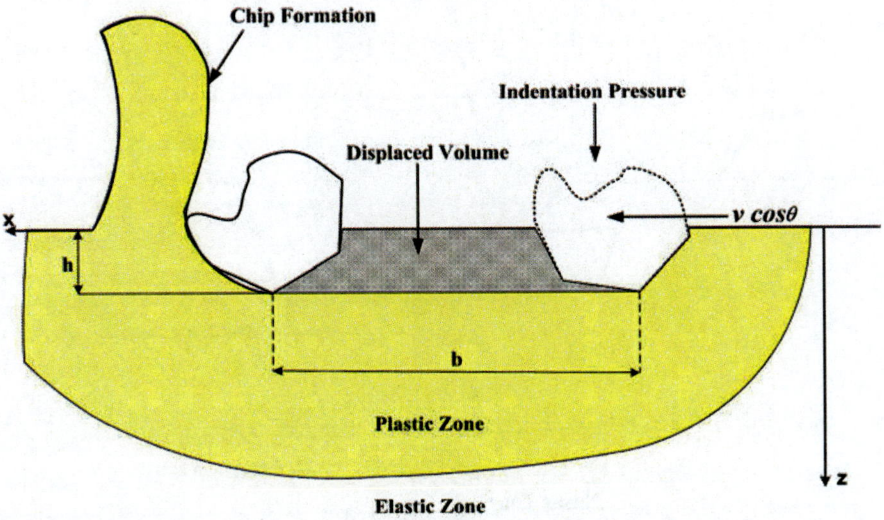

Fig. 9.6 Material removal by tangential (scratching) abrasive motion

where k_n and k_t are experimental coefficients, and v_n and v_t are threshold velocities below which no material removal can occur. The erosion model in (9.19) is similar to that proposed by Neilson and Gilchrist [70], which is a generalization of earlier erosion models by Finnie [71] and Bitter [72, 73]. There are many other ductile erosion models, e.g. [74], as reviewed in [31].

9.3.2 *Brittle Erosion*

When a target material is not capable of significant plastic deformation, it responds to impact loads by fracture, forming cracks, and fragmentation of the affected zone. Cracks can be of the ring type (due to elastic stress waves [75]), conical type (due to quasi-static tensile failure [76]), radial, median, and lateral type (due to stress redistribution in the plastic zone at the crack-tip during loading and unloading) [77–79], as shown in Fig. 9.7. The estimation of those brittle fracture mechanisms require some assumptions about flaws sizes and their distribution; the location, size, and growth rate of the various cracks; and the interactions among various cracks. Such models eventually resort to experimental data for quantitative verification.

All brittle erosion models can be reduced to a causality formula of the form:

$$V_r \propto \rho_p^{q_1}\ d^{q_2}\ v^{q_3}\ H^{q_4}\ K^{q_5}\ E^{q_6}, \tag{9.20}$$

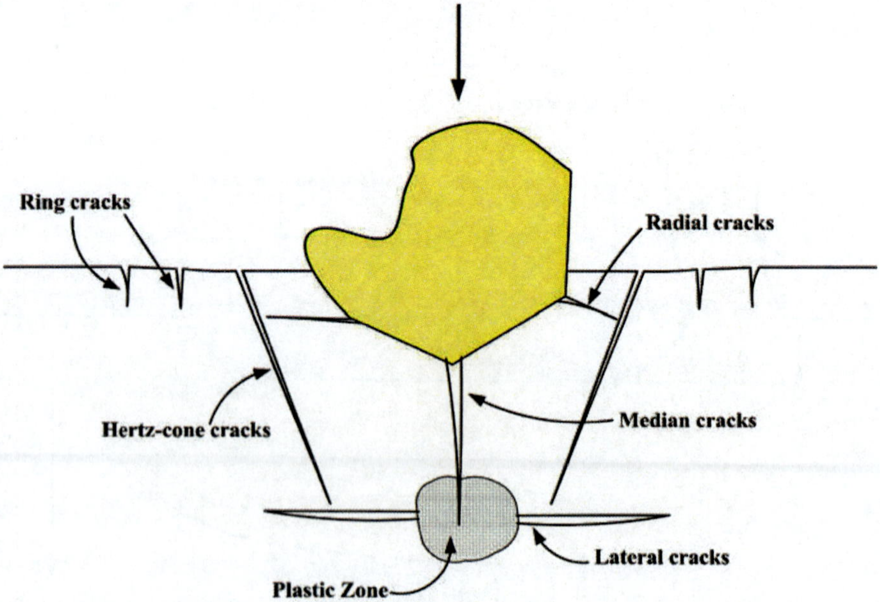

Fig. 9.7 Brittle erosion by cracking and fragmentation

where ρ_p is particle density, d is particle size, v its impact velocity, and H, K and E are hardness, fracture toughness and Young's modulus of the target material, respectively. Depending on the specific fracture model being assumed, the exponents in (9.20) are found to be: $q_1 = 0$–1.6, $q_2 = 3$–4.2, $q_3 = 2$–3.2, $q_4 = -1.4$ to 1, $q_5 = -2$ to -1, and $q_6 = 1$–1.25. For quantitative accuracy, it is appropriate to evaluate those exponents experimentally for each specific particle-target materials combination.

9.3.3 Unified Erosion Model

It is noted that all ductile erosion models, e.g. (9.19), predict that $V_r \propto v^2$, while brittle erosion models, e.g. (9.20), predict that $V_r \propto v^3$. This is in disagreement with experimental observations, where it is found that the velocity exponent is dependent on material property, and is 2.2–2.5 for metals and 2.6–3 for ceramics. The deviation between models and experiment is understood, because of the fact that no real material is absolutely ductile or absolutely brittle, and some interpolation between the two extreme idealizations is needed. As impact velocity is the most important variable in impact erosion, error in estimating the velocity exponent can degrade the accuracy of all those models.

Further, it is observed that maximum erosion occurs at an impact angle, θ_{max}, of near 90° for brittle materials at higher velocity or using coarser abrasive particle (Fig. 9.8). The peak erosion angle is 20–30° for ductile materials at lower impact velocity or using finer particles. The above models are unable to explain this peak shifting. As impact angle is the second most important factor affecting material removal, this represents an added source of inaccuracy.

In a yet to be published work, the authors present a unified erosion model that is applicable to ductile and brittle materials alike. It is able to interpolate between the extreme cases of ideal plasticity and ideal fracture, and provide a more realistic representation of real engineering materials. The model can explain the variable velocity exponent, the peak erosion angle shifting, the effect of various material properties, the effect of particle size and shape, and other effects such as ductile-brittle transition and size effect. According to such a model

$$\frac{V_r}{m_p} = (A_1 \sin^2\theta + A_3 \cos^2\theta)v^2 + (A_2 \sin^3\theta + A_4 \cos^3\theta)v^3, \tag{9.21}$$

where

$$A_1 = k_1 \frac{10^4 \Delta}{\alpha + \beta H}, \quad A_2 = k_2 (1 - 10^4 \Delta) \left(\frac{d}{2H\Delta^2} \right)^{3/2}, \quad A_3 = A_1 \frac{k_3}{k_1}, \quad A_4 = A_2 \frac{k_4}{k_2},$$

$$\Delta = \frac{K}{\sqrt{2EH(1 + 2\varepsilon_f/\varepsilon_e)}}, \tag{9.22}$$

α and β are material constants relating dynamic to quasi-static hardness, ε_f and ε_e are fracture and elastic strains (material properties of the target material), and k_1–k_4 are experimental parameters. For specific abrasive particle-target materials combinations, it is better to use (9.21), with A_1–A_4 determined experimentally.

9.3.4 Material Removal by Abrasive Jet

For ductile materials, Hashish [80] proposed an AWJ machining model based on the initial Finnie model [71]. This was later improved [81] based on the revised model by Finnie and Sheldon [82]. Numerous models (e.g. [83–85]) have been proposed over the years, and they are all fundamentally similar, and are based on some modification of (9.19). For brittle materials, various models for AWJ machining have been proposed (e.g. [86–88]), and they are all based on the quasi-static fracture models developed in [77–79], as generalized by (9.20). All such AWJ machining models are subject to limitations imposed by their parent impact erosion models, as discussed in Sect. 9.3.3. Specifically, these models cannot predict the material-dependent velocity exponent, erosion peak angle shifting due to particle size, nor the ductile-brittle transition experience in real material behaviour, e.g. Fig. 9.8.

The average velocity of the abrasive particle, as it exits the nozzle, can be estimated from (9.13). The average target material removal due to impact by the abrasive particle can be estimated from (9.21). For a jet stream containing

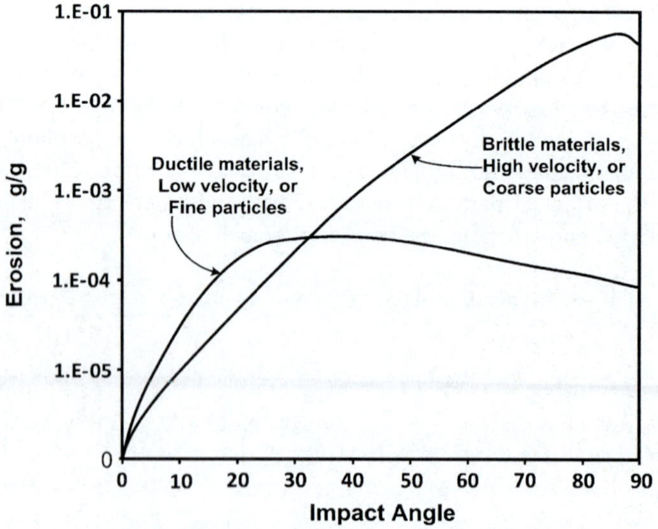

Fig. 9.8 Peak erosion angle shifting with material and impact conditions

numerous abrasive particles, it is suitable to take the time derivative of (9.21), and express the material removal rate per abrasive particle mass flow rate, as

$$E = \frac{\dot{V}_r}{\dot{m}_p} = \frac{\delta D U}{\dot{m}_p}, \tag{9.23}$$

where δ is depth of cut, D is nozzle diameter (approximately equal to width of cut, for small standoff distance), and U is traverse speed of the jet on the surface of the target material. Typical removal rates using AWJ cutting are listed in Table 9.2.

For a given nozzle diameter and abrasive flow rate (as defined in Sect. 9.2.3), it is seen that volume removal rate is conserved. There is direct trade-off between depth of cut and traverse speed: increasing traverse speed decreases the depth of cut. The only way to increase depth of cut at constant traverse speed, abrasive flow rate, and nozzle diameter is by increasing the erosion rate itself, as dictated by (9.21). For a given target material, maximizing erosion is achieved by either maximizing the impact velocity, v in terms of (9.13), or by selecting the optimum impact angle for peak erosion, as in Fig. 9.8.

Using (9.23), it is possible to estimate a typical depth of cut due to AWJ, using any jet parameters, and for any target material. Real material removal by AWJ is more complicated than described above. As the AWJ traverses across a target workpiece, a kerf is gradually developed. This process was extensively studied using high-speed camera visualization techniques [89, 90], and is outlined in Fig. 9.9. Initially, abrasive particles impact normal to the workpiece surface, and material removal is similar to Fig. 9.5. However, as the jet proceeds and some material is removed, a ramp is formed: abrasives then approach tangential impact as in Fig. 9.6. As the jet continues its traverse, a step is formed, and a condition of normal impact is restored. The cycle continues repeatedly, until a steady state condition is reached; a kerf profile, having a step-wise ramp shape, is maintained for the rest of the cut.

As can be seen from the right-most view in Fig. 9.9, under steady state conditions, some particles impact on the material at near normal (indentation-like), while other impact at near tangential (scratching-like) direction. For historic reasons, Hashish [80, 89] refers to the first condition as "deformation wear" and to the latter as "cutting wear". In reality, material is removed during both cases, but the mechanism varies, as shown in Figs. 9.5 and 9.6. The importance of kerf formation can be understood from inspection of Fig. 9.8. It is seen that material removal rate is maximized for brittle

Table 9.2 Typical removal rates by AWJ for depth $\delta = 25$ mm

Material	Speed, U (mm/s)	Material	Speed, U (mm/s)	Material	Speed, U (mm/s)
Carbon steels	0.5–2	Aluminum	0.3–3.5	Glass	35
Alloy steels	0.2–1	Titanium	0.2–2	Quartz	0.3
Stainless steels	0.2–1.5	Nickel alloys	0.04–0.15	Alumina	0.05–0.3

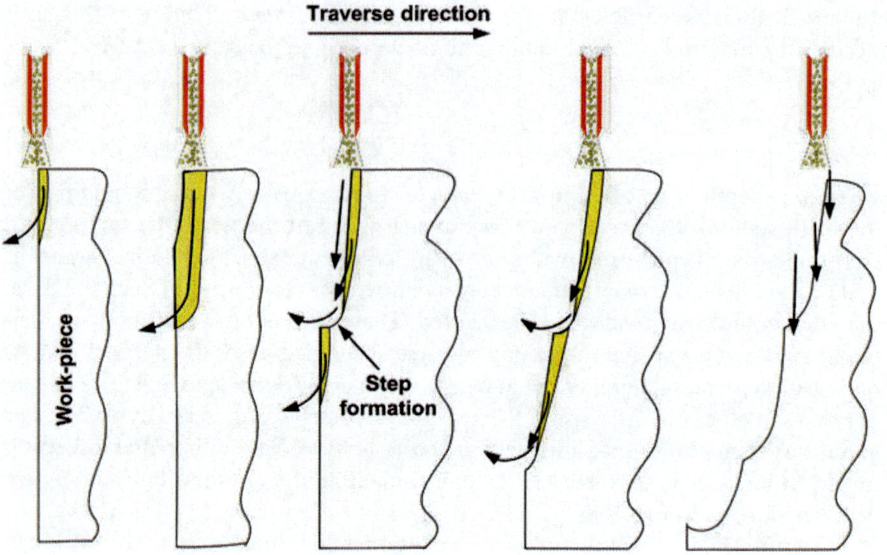

Fig. 9.9 Step formation during abrasive jet traverse

materials when particles impact near normal. For ductile materials, material removal is maximized when particles impact at 20–30° to the surface.

The geometry in Fig. 9.9 results in particles impacting at angles whose average is the optimum angle for neither ductile nor brittle materials. At present, there is no satisfactory theory to estimate the equivalent average impact angle and kerf formation, in terms of process variables. However, from geometry, the kerf can have a steeper ramp when the traverse speed is lower. This, in turn, reduces the average impact angle, and brings it closer to the scratching mode. Therefore, it is more efficient to cut ductile materials at slower traverse speed (smaller effective impact angle), while faster traverse speed (larger effective impact angle) is more productive for machining brittle material. However, as seen from (9.23), there is interdependence between traverse speed and depth of cut. A better alternative to enhance (and control) the effective impact angle by tilting the nozzle, as will be explained in Sect. 9.4.1.

From the circular cross-section of the jet, it is clear that more abrasive particles impact near centerline of the cutting path than near its edge. Moreover, there is greater loss of kinetic energy near edge of the jet, due to abrasive particle deflections, inter-particle collisions during rebound, and interference from the fluid jet. This results in tapering of the cut groove, as shown in Fig. 9.10, and the final cut can be divided into three distinct zones [91, 92]:

- Smooth cut zone: where the tapering effect is relatively small and surface roughness of the produced wall is reasonably fine.
- Rough or striation cut zone: where the tapper angle is significantly larger, and significant scratch marks and rough wall surface is observed.
- Pocket formation: where material removal is due to random collisions from rebounding particles.

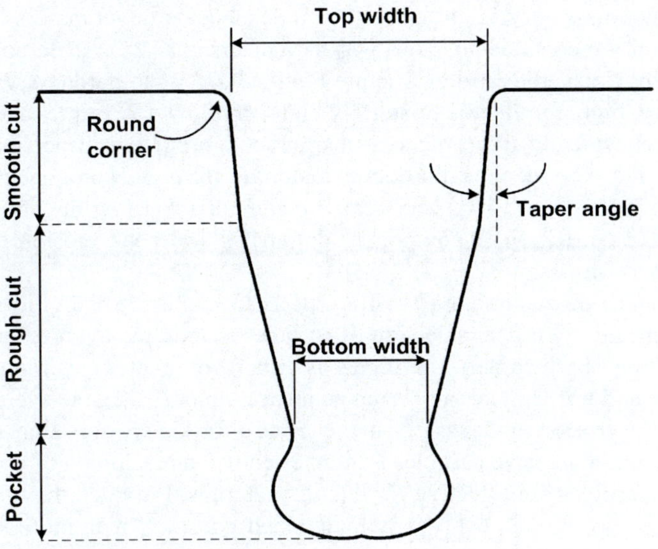

Fig. 9.10 Cross section of machined groove

The pocket formation stage is practically useless for any productive machining application. The rough cut is useful only during crude applications, when the only purpose is to cut the work material into parts at the fastest possible rate, regardless of dimensional tolerances or quality of the produced surfaces. For all precision applications of AWJ, the smooth depth of cut represents the practically usable range of material removal.

Due to the random nature of particles impact and collisions, it is practically impossible to estimate the exact shape of the cut groove from process parameters alone. There is currently no satisfactory model for estimating the exact geometry of the kerf, taper angle, or smooth depth of cut. The best option seems to be semi-empirical in nature, where the qualitative behavior is established from dimensional analysis and energy conservation considerations, and the exact quantitative behavior is obtained by fitting to experimental data. Models for estimating the smooth depth of cut, rough depth of cut, taper angle, and surface roughness of the finished surface can be found in [93–99].

9.3.5 *Design Rules*

To maximize material removal rate by AWJ machining, the most important factor is to maximize particle impact velocity. This is achieved by proper selection of parameters in (9.13) according to the general guidelines given in Sect. 9.2.3. Next, it is important to maximize the erosion rate per impacting particle, by selecting the appropriate impact angle, θ, in (9.21).

This selection is not straightforward, as it depends on target material properties, whose effects are contained in the A_1–A_4 parameters in (9.22). The crucial factor in material behavior is the ratio of fracture toughness, K_{IC}, to hardness, H. The ratio K/H is very high for ductile materials and very low for brittle materials (the difference is typically two orders of magnitude). From the inspection of (9.21) and. (9.22), it can be seen that for ductile materials, the $cos^2\theta$ component dominates (because, typically $k_3 \approx 3\ k_1$) and peak erosion occurs at low impact angles. For ductile materials, the $sin^3\theta$ component dominates and peak erosion occurs near normal impact angles.

Other means of maximizing erosion rate is to maximize the efficiency of the abrasive particles by proper selection of particles shapes and average size. A sharp particle, having sharp corners and angles, is capable of greater material removal for both ductile and brittle materials. From an inspection of (9.22), the average particle size enters the erosion model as $d^{3/2}$ in the terms of brittle erosion. Hence, it is clear that using coarser abrasive particles is more effective in machining brittle materials, while finer particles are effective for ductile materials. Provided the total abrasive flow rate (i.e. jet kinetic energy) is maintained constant, it is more effective to machine brittle materials using coarser particles. However, selection of particle size must be subject to the constraint imposed in (9.10), (9.11) and (9.13), where it is found that the fluid drag coefficient cannot be maintained high enough while both particle size and jet pressure are increased. In other words, increasing particle size may come the expense of reduced particle velocity (even for the same abrasive flow rate); thus, negating the advantage of the size effect.

To maximize erosion, it is also important that the abrasive particles remain intact and effectively rigid. Clearly, any energy consumed in deforming or fracturing the abrasive particle, instead of the target workpiece, is a waste of resources and should be minimized. A reasonable specification is to require that fracture toughness of the abrasive particle be greater than that of the target ($K_p > K_t$), and their relative hardness is $H_p > 2.5\ H_t$. This latter constraint can be expensive to achieve when machining very hard materials. It can be practically reduced to $H_p > 1.6\ H_t$ for hard materials. However, it is to be noted that $H_p > 1.25\ H_t$ is the absolute minimum requirement for abrasive particles, and below that limit, machining will be very wasteful, with most deformation occurring on the particle instead of the target workpiece.

9.4 Process Improvement

From the discussion in Sect. 9.3.4, there are three issues about material removal by AWJ that need to be addressed:

1. Due to the traversing motion of the nozzle, abrasive particles may impact the workpiece surface at non-optimum angles; thus, reducing the opportunity for maximizing material removal rate.

2. Due to energy loss with depth, a tapered cut is formed; thus, reducing the ability to achieve tight dimensional and geometrical tolerance and accuracy.
3. Due to both factors above, surface finish is significantly degraded (striation formations) at greater depth of cut, and may be unacceptable at the rough zone of cut.

One of the main strengths of AWJ is its flexibility and adaptability to various situations. As seen in Fig. 9.1, the cutting head can be held by a robot arm, or similar multi-degrees-of-freedom manipulator. Hence, there is a greater ability to alter the process, by simple manipulation of the nozzle kinematics. In this section, how managing the nozzle kinematics can result in minimizing the drawbacks mentioned above is discussed.

9.4.1 Nozzle Planar Tilting

There is no requirement to maintain an orthogonal relationship between the nozzle and the workpiece surface. The AWJ nozzle can be oriented along any direction that can optimize kerf formation and maximize material removal. Figure 9.11 shows three such possibilities of tilting nozzle in the cutting plane:

(a) The nozzle is tilted sharply forward, so that abrasive particles impact the workpiece at near normal angle, which is more effective for brittle materials. The cut groove is smooth and narrow, and depth of cut is controllable by the nozzle tilt angle and standoff distance.
(b) The nozzle is tilted slightly forward, 5–10°, so that abrasive particles impact at an effective angle 20–30° to the surface, which is optimum for cutting some materials, such ductile materials. The cut groove is narrow, but the depth of cut can be so large that large taper and the deep rough-cut zone are formed.

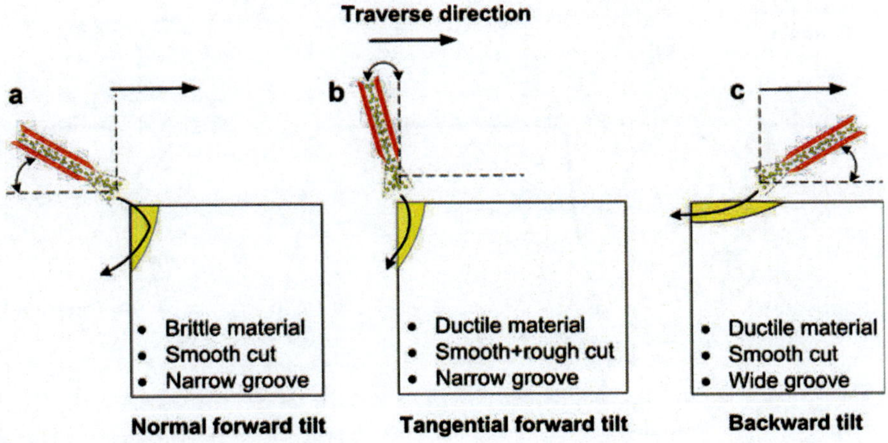

Fig. 9.11 Nozzle tilting configurations

(c) The nozzle is tilted backward so that abrasives impact at 20–30° to the *top* surface of the workpiece. This arrangement approximates the conditions in surface grinding, and is ideal for ductile materials. The cut groove can be smooth, with controlled depth, but taper angle is very large, with the cut groove having almost no flat bottom.

In practice, nozzle tilting in the configuration (b) is the most commonly adopted for AWJ cutting of various materials [59, 95, 97], particularly for ductile materials, and has been found [100] to increase the smooth depth of cut, total depth of cut, and even reduce surface roughness of the cut groove [59].

9.4.2 *Nozzle Lateral Tilting*

For the purpose of dimensional accuracy, it is practically important that taper angle is reduced to a minimum. Ideally, the cut groove should be square, with uniform width throughout the depth of cut. This is not achievable, in reality, due to reduced number of abrasive particles and kinetic energy near the wall of the groove. A remedy to this effect is achieved by slight lateral tilting of the nozzle (Fig. 9.12). By aiming the jet towards the walls of the groove, it is possible to remove more material from the side and, in principle, to correct for the taper angle and deliver a square cut, at least within the smooth cut zone. Experiments

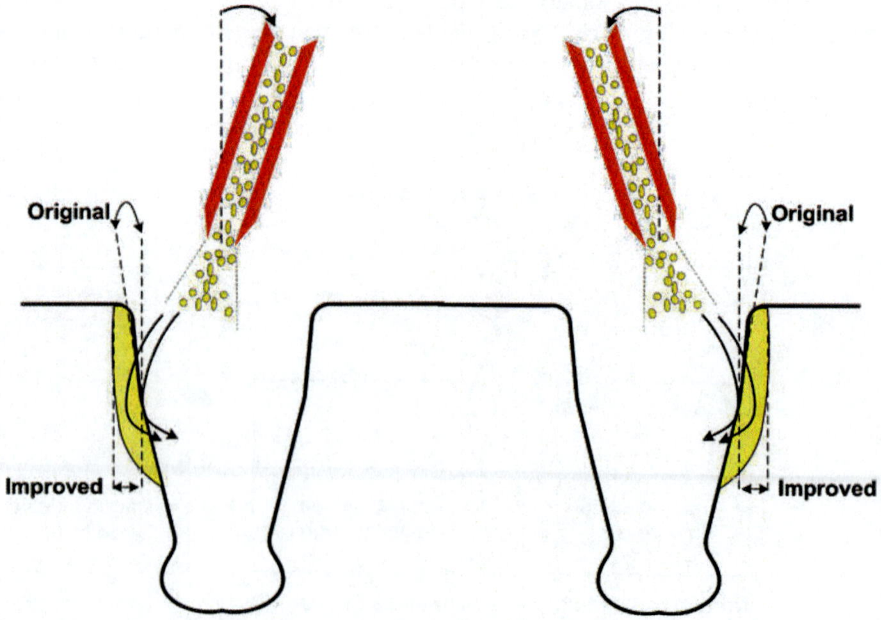

Fig. 9.12 Taper angle reduction by nozzle tilt and oscillation

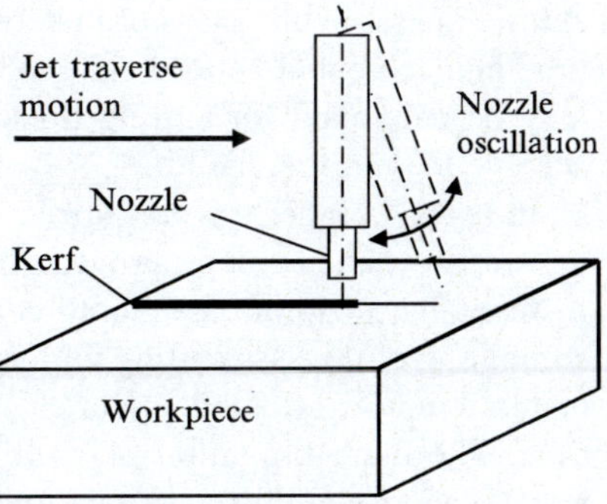

Fig. 9.13 Schematic of controlled nozzle oscillation

conducted on alumina ceramics [101] show that the taper can be completely eliminated by tilting the nozzle 4–5° against the wall.

9.4.3 Controlled Nozzle Oscillation

Oscillating the nozzle or cutting head, as shown in Fig. 9.13, has been found to be an effective means to improve AWJ cutting process. It was found [97, 102] that by oscillating the nozzle in the cutting plane, the depth of smooth zone, and total depth of cut, kerf taper, surface roughness can all be improved. The effect, however, is dependent upon the magnitude and frequency of oscillation, and may have negative effect if those two parameters are wrongly selected. Dimensional analysis [97, 102] provides some guidance on the effect of various parameters; yet, experimental correlations are needed on a case-by-case, as there is no confirmed theory to explain those effects. A comprehensive review of the technique and its benefits to the cutting process is given in [103].

9.4.4 Multi-pass Cutting

As seen from (9.23), the material removal rate (product of depth of cut by traverse speed) is conserved, if impact kinetics of the particle is determined. For example, it is possible to cut 15 mm thickness of a plate material at 10 mm/s, in about the same time it takes to cut three passes on a 5 mm thick sheet of the same material at 30 mm/s: volume removal rate is the same. However, in the second case, depth of cut per pass is smaller. This results in a smoother wall roughness and smaller taper angle [104], without sacrificing on production time. In a multi-pass cutting operation, the nozzle travels over the same kerf a number of times to achieve the cutting result. It has been found [97, 105] that a multi-pass cutting at a faster nozzle traverse speed for each pass can yield a deeper cut with good kerf quality than a single pass cutting at a slower speed, within the same cutting time. This result indicates that to cutting the same thickness of material, multi-pass can be used to reduce the cutting time and achieve a better kerf quality. Furthermore, multi-pass cutting operations may be employed for cutting thick materials where a single pass cutting cannot achieve the required depth.

In reality, higher traverse speed reduces the risk of inter-particle collisions and interference, and some improvement in removal rate is expected when the process is maintained within the smooth cut depth. A further enhancement is achieved by combining multi-pass cutting with nozzle oscillation, where it was found [105] that material removal (total depth of cut) can be increased by 50% within the same cycle time. The main limitation of multi-pass stems from positioning accuracy of the nozzle, and its ability to retrace the same path with high accuracy. Moreover, as each pass of material is removed, the effective nozzle standoff distance increases, and consequently, the effective jet width increases. There is a trade-off between the number of passes and the overall accuracy and quality of the finished cut. In practice, it is found that 2–3 passes seem to be the optimum range for cutting alumina ceramics [105].

In conclusion, with judicious choice of AWJ parameters, multi-pass cutting, nozzle planar (forward or backward) and lateral tilting, nozzle oscillation and their combinations can be employed to achieve smooth, square cuts, with accuracy and precision comparable with those achievable by other machining process, but at a higher material removal rate.

9.5 Machining Operations

For all practical purposes, the jet produced by AWJ may be viewed as a thin (typically ~1 mm in diameter) cutting tool, which is able to cut along any direction and in any orientation. The advantages of AWJ over other machining processes are [59]:

- Versatile: the same cutting tool can be used for drilling, milling, turning, contouring, etc, of virtually any geometry; possibly, replacing an entire workshop of machine tools. When the initial capital cost of an AWJ machine is compared with that of a corresponding machining centre, it is also favourable and more competitive.
- Flexible: the AWJ machine is able to cut any material (metals, ceramics, plastics, glass) with equal ease and convenience, without modification or special fixtures or tooling.
- Easy cutting: while the impact pressure by an individual particle is very high, the average force applied on the workpiece is tiny. Small forces mean that the workpiece and machine structure deflections are minimal. Workpiece fixture and setup time are also minimal, because the workpiece may be held in position even by its own weight.
- Cool cutting: while temperature at the particle impact zone can be very high, the bulk temperature is modest [106, 107]. The presence of a high speed waterjet

acts as very effective convective coolant as well. Hence, (as opposed to laser or plasma cutting, for example) AWJ is most suitable for machining heat-sensitive and hard-to-machine materials.

- Green cutting: an AWJ is simply composed of water and some naturally occurring minerals (some form of sands). It does not require synthetic oils or other chemicals for lubrication and cooling. All components of an AWJ are environmentally friendly and recyclable.

In the following sections, the versatility of AWJ is demonstrated, by its application to various machining operations.

9.5.1 Milling and Its Siblings

The abrasive jet can be adapted to any milling operation. The produced surfaces and boundaries can be linear or curved, with accuracy and resolution depending on those of the robotic arm. For simplicity, only planner linear milling is discussed here. Figure 9.14 shows the types of AWJ milling operations, classified by the number of bounding walls (from one edge, to four containing edges), and by whether the cut is through the entire workpiece thickness or only to a partial depth of cut. The case of two-walls through cutting (channel slicing) is the same as contouring and route cutting in sheets and plates, or straight cutting using a saw. The case of four walls through cutting (slot boring) is similar to drilling, internal boring or punching of plates. By a suitable combination of those geometries, any three-dimensional shape may be produced.

The accuracy, precision, and the smallest feature size that can be reproduced depend primarily on the robotic motion control, and the jet size. As typical nozzle diameters are ~0.75 mm, jet diameter at the point of cutting is ~1 mm, but can be made bigger by increasing standoff distance. Smaller nozzles and finer abrasive particles can be used to produce smaller features, at the expense of reduced material removal rate. Therefore, it seems that the narrowest channel that can be practically produced by AWJ is 0.5 mm. The depth of cut is controllable by varying the traverse speed, nozzle tilt, and other process variable. With careful process control, depth of cut can be maintained within 0.1 mm tolerance during multi-pass cutting.

The removal of volumes greater than a thin channel would require the AWJ to sweep a surface area, as shown by the shaded area in Fig. 9.14. The jet sweeping motion can follow any number of common patterns, as shown in Fig. 9.15, and at different sweeping speeds (also known as cross-feed or overlap factor). The additional process variable, cross-feed, plays an important role in determining not only material removal rate, but also surface roughness and quality of the finished product. For example, the lapping pattern can produce walls with relatively smooth surface finish, but rough linear pattern at the bottom of a channel. The spiral pattern is particularly favourable for drilling large or very deep holes that would be difficult to produce using a conventional drill.

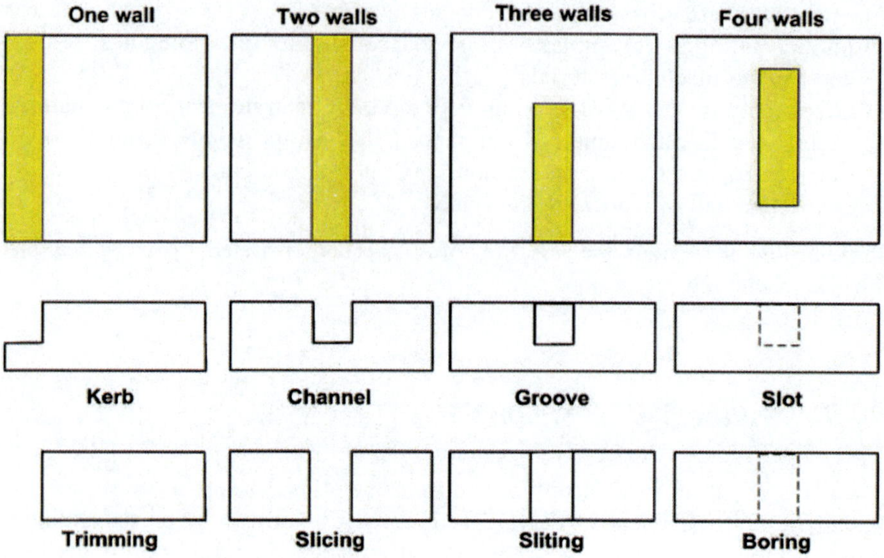

Fig. 9.14 Classification of AWJ milling, by the bounding surface

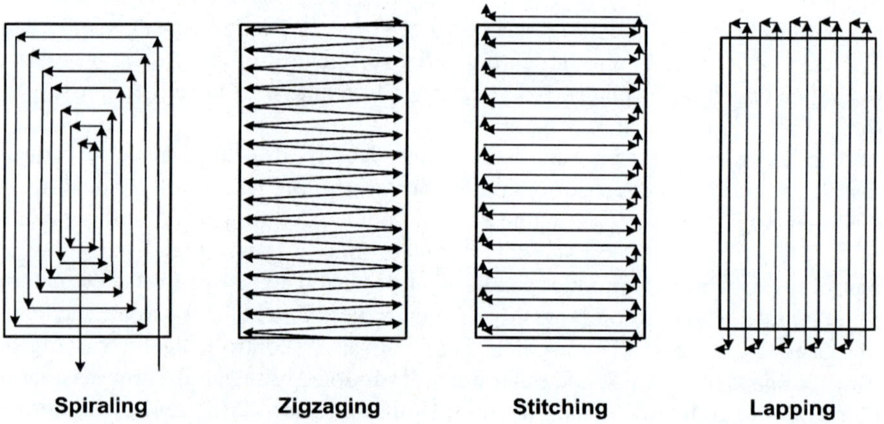

Fig. 9.15 Area sweeping patterns for AWJ milling

Examples of the research efforts into AWJ milling include [108–114]. The choice of the most suitable sweeping pattern is complicated by the planar/lateral tilting and oscillation of the nozzle. At present, there is neither theory nor comprehensive experimental data on the effect of various process parameters; correlation between the complex nozzle kinematics and product characteristics, such as dimensional accuracy and surface roughness, is a shop-floor art or a trial-and-error exercise, and there is a need for a more comprehensive systematic treatment.

9.5.2 Turning

The advantages of AWJ over other machining processes are particularly beneficial to turning. It is possible to use a single tool (the abrasive jet) to achieve numerous axi-symmetric shapes that would otherwise require various tools, and long setup time on a conventional lathe. There is no other machining process where it is possible to turn cylindrical profiles, cut threads, add fillets, chamfers, keyways, pin holes, either externally or internally, without changing the tool, and possibly even without stopping the machine. In addition to all the jet generation parameters and nozzle kinematics variables discussed above, turning introduces three additional process variables: (1) rotational speed, N, (2) spin direction (up/down, clockwise/counter clockwise), and (3) radial in-feed, r. Along with cross-feed, this makes the kinematics of AWJ turning very rich. Three unique AWJ turning modes can be identified in Fig. 9.16.

During the radial mode of turning, abrasive particle impact near normal to the surface, when rotational speed is very high, and near tangential, when N is very low (see Figs. 9.9 and 9.11 for details of material removal). Hence, by varying the rotational speed, it is possible to not only control the depth of cut, but also modify the erosion process, to make it more effective for ductile or brittle materials. In this turning mode, the direction of rotation has no effect, and the effect of in-feed is the same as in multi-pass linear cutting. Surface roughness and tolerances depend on rotational speed, and are, generally, rough and harder to control, but high material removal rate is achievable.

Offset mode turning is achieved by shifting the jet velocity vector away from centreline of the workpiece. This is equivalent to forward/backward tilting of the nozzle during linear cutting. In this mode, direction of rotation is important. It can be seen that up-cutting is equivalent to steep backward tilting of the nozzle, which is advantageous for ductile material. On the other hand, down-cutting is equivalent to steep forward tilting of the nozzle, which is beneficial for turning brittle materials.

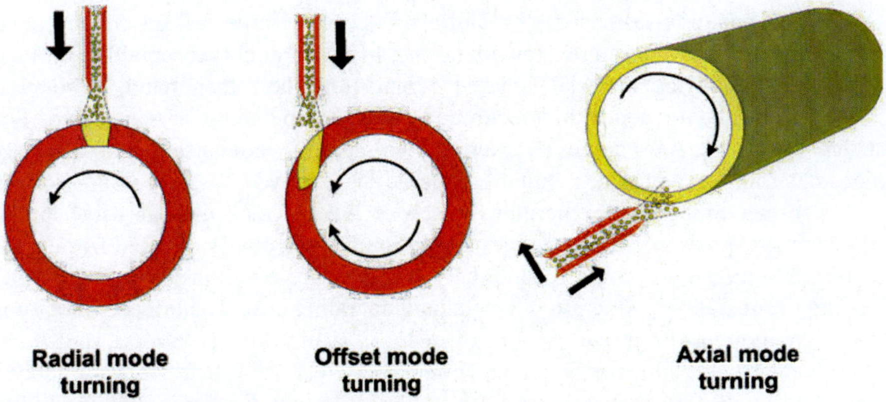

Fig. 9.16 Three modes of AWJ turning

The depth of cut and material removal rate depend on the rotational speed and on the in-feed, where the depth of cut can be precisely controlled by in-feed positioning of the jet. Therefore, this mode of turning is likely to produce better surface finish and greater dimensional accuracy.

Axial mode turning is essentially the same as kerb or channel cutting during milling (Fig. 9.14), with rotational speed having the effect of traverse speed while in-feed has the effect of multi-pass cutting. Material removal rate can be high, while surface finish and dimensional accuracy is maintained. However, there is a risk of collision between the nozzle and workpiece, if nozzle motion is maintained purely axial. This turning mode may be useful for short workpiece lengths (e.g. turning a thin disk).

It is interesting to note that the taper angle, experienced in linear cutting (Fig. 9.10) is not important during those modes of turning, because such a pattern is destroyed by workpiece rotation. Therefore, nozzle lateral tilting is not required during turning. All the issue discussed above, however, remain qualitative, as there is no comprehensive investigation of AWJ turning yet. Undoubtedly, AWJ turning is still in its infancy, and once its full potential is explored and exploited, it will become the method of choice for turning many engineering components, especially those that are hard or expensive to produce using other methods. Examples of efforts exploring AWJ turning can be found in [115–119].

9.5.3 Micro-machining

Because AWJ was originally aimed at cutting difficult-to-machine materials, where material removal rate by other methods is very low, early AWJ process designs were focused on maximizing this effect. As AWJ became a mainstream machining process, the desire for such aggressive material removal appears not always necessary, especially when it comes at the expense of precision. There are a number of emerging applications (optical, electronic, and biomedical industries, as examples) where the ability to produce fine features is more important than production time. For those "sensitive-to-machine" materials, excessive forces (as in conventional machining) and temperatures (as in laser cutting) are not acceptable, as they damage the subject material beyond repair. On the other hand, producing such features using deposition techniques may be too slow or expensive. For such applications, AWJ comes as a saviour, with its soft, cool cutting, and relatively high material removal rate, if suitable accuracy is achieved.

Micro-machining is an emerging branch of AWJ, which uses abrasive air jet (AAJ) or low pressure AWJ, so that particle impact velocity is reduced from ~500 to 100–200 m/s. This would reduce the water pressure by an order of magnitude, leading to substantial reduction in capital and operating cost. The impact velocity is still high enough to remove <1 mm of material at reasonable traverse speed and cross feed. Meanwhile, nozzle diameter and abrasive particle size are also reduced significantly, so that part features in the order of hundreds of microns width and depth can be reproduced accurately [120].

References

1. Hutchings IM (1992) Tribology: friction and wear of engineering materials. Edward Arnold, London
2. Ludema KC (1996) Friction, wear, lubrication: a textbook in tribology. CRC Press, Boca Raton
3. Rabinowicz E (1995) Friction and wear of materials. Wiley-Interscience, New York
4. Stachowiak GW, Batchelor AW (2005) Engineering tribology. Elsevier Butterworth-Heinemann, Amsterdam
5. Blazynski TZ (1983) Explosive welding, forming, and compaction. Applied Science, London
6. Ezra AA (1973) Principles and practice of explosive metalworking. Industrial Newspapers, London,
7. Ritter JE (1992) Erosion of ceramic materials. Trans Tech Pub., Zurich
8. Kleis I, Kulu P (2008) Solid particle erosion: occurrence, prediction and control. Springer, London
9. Engel PA (1976) Impact wear of materials. Elsevier, Amsterdam
10. Springer GS (1976) Erosion by liquid impact. Scripta Pub., Washington
11. Preece CM (1979) Erosion. Academic Press, New York
12. Melosh HJ (1989) Impact cratering. Oxford University Press, New York
13. Kaczmarek J (1976) Principles of machining by cutting, abrasion and erosion. P. Peregrinus, Stevenage, England
14. Marinescu ID, Rowe WB, Dimitrov B, Inasaki I (2004) Tribology of abrasive machining processes. William Andrew, Norwich
15. Marinescu ID, Rowe WB, Dimitrov B, Inasaki I (2007) Handbook of machining with grinding wheels. CRC, Boca Raton
16. Malkin S, Guo C (2008) Grinding technology: theory and application of machining with abrasives. Industrial Press, New York
17. Shaw MC (1996) Principles of abrasive processing. Clarendon Press, Oxford
18. Flow International. http://www.flowcorp.com/. Accessed 7/2009
19. Haskel Pumps. http://www.haskel.com.au/. Accessed 7/2009
20. Hashish M (2009) Trends and cost analysis of AWJ operation at 600 MPa pressure. J Press Vessel Technol Trans ASME 131: 021410: 1-7
21. Labus TJ (2001) High pressure equipment and systems. In: Labus TJ, Savanick GA (eds) An overview of waterjet fundamentals and applications, 5th Ed. Waterjet Technology Assoc., Saint Louis
22. Waterjets.Org. http://www.waterjets.org/. Accessed 7/2009
23. Water Jetting Directory. http://www.waterjettingdirectory.com/waterjet.htm. Accessed 7/2009
24. WaterJet Technology Association. http://www.wjta.org/. Accessed 7/2009
25. Water Jetting Association. http://www.waterjetting.org.uk/. Accessed 7/2009
26. Al-Obaid YF (1995) Shot peening mechanics: experimental and theoretical analysis. Mech Mater 19: 251–260
27. Kobayashi M, Matsui T, Murakami Y (1998) Mechanism of creation of compressive residual stress by shot peening. Int J Fatigue 20: 351–357
28. Wang S, Li Y, Yao M, Wang R (1998) Compressive residual stress introduced by shot peening. J Mater Process Technol 73: 64–73
29. Ramulu M, Kunaporn S, Arola D, Hashish M, Hopkins J (2000) Waterjet machining and peening of metals. J Press Vessel Technol Trans ASME 122: 90–95
30. Kunaporn S, Ramulu M, Hashish M (2005) Mathematical modeling of ultra-high-pressure waterjet peening. J Eng Mater Technol Trans ASME 127: 186–191
31. Ellermaa RRR (1993) Erosion prediction of pure metals and carbon steels. Wear 162–164: 1114–1122

32. Unified Abrasives Manufacturers' Association. http://www.uama.org/. Accessed 7/2009
33. FEPA. http://www.fepa-abrasives.org/. Accessed 7/2009
34. British Abrasives Federation. http://www.thebaf.org.uk/. Accessed 7/2009
35. Abrasive Engineering Society. http://www.abrasiveengineering.com/. Accessed 7/2009
36. Fox RW, McDonald AT, Pritchard PJ (2009) Introduction to fluid mechanics, 7th Ed. Wiley, Hoboken
37. Momber AW, Kovacevic R (1998) Principles of abrasive water jet machining. Springer, London
38. Korobkin A (2004) Analytical models of water impact. Eur J Appl Math 15: 821–838
39. Korobkin AA (1997) Asymptotic theory of liquid-solid impact. Philos Trans R Soc A Math Phys Eng Sci 355: 507–522
40. Korobkin AA, Pukhnachov VV (1988) Initial stage of water impact. Annu Rev Fluid Mech 20: 159–185
41. Battistin D, Iafrati A (2003) Hydrodynamic loads during water entry of two-dimensional and axisymmetric bodies. J Fluids Struct 17: 643–664
42. May A, Woodhull JC (1950) The virtual mass of a sphere entering water vertically. J Appl Phys 21: 1285–1289
43. May A, Woodhull JC (1948) Drag coefficients of steel spheres entering water vertically. J Appl Phys 19: 1109–1121
44. Shiffman M, Spencer DC (1945) The force of impact on a sphere striking a water surface, AMP Report 42. 2R. AMG-New York University, New York
45. Chow C-Y (1979) An introduction to computational fluid mechanics. Wiley, New York
46. Nguyen T, Shanmugam DK, Wang J (2008) Effect of liquid properties on the stability of an abrasive waterjet. Int J Mach Tools Manuf 48: 1138–1147
47. Labus TJ, Neusen KF, Alberts DG, Gores TJ (1991) Factors influencing the particle size distribution in an abrasive waterjet. J Eng Ind Trans ASME 113: 402–411
48. Vijay MM (2001) Fluid mechanics of jets. In: Labus TJ, Savanick GA (eds) An overview of waterjet fundamentals and applications, 5th Ed. Waterjet Technology Assoc., Saint Louis
49. Momber AW, Kovacevic R (1999) An energy balance of high-speed abrasive water jet erosion. Proc Inst Mech Eng Part J J Eng Tribol 213: 463–472
50. Whiting CE, Graham EE, Ghorashi B (1990) Evaluation of parameters in a fluid cutting equation. J Eng Ind Trans ASME 112: 240–244
51. Lesser MB (1981) Analytic solutions of liquid-drop impact problems. Proc R Soc A Math Phys Eng Sci 377: 289–308
52. Pack DC, Evans WM (1951) Penetration by high-velocity ('Munroe') jets: I. Proc Phys Soc B 64: 298–302
53. Wood B (2001) Cleaning applications for waterjets. In: Labus TJ, Savanick GA (eds) An overview of waterjet fundamentals and applications, 5th Ed. Waterjet Technology Assoc., Saint Louis
54. Summers DA (1995) Waterjetting technology. E & FN Spon, London
55. Hashish M (1994) Observations of wear of abrasive-waterjet nozzle materials. J Tribol-Trans ASME 116: 439–444
56. Li HZ, Wang J, Fan JM (2009) Analysis and modelling of particle velocities in micro-abrasive air jet. Int J Mach Tools Manuf 49: 850–858
57. Shimizu S, Wu Z-L (1996) Acceleration of abrasive particles in premixed abrasive water jet nozzle. JSME Int J 39: 562–567
58. Tazibt A, Parsy F, Abriak N (1996) Theoretical analysis of the particle acceleration process in abrasive water jet cutting. Comput Mater Sci 5: 243–254
59. Wang J (2003) Abrasive waterjet machining of engineering materials. Trans Tech Publications, Uetikon-Zuerich, Switzerland
60. Liu H, Wang J, Kelson N, Brown RJ (2004) A study of abrasive waterjet characteristics by CFD simulation. J Mater Process Technol 153–154: 488–493

61. Wang J (2009) Particle velocity models for ultra-high pressure abrasive waterjets. J Mater Process Technol 209: 4573–4577
62. Prisco U, D'Onofrio MC (2008) Three-dimensional CFD simulation of two-phase flow inside the abrasive water jet cutting head. Int J Comp Meth Eng Sci Mech 9: 300–319
63. Ye J, Kovacevic R (1999) Turbulent solid-liquid flow through the nozzle of premixed abrasive water jet cutting systems. Proc Inst Mech Eng B J Eng Manuf 213: 59–67
64. Yong Z, Kovacevic R (1997) Simulation of chaotic particle motion in particle-laden jetflow and application to abrasive waterjet machining. J Fluids Eng Trans ASME 119: 435–442
65. Johnson KL (1985) Contact mechanics. Cambridge University Press, Cambridge
66. Galin LA (1961) Contact problems in the theory of elasticity. Dept. of Mathematics, North Carolina State College, Raleigh
67. Gladwell GML (1980) Contact problems in the classical theory of elasticity. Sijthoff & Noordhoff, Alphen aan den Rijn, The Netherlands
68. Hamilton GM (1983) Explicit equations for the stresses beneath a sliding spherical contact. Proc Inst Mech Eng C 197: 53–59
69. Sackfield A, Hills D (1983) A note on the hertz contact problem: a correlation of standard formulae. J Strain Anal 18: 195–197
70. Neilson JH, Gilchrist A (1968) Erosion by a stream of solid particles. Wear 11: 111–122
71. Finnie I (1960) Erosion of surfaces by solid particles. Wear 3: 87–103
72. Bitter JGA (1963) A study of erosion phenomena, Part I. Wear 6: 5–21
73. Bitter JGA (1963) A study of erosion phenomena, Part II. Wear 6: 169–190
74. Beckmann G, Gotzmann J (1981) Analytical model of the blast wear intensity of metals based on a general arrangement for abrasive wear. Wear 73: 325–353
75. Bowden FP, Field JE (1964) The brittle fracture of solids by liquid impact, by solid impact, and by shock. Proc R Soc A Math Phys Eng Sci 282: 331–352
76. Sheldon GL, Finnie I (1966) The mechanism of material removal in the erosive cutting of brittle materials. J Eng Ind Trans ASME 88: 393–400
77. Wiederhorn SM, Lawn BR (1977) Strength degradation of glass resulting from impact with spheres. J Am Ceram Soc 60: 451–458
78. Evans AG, Wilshaw TR (1977) Dynamic solid particle damage in brittle materials: an appraisal. J Mater Sci 12: 97–116
79. Evans AG, Gulden ME, Rosenblatt M (1978) Impact damage in brittle materials in the elastic-plastic response regime. Proc R Soc A Math Phys Eng Sci 361: 343–365
80. Hashish M (1984) A modeling study of metal cutting with abrasive waterjets. J Eng Mater Technol Trans ASME 106: 88–100
81. Hashish M (1989) A model for abrasive-waterjet (AWJ) machining. J Eng Mater Technol Trans ASME 111: 154–162
82. Finnie I, McFadden DH (1978) On the velocity dependence of the erosion of ductile metals by solid particles at low angles of incidence. Wear 48: 181–190
83. Lemma E, Deam R, Chen L (2005) Maximum depth of cut and mechanics of erosion in AWJ oscillation cutting of ductile materials. J Mater Process Technol 160: 188–197
84. Paul S, Hoogstrate AM, Van Luttervelt CA, Kals HJJ (1998) Analytical and experimental modelling of the abrasive water jet cutting of ductile materials. J Mater Process Technol 73: 189–199
85. El-Domiaty AA, Shabara MA, Abdel-Rahman AA, Al-Sabeeh AK (1996) On the modelling of abrasive waterjet cutting. Int J Adv Manuf Technol 12: 255–265
86. Ness E, Zibbell R (1996) Abrasion and erosion of hard materials related to wear in the abrasive waterjet. Wear 196: 120–125
87. El-Domiaty AA, Abdel-Rahman AA (1997) Fracture mechanics-based model of abrasive waterjet cutting for brittle materials. Int J Adv Manuf Technol 13: 172–181
88. Zeng J, Kim TJ (1996) An erosion model of polycrystalline ceramics in abrasive waterjet cutting. Wear 193: 207–217

89. Hashish M (1988) Visualization of the abrasive-waterjet cutting process. Exp Mech 28: 159–169
90. Hashish M (1991) Characteristics of surfaces machined with abrasive-waterjets. J Eng Mater Technol Trans ASME 113: 354–362
91. Chen L, Siores E, Wong WCK (1996) Kerf characteristics in abrasive waterjet cutting of ceramic materials. Int J Mach Tools Manuf 36: 1201–1206
92. Chen FL, Wang J, Lemma E, Siores E (2003) Striation formation mechanisms on the jet cutting surface. J Mater Process Technol 141: 213–218
93. Wang J, Wong WCK (1999) A study of abrasive waterjet cutting of metallic coated sheet steels. Int J Mach Tools Manuf 39: 855–870
94. Wang J (1999) Machinability study of polymer matrix composites using abrasive waterjet cutting technology. J Mater Process Technol 94: 30–35
95. Wang J (1999) Abrasive waterjet machining of polymer matrix composites – cutting performance, erosive process and predictive models. Int J Adv Manuf Technol 15: 757–768
96. Wang J, Guo DM (2002) A predictive depth of penetration model for abrasive waterjet cutting of polymer matrix composites. J Mater Process Technol 121: 390–394
97. Wang J (2007) Predictive depth of jet penetration models for abrasive waterjet cutting of alumina ceramics. Int J Mech Sci 49: 306–316
98. Kovacevic R (1991) Surface texture in abrasive waterjet cutting. J Manuf Syst 10: 32–40
99. Kovacevic R, Fang M (1994) Modeling of the influence of the abrasive waterjet cutting parameters on the depth of cut based on fuzzy rules. Int J Mach Tools Manuf 34: 55–72
100. Hashish M (1993) Effect of beam angle in abrasive-waterjet machining. J Eng Ind Trans ASME 115: 51–56
101. Shanmugam DK, Wang J, Liu H (2008) Minimisation of kerf tapers in abrasive waterjet machining of alumina ceramics using a compensation technique. Int J Mach Tools Manuf 48: 1527–1534
102. Xu S, Wang J (2006) A study of abrasive waterjet cutting of alumina ceramics with controlled nozzle oscillation. Int J Adv Manuf Technol 27: 693–702
103. Wang J (2009) A focused review on enhancing the abrasive waterjet cutting performance by using controlled nozzle oscillation. Key Eng Mater 404: 33–44
104. Wang J, Guo DM (2003) The cutting performance in multipass abrasive waterjet machining of industrial ceramics. J Mater Process Technol 133: 371–377
105. Wang J, Zhong Y (2009) Enhancing the depth of cut in abrasive waterjet cutting of alumina ceramics by using multipass cutting with Nozzle oscillation. Mach Sci Technol 13: 76–91
106. Kovacevic R, Mohan R, Beardsley HE (1996) Monitoring of thermal energy distribution in abrasive waterjet cutting using infrared thermography. J Manuf Sci Eng Trans ASME 118: 555–563
107. Ohadi MM, Ansari AL, Hashish M (1992) Thermal energy distributions in the workpiece during cutting with an abrasive waterjet. J Eng Ind Trans ASME 114: 67–73
108. Fowler G, Pashby IR, Shipway PH (2009) The effect of particle hardness and shape when abrasive water jet milling titanium alloy Ti6Al4V. Wear 266: 613–620
109. Hashish M (1998) Controlled-depth milling of isogrid structures with AWJs. J Manuf Sci Eng Trans ASME 120: 21–27
110. Hocheng H, Tsai HY, Shiue JJ, Wang B (1997) Feasibility study of abrasive-waterjet milling of fiber-reinforced plastics. J Manuf Sci Eng Trans ASME 119: 133–142
111. Paul S, Hoogstrate AM, Van Luttervelt CA, Kals HJJ (1998) An experimental investigation of rectangular pocket milling with abrasive water jet. J Mater Process Technol 73: 179–188
112. Hashish M (1989) Investigation of milling with abrasive-waterjets. J Eng Ind Trans ASME 111: 158–166
113. Feng YX, Huang CZ, Wang J, Hou RG, Lu XY (2007) An experimental study on milling Al2O3 ceramics with abrasive waterjet. Key Eng Mater 339: 500–504
114. Zeng J, Kim TJ (1996) An erosion model for abrasive waterjet milling of polycrystalline ceramics. Wear 199: 275–282

115. Axinte DA, Stepanian JP, Kong MC, McGourlay J (2009) Abrasive waterjet turning-An efficient method to profile and dress grinding wheels. Int J Mach Tools Manuf 49: 351–356
116. Manu R, Babu NR (2009) An erosion-based model for abrasive waterjet turning of ductile materials. Wear 266: 1091–1097
117. Ansari AI, Hashish M, Ohadi MM (1992) Flow visualization study of the macromechanics of abrasive-waterjet turning. Exp Mech 32: 358–364
118. Ansari AI, Hashish M (1995) Effect of abrasive waterjet parameters on volume removal trends in turning. J Eng Ind Trans ASME 117: 475–484
119. Zhong ZW, Han ZZ (2002) Turning of glass with abrasive waterjet. Mater Manuf Process 17: 339–349
120. Wang J, Nguyen T, Pang KL (2009) Mechanisms of microhole formation on glasses by an abrasive slurry jet. J Appl Phys 105(4): 044906.1–044906.4

Index